Handbook of

Terahertz Technologies

Handbook of
Terahertz Technologies
DEVICES AND APPLICATIONS

edited by
Ho-Jin Song
Tadao Nagatsuma

CRC Press
Taylor & Francis Group
Boca Raton London New York

CRC Press is an imprint of the
Taylor & Francis Group, an **informa** business

Handbook of Terahertz Technologies: Devices and Applications

First published in 2015 by Pan Stanford Publishing Pte. Ltd.

Published 2022 by CRC Press
Taylor & Francis Group
6000 Broken Sound Parkway NW, Suite 300
Boca Raton, FL 33487-2742

ISBN-13: 978-981-4613-08-8 (hbk)

Publisher's Note
The publisher has gone to great lengths to ensure the quality of this reprint but points out that some imperfections in the original copies may be apparent.

British Library Cataloguing-in-Publication Data
A catalogue record for this book is available from the British Library.

Visit the Taylor & Francis Web site at
http://www.taylorandfrancis.com

and the CRC Press Web site at
http://www.crcpress.com

Contents

Preface

Electromagnetic waves with frequencies from 0.1 THz (100 GHz) to 10 THz are referred as terahertz (THz) waves, and their frequencies lie between those of microwaves and infrared light waves. THz waves have long been a big concern, especially in astronomy, because 98% of the total photons emitted in the history of the universe since the Big Bang lie in the THz region of electromagnetic waves.

THz waves exhibit various interesting unique properties such as high absorption in water; high transparency over paper, plastic, and cloth; correspondence to many molecular absorption lines; and no harmful ionization effects on biological tissues. Because of lack of devices and techniques for their generation and detection, the THz region was untapped in industrial applications up till the middle of 1980s.

However, enormous efforts have been made over the last couple of decades in research and development of THz technologies, and it has been proved that, unlike conventional radio waves and light waves, THz waves have lots of capabilities that can be applied in various fields, such as imaging, spectroscopy, security, and communications. Historically, use of pulsed THz waves, which were generated and detected with femtosecond-pulse lasers, opened the first practical and powerful applications of THz waves both in spectroscopy and imaging in the early 1990s. The advent of semiconductor device technologies have brought about new components and applications, for example, THz lasers, THz cameras, and THz communications systems in the 2000s. The measurement and testing technologies such as THz network analyzers and spectrum analyzers have also accelerated the progress of integrated circuit technologies, pushing THz technologies into compact and cost-effective systems in 2010s.

This book covers some of those technical breakthroughs in terms of device technologies. It not only describes the theoretical details and typical features of the state-of-art technologies, but also discusses some issues and challenges related to them. It also shows what we can actually do with THz-wave technologies, suggested through reviews of their usage in several successful applications, such as wireless communications, industry, remote sensing, chemical analysis, and two-dimensional/tomographic imaging.

We hope this book will be a useful resource for researchers and engineers in the field of THz-wave technologies who are searching for current research trends and technology needs. We also wish that the several examples of some successful applications of THz technologies given in the book are invaluable for those who are seeking new approaches or technologies for their own systems. Finally, we would like to give a big thanks to all authors for their valuable contributions and support in this book project.

Ho-Jin Song

Tadao Nagatsuma

Spring 2015

Chapter 1

Photoconductive Antenna vs. Photomixer for Broadband Terahertz Wave Generation and Detection

Kyung Hyun Park,[a] Kiwon Moon,[a] Han-Cheol Ryu,[b] Namje Kim,[a] and Sang-Pil Han[a]

[a]*THz Photonics Creative Research Center, ETRI, Daejeon, 305-700, Korea*

[b]*Department of Car-Mechatronics, Sahmyook University, Seoul, 139-742, Korea*

khp@etri.re.kr

This paper reviews the main building blocks for the generation and detection of a broadband Terahertz (THz) wave. The role played by GaAs and InGaAs grown at low temperature in frequency down-conversion is described, including material growth and device performance. A photonics-based beating source is also surveyed for the realization of a compact and cost-effective THz spectrometer. In addition to the material and device, the THz field applications using THz devices are described.

Handbook of Terahertz Technologies: Devices and Applications
Edited by Ho-Jin Song and Tadao Nagatsuma

ISBN 978-981-4613-08-8 (Hardcover), 978-981-4613-09-5 (eBook)
www.panstanford.com

1.1 Introduction

Terahertz (THz) technology has shown potential in a wide variety of applications, including spectroscopy, imaging, and sensing, because of its unique properties [1–3]. To utilize these properties, a high-performance source and detector are required. As such, continuous efforts to produce high-performance THz components are being made. THz systems based on optical technologies have prevailed from the point of view of THz spectroscopy systems. Among such optical-based THz systems, time domain spectroscopy (THz-TDS) has had remarkable success in terms of its high resolution, moderate size, broad band, and high signal-to-noise ratio (SNR). Although it delivers a great advantage in investigating the properties of materials with a high SNR, the performance of THz-TDS is strongly limited by the characteristics of the femtosecond pulse laser and the quality of the photoconductive (PC) switch.

In comparison to THz-TDS, frequency domain THz continuous-wave (CW) systems have superior advantages in terms of frequency resolution, compactness, and their wide range of field applications [4, 5]. Compact, lightweight, and cost-effective THz spectroscopy and imaging systems have gradually become the preferred choice for use outdoors or in mobile situations in the fields of security, noninvasive testing, food and agricultural goods quality control, and environmental monitoring among others. Fiber-coupled THz systems are one of the possible solutions, as they offer lower cost and higher stability and are more portable compared to free-space THz systems, largely because they have movable THz emitters and detectors [6–8]. The frequency domain THz system, which consists of a combination of beating sources and a broadband photomixer, shows the way to differentiate the performance from THz-TDS. The good thermal conductivity of the photomixer, which is the frequency down-converter, is essential for its high performance. Recently, many research groups have been trying to develop a high-performance photomixer that can be used in a long wavelength range. The use of a long-wavelength beating source can bring the connection between the THz and InP-based communication technologies, since well-developed InP-based optoelectronic technologies are expected to enable the integration of tunable laser diode (LD) sources with other optical components such as semiconductor optical amplifiers (SOAs), electroabsorption modulators, and waveguide-type THz photomixers [9]. Many CW-THz spectrometers have been commercialized using these long-wavelength components.

This chapter gives a brief perspective on the main building blocks for broadband THz generation and detection in both time and frequency domain THz systems.

1.2 Photoconductive Antenna

1.2.1 Emission and Detection of THz Pulses by a Photoconductive Antenna

A PC antenna is composed of two metallic electrodes facing each other, patterned onto the surface of a semi-insulating semiconductor surface. Usually, the shapes of the two electrodes take the form of a simple dipole antenna with a microscale gap between them (Fig. 1.1). For THz pulse generation, electron–hole pairs are locally generated at the PC gap by ultrashort laser pulses and accelerated by applied electric field between the two electrodes. This induces an ultrafast surge current, which causes THz pulse radiation [10, 11].

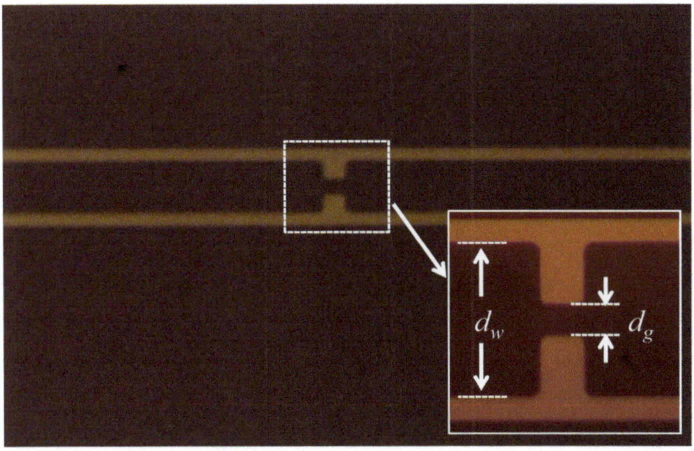

Figure 1.1 Typical photoconductive antenna fabricated on a semiconductor substrate. The photoconductive gap (d_g) and the distance between the two electrodes (d_W) were set at 5 μm and 30 μm, respectively.

There have been many studies on the THz emission mechanism, which is briefly introduced in Refs. [12–14]. The relationship between the transient photocurrent $i_{PC}(t)$ and the radiated electric field E_{THz} is expressed with the following equation:

$$E_{THz}(r,\theta,t)=\frac{l_e}{4\pi\varepsilon c^2 r}\frac{\partial i_{PC}(t)}{\partial t}\sin\theta \tag{1.1}$$

where r is the distance from the dipole, θ is the angle relative to the dipole axis, l_e is the effective antenna length, and c is the speed of light in a vacuum. The transient current $i_{PC}(t)$ is determined by the time-dependent carrier density $n(t)$ and the electron velocity $v_e(t)$ by

$$i_{PC}(t)\propto P_{pulse}(t)\otimes(n(t)v_e(t)) \tag{1.2}$$

where $\otimes$ denotes the convolution integral. The dynamics of $n(t)$ are expressed as

$$\frac{dn}{dt}=\eta\frac{P_{pulse}(t)}{h\nu}-\frac{n}{\tau_c} \tag{1.3}$$

and by adopting a 1D Drude–Lorentz model, the electron velocity $v_e(t)$ can be expressed by

$$\frac{dv_e}{dt}=-\frac{v_e}{\tau_s}+\frac{e}{m^*}E \tag{1.4}$$

where η is the absorption efficiency and $P_{pulse}(t)$ represents the time evolution of the ultrashort laser power. τ_c, τ_s, and m^* represent the carrier lifetime, the momentum relaxation time, and the electron effective mass, respectively. E represents the local electric field as seen by the photogenerated electrons.

To achieve an ultrafast surge current, it is desirable for the pulse duration and the carrier lifetime to be less than 200 fs and 1 ps, respectively. In addition, the carrier mobility of the substrate should be as high as possible since electron velocity also plays an important role, as shown in Eq. 1.2.

As shown in Eq. 1.1, the time evolution of the emitted electric field is proportional to the time derivative of the photocurrent. Therefore, the surge current generates a bipolar-shaped THz pulse having a positive peak and a subsequent negative peak. Detailed studies on the dynamics of $i_{PC}(t)$ can be found in Refs. [12–14].

The PC antenna is used for THz detection as well [10, 15]. Typically, the THz pulse is focused on the PC antenna from the substrate side, and the femtosecond pulse, which is split from the THz generating pulse, is focused at the gap. The gap becomes conductive at the moment of probe pulse incidence, permitting an instantaneous current flow that is proportional to the local electric bias provided by the incident THz pulse. Therefore, the transient of the THz pulse is measured in the time domain by scanning the phase difference between the THz pulse and the probe femtosecond pulse. An example of the THz-TDS system is illustrated in Fig. 1.2.

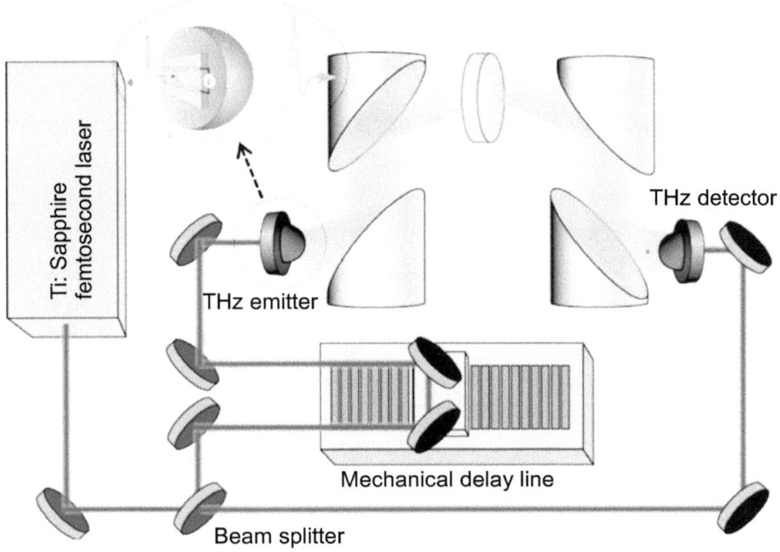

Figure 1.2 A schematic of the THz-TDS system using a femtosecond ultrafast laser and two PC antennas for the emission and detection of the THz pulse.

1.2.2 The Growth and Characterization of Photoconductive Materials

The key to broadband THz generation and detection is the semi-insulating substrate of the subpicosecond carrier lifetime having high carrier mobility. This is typically achieved by developing a nonstoichiometric compound semiconductor material with molecular beam epitaxy (MBE) and then by carrying out a thermal annealing process to recover the crystalline quality and generate metallic precipitates that act as carrier-trapping centers [16–20].

It is well known that low-temperature-grown GaAs (LTG GaAs) and LTG InGaAs have a high density of arsenic antisites, which reduce the carrier lifetime by trapping excess carriers [19, 20]. However, they are good electric conductors due to the conduction by the nearest-neighbor hopping mechanism occurring in the deep donor band created by the arsenic antisites [16–18]. After the subsequent annealing process, arsenic precipitates are generated. This decreases the deep donor band and generates metallic islands in the layer, which are regarded as buried Schottky barriers [16]. As the density of the precipitates increases, the depletion region overlaps, leading to the formation of a high-resistivity layer.

The density of the arsenic antisite is a function of the growth temperature. Studies on the effects of growth conditions on layer properties and THz emission have been presented in Ref. [18]. LTG GaAs layers were grown at various temperatures ranging from 200°C to 250°C, while the growth rate was fixed at 1 μm/hr. The lattice mismatch increased as the growth temperature decreased, as shown in Fig. 1.3a. The effect of the subsequent thermal annealing can be observed by X-ray diffraction (XRD) and resistivity measurement. The XRD rocking curves were measured before and after the annealing process (Fig. 1.3b), which was performed for 10 minutes at temperatures between 500°C and 620°C.

Mismatch between the as-grown layer and the GaAs substrate indicates the incorporation of excess arsenic. The mismatch was relaxed after thermal annealing, indicating the formation of arsenic precipitates [17, 18]. The thickness fringe patterns in the rocking curves indicate excellent macroscopic crystal quality, although the growth temperature deviates from the optimum value. The subpicosecond lifetime was confirmed by transient reflectivity measurement (Fig. 1.4a), and sheet resistance increases more than 1000 MΩ/sq. under dark conditions (Fig. 1.4b).

1.2.3 Generation and Detection of THz Pulses

A PC antenna fabricated on an LTG GaAs layer and grown at a growth temperature of T_g = 230°C and an annealing temperature of

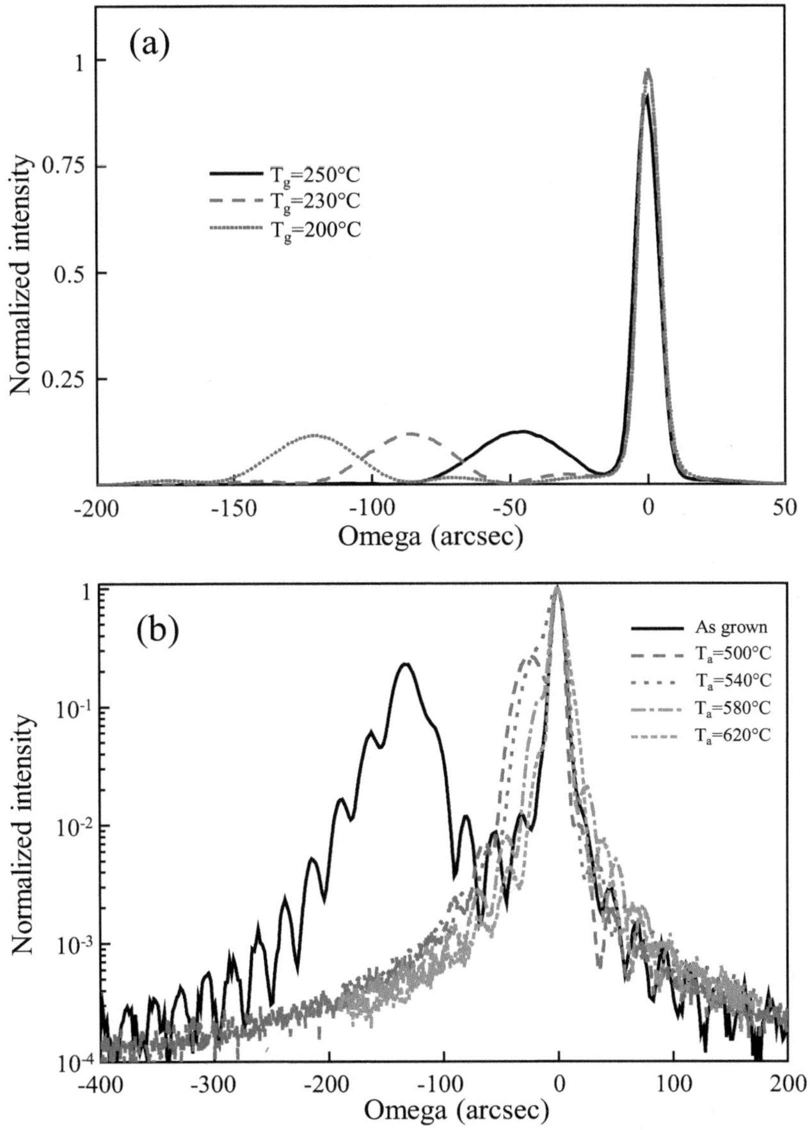

Figure 1.3 Measured XRD rocking curves. (a) XRD rocking curves of as-grown samples. The sample thickness was 500 nm. The measured mismatch corresponds to the excess arsenic density of 1%, 0.7%, and 0.35% for the growth temperatures of 200°C, 230°C, and 250°C, respectively. (b) Measured rocking curves after annealing for a sample grown at 200°C. Thickness of the sample was 1 μm. Thickness fringes indicate macroscopic quality of the crystal.

T_a = 580°C was used for the generation and detection of THz pulses. The gap between the two electrodes (d_g) and the distance between the coplanar lines (d_w) were 5 μm and 30 μm, respectively (see Fig. 1.1). THz pulse radiation from an InAs crystal was measured using the fabricated PC antenna to estimate its performance as a detector. A coherent MIRA 900 fs laser was used for the experiment. Optical pulse duration, repetition rate, and average output power were 120 fs, 80 MHz, and 600 mW, respectively. The measured THz pulse and its Fourier amplitude spectrum are presented in Fig. 1.5a,b. A bandwidth of more than 2.5 THz was obtained, and the water absorption peaks were clearly observed.

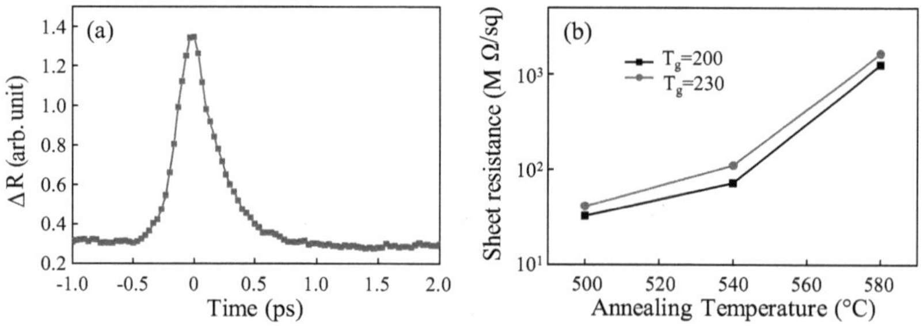

Figure 1.4 (a) Transient reflectivity of a sample grown at 200°C and annealed at 580°C and (b) measured sheet resistance as a function of annealing temperature (T_a).

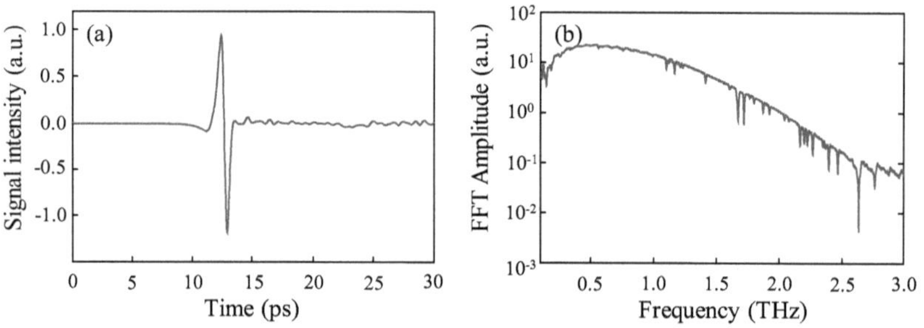

Figure 1.5 InAs-emitted THz pulse measured by the fabricated PC antenna. (a) Time domain signal and (b) fast Fourier transform (FFT) spectrum.

The PC antenna was also used for THz pulse emission. The excitation optical power was 30 mW, and a 25 V DC bias was applied between the two electrodes for THz generation. The results are shown in Fig. 1.6. It should be noted that the emitted pulse intensity was comparable to the emission from the InAs crystal at the optical excitation of 30 mW. Although relatively high THz pulse power is delivered by the PC antenna, improving its efficiency is an important issue for commercial applications.

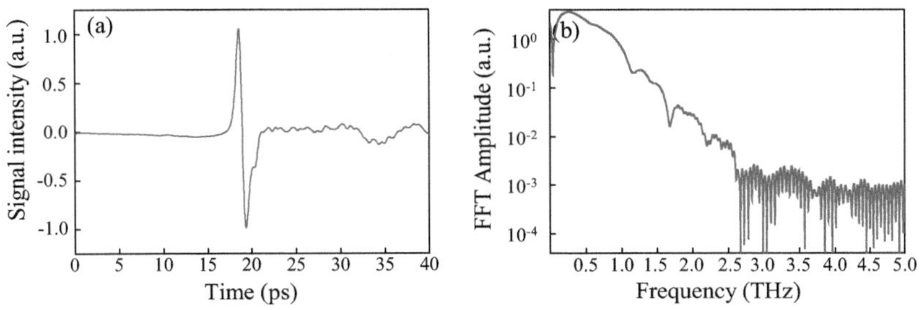

Figure 1.6 PC-antenna-emitted THz pulse measured by the fabricated PC antenna. (a) Time domain signal and (b) FFT spectrum.

On the other hand, InGaAs-based PC antennas have been studied to utilize the well-established fiber optic communication technology [19, 20]. In much the same way as LTG GaAs, the LTG InGaAs layer contains a high density of arsenic antisites, thus generating arsenic clusters after thermal annealing. To compensate spontaneous n-doping, beryllium counterdoping is usually employed [21]. A 1.2 μm thick LTG InGaAs layer was grown by MBE. The XRD rocking curve showed good crystalline quality, and the carrier lifetime was measured to be about 1 ps (Fig. 1.7). The sheet resistance of LT-InGaAs was about 1–10 kΩ/sq.

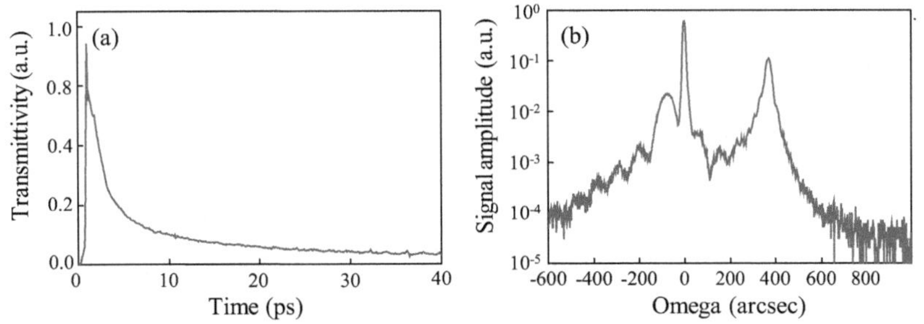

Figure 1.7 Characteristics of the LTG-InGaAs layer. (a) Transient transmittivity measurement. By double exponential fitting, the carrier decay time was estimated to be 1.3 ps. (b) XRD rocking curve of the LTG-InGaAs layer.

A log-periodic PC antenna was fabricated on the surface of the LTG InGaAs layer. To prevent dark current, a dielectric insulating layer was inserted between the metal layer and the surface of the semiconductor, except for the active gap area. The THz emission characteristics were measured by a fiber-optic-based THz-TDS system, as shown in Fig. 1.8.

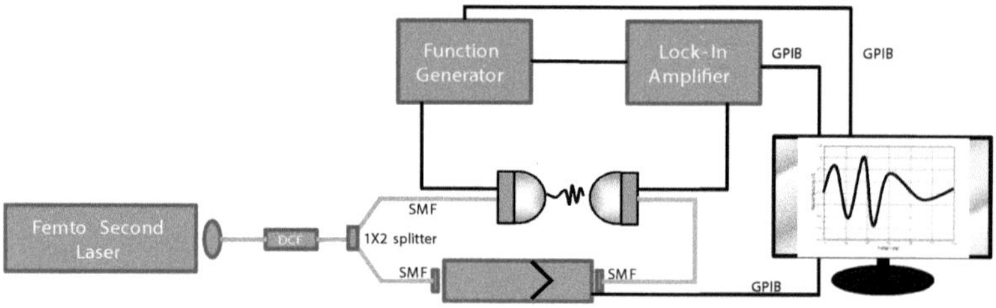

Figure 1.8 Fiber-optic-based THz-TDS system. The center wavelength, duration, and average power of the laser are 1550 nm, 70 fs, and 20 mW, respectively. A dispersion compensation fiber is used to adjust the pulse duration to about 120 fs at the PC antenna. Reproduced from Ref. [6] by permission of Optical Society of America, © 2011.

The measured time domain curve and the spectrum are depicted in Figs. 1.9a and 1.9b, respectively. Note that the emitted THz pulse shows negative dispersion compared to the dipole emission shown in Fig. 1.5. This is due to the frequency response of the log-periodic antenna, which is inappropriate for the pulse spectroscopy. However, the log-spiral antenna has significantly improved the signal intensity, which is important for the CW terahertz system.

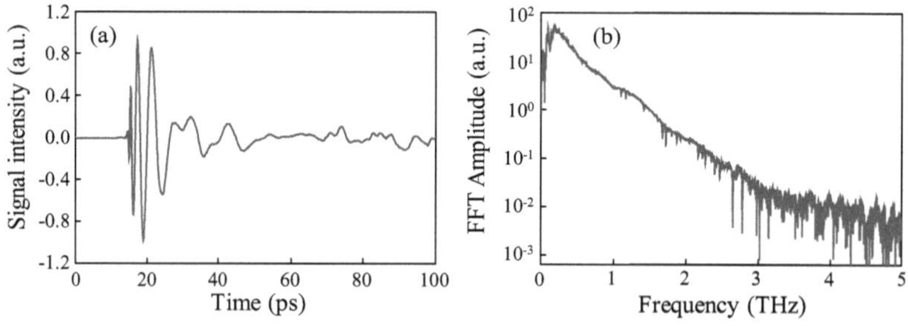

Figure 1.9 THz emission from a log-periodic PC antenna fabricated on an LTG-InGaAs wafer. (a) Time domain signal and (b) FFT amplitude. Reproduced from Ref. [22] by of permission of © 2012 Optical Society of America.

1.3 Photomixers and Beating Sources

1.3.1 Photomixers for Continuous-Wave Terahertz Applications

A photomixer is a type of frequency down-converter. It receives the optical beat signal generated by two laser beams and emits CW THz radiation through an integrated antenna. When the optical beat signal irradiates the active layer of the photomixer, a high-frequency AC current with the same frequency as that of the optical beat signal is generated. This AC current is then applied to an integrated antenna, which emits the THz radiation into free space. Figure 1.10 shows a typical homodyne configuration used for CW THz emission and detection.

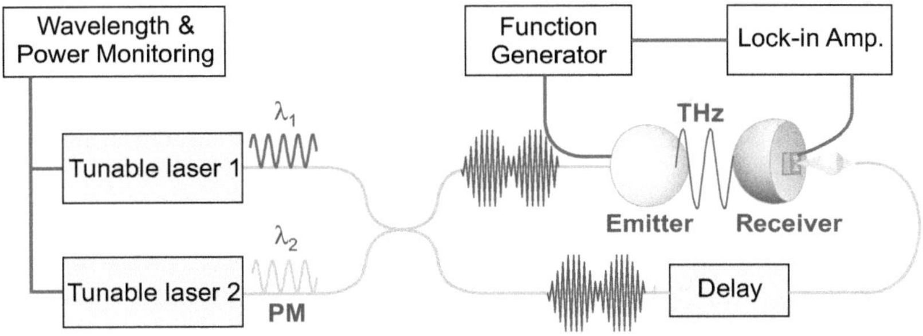

Figure 1.10 A typical fiber-coupled CW THz system with homodyne configuration (PM: polarization-maintaining fiber).

The emitted power of CW THz radiation and the characteristics of the photomixer are described by using the equivalent circuit; the output power obtained by using the photomixing method is as follows:

$$P_L = \frac{1}{2}\left(\frac{e\eta}{hv}\right)^2 P_1 P_2 \times \frac{1}{1+(\Delta\omega\tau)^2} \times \frac{G_L}{(G_L+G_P)^2+(\Delta\omega C_P+B_L)^2} \tag{1.5}$$

where P_1 and P_2 are the optical powers of the lasers, η is the external quantum efficiency, τ is the carrier lifetime (assuming that $\tau_e = \tau_h$; C_P is the capacitance), G_P is the differential resistance, and G_L and B_L are the real and imaginary parts of the complex admittance, respectively [23].

Therefore, the active layer in the photomixer needs to have high-frequency operation capability, high output current, and high dark resistance. High-frequency operation capability means a short carrier lifetime in LTG III–V materials and a short carrier transit time in photodiode (PD)-based photomixers. For high current operation, the carrier mobility should be high, whereas for a high applied voltage, a high dark resistance is preferable. Thus far, photomixers have been categorized into two groups: one group is based on LTG III–V-based materials such as LTG GaAs and InGaAs, which are characterized by a short carrier lifetime due to metallic precipitates and a wide bandwidth of over 2 THz; the other consists of high-speed PD-based photomixers.

1.3.2 Low-Temperature-Grown III–V Material–Based Photomixers

LTG III–V materials act as a key component consisting of an active layer of photomixers for generating CW THz radiation using the photomixing method. Historically, LTG III–V materials have, since the beginning of the 1990s, been intensively researched as a buffer layer for high-speed electronic devices and for generating short electrical pulses [24–26]. It opens up a new field of epitaxy called defect engineering. Single-crystal growth with high resistivity was the main focus at that time. Carrier lifetime, mobility, and dark resistance are important figures of merit for describing and characterizing the quality of LTG III–V materials. Metallic precipitates in host III-V materials are the physical origin of a short carrier lifetime. However, their carrier mobility was reduced due to a simultaneous increase in scattering probability. The trade-off between carrier lifetime and mobility in LTG III–V materials, such as in GaAs and InGaAs, makes CW THz applications difficult. To overcome the trade-off relation, it was proposed that the carrier-trapping layer be separated from the absorption layer. The ErAs/InGaAs system utilizes separated metallic nanoparticles [27]. The InGaAs/InAlAs multilayer structure also utilizes the separate carrier-trapping layer [28]. A relatively high-quality absorbing layer consisting of InGaAs ensures high carrier mobility, and the carrier-trapping layer of InAlAs shortens the carrier lifetime.

Dark resistance also poses a critical problem in CW THz applications. The low band gap of InGaAs increases the dark current and decreases the breakdown voltage, which limits the applied voltage and the THz output power.

Although LTG III–V materials have acted as an important cornerstone from the viewpoint of a CW THz emitter, the low conversion efficiency of infrared (IR) to CW THz power was simultaneously an obstacle. The thermal issue due to high dark current is an important cause of low conversion efficiency [8].

Peytavit et al. reported a milliwatt-level THz output by using wafer-bonding technology. They used an LTG GaAs vertical resonant photoconductor and a metallic mirror Fabry–Pérot cavity [29]. Meanwhile, Tanoto et al. reported that tip-to-tip nanogap electrodes can greatly enhance the THz output due to a highly significant enhancement of the THz field [30]. They demonstrated an enhancement of 2 orders of magnitude in THz power emission compared to that emitted by a conventional photomixer.

1.3.3 High-Speed Photodiode-Based Photomixers

Since the mid-1990s, high-speed and high-output power PDs have been researched intensively for optical communication with a bit rate of over 40 Gbps. Typical surface-illuminated PDs have suffered due to the trade-off between responsivity and speed. Waveguide-type PDs allow high responsivity and high-speed operation by using the thin absorption layer. However, the saturation effect in the first few microns of the absorption layer limits the speed of waveguide-type PDs and also reduces the saturation input power. The evanescent PD solves these problems as it has a uniform carrier distribution inside the absorption layer [31]. For THz applications, photomixers based on a PD structure operating at 1.3 μm and 1.55 μm are technically important because we can connect the THz technology to well-developed photonics technology. Small and compact active devices such as tunable LDs, modulators, and optical amplifiers are commercially available. Furthermore, fiber-based passive devices also have their own merits in making THz systems. There are many reports on the merits of PD-based photomixers. Notably, uni-travelling-carrier photodiodes (UTC PDs) have been reported [32].

These UTC PDs utilize a p-doped absorption layer, which excludes holes with low mobility in carrier transport. They only use electrons as electrons have high mobility compared to holes. (For example, electron mobility is 11,200 cm^2/Vs, whereas hole mobility is 300 cm^2/Vs for bulk InGaAs.) Therefore, the 3dB bandwidth exceeds 310 GHz. CW THz output power is over 10 μW [33]. This high 3dB bandwidth is appropriate for THz applications. Although UTC PDs show high THz output power and high saturation input power, surface-illuminated UTC PDs suffer from low responsivities below 0.1 A/W. Recently, Rouvalis et al. solved the low responsivity problem by introducing the travelling-wave PD structure while maintaining the UTC PD epitaxial layers [34]. They reported a CW THz power of 148 μW at 457 GHz in the case of a resonant antenna with a high responsivity of 0.14–0.32 A/W. Sartorius et al. studied high-speed p-i-n PD-based photomixers [28] and reported a THz output power of 10 μW at 200 GHz with an IR input power of 30 mW and a wide THz bandwidth of over 1.5 THz.

Although PD-based photomixers show superior physical properties compared to LTG III–V-based photomixers with respect to THz output power, the IR to THz power conversion efficiency is still of the order of 10^{-3}. We therefore need a high optical input power of several tens or even hundreds of milliwatts. Low conversion efficiency is clearly the main obstacle in realizing a single-chip CW THz emitter, which exploits the wide tunability of the photomixing method.

1.3.4 Photomixer Module

LTG InGaAs–based and PD-based photomixers have a unique advantage in that they connect THz technologies to InP-based photonics technology, as mentioned above. However, the high dark current associated with InGaAs-based photomixers impedes efficient THz emission due to background thermal radiation. The photonics technology and related package method can help to realize a compact photomixer module and reduce the thermal problem. In this section, a compact InGaAs photomixer module is described.

A number of compact THz emission and detection systems and their component modules have been developed. Regarding such systems, there have been various descriptions of fiber-coupled THz spectrometers. Owing to the versatility of their long-wavelength optical components, compact size, and cost-effectiveness, InGaAs-based fiber-coupled THz spectroscopy systems such as those presented in Refs. [6, 7, 35] are considered promising. A number of fiber-connected photomixer modules for these systems have been developed: one is a butterfly-type fiber-pigtailed THz transmitter module measuring 1 in. × 0.5 in. × 0.5 in. and weighing a couple of ounces [36]; another comprises a miniaturized THz generator and detector heads whose electrical and optical parts are fixed by epoxy in the limited volume of the module [37]; the third has two input optical fibers connected to the THz emitter and receiver, respectively, but they can be adjusted using built-in adjustment screws [38]; and, finally, the CAN-type module has an effective heat dissipation solution as well as a module volume of less than 0.7 cc [8].

Figures 1.11b and 1.11c show a photograph of the module that is based on CAN-type packaging and its cross-sectional assembly process, respectively. The fiber-pigtailed photomixer (FPP) module consists of a log-spiral antenna-integrated LTG InGaAs photomixer chip, as shown in Fig. 1.11a, a heat-conductive submount, a high-resistivity hyperhemispherical Si lens, a single-mode fiber assembly with an aspherical glass lens (A-lens), a miniature coaxial cable, and a module housing [8]. The FPP module was assembled as follows: First, the photomixer chip and the heat-conductive submount were bonded together using a flip-chip bonding process. Then, the submount was attached to the body of the module housing, and the bias pads of the submount were connected with the miniature coaxial cable. The Si lens was placed on the bottom surface of the photomixer chip and fixed with the cover screw of the module housing. Finally, the entire module housing was actively aligned and welded to the fiber assembly with the A-lens, using a laser welding technique. Note that the active alignment between the active area of the photomixer chip and the fiber core of the fiber assembly must be precise before commencing laser welding. The active alignment was necessary to monitor and find the maximum value of the photomixer photocurrent when a bias voltage of about 0.1 V was applied and an optical beam with a power of about 5 dBm was adjusted and focused to the active area of the photomixer chip.

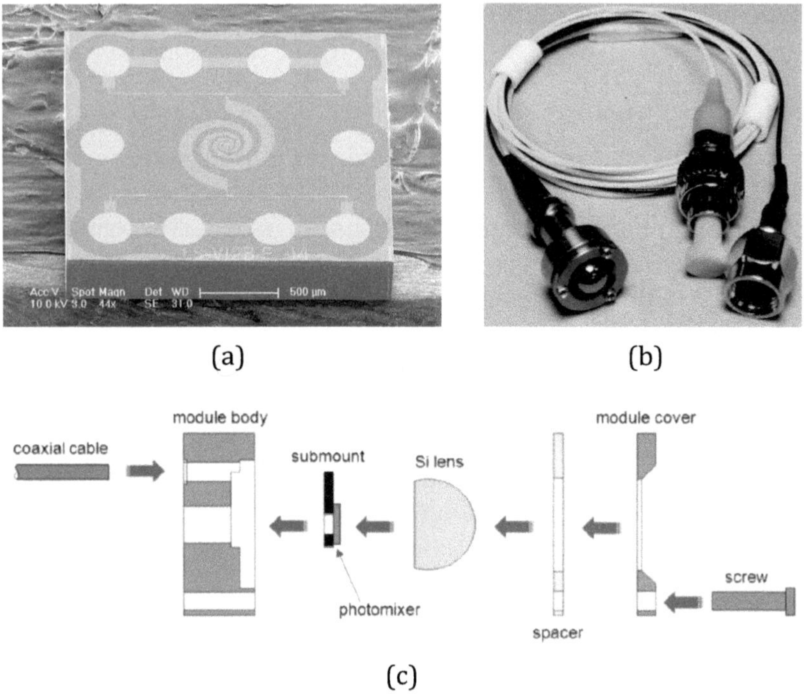

Figure 1.11 Compact photomixer module. (a) SEM image of an LTG InGaAs photomixer, (b) a photograph of a fiber-pigtailed photomixer module, and (c) the assembly process thereof. Reproduced from Ref. [8] by permission of Optical Society of America, © 2012.

Heat generated in the active area of the photomixer must be dissipated to increase the SNR and performance of the THz systems, since inefficient heat dissipation from the photomixer will result in thermal degradation. Thermal degradation is one of the factors that limit photomixer performance. The sources of this heat are the optical energy absorbed by the active area of the photomixer and joule heating due to the photocurrent. Furthermore, in narrow-band-gap materials such as InGaAs, the THz output power of the photomixer can also be limited by joule heating from the dark current [21]. Therefore, heat dissipation is vital for InGaAs photomixers because the bias voltage applied at the photomixer is limited by the heat generated due to the electrical current [39].

1.3.5 Optical Beat Sources

The quest to increase the conversion efficiency of photomixers has led to a continuous increase in the demand for efficient and compact optical beat sources. The role of an optical beat source in a photomixing system is not limited to its use as an optical pumping source to generate electron–hole pairs. The spectral characteristics of THz radiation are governed by the spectral purity of the optical beat signal, such as spectral linewidth, relative intensity noise (RIN), polarization matching, power balance, and spatial overlap. First, polarization matching and spatial overlap determine the mode beat efficiency directly. Therefore, a polarization-maintaining fiber is typically used in CW THz systems containing an optical fiber. Second, the linewidth of the individual laser beam and the phase noise between the two lasers are reflected in the spectral linewidth of the generated CW THz radiation. Because the frequency of the THz radiation is equal to the difference in frequency of the two laser beams, fluctuations of each laser beam in the frequency domain result in linewidth broadening of the THz radiation. Finally, the RIN of the laser also causes fluctuations in the output power of the THz radiation. The spectral purity of the two laser beams should be high in order to generate an efficient and clean mode beat signal.

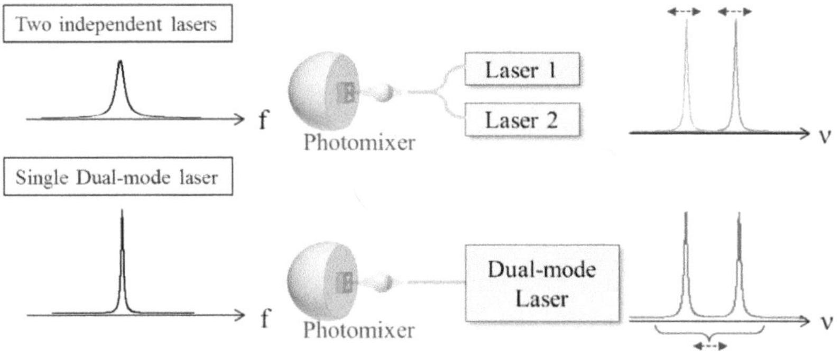

Figure 1.12 Noise and beat source characteristics.

From this viewpoint, there have been many efforts to utilize the two lasing modes from a single optical resonator. If one laser emits two lasing modes simultaneously, the linewidth of the THz radiation would be minimized to maintain the wavelength difference from environmental noise sources such as temperature variation, current source noise, external vibration, etc. This effect is well known as the common mode noise rejection effect [40] and has been clearly demonstrated in experiments in the field of microwave generation using the photonic method [41]. The system configuration is also greatly simplified when it is combined with a heterodyne receiver, such as a Schottky diode [42] and field-effect transistors [43]. However, the development of a widely tunable dual-mode laser with a single cavity structure is technically challenging. Because typical tuning methods, such as thermal and electrical tuning via the plasma effect, affect both lasing modes, the beat frequency cannot be tuned over a wide range of wavelengths. Furthermore, the gain competition between the two lasing modes also prevents the optical beat source with a single active medium from performing dual-mode operation. Optical beat sources have been widely reported for CW THz systems. Fiber-based laser systems have been developed to have a wide tuning range [44]. However, fiber lasers have problems in longitudinal multimodes due to their long cavity length and the gain competition of the erbium-doped fiber. Longitudinal multimodes generate a low-frequency envelope in the mode beat signal, which has a relatively high THz conversion efficiency in the photomixer. The second important category consists of the external cavity tunable LDs [45, 46]. They exhibit a wide tuning range and high output power. However, external cavity lasers need mechanical moving parts that are not appropriate for field applications. Finally, it must be noted that semiconductor-based monolithically integrated dual-mode laser diodes (DMLs) have been reported. For single-frequency THz applications, the two-color Fabry–Pérot LD was reported [47]. They selected two lasing modes by using the spatially varying effective refractive index profile along the cavity. Klehr et al. reported dual-mode distributed feedback (DFB) LDs with high output power [48]. They utilize the first and second lateral modes. Furthermore, by using the mode-hopping mechanism, it can be operated at two distinct frequencies of 150 GHz and 350 GHz. The output power exceeds 435 mW. However, these kinds of devices do not have the ability of continuous tuning, which indicates the loss of an important merit of the photomixing method. Price et al. integrated two distributed Bragg reflector laser diodes (DBR LDs) in parallel geometry [49]. It exhibits a wide tuning range of over 4 THz. Although it is compact and has an easy operation principle, it does not have single-cavity geometry and there is no mode correlation between the two modes.

Recently, widely tunable monolithically integrated DMLs have been demonstrated [50, 51]. Two DFB LDs and the phase section, which controls the interaction between the two DFB LDs, have been integrated. Integrated microheaters (μ-heaters) provide wavelength tuning without mode hopping. The DML has single-cavity geometry and realizes the compact and widely tunable optical beat source. It can be integrated with other optical devices such as semiconductor optical amplifiers, modulators, and photomixers. If a DML is integrated with PD-based photomixers, the widely tunable single chip THz emitter will be realized.

Figure 1.13 shows the scanning electron microscope (SEM) image of a monolithically integrated 1.3 μm phase-shifted dual-mode laser diode (1.3-pDML). It consists of two DFB LDs with a cavity length of 400 μm and one phase section with a length of 50 μm. The total cavity length, including the spacer between each section, is 880 μm. Quarter-wavelength-shifted DFB LDs offer stable single-mode operation characteristics for each of the DFB LDs [52]. Because simple DFB LDs with an index-coupled grating show phase-sensitive operating characteristics due to the competition between −1 and +1 modes, the phase-shifted grating is preferable. The operating wavelengths of each of the DFB LDs are set to be 1300 nm and 1304 nm to obtain an initial operating mode beat frequency of around 700 GHz. The μ-heaters on each DFB LD control the operating wavelength of the DFB LDs independently. Because the μ-heaters only control the temperature of local area below the μ-heaters, it is possible to achieve independent wavelength tuning. The tuning speed is much faster than the conventional method that controls the temperature of the submount and laser cavity with a thermoelectric controller [53]. Therefore, the DMLs are able to tune the mode beat frequency over a wide range.

Figure 1.14 shows the initial operating state of the pDML. The side mode suppression ratio (SMSR) is over 50 dB, and clear four-wave mixing (FWM) signals are observed. Physically, the FWM signals indicate the degree of the mode correlation. When the spacing between the two modes is over 100 GHz, the carrier density pulsation along the cavity is the main mechanism of the FWM signal [41]. This means that the phase of the two modes is stable and well correlated.

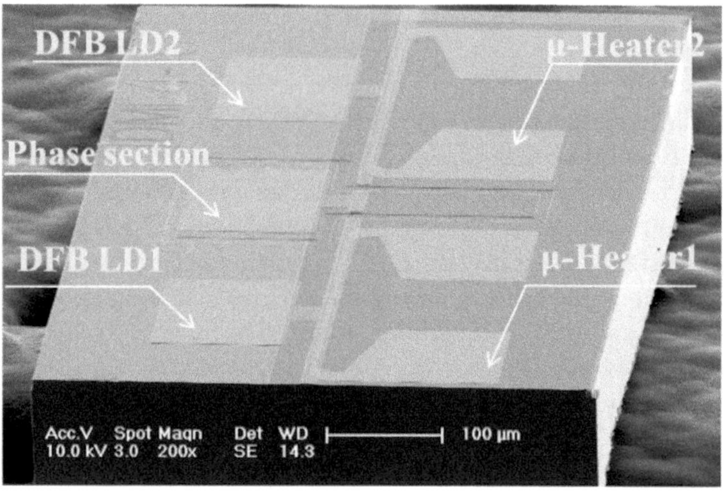

Figure 1.13 SEM image of a 1.3-pDML.

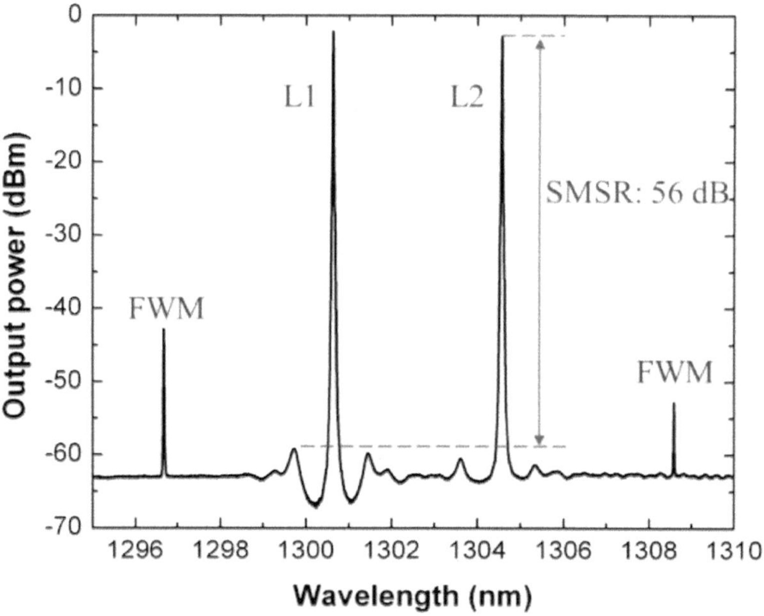

Figure 1.14 Initial operating state of the 1.3-pDML. The mode beat frequency corresponds to 693 GHz. Reproduced from Ref. [51] by permission of IOP Publishing & Deutsche Physikalische Gesellschaft. CC BY-NC-SA, © 2013.

Another important feature in the operating principle of the 1.3-pDML is the role of the phase section. When the operating currents are applied to two DFB LDs simultaneously, the compound cavity modes are activated. When the difference in wavelength between the two lasing modes is smaller than the width of the stopbands of the DFB LDs, the mutual interaction caused by the strong optical feedback leads to compound cavity mode operation, which obtains optical gain from the entire optical cavity. These modes are very phase sensitive, and the number of lasing modes and the wavelength depend on the operating condition of the phase section. However, when the wavelength difference between the two lasing modes is wide enough, the pDML acts as an SOA-integrated DFB LD. At first, one of the DFB LDs lases and the other DFB LD just amplifies the single lasing mode. Therefore, the optical gain of the compound cavity modes should be controlled without sacrificing the optical gain of each lasing mode [54]. The phase section with reverse bias controls the round-trip optical gain by controlling the optical absorption and successfully suppresses the compound cavity modes.

After obtaining the dual-mode operation, as shown in Fig. 1.14, the operating wavelength of each DFB LD can be tuned independently through integrated μ-heaters, as shown in Fig. 1.15. By passing a current through the μ-heater for the lasing mode L1, the mode beat frequency is decreased from 693 GHz to 290 GHz. In this case, the overlap of the stopbands limits the mode beat frequency tuning range. When the current is passed through the μ-heater for the lasing mode of L2, the mode beat frequency is increased up to 1.2 THz. The tuning spectra of the entire tuning range are shown in Fig. 1.15.

CW THz radiation was also demonstrated by combining LTG InGaAs photomixers and 1.3-pDML, as shown in Fig. 1.16. A typical homodyne configuration was used [55]. The 1.3 μm SOA was used to amplify the output from the 1.3-pDML. The photomixers were fabricated using the LTG InGaAs bulk layer grown by MBE. The growth temperature was 160°C, and the thickness of InGaAs was 1.2 μm. Ex situ annealing at 560°C was performed. An SNR of 48 dB at 300 GHz was measured. The experimental results indicate that the pDML is a promising key component of the single-chip THz emitter using a photomixing system.

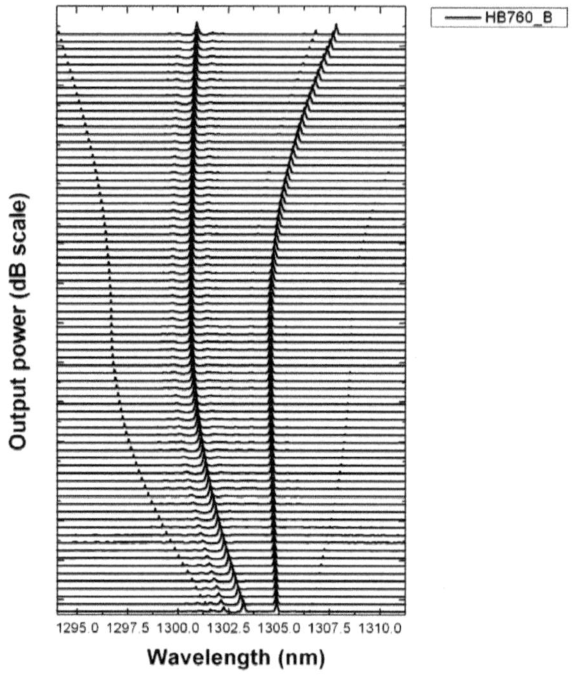

Figure 1.15 The tuning spectra of the 1.3-pDML. The mode beat frequency tuning ranges from 0.29 THz to over 1.215 THz. Clear FWM signals are observed throughout the tuning range.

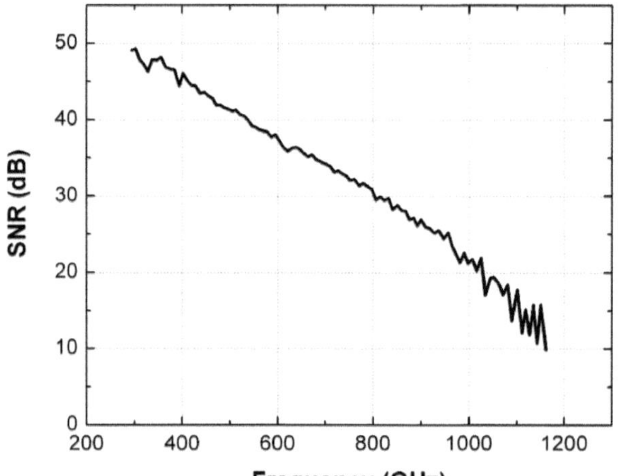

Figure 1.16 CW THz spectrum measured with LTG InGaAs photomixers and 1.3-pDML.

1.4 Applications

1.4.1 Cost-Effective Coherent Homodyne CW THz System

THz technology has shown potential in a wide variety of applications such as spectroscopy, imaging, and sensing because of its unique properties [1–3]. CW THz systems based on photomixing have recently received considerable attention as an important candidate for industrialization because of their lower cost, smaller size, and higher frequency resolution compared to pulsed THz systems [4, 5]. A coherent homodyne CW THz system setup is shown in Fig. 1.17. The system consists of a $\lambda/4$ phase-shifted DML and an all-fiber path for the laser beams in order to increase flexibility and compactness. Two DFB lasers operating at slightly different wavelengths with a fiber optic 50:50 combiner can be used instead of one DML. The optical beat signal emitted from the DML package was amplified by an erbium-doped fiber amplifier (EDFA). The amplified spontaneous emission generated by the EDFA was filtered out by an optical bandpass filter (BPF). The output signal from the BPF was coupled to a fiber optic 50:50 splitter and divided into two integrated photomixers/antennas. They play the roles of a THz transmitter and receiver, respectively, in the coherent homodyne CW THz system.

1.4.2 Measurement of Thickness Using a Coherent Homodyne CW THz System

The detected photocurrent in the coherent homodyne CW THz system is proportional to the THz electric field E_{THz}, and its phase depends on the phase difference between the THz CW and the laser beat signals for the transmitter and receiver, which can be written as

$$I_{\mathrm{THz}} \propto E_{\mathrm{THz}} \cdot \cos\left(\frac{2\pi f}{c} \cdot \Delta L + \varphi\right) \quad (1.6)$$

$$\text{where } \Delta L = \left(L_{\mathrm{Tx}} + L_{\mathrm{THz}}\right) - L_{\mathrm{Rx}}$$

The phase offset φ is frequency independent. The path length difference ΔL can be changed by scanning the THz CW path (L_{THz}) or the optical path for the THz transmitter (L_{Tx}) or receiver (L_{Rx}), as shown in Fig. 1.17. Especially when ΔL is zero, the normalized THz photocurrents are constant throughout the entire frequency band. Furthermore, the length of the THz path depends on the refractive index and thickness of a sample. Thus, the thickness of a sample having a known refractive index can be calculated exactly by determining the optical path length that maintains a constant normalized THz photocurrent in the frequency band. This optical path length can be found by simply scanning the optical delay line inserted into the optical path. The optical path length for the receiver is controlled by scanning a precise optical fiber delay line located between the laser source and the THz receiver, while the optical path length for the transmitter is fixed, as shown in Fig. 1.17. The optical path lengths for the receiver, which yield constant THz photocurrents in the measured beat frequency band, are expressed as follows:

$$0 = (L_{\mathrm{Tx}} + L_{\mathrm{THz\_ref}}) - L_{\mathrm{Rx\_ref0}}$$
$$0 = (L_{\mathrm{Tx}} + L_{\mathrm{THz\_sam}}) - L_{\mathrm{Rx\_sam0}} \quad (1.7)$$

where $L_{\mathrm{Rx\_ref0}}$ and $L_{\mathrm{Rx\_sam0}}$ are the optical path lengths for the reference and the sample, respectively, when the THz photocurrents are constant in the measured frequency band. For the reference data, the THz photocurrents could be measured through air without any sample in the frequency band. The refractive index of a sample is the same as its group refractive index if the measured frequency band is narrow enough for the sample not to be dispersive. Therefore, the relationship between the THz path lengths for the reference and the sample is

$$L_{\mathrm{THz\_sam}} = L_{\mathrm{THz\_ref}} + (n - 1)\cdot d \quad (1.8)$$

where d and n are the sample's thickness and refractive index, respectively. The thickness of a sample can be calculated using Eqs. 1.7 and 1.8. With this method of measuring thickness, there is no modulo 2π ambiguity, which can be a problem in methods that use exact frequency scanning at fixed optical path lengths [56, 57]. Moreover, this method does not require precise control of the beat frequency; only accurate scanning of a simple optical delay line is needed [53].

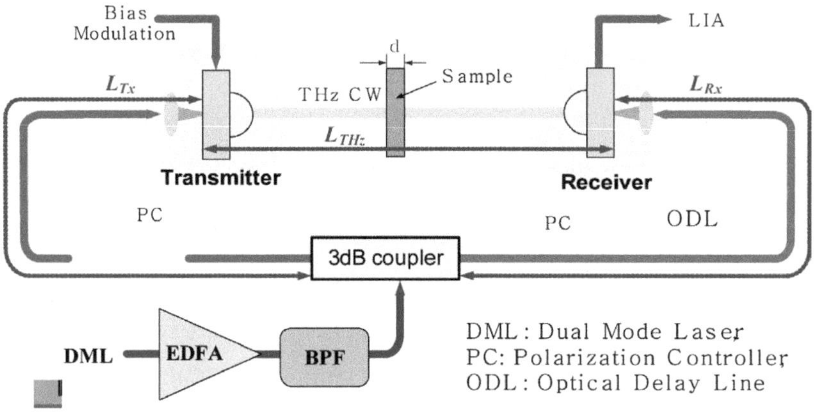

Figure 1.17 Schematic diagram of a coherent homodyne CW THz system setup. Reproduced from Ref. [53] by permission of Optical Society of America, © 2012.

Figure 1.18 shows the normalized THz photocurrents measured without and with the polyethylene (PE) pellet according to the variations in the optical delay in the frequency range from 448 GHz to 610 GHz at intervals of about 4 GHz. The lock-in time constant was set to 100 ms to obtain every time trace for a thickness measurement. The total number of measured frequencies was 37. The step in the delay line was 0.05 ps, which determines the resolution of the thickness measurement and measuring time.

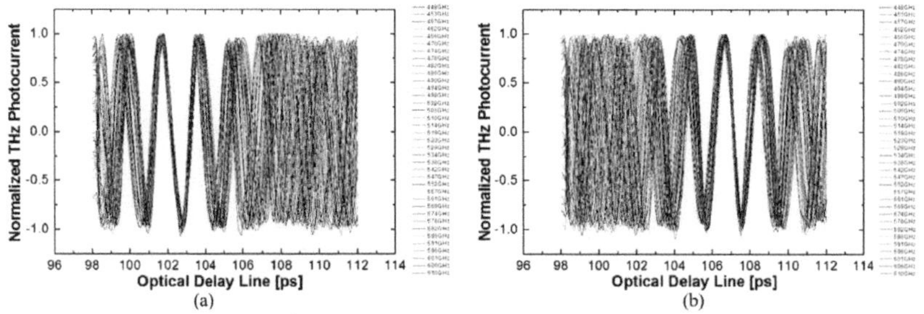

Figure 1.18 Normalized THz photocurrent measured (a) without and (b) with the PE pellet according to the variations in the optical delay in the frequency range from 448 GHz to 610 GHz at intervals of about 4 GHz. Reproduced from Ref. [53] by permission of Optical Society of America, © 2012.

The equiphase positions of the delay line for the reference and the PE sample, at which the normalized THz photocurrents are constant in the measured frequency band, are located between 101.75 ps and 102.75 ps and between 106.65 ps and 107.55 ps, respectively. The equiphase positions can be easily obtained by adding all the time traces measured at all 37 frequencies, as shown in Fig. 1.19. The displacement of the maximum peak after THz CW penetration through the PE sample represents the phase delay of the THz CW through the PE sample because the phase offset φ is frequency independent. The delay time of the THz CW caused by the PE sample was between 4.8 ps and 4.9 ps. The refractive index of the PE measured using our THz-TDS was 1.46. Therefore, the deduced thickness of the PE sample was between 3.13 mm and 3.19 mm. The measurement deviation was caused by the intensity fluctuation of the measured THz signal due to the variation of the optical polarization and phase during the propagation through the single-mode fiber. The measurement by the THz CW system agreed well with that obtained using a standard mechanical micrometer within the error range of the micrometer.

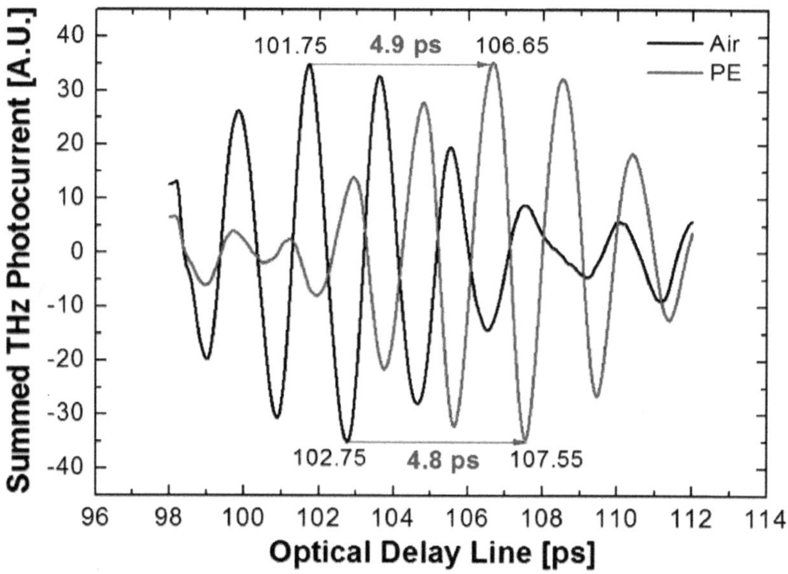

Figure 1.19 Summations of time traces measured at 37 frequencies without and with PE pellet. Reproduced from Ref. [53] by permission of Optical Society of America, © 2012.

The equiphase positions of the delay line can also be found by adding only three time traces measured at three frequencies. The three time traces were used to find the equiphase positions exactly, even though two time traces could be used to find the equiphase positions in principle. Figures 1.20a and 1.20b show the summations of the three time traces measured at three frequencies without and with the PE sample, respectively, for various frequency spans from 20 GHz to 160 GHz. The central frequency was 523 GHz. The difference between the values of the maximum and second peaks decreased as the frequency span decreased from 160 GHz to 20 GHz. The difference between the adjacent frequencies determines the measurable maximum thickness of a sample. These results show that only three beat frequencies are sufficient for measuring the thickness of a sample and that they do not need to be exactly controlled.

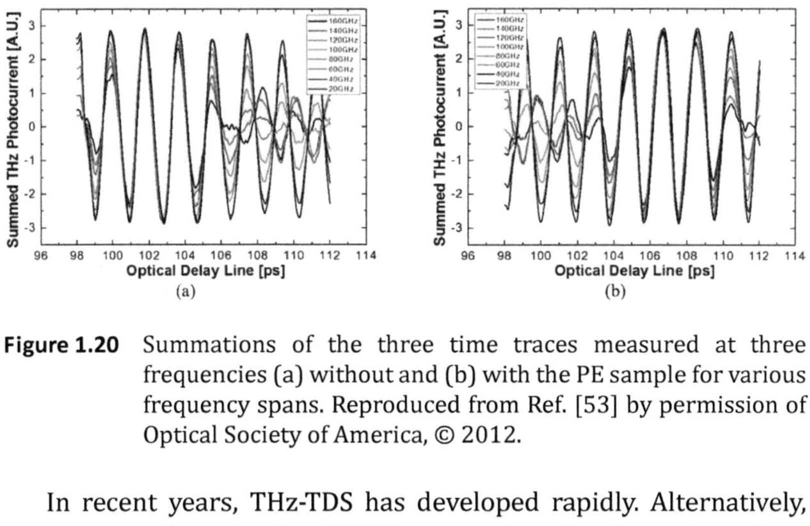

Figure 1.20 Summations of the three time traces measured at three frequencies (a) without and (b) with the PE sample for various frequency spans. Reproduced from Ref. [53] by permission of Optical Society of America, © 2012.

In recent years, THz-TDS has developed rapidly. Alternatively, spectroscopy by using the CW THz system has also shown its unique advantages in terms of frequency resolution, size, and cost. Both the real and imaginary parts of the complex dielectric function have been determined reliably by measuring both the amplitude and the phase of the THz wave in the coherent THz system, as shown in Fig. 1.21. The measured photocurrent in the CW THz system depends both on the amplitude of the THz electric field and on the phase difference that can be controlled by scanning either the THz frequency or the optical delay length. When a sample is introduced into the THz beam path, the THz wave is delayed and attenuated, and the detected current can be expressed as

$$I_{\mathrm{THz}} \propto E_{\mathrm{THz}} e^{-\pm \mathrm{d}} \cdot \cos\left(\frac{2\pi f}{c} \cdot \Delta L + \varphi\right)$$
$$\text{where } \Delta L = \left(L_{\mathrm{Tx}} + L_{\mathrm{THz}}\right) - L_{\mathrm{Rx}} \tag{1.9}$$
$$L_{\mathrm{THz}} = \left(n(f) - 1\right)d$$

where α is the absorption coefficient of the amplitude of the THz CW and $n(f)$ is the real part of the refractive index of a sample that is frequency dependent. Then, the complex dielectric characteristics of a sample can be calculated by using the relation between Eqs. 1.6 and 1.9. Figure 1.21 shows the measured dielectric characteristics of a 17 mm thick high-resistive silicon sample by using the coherent homodyne THz CW system [57]. The phase shift and attenuation shown in Fig. 1.21a are due to the delay of the THz path and the absorption and the Fresnel reflection, respectively. The modulo 2π ambiguity can be removed by the linear relationship between the absolute phase shift and the frequency, as with THz-TDS.

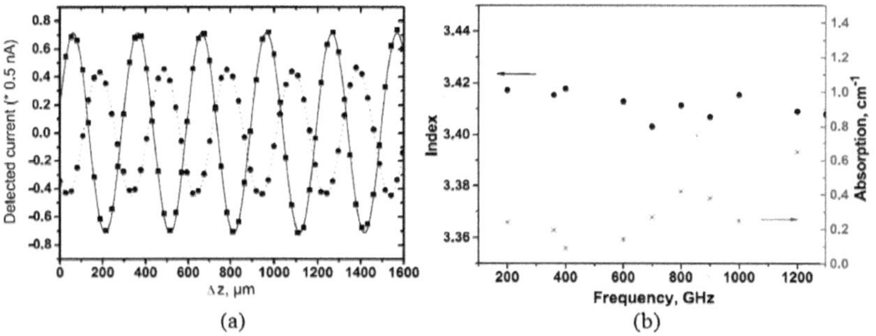

Figure 1.21 (a) Detected electric field of THz radiation vs. position of the delay line at 1000 GHz. The black line is the reference curve, and the dashed line is the transmission of the THz electric field of high-resistive silicon. (b) Index of refraction and absorption coefficient of high-resistive silicon. Reproduced with kind permission from Springer Science+Business Media [57].

The complex dielectric characteristics of a sample can be analyzed on the basis of the scanning of an interference pattern in a frequency, when the beat source is precisely controlled using an interferometric frequency controller. This method precisely scans only a beat frequency while fixing a delay line to change the phase difference of a detected photocurrent. The detected photocurrent oscillates with frequency, and the oscillation period is determined by the phase difference of the photocurrent. Comparing the reference and sample data, the refractive index can be calculated and the sample thickness determined independently [56]. Figure 1.22 shows the detected THz photocurrent with and without the lactose sample and the calculated complex dielectric characteristics of the lactose.

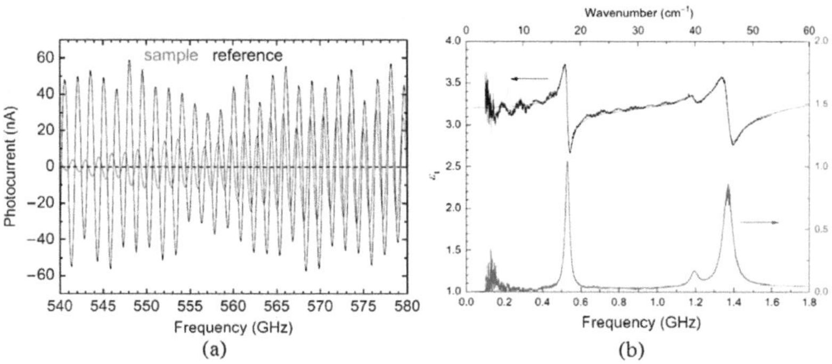

Figure 1.22 (a) The detected THz photocurrent with and without the lactose sample and (b) the real (black) and imaginary (blue) parts of the dielectric characteristics of the lactose. Reproduced from Ref. [56] by permission of IOP Publishing & Deutsche Physikalische Gesellschaft. CC BY-NC-SA, © 2010.

1.5 Conclusion

In this chapter, we reviewed the current status and prospects of photonics-based broadband THz generation technologies. Even though there emerge some commercialized CW THz spectroscopy systems, for more practical applications at hand, there still remain technical issues that need innovative progress in related technologies. Although there are time-requiring technical hurdles in the sizes, power efficiencies, and prices of the core modules or devices for THz technologies to become general industrial ones, the CW THz system still remains very attractive as a realizable candidate for the core of human-centric technologies such as health monitoring or any other kinds of communications between humans and smart devices. Although not fully introduced in this chapter, to enhance the optoelectric conversion efficiency for the generation and detection of photonics-based broadband THz technologies, various approaches such as using nanostructures or microcavities, hybridizations of an LTG PC film with a signal line, and guided-type photodetectors have been proposed. We expect that a cost-effective compact CW THz system would be available in the near future and be used for many industrial applications and human-centric technologies.

References

1. P. U. Jepsen, D. G. Cooke, and M. Koch, Terahertz spectroscopy and imaging: modern techniques and applications, *Laser Photon. Rev.* **5**(1), 124–166 (2011).
2. H. B. Liu, H. Zhong, N. Karpowicz, Y. Chen, and X.-C. Zhang, Terahertz spectroscopy and imaging for defense and security applications, *Proc. IEEE* **95**(8), 1514–1527 (2007).
3. M. Tonouchi, Cutting-edge terahertz technology, *Nat. Photonics* **1**, 97–105 (2007).
4. S. Verghese, K. A. McIntosh, and E. R. Brown, Highly tunable fiber-coupled photomixers with coherent terahertz output power, *IEEE Trans. Micro. Theo. Tech.* **45**(8), 1301–1309 (1997).
5. I. S. Gregory, C. Baker, W. R. Tribe, I. V. Bradley, M. J. Evans, E. H. Linfield, A. G. Davies, and M. Missous, Optimization of photomixers and antennas for continuous-wave terahertz emission, *IEEE J. Quantum Electron.* **41**, 717–728 (2005).
6. S.-P. Han, H. Ko, N. Kim, H.-C. Ryu, C. W. Lee, Y. A. Leem, D. Lee, M. Y. Jeon, S. K. Noh, H. S. Chun, and K. H. Park, Optical fiber-coupled InGaAs-based terahertz time-domain spectroscopy system, *Opt. Lett.* **36**(16), 3094–3096 (2011).
7. B. Sartorius, H. Roehle, H. Künzel, J. Böttcher, M. Schlak, D. Stanze, H. Venghaus, and M. Schell, All-fiber terahertz time-domain spectrometer operating at 1.5 microm telecom wavelengths, *Opt. Express* **16**(13), 9565–9570 (2008).
8. S.-P. Han, N. Kim, H. Ko, H.-C. Ryu, C. W. Lee, Y.-J. Yoon, J.-H. Shin, D. Lee, S.-H. Park, S.-H. Moon, S.-W. Choi, H. S. Chun, and K. H. Park, Compact fiber-pigtailed InGaAs photomixer module for terahertz-wave generation and detection, *Opt. Express,* **20**(16), 18432–18439 (2012).
9. N. Kim, J. Shin, E. Sim, C. W. Lee, D.-S. Yee, M. Y. Jeon, Y. Jang, and K. H. Park, Monolithic dual-mode distributed feedback semiconductor

laser for tunable continuous-wave terahertz generation, *Opt. Express* **17**(16), 13851–13859 (2009).

10. B. Ferguson and X.-C. Zhang, Materials for terahertz science and technology, *Nat. Mater.*, **1**, 26–33 (2002).
11. D. H. Auston, K. P. Cheung, and P. R. Smith, Picosecond photoconducting Hertzian dipoles, *Appl. Phys. Lett.* **45**, 284–286 (1984).
12. L. Duvillaret, F. Garet, J.-F. Rous, and J.-L. Coutaz, Analytical modeling and optimization of terahertz time-domain spectroscopy experiments, using photoswitches as antennas, *IEEE J. Sel. Top. Quantum Electron.* **7**, 615–623 (2001).
13. P. U. Jepsen, R. H. Jacobsen, and S. R. Keiding, Generation and detection of terahertz pulses from biased semiconductor antennas, *J. Opt. Soc. Am. B* **13**, 2424–2436 (1996).
14. Z. Piao, M. Tani, and K. Sakai, Carrier dynamics and terahertz radiation in photoconductive antennas, *Jpn. J. Appl. Phys.* **39**, 96–100 (2000).
15. D. Grischkowsky, S. Keiding, M. Exter, and Ch. Fattinger, Far-infrared time-domain spectroscopy with terahertz beams of dielectrics and semiconductors, *J. Opt. Soc. Am. B* **7**, 2006–2015 (1990).
16. K. Mahalingam et al., Substrate temperature dependence of arsenic precipitate formation in AlGaAs and GaAs, *J. Vac. Sci. Technol. B* **9**, 2328–2332 (1991).
17. M. R. Melloch et al., Low-temperature grown III-V materials, *Annu. Rev. Mater. Sci.* **25**, 547–600 (1995).
18. I. S. Gregory et al., High resistivity annealed low-temperature GaAs with 100 fs lifetimes, *Appl. Phys. Lett.* **83**, 4199–4201 (2003).
19. R. A. Metzger, A. S. Brown, L. G. McCray, and J. A. Henige, Structural and electrical properties of low temperature GaInAs, *J. Vac. Sci. Technol. B* **11**, 798–801 (1992).
20. Y. Chen et al., Trapping and recombination dynamics of low-temperature-grown InGaAs/InAlAs multiple quantum wells, *Appl. Phys. Lett.* **72**, 439–441 (1998).
21. K. Namje et al., Tunable continuous-wave terahertz generation/detection with compact 1.55 μm detuned dual-mode laser diode and InGaAs based photomixer, *Opt. Express* **19,** 146541 (2011).
22. S.-P Han, H. Ko, J.-W. Park, N. Kim, Y.-J. Yoon, J.-H. Shin, D. Y. Kim, D. H. Lee, and K. H. Park, InGaAs Schottky barrier diode array detector for a real-time compact terahertz line scanner. *Opt. Express* **21**(22), 25874–25882 (2012).

23. E. R. Brown, THz Generation by photomixing in ultrafast photoconductors, *Int. J. High Speed Electron. Syst.* **13**(2) 497–545 (2003).

24. S. Gupta, J. F. Whitaker, and G. A. Mourou, Ultrafast carrier dynamics in III-V semiconductors grown by molecular-beam epitaxy at very low substrate temperatures, *IEEE J. Quantum Electron.* **28**(10), 2464–2472 (1992).

25. H. Künzel, J. Böttcher, R. Gibis, and G. Urmann, Material properties of Ga0.474In0.53As grown on InP by low-temperature molecular beam epitaxy, *Appl. Phys. Lett.* **61**, 1347–1349 (1992).

26. M. R. Melloch, J. M. Woodall, E. S. Harmon, N. Otsuka, F. H. Pollak, D. D. Nolte, R. M. Feenstra, and M. A. Lutz, Low-temperature grown III-V materials, *Annu. Rev. Mater. Sci.* **25**, 547–600 (1995).

27. M. Sukhotin, E. R. Brown, A. C. Gossard, D. Driscoll, M. Hanson, P. Maker, and R. Muller, Photomixing and photoconductor measurements on ErAs/InGaAs at 1.55 um, *Appl. Phys. Lett.* **82**, 18, 3116–3118 (2003).

28. B. Sartorius, M. Schlak, D. Stanze, H. Roehle, H. Künzel, D. Schmidt, H.-G. Bach, R. Kunkel, and M. Schell, Continuous wave terahertz systems exploiting 1.5 um telecom technologies, *Opt. Express* **17**, 15001–15007 (2009).

29. E. Peytavit, S. Lepilliet, F. Hindle, C. Coinon, T. Akalin, G. Ducournau, G. Mouret, and J.-F. Lampin, Milliwatt level output power generated by photomixing in a GaAs photoconductor, IRMMW-THz 2012, Thu-C-1-1 (2012).

30. H. Tanoto, J. H. Teng, Q. Y. Wu, M. Sun, Z. N. Chen, S. A. Maier, B. Wang, C. C. Chum, G. Y. Si, A. J. Danner, and S. J. Chua, Greatly enhanced continuous-wave terahertz emission by nano-electrodes in a photoconductive photomixer, *Nat. Photon.* **6**, 121–126 (2012).

31. N. Michel, V. Magnin, J. Harari, A. Marceaux, O. Parillaud, D. Decoster, and N. Vodjdani, High-power evanescently-coupled waveguide photodiodes, *EEE Proc. Optoelectron.* **153**(4), 199–204 (2006).

32. H. Ito, T. Furuta, S. Kodama, and T. Ishibashi, InP/InGaAs uni-travelling-carrier photodiode with 310 GHz bandwidth, *Electron. Lett.* **36**, 1809 (2000).

33. H. Ito, F. Nakajima, T. Furuta, and T. Ishibashi, Continuous THz-wave generation using antenna-integrated uni-travelling-carrier photodiodes, *Semicond. Sci. Technol.* **20**, S191 (2005).

34. E. Rouvaliz, C. C. Renaud, D. G. Moodie, M. J. Robertson, and A. J. Seeds, Traveling-wave uni-traveling carrier photodiodes for continuous wave THz generation, *Opt. Express* **18**(11), 11105–11110 (2010).

35. C. Baker, I. S. Gregory, M. J. Evans, W. R. Tribe, E. H. Linfield, and M. Missous, All-optoelectronic terahertz system using low temperature-grown InGaAs photomixers, *Opt. Express* **13**(23), 9639–9644 (2005).

36. D. Zimdars, J. V. Rudd, and M. Warmuth, A compact, fiber-pigtailed, terahertz time domain spectroscopy system, *Proc. ISSTT*, 414–423 (2000).

37. Y. B. Ji, E. S. Lee, S.-H. Kim, J.-H. Son, and T.-I. Jeon, A miniaturized fiber-coupled terahertz endoscope system, *Opt. Express* **17**(19), 17082–17087 (2009).

38. D. Stanze, A. Deninger, A. Roggenbuck, S. Schindler, M. Schlak, and B. Sartorius, Compact CW Terahertz spectrometer pumped at 1.5 μm wavelength, *J. Infrared Milli. Terahz. Waves* **22** (2010).

39. N. S. Daghestani, S. Persheyev, M. A. Cataluna, G. Ross, and M. J. Rose, THz generation from a nanocrystalline silicon-based photoconductive device, *Semicond. Sci. Technol.* **26**, 075015 (2011).

40. M. Tani et al., Generation of terahertz radiation by photomixing with dual- and multiple-mode lasers, *Semicond. Sci. Technol.* **20**(7), S151–S163 (2005).

41. J. Renaudier, G.-H. Duan, J.-G. Provost, H. Debregeas-Sillard, and P. Gallion, Phase correlation between longitudinal modes in semiconductor self-pulsating DBR lasers, *IEEE Photon. Technol. Lett.* **17**(4), 741–743 (2005).

42. N. Karpowicz, H. Zhong, C. Zhang, K.-I. Lin, J.-S. Hwang, J. Xu, and X.-C. Zhang, Compact continuous-wave subterahertz system for inspection applications, *Appl. Phys. Lett.* **86**(5), 054105 (2005).

43. B. Gershgorin, V. Yu. Kachorovskii, Y. V. Lvov, and M. S. Shur, Field effect transistor as heterodyne terahertz detector, *Electron. Lett.* **44**(17), 1036–1037 (2008).

44. M. Y. Jeon, N. Kim, J. Shin, J. S. Jeong, S.-P. Han, C. W. Lee, Y. A. Leem, D.-S. Yee, H. S. Chun, and K. H. Park, Widely tunable dual-wavelength Er3+-doped fiber laser for tunable continuous-wave terahertz radiation, *Opt. Express* **18**(12), 12291–12297 (2010).

45. S. Hoffmann, M. Hofmann, M. Kira, and S. W. Koch, Two-colour diode lasers for generation of THz radiation, *Semicond. Sci. Technol.* **20**(7), S205–S210 (2005).

46. M. Scheller, S. W. Koch, and J. V. Moloney, Multi-watt femtosecond and continuous wave semiconductor disc lasers for terahertz systems, IRMMW-THz 2012, Tue-B-1-2 (2012).

47. S. Osborne, S. O'Brien, E. P. O'Reilly, P. G. Huggard, and B. N. Ellison, Generation of CW 0.5 THz radiation by photomixing the output of a two-colour 1.49 μm Fabry–Pérot diode laser, *Electron. Lett.* **44**(4), 296 (2008).

48. A. Klehr, J. Fricke, A. Knauer, G. Erbert, M. Walther, R. Wilk, M. Mikulics, and M. Koch, High-power monolithic two-mode DFB laser diode for the generation of THz radiation, *IEEE J. Sel. Top. Quantum Electron.* **14**(2), 289–294 (2008).

49. R. K. Price, V. B. Verma, K. E. Tobin, V. C. Elarde, and J. J. Coleman, Y-branch surface-etched distributed Bragg reflector lasers at 850 nm for optical heterodyning, *IEEE Photon. Technol. Lett.* **19**(20), 1610–1612, (2007).

50. N. Kim, J. Shin, E. Sim, C. W. Lee, D.-S. Yee, M. Y. Jeon, Y. Jamg, and K. H. Park, Monolithic dual-mode distributed feedback semiconductor laser for tunable continuous-wave terahertz generation, *Opt. Express* **17**(6), 13851–13859 (2009).

51. N. Kim, H.-C. Ryu, D. Lee, S.-P. Han, H. Ko, K. Moon, J.-W. Park, M. Y. Jeon, and K. H. Park, Monolithically integrated optical beat sources toward a single-chip broadband terahertz emitter, *Laser Phys. Lett.* **10**, 085805 (2013).

52. J. E. A. Whiteaway. G. H. B. Thompson, A. J. Collar, and C. J. Armistead, The design and assessment of λ/4 phase-shifted DFB laser structures, IEEE J. *Quantum Electron.* **25**(6) 1261 (1989).

53. G. Mouret, S. Matton, R. Bocquet, D. Bigourd, F. Hindle, A. Cuisset, J. F. Lampin, K. Blary, and D. Lippens, THz media characterization by means of coherent homodyne detection, results and potential applications, *Appl. Phys. B* **89**, 395–399 (2007).

54. N. Kim, Y. A. Leem, H. Ko, M. Y. Jeon, C. W. Lee, S.-P. Han, D. Lee, and K. H. Park, Widely tunable 1.55-um detuned dual-mode laser diode for compact continuous-wave THz emitter, *ETRI J.* **33**(5), 810–813 (2011).

55. N. Kim, S.-P. Han, H. Ko, Y. A. Leem, H.-C. Ryu, C. W. Lee, D. Lee, M. Y. Jeon, S. K. Noh, and K. H. Park, Tunable continuous-wave terahertz generation/detection with compact 1.55 μm detuned dual-mode laser diode and InGaAs based photomixer, *Opt. Express* **19**(16), 15397–15403 (2011).

56. A. Roggenbuck, H. Schmitz, A. Deninger, I. Cámara Mayorga, J. Hemberger, R. Güsten, and M. Grüninger, Coherent broadband continuous-wave terahertz spectroscopy on solid-state samples, *New J. Phys.* **12**(4), 043017 (2010).

57. H.-C. Ryu, N. Kim, S.-P. Han, H. Ko, J.-W. Park, K. Moon, and K. H. Park, Simple and cost-effective thickness measurement terahertz system based on a compact 1.55 μm $\lambda/4$ phase-shifted dual-mode laser, *Opt. Express* **20**(23), 25990–25999 (2012).

Chapter 2

Nonlinear Crystal for Broadband Terahertz Pulse Generation and Detection

Masaya Nagai
Graduate School of Engineering Science, Osaka University, Toyonaka, Osaka 560-8531, Japan
mnagai@mp.es.osaka-u.ac.jp

A nonresonant electro-optic (EO) process is one of the simplest methods for both THz generation with optical rectification and detection using the EO sampling technique. It is because the optimization of nonlinear crystals is well established, even in the long-wavelength region. Nonlinear crystals are widely used for emitters and detectors using two narrowband optical beams [1]. However, I focus on broadband THz pulse generation and detection using ultrashort optical pulses here. The recent development of laser technologies provides extremely intense optical pulses or extremely ultrashort (pulse duration is less than 10 fs) optical pulses. By using materials with an instantaneous optical nonlinear response, it may be possible to capture much of this enormous bandwidth above 100 THz [2]. Moreover, high conversion efficiency of emitted THz radiation is retained even under significantly high-power excitation.

Handbook of Terahertz Technologies: Devices and Applications
Edited by Ho-Jin Song and Tadao Nagatsuma

ISBN 978-981-4613-08-8 (Hardcover), 978-981-4613-09-5 (eBook)
www.panstanford.com

The field amplitude of output THz pulses above 1 MV/cm is reported using optimized nonlinear crystals [3]. In this chapter, I summarize the fundamentals of THz pulse generation and detection using nonlinear crystals and the ultrashort optical pulse. In Section 2.1, I review the techniques of THz pulse generation and introduce that the phase-matching condition is critical for the generation efficiency and its bandwidth of the THz pulse. In Section 2.2, I show the details of the EO sampling technique.

2.1 THz Pulse Generation from a Nonresonant Nonlinear Crystal

In a transparent medium, nonresonant nonlinear responses are apparent instantaneously using ultrashort optical pulses. Second-order nonlinear polarization $P^{(2)}(t)$ is described in the time domain as $P^{(2)}(t) = \chi^{(2)}E(t)^2$, where $\chi^{(2)}$ is second-order electric susceptibility and the incident optical pulse is $E(t)$. The emitted THz pulse is the second-order derivative of nonlinear polarization. In the frequency domain, the optical pulse is assumed as the superposition of Fourier components with different frequencies. If two optical components $E(\omega_1)$ and $E(\omega_2)$ in the ultrashort optical pulse are incident with the frequencies of ω_1 and ω_2, induced nonlinear polarization $P^{(2)}(\omega_3 = \omega_1 - \omega_2)$ is generated with the frequency ω_3 as follows:

$$P^{(2)}(\omega_3) = \int \chi^{(2)}(\omega_3;\omega_1,\omega_2) E(\omega_1) E^*(\omega_2) d\omega_1 \tag{2.1}$$

THz pulse generation in a nonlinear crystal has been demonstrated for the first time by Yajima and Takeuchi [4]. They observed far-infrared (IR) pulse generation when a mode-locked glass laser was incident on $LiNbO_3$, ZnTe, ZnSe, CdS, and quartz crystals. Coincidently, Yang et al. demonstrated far-IR pulse generation using a $LiNbO_3$ crystal [5]. After the ultrashort optical pulse technologies based on the Ti: sapphire laser were developed in the 1990s, a lot of demonstrations for THz pulse generation from various nonlinear crystals have been reported.

2.1.1 The Optimization of the Nonlinear Crystal

For the high efficiency of THz pulse generation in the nonlinear crystal, four conditions are required: (1) the damage threshold for the incident excitation pulse is high, (2) the $\chi^{(2)}$ value is high, (3) the phase-matching condition is satisfied, and (4) the absorption coefficients in the crystal are low. For the generation efficiency of the THz pulse, (3) the phase-matching condition is the most critical. In general, the generation efficiency of THz emission from the individual nonlinear polarization is not so large. However, the excitation pulse propagating in the nonlinear crystal induces a long interaction length in the crystal. If these components of the emitted THz pulse interfere constructively, we can extract the intense THz pulse out of the nonlinear crystal.

The phase-matching condition can be derived from the zero phase term in the Maxwell's propagation equation with the nonlinear polarization [6]. Here we assume nonlinear polarization with the THz frequency ω_3 induced by the two optical components with the frequency ω_1 and ω_2. These components lie within the spectrum of the optical excitation pulse. Using a corresponding wavenumber $k_i = (n_i/c)\ \omega_i$, the phase-matching condition for the nonlinear process is expressed as follows:

$$k_1 - k_2 = k_3 \tag{2.2}$$

Since ω_3 is usually much smaller than the optical frequency ω_1 and ω_2, the left term in Eq. 2.2 is approximated as $k_1 - k_2 = dk/d\omega\ \omega_3 = n_g/c\ \omega_3$, where n_g is the index of the group velocity in the optical frequency region. Therefore the phase-matching condition is satisfied when the index of the group velocity for the optical excitation pulse, n_g, is equal to the index of the phase velocity in the THz frequency region, n_3.

Figure 2.1a shows the refractive index in the widely used nonlinear crystal ZnTe. The refractive index n_3 in the THz frequency region is different from the index n_1 at 0.8 μm, but the index at 2 THz is matched to the group velocity index n_g [7]. Figure 2.1b shows the time profile of the THz pulse emitted from a 1 mm thick (110)-oriented ZnTe crystal. The bandwidth of the emitted THz pulse is 2.5 THz, which is determined by the phase-matching condition. Zinc-blende crystals have a cubic structure with a point group 43 m and have only one independent nonvanishing second-order nonlinear optical coefficient, $d_{14} = d_{25} = d_{36}$. The field amplitude of the emitted THz pulse can be expressed as follows:

$$|E_{THz}| \propto d_{14}E^2 \sin\theta\sqrt{1-3\cos^2\theta} \tag{2.3}$$

where θ is the azimuthal angle between the [100] direction and the polarization direction of the incident excitation field E. The generation efficiency is maximum when the polarization direction of the excitation pulse is $\theta = 55°$, which is parallel to the [111] direction [8]. In this case, the polarizations of the THz pulse and the incident pulse are the same.

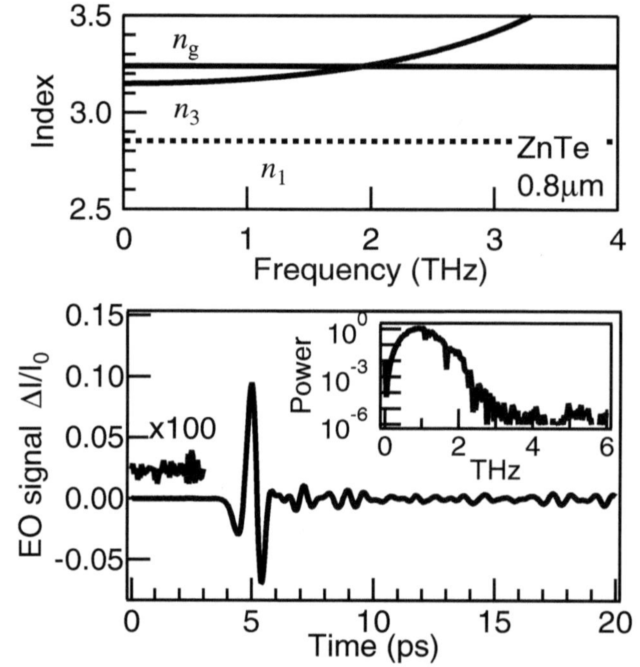

Figure 2.1 (a) Indexes of n_1, n_g, and n_3 in ZnTe. (b) The time profile of a THz pulse emitted from a 1 mm thick (110)-oriented ZnTe crystal. The inset shows the corresponding power spectrum.

When a nonlinear crystal has threefold rotational symmetry, the polarization of the emitted THz pulse is changed, but its amplitude does not depend on the azimuthal angle of the linearly polarized optical pulse. In this situation, the azimuthal angle of the THz wave is twice of that of the incident excitation pulse. Using this character, the THz vector beam can be generated using a segmented (111)-oriented zinc-blende crystal [9].

A THz pulse can also be generated using different phase-matched materials. The characteristic parameter for the optical rectification process is the coherent length l_c. It can be expressed as a function of frequencies ω_1 and ω_3 by

$$l_c = \frac{\pi c}{\omega_3 \left| n_g - n_3 \right|} \tag{2.4}$$

where c is the light speed. Figure 2.2 shows the coherent length l_c in some binary semiconductors with a zinc-blende structure as a function of optical wavelength [10]. One can see the enhancement of the coherent length at 0.8 μm, indicating that ZnTe has been determined to be the most suitable EO crystal for an emitter and a detector using a 0.8 μm laser. At a wavelength of 1.0 μm, coherent lengths of CdTe and GaP crystals are enhanced. While CdTe has almost the same static dielectric constant as that of ZnTe, its narrower band gap modulates the position of the group velocity matched to the phase velocity in the THz frequency region toward a longer wavelength. Experimentally it has been demonstrated by Xie et al. [11] and Shoji et al. [12]. This phase-matched wavelength can be tuned by controlling the dispersion of ternary material. Liu et al. measured a THz wave with ternary zinc-blende semiconductors of $Zn_xCd_{1-x}Te$ using a 0.82 μm laser and concluded that net efficiency of THz wave generation from an EO crystal of $Zn_{0.95}Cd_{0.05}Te$ is the highest [13].

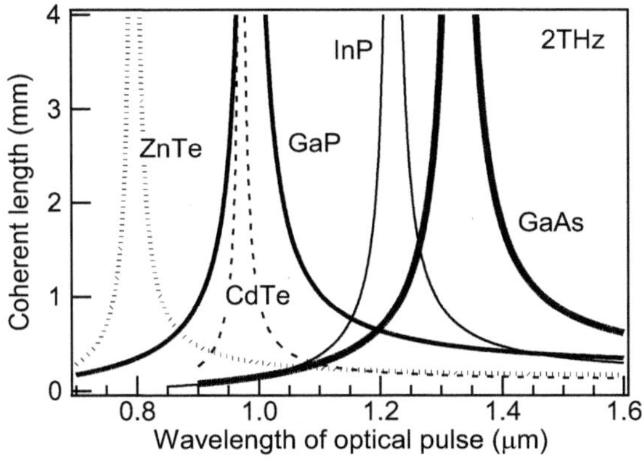

Figure 2.2 Coherent length of THz pulse generation at 2 THz as a function of the wavelength of the excitation pulse in various binary semiconductors with a zinc-blende structure.

A 1.55 μm laser, such as an Er-doped fiber laser, is especially attractive from the viewpoint of coupling with advanced optical communication techniques. Figure 2.2 implies that GaAs is the longest in binary zinc-blende semiconductors [10]. Another candidate of the nonlinear crystal for a 1.55 μm laser system is the organic crystal 4-*N*,*N*-dimethylamino-4′-*N*′-methyl stilbazolium tosylate (DAST). It has large hyperpolarizability, and good alignment of the charge transfer axes in the crystal greatly increases the nonlinear optical coefficients ($\chi^{(2)}{}_{111}$ = 490 pm/V at λ = 1.535 μm). Fortunately the phase-matching condition is satisfied at 1.55 μm [14]. The spectral bandwidth of a THz pulse from a DAST crystal is significantly enhanced at higher frequencies compared to that from a photoconductive switch that is pumped at 0.78 μm [15]. Therefore a DAST crystal is used in the fiber laser–based broadband THz spectrometer.

The next important criterion for nonlinear crystals is the low absorption coefficient in the THz frequency region. An important factor for absorption in the THz frequency region is free carrier absorption. Lui et al. concluded from doped ZnCdTe measurements that the resistivity of a crystal for both emitter and sensor applications should be greater than 100 W·cm [13]. Therefore, a semi-insulating nonlinear crystal is required for a high-efficiency THz emitter. Another dominant absorption at the low-frequency region is due to phonon resonances. In the case of a ZnTe crystal low-frequency resonance lies at 5.4 THz, which is expected to attenuate the high-frequency components of the THz radiation. A GaP crystal, which has a smaller value of $\chi^{(2)}$, has its first transverse-optical (TO) phonon resonance near 10.8 THz. Owing to the phase-matching condition for a 0.8 μm excitation pulse, a thin GaP crystal is an emitter with a broad bandwidth of 7 THz [16].

2.1.2 Broadband THz Pulse Generation above 10 THz

These nonlinear crystals have some phonons at a few THz, which limit the bandwidth of the emitted THz pulse. Of course, THz pulse generation is possible above the phonon frequency, but the response bandwidth of an isotropic nonlinear crystal such as ZnTe and GaP is limited by the mismatch between the THz phase velocity and the group velocity of the optical excitation pulse. Above the 10 THz region, the birefringent nonlinear crystals satisfy the phase-matching condition. GaSe crystals have a hexagonal structure with a point group 62 m, and a z-cut naturally cleaved crystal can be obtained. It has a large $\chi^{(2)}$ value (d_{22} = 54 pm/V) and is widely used for mid-IR pulse generation with cascading optical parametric amplification and differential frequency generation. However, mixing the components within the broad spectrum of individual excitation pulses is significantly simpler. Kaindl et al. demonstrated mid-IR pulse generation using a GaSe crystal and near-IR 13 fs pulses [17, 18]. Huber et al. also demonstrated broadband THz pulse generation up to 41 THz with a thin (30 μm) GaSe crystal [19]. Using an ultrathin GaSe crystal and an ultrabroadband excitation pulse (its duration is 10 fs), the bandwidth of the emitted THz pulse is beyond 120 THz [20, 21].

Additional to the GaSe crystal, a few nonlinear crystals are desired for broadband THz pulse generation. Zentgraf et al. succeeded in 50–130 THz pulse generation using a $LiIO_3$ crystal [22]. Katayama et al. demonstrated a broad band using a DAST crystal and 5 fs excitation pulses [23]. Figure 2.3 shows the power spectrum of a THz pulse from a DAST crystal using a 5 fs Ti:sapphire laser. It was evaluated with a monochromator equipped with a HgCdTe detector. Its spectral component is expanded to 180 THz, which is close to the optical telecommunication band. The spectrum of the excitation pulse is added in Fig. 2.3, and it covers from 0.65 μm (461 THz) to 1.05 μm (286 THz). Therefore, it is very promising for applications in the next IR spectrometers.

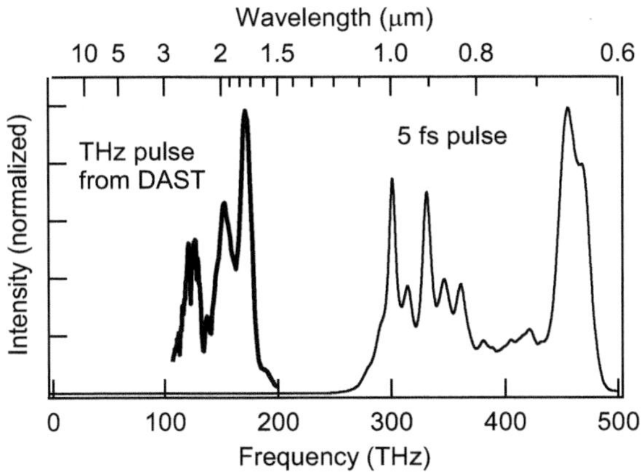

Figure 2.3 Power spectra of a THz pulse from a DAST crystal (bold curve) and a 5 fs Ti: sapphire laser (thin curve).

2.1.3 Intense THz Pulse Generation

The recent development of laser technologies provides us with intense ultrashort pulses, so intense THz pulse generation is possible via frequency conversion. A few-cycle pulse with the center frequency above 10 THz can be easily obtained using a thin GaSe crystal. The pulse with a short wavelength can be easily focused on a small spot size, so the maximum electric field is above 100 MV/cm [24]. An intense high-frequency THz pulse can also be generated using a DAST crystal, and its field amplitude approaches a few hundreds kV/cm [25, 26]. These field amplitudes are intense enough to perform nonlinear spectroscopy in the THz frequency region, such as 2D spectroscopy.

For the intense THz pulse generation below 2 THz, the ZnTe crystal is one candidate crystal for intense THz pulse generation. Blanchard et al. demonstrate 1.5 μJ THz pulse generation using a 75 mm diameter ZnTe single crystal wafer and 48 mJ, 0.8 μm, 30 fs excitation pulses [27]. In this experiment, the energy conversion efficiency is limited to $\sim 3 \times 10^{-5}$. The phase-matching condition in typical nonlinear crystals is satisfied at the optical frequency near the band gap (2.26 eV for a ZnTe crystal), so the generation efficiency of THz pulse generation is saturated due to the free carrier generation via two-photon absorptions of excitation pulses.

$LiNbO_3$ is also a good candidate nonlinear crystal for low-frequency THz pulse generation. It has a large band-gap energy (3.8 eV) and a large $\chi^{(2)}$ material (d_{33} = 168pm/V). However, a huge index dispersion ($n_g \approx 2.2$ in the near-IR region and $n_3 \approx 5$ in the THz frequency region) implies that the phase-matching condition is not satisfied in the colinear configuration [28]. Experimentally, the excitation pulse is focused on the $LiNbO_3$ crystal, and intense radiation is emitted in the noncolinear geometry after the excitation pulse passing. This scheme is similar to Cherenkov radiation. When the charged particles propagates in the medium faster than the speed of light in the medium, the electromagnetic field surrounding a charged particle is left behind, and the wavefront of an electromagnetic wave via the noncoaxial overlap of the waves behaves like a shock wave. In the case of THz radiation, the excitation pulse is equivalent to the charged particle. Then, a THz pulse is emitted in the radiation angle of $\cos\theta = n_g/n_3$. If the beam size of the excitation pulse is larger than the wavelength of the THz pulse, the lateral emission is strongly suppressed due to the destructive interference in the beam size. The scheme is illustrated in Fig. 2.4a. It has been experimentally demonstrated by Auston [29, 30], and broadband THz pulse generation is demonstrated with a $LiNbO_3$ waveguide coupled with a prism [31, 32]. For a collinear phase-matching condition, a periodically poled crystal is designed for narrowband THz pulse generation [33, 34].

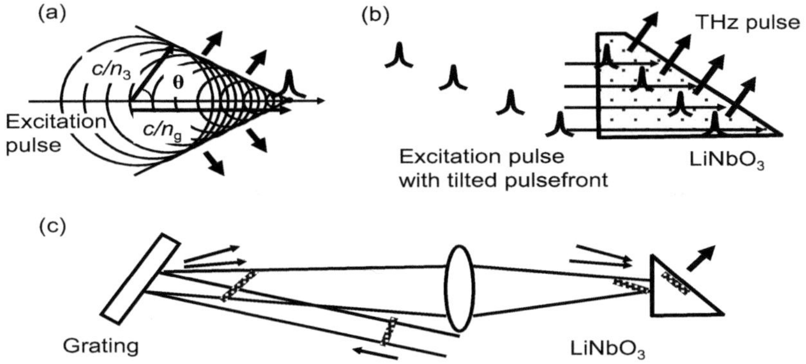

Figure 2.4 (a) Scheme of Cherenkov-like radiation from a moving source. (b) THz pulse generation in $LiNbO_3$ using an excitation pulse with a tilted pulse front.

Recently an intense THz wave generation technique has been demonstrated using a prism-shaped $LiNbO_3$ crystal [35]. The scheme is shown in Fig. 2.4b. Pulse front tilting of the ultrashort excitation pulse permits phase matching in the direction of THz pulse propagation, even for broadband pulse excitation. This was proposed by Hebling the first time [36, 37], and later many groups experimentally demonstrated it. Figure 2.4c shows the schematic of the experimental setup for the THz electric field measurement. A 2000 lines/mm grating is applied to tilt the intensity front of the pump pulses, and two lenses are used to image the spot of the pump beam on the grating into the $LiNbO_3$ crystal. They control lateral dispersion of the excitation wavepacket and simultaneously control spectral angular dispersion matched to the phase-matching condition. This optical design strongly influences the distortion of the pulse front for the excitation pulses [38–40], which suppresses the generation efficiency with the intense excitation pulse with a large spot size. The emitter is a stoichiometric $LiNbO_3$ crystal with a Mg concentration. Reduced Mg doping (less than a few percent) is still sufficient to prevent photorefractive damage [41]. To extract the THz pulse out of the crystal, one surface is cut with an angle of 62°.

Figure 2.5a shows the temporal evolution of the electric field evaluated from the EO signal [42]. The observed electric field approaches 380 kV/cm. In this measurement, although we used a sampling pulse with a duration of 200 fs, the delay time between the maximum and the minimum electric field is only 300 fs. This implies that the temporal profile of the electric field becomes dull. By deconvoluting the sampling pulse, we found that the prospective electric field exceeds 500 kV/cm. The output pulse energy 1.1 μJ is evaluated using a thermal sensor. The input pulse energy was measured to be 520 μJ using the same sensor, meaning that the conversion efficiency from these pulse energies reached 0.21%. The inset shows the power spectrum. The weighted center frequency of the emitted THz pulse is estimated to be 1.1 THz. The corresponding quantum efficiency is 0.54, and the net quantum efficiency exceeds the one inside the crystal.

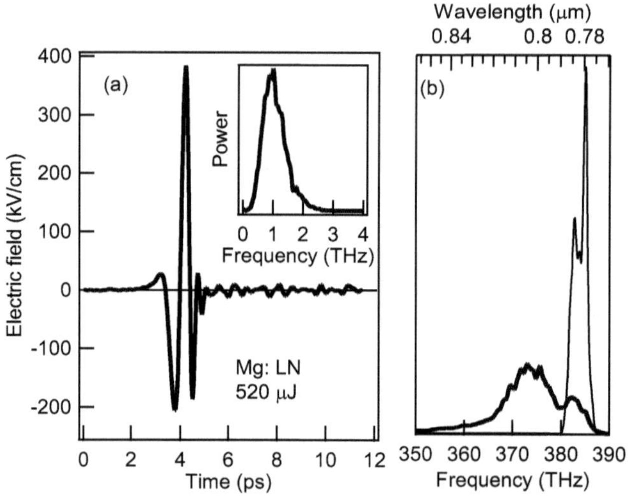

Figure 2.5 (a) Temporal profile of a THz pulse emitted from a $LiNbO_3$ crystal. The power of the excitation pulse is 520 μJ. Inset shows the power spectrum. (b) Spectra of the excitation pulses before (thin curve) and after (bold curve) passing through the $LiNbO_3$ crystal.

Many groups have demonstrated intense THz pulse generation using this technique [43–48]. The maximum amplitude of the electric field exceeds 1 MV/cm using a 4 mJ Ti:sapphire laser [39], and the maximum output power is 125 μJ using a 70 mJ excitation pulse [18].

Obeying the Manley–Rowe limit, the energy conversion rate in the $\chi^{(2)}$ process should be low (~0.1%) because of the tiny photon quantum energy of the THz wave. However, an excitation photon emitting one THz photon beam converts into an optical beam with a small low-frequency shift, which will promote further the optical rectification process. Figure 2.5b shows the spectrum of the excitation pulse before (thin curve) and after (bold curve) passing through the $LiNbO_3$ crystal. Strong spectral modulation of the excitation beam induced by THz wave generation indicates the distortion of the excitation pulse shape. It is due to two reasons. One is the cascaded $\chi^{(2)}$ processes where the emitted THz pulse modulates the excitation pulse. Another reason is the stimulated Raman scattering process [42]. This is a third-order nonlinear process, and it is widely used for the spectral broadening of ultrashort pulses. $LiNbO_3$ has also a large $\chi^{(3)}$ value related to the stimulated Raman scattering process and the spectral broadening of the excitation pulse to be accelerated. Such higher-order nonlinear processes are critical for the high generation efficiency of a THz pulse in a $LiNbO_3$ crystal [49]. If the pulse duration of the excitation pulse is shortened via the generated intense THz wave, the conversion efficiency of THz wave generation is enhanced and the bandwidth of the THz wave becomes broadened. Actually broadband THz pulse generation beyond the limitation of the incident pulse profile is succeeded using a Yb-doped fiber laser pulse with a 1.04 μm wavelength and a 600 fs pulse duration [50].

2.2 Broadband THz Pulse Detection in the Nonresonant Nonlinear Crystal

For THz time domain spectroscopy, we usually detect the electric field amplitude using a photoconductive antenna or an EO crystal with optical sampling methods. The former is based on the current measurement of a photoconductive switch with a gated optical pulse. The latter is based on the birefringence measurement induced by the electric field of THz pulses. Either signal is proportional to the electric field amplitude of the THz pulse at the time position of the sapling pulse, and by varying the time delay between a THz pulse and a sampling pulse, the electric field amplitude of THz pulses $E(t)$ can be detected as a function of time. The merit of the EO sampling technique is that the absolute value of the electric field amplitude can be evaluated from the polarization change of the sampling pulse [7, 51].

Figure 2.6 shows the concept of the sampling methods with the EO effect [52]. If the linearly polarized sampling pulse is incident on the EO crystal, the polarization of the sampling pulse is modulated by the index anisotropy induced by the electric field. The polarization change induced by the THz electric field is not influenced by the thermal noise, which allows for sensitive detection of THz pulse, even at room temperature.

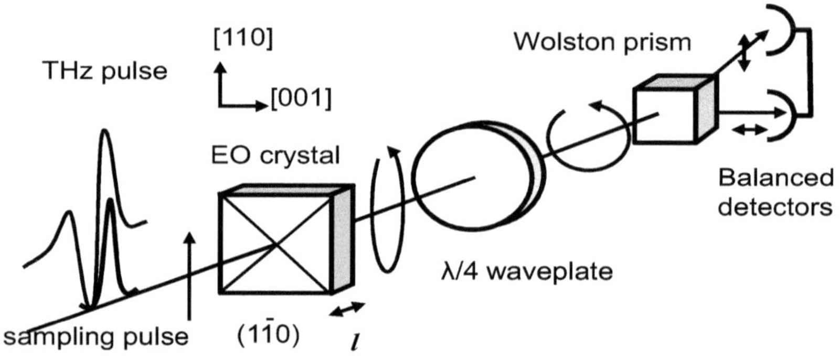

Figure 2.6 The sampling method with the EO effect.

Let us assume (110)-oriented zinc-blende semiconductors with index n_0 and EO coefficient r_{41}. The polarization directions of the THz field and sampling pulse incident on the crystal are parallel to the [110] direction. The refractive index of the EO crystal is obeyed along the equation of the index ellipsoid in the presence of an electric field as follows:

$$\left(\frac{1}{n_0^2}-r_{41}E\right)x'^2+\left(\frac{1}{n_0^2}+r_{41}E\right)y'^2+\frac{1}{n_0^2}z'^2=1 \tag{2.5}$$

Here we introduce a new optical axis of x' //$[1/2,1/2,1/\sqrt{2}]$ and y'// $[-1/2, 1/2, 1/\sqrt{2}]$. The lengths of the new axes of the ellipsoid are $n_{x'}$ and $n_{y'}$, where $1/n_{x'}^2 = 1/n_0^2 - r_{41}E$, $1/n_{y'}^2 = 1/n_0^2 + r_{41}E$, and $1/n_{z'}^2 = 1/n_0^2$. Assuming $r_{41}E << 1/n_0^2$, we obtain the refractive index, $n_{x'} \approx n_0 + \Delta n$, $n_{y'} \approx n_0 - \Delta n$, and $n_{z'} = n_0$ with $\Delta n = n_0^3 r_{41} E/2$.

If a linearly polarized optical pulse is incident on the EO crystal, the polarization is distorted due to the phase difference between x' and y' components. The phase retardation at the output plane $z = l$ is

$$\Gamma = \phi_{y'} - \phi_{x'} = \frac{2\omega\Delta n}{c} \cdot l = \frac{2\pi}{\lambda} n_0^3 r_{41} l E \quad (2.6)$$

where λ is the wavelength of the sampling pulse. Equation 2.6 shows that the elliptical polarization becomes changed with the phase difference Γ between those in the directions of x' and y' after the linearly polarized ultrashort pulse passes through the electric-field-induced EO crystal. Hence the ratio ΔI of the vertically and horizontally polarized components is simply expressed as

$$\frac{\Delta I}{I} \cong \Gamma \quad (2.7)$$

Since Γ is proportional to the electric field E, the absolute value of the electric field can be evaluated from the measurements of ΔI.

In the experiments, we have to note two factors for the evaluation of the electric field. One is the electric field reduced by Fresnel loss at the surface of the EO crystal. The other is the phase-matching condition in the EO crystal for detection. Thus the complex Fourier component of the electric field has to be divided by a sensitivity factor of $2/(n_3 - 1)$ $(e^{i\Delta kL} - 1)/i\Delta k$ in GaP determined by Fresnel loss and phase mismatching $\Delta k = (n_g - n_3)\omega_3/c$ for obtaining the true time evolution of the electric filed.

In the previous section, we introduced that ZnTe is the most desired nonlinear crystal in the zinc-blende binary semiconductors for a 0.8 μm laser pulse. It is also applicable to the detection process. A ZnTe crystal has the parameters of $n_0 = 3.2$ and $r_{41} = 4.0$ pm/V. The THz pulse with a field amplitude of $E = 100$V/m is incident on a 1 mm thick ZnTe crystal, $\Delta n/n_0 \approx (1/2)n_0^2 r_{41} E \approx 2 \times 10^{-9}$, so phase retardation is $\Gamma \approx 1 \times 10^{-4}$. For the detection of the tiny birefringence high signal-to-noise detection is required.

Using the EO sampling technique, we can detect a broadband THz pulse coherently because the detectable bandwidth is limited by the phase-matching condition. Therefore, several tens of THz pulses can be detected using an ultrathin ZnTe or GaP crystal [53]. In the mid-

IR region, a GaSe crystal is usually used for coherent detection [20]. Figure 2.7 shows the temporal evolution of a THz pulse using a 20 μm thick GaSe crystal [54]. An ultrabroadband IR pulse is generated through a plasma by focusing a hollow-fiber compressed, intense 10 fs pulse and its second harmonic in air. Due to the phase-matching condition, the detectable bandwidth depends on the crystal angle. A signal up to 100 THz is detected with a crystal angle of 50°.

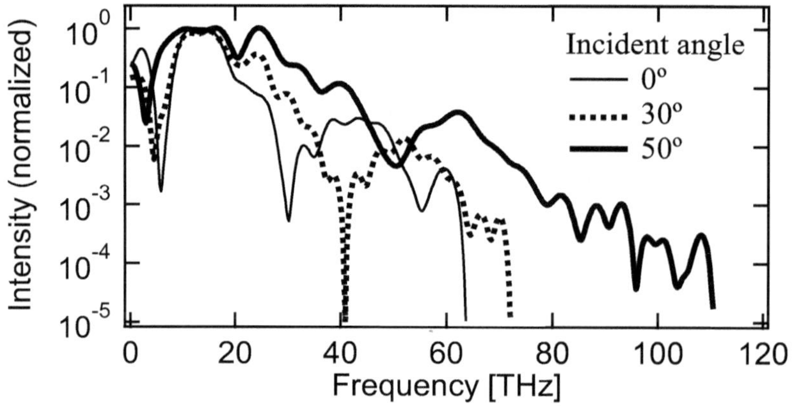

Figure 2.7 Spectrum of an ultrabroadband THz pulse detected with the EO sampling technique. As the EO crystal, a 20 μm thick GaSe crystal is used.

The excellent conversion from the THz field amplitude to an optical signal in the EO sampling technique allows great real-time, multichannel detection imaging [55]. Array detection such as charge-coupled devices (CCDs) and digital-signal-processing technologies are well developed, which bring in the spatial mapping of the THz image. Wu et al. demonstrated it using the multichannel EO sampling technique, and such an imaging system offers the ability to measure the 2D spatially distributed phase and amplitude information of a THz wave in a true real-time mode.

In addition to imaging, multichannel EO detection permits two single-shot EO measurement techniques of the temporal wave form of a THz pulse. One is the time-to-spectrum conversion technique [56, 57]. Using a linearly chirped sampling pulse, a temporal wave form of a copropagating THz field is linearly encoded onto the frequency spectrum of the sampling pulse and then decoded by a multichannel optical detector with a spectrometer. Another is the time-to-space conversion technique [58]. The method involves transcribing the temporal profile of a THz pulse onto the transverse spatial profile of the optical sampling beam through the use of noncollinear geometry. Thus acquisition of picosecond THz field pulses is realized without any mechanical time delay device. These techniques are useful to perform computer tomography [59, 60], which strongly requires a rapid scanning time for time and space.

References

1. K. Kawase, H. Minamide, K. Imai, J. Shikata, and H. Ito, Injection-seeded terahertz-wave parametric generator with wide tunability, *Appl. Phys. Lett.* **80,** 195–197 (2002).
2. R. Huber, B. A. Schmid, R. A. Kaindl, and D. S. Chemla, Femtosecond THz studies of intra-excitonic transitions, *Phys. Status Solidi B* **245,** 1041–1048 (2008).
3. K. Tanaka, H. Hirori, and M. Nagai, THz nonlinear spectroscopy of solids, *IEEE Trans. THz Sci. Tech.* **1,** 301–312 (2011).
4. T. Yajima and N. Takeuchi, Far-infrared difference-frequency generation by picosecond laser pulses, *Jpn. J. Appl. Phys.* **9,** 1361–1371 (1970).
5. K. H. Yang, P. L. Richards, and Y. R. Shen, Generation of far infrared radiation by picosecond light pulses in $LiNbO_3$, *Appl. Phys. Lett.* **19,** 320–323 (1971).
6. Y. R. Shen, *The Principles of Nonlinear Optics* (Wiley Interscience, 1984).
7. A. Nahata, A. S. Weling, and T. F. Heinz, A wideband coherent terahertz spectroscopy system using optical rectification and electro-optic sampling, *Appl. Phys. Lett.* **69,** 2321–2323 (1996).
8. A. Rice, Y. Jin, X. F. Ma, X.-C. Zhang, D. Bliss, J. Larkin, and M. Alexander, Terahertz optical rectification from 110 zinc-blende crystals, *Appl. Phys. Lett.* **64,** 1324–1326 (1994).
9. R. Imai, N. Kanda, T. Higuchi, Z. Zheng, K. Konishi, and M. Kuwata-Gonokami, Terahertz vector beam generation using segmented nonlinear optical crystals with threefold rotational symmetry, *Opt. Express* **20,** 21896–21904 (2012).
10. M. Nagai, K. Tanaka, H. Ohtake, T. Bessho, T. Sugiura, T. Hirosumi, and M. Yoshida, Generation and detection of terahertz radiation by electro-optical process in GaAs using 1.56 μm fiber laser pulses, *Appl. Phys. Lett.* **85,** 3974–3976 (2004).

11. X. Xie, J. Xu, and X.-C. Zhang, Terahertz wave generation and detection from a CdTe crystal characterized by different excitation wavelengths, *Opt. Lett.* **31,** 978–980 (2006).
12. A. Syouji, S. Saito, K. Sakai, M. Nagai, K. Tanaka, H. Tanaka, T. Bessho, T. Sugiura, T. Hirosumi, and M. Yoshida, Evaluation of a terahertz wave spectrum and construction of a terahertz wave-sensing system using a Yb-doped fiber laser, *J. Opt. Soc. Am. B* **24,** 2006–2012 (2007).
13. K. Liu, H. S. Kang, T. K. Kim, and X.-C. Zhang, Study of ZnCdTe crystals as terahertz wave emitters and detectors, *Appl. Phys. Lett.* **81,** 4115–4117 (2002).
14. A. Schneider, M. Stillhart, and P. Günter, High efficiency generation and detection of terahertz pulses using laser pulses at telecommunication wavelengths, *Opt. Express,* **14,** 5376–5384 (2006).
15. D. Stehr, C. M. Stehr, C. Schmidt, and M. S. Sherwin, High-performance fiber-laser-based terahertz spectrometer, *Opt. Lett.* **35,** 3799–3801 (2010).
16. Q. Wu and X.-C. Zhang, 7 terahertz broadband GaP electro-optic sensor, *Appl. Phys. Lett.* **70**, 1785–1787 (1997).
17. R. A. Kaindl, D. C. Smith, M. Joschko, M. P. Hasselbeck, M. Hasselbeck, and T. Elsaesser, Femtosecond infrared pulses tunable from 9 to 18 μm at an 88-MHz repetition rate, *Opt. Lett.* **23,** 861–863 (1998).
18. R. A. Kaindl, F. Kaindl, M. Woerner, and T. Elsaesser, Broadband phase-matched difference frequency mixing of femtosecond pulses in GaSe: experiment and theory, *Appl. Phys. Lett.* **75,** 1060–1062 (1999).
19. R. Huber, A. Brodschelm, F. Tauser, and A. Leitenstorfer, Generation and field-resolved detection of femtosecond electromagnetic pulses tunable up to 41 THz, *Appl. Phys. Lett.* **76,** 3191–3193 (2000).
20. C. Kübler, R. Huber, S. Tübel, and A. Leitenstorfer, Ultrabroadband detection of multi-terahertz field transients with GaSe electro-optic sensors: approaching the near infrared, *Appl. Phys. Lett.* **85,** 3360–3362 (2004).
21. A. Sell, R. Scheu, A. Leitenstorfer, and R. Huber, Field-resolved detection of phase-locked infrared transients from a compact Er:fiber system tunable between 55 and 107 THz, *Appl. Phys. Lett.* **93,** 251107–251109 (2008).
22. T. Zentgraf, R. Huber, N. C. Nielsen, D. S. Chemla, and R. A. Kaindl, Ultrabroadband 50-130 THz pulses generated via phase-matched difference frequency mixing in $LiIO_3$, *Opt. Express,* **15,** 5775–5781 (2007).

23. I. Katayama, R. Akai, M. Bito, H. Shimosato, K. Miyamoto, H. Ito, and M. Ashida, Ultrabroadband terahertz generation using 4-N,N-dimethylamino-4′-N′-methyl-stilbazolium tosylate single crystals, *Appl. Phys. Lett.* **97,** 021105–021107 (2010).

24. A. Sell, A. Leitenstorfer, and R. Huber, Phase-locked generation and field-resolved detection of widely tunable terahertz pulses with amplitudes exceeding 100 MV/cm, *Opt. Lett.* **33,** 2767–2769 (2008).

25. C. Ruchert, C. Vicario, and C. P. Hauri, Scaling submillimeter single-cycle transients toward megavolts per centimeter field strength via optical rectification in the organic crystal OH1, *Opt. Lett.* **37,** 899–901 (2011).

26. C. P. Hauri, C. Ruchert, C. Vicario, and F. Ardana, Strong-field single-cycle THz pulses generated in an organic crystal, *Appl. Phys. Lett.* **99,** 161116–161119 (2011).

27. F. Blanchard, L. Razzari, H.-C. Bandulet, G. Sharma, R. Morandotti, J.-C. Kieffer, T. Ozaki, M. Reid, H. F. Tiedje, H. K. Haugen, and F. A. Hegmann, Generation of 1.5 μJ single-cycle terahertz pulses by optical rectification from a large aperture ZnTe crystal, *Opt. Express* **15,** 13212–13220 (2007).

28. E. D. Palik (Ed.) *Handbook of Optical Constants of Solids* (Academic Press) (1997).

29. D. H. Auston, Subpicosecond electro-optic shock waves, *Appl. Phys. Lett.* **43,** 713–715 (1983).

30. D. H. Auston, K. P. Cheung, J. A. Valdmanis, and D. A. Kleinman, Cherenkov radiation from femtosecond optical pulses in electro-optic media, *Phys. Rev. Lett.* **53,** 1555–1558 (1984).

31. M. Theuer, G. Torosyan, C. Rau, R. Beigang, K. Maki, C. Otani, and K. Kawase, Efficient generation of Cherenkov type terahertz radiation from a lithium niobate crystal with a silicon prism output coupler, *Appl. Phys. Lett.* **88,** 071122–071124 (2006).

32. S. B. Bodrov, A. N. Stepanov, M. I. Bakunov, B. V. Bakunov, I. E. Ilyakov, and R. A. Akhmedzhanov, Highly efficient optical-to-terahertz conversion in a sandwich structure with $LiNbO_3$ core, *Opt. Express,* **17,** 1871–1879 (2009).

33. J. A. L'huillier, G. Torosyan, M. Theuer, Y. Avetisyan, and R. Beigang, Generation of THz radiation using bulk, periodically and aperiodically poled lithium niobate: part 1; theory. *Appl. Phys. B* **86,** 185–196 (2007).

34. J. A. L'huillier, G. Torosyan, M. Theuer, C. Rau, Y. Avetisyan, and R. Beigang, Generation of THz radiation using bulk, periodically and

aperiodically poled lithium niobate: part 2; experiments, *Appl. Phys. B* **86,** 197–208 (2007).

35. J. Hebling, K.-L. Yeh, M. C. Hoffmann, B. Bartal, and K. A. Nelson, Generation of high-power terahertz pulses by tilted-pulse-front excitation and their application possibilities, *J. Opt. Soc. Am. B* **25,** 6–19 (2008).

36. J. Hebling, G. Almási, and I. Z. Kozma, Velocity matching by pulse front tilting for largearea THz-pulse generation, *Opt. Express* **10,** 1161–1166 (2002).

37. J. Hebling, A. G. Stepanov, G. Almási, B. Bartal, and J. Kuhl, Tunable THz pulse generation by optical rectification of ultrashort laser pulses with tilted pulse fronts, *Appl. Phys. B* **78,** 593–599 (2004).

38. J. A. Fülöp, L. Pálfalvi, G. Almási, and J. Hebling, Design of high-energy terahertz sources based on optical rectification, *Opt. Express* **18,** 12311–12327 (2010).

39. H. Hirori, A. Doi, F. Blanchard, and K. Tanaka, Single-cycle terahertz pulses with amplitudes exceeding 1 MV/cm generated by optical rectification in $LiNbO_3$, *Appl. Phys. Lett.* **98,** 091106–091108 (2011).

40. J. A. Fülöp, L. Pálfalvi, G. Almási, M. C. Hoffmann, and J. Hebling, Towards generation of mJ-level ultrashort THz pulses by optical rectification, *Opt. Express* **19,** 15090–15097 (2011).

41. L. Pálfalvi, J. Hebling, G. Almási, Á. Péter, K. Polgár, K. Lengyel, and R. Szipöcs, Nonlinear refraction and absorption of Mg doped stoichiometric and congruent $LiNbO_3$, *J. Appl. Phys.* **95,** 902 (2004).

42. M. Nagai, E. Matsubara, and M. Ashida, High-efficiency terahertz pulse generation via optical rectification by suppressing stimulated Raman scattering process, *Opt. Express* **85,** 011105–011110 (2012).

43. G. Stepanov, J. Kuhl, I. Z. Kozma, E. Riedle, G. Almási, and J. Hebling, Scaling up the energy of THz pulses created by optical rectification, *Opt. Express* **13,** 5762–5768 (2005).

44. K.-L. Yeh, M. C. Hoffmann, J. Hebling, and K. A. Nelson, Generation of 10 μJ ultrashort terahertz pulses by optical rectification, *Appl. Phys. Lett.* **90,** 171121–171123 (2007).

45. M. C. Hoffmann, K.-L. Yeh, J. Hebling, and K. A. Nelson, Efficient terahertz generation by optical rectification at 1035 nm, *Opt. Express* **15,** 11706–11713 (2007).

46. M. C. Hoffmann, K.-L. Yeh, H. Y. Hwang, T. S. Sosnowski, B. S. Prall, J. Hebling, and K. A. Nelson, Fiber laser pumped high average power single-cycle terahertz pulse source, *Appl. Phys. Lett.* **93,** 141107–141109 (2008).

47. A. G. Stepanov, S. Henin, Y. Petit, L. Bonacina, J. Kasparian, and J.-P. Wolf, Mobile source of high-energy single-cycle terahertz pulses, *Appl. Phys. B* **101,** 11–14 (2010).

48. J. A. Fülöp, L. Pálfalvi, S. Klingebiel, G. Almási, F. Krausz, S. Karsch, and J. Hebling, Generation of sub-mJ terahertz pulses by optical rectification, *Opt. Lett.* **37,** 557–559 (2012).

49. M. Jewariya, M. Nagai, and K. Tanaka, Enhancement of terahertz wave generation by cascaded $\chi^{(2)}$ processes in $LiNbO_3$, *J. Opt. Soc. Am. B* **26,** A101–A106 (2009).

50. M. Nagai, M. Jewariya, Y. Ichikawa, H. Ohtake, T. Sugiura, Y. Uehara, and K. Tanaka, Broadband and high power terahertz pulse generation beyond excitation bandwidth limitation via $\chi^{(2)}$ cascaded processes in $LiNbO_3$, *Opt. Express* **13,** 11543–11549 (2008).

51. Q. Wu and X.-C. Zhang, Ultrafast electro-optic field sensors, *Appl. Phys. Lett.* **68,** 1604–1606 (1996).

52. A. Yariv, *Quantum Electronics*, 3rd ed. (John Wiley & Sons, 1989).

53. A. Leitenstorfer, S. Hunsche, J. Shah, M. C. Nuss, and W. H. Knox, Detectors and sources for ultrabroadband electro-optic sampling: experiment and theory, *Appl. Phys. Lett.* **74,** 1516–1518 (1999).

54. E. Matsubara, M. Nagai, and M. Ashida, Ultrabroadband coherent electric field from far infrared to 200 THz using air plasma induced by 10 fs pulses, *Appl. Phys. Lett.* **101,** 011105–011107 (2012).

55. Q. Wu, T. D. Hewitt, and X.-C. Zhang, Two-dimensional electro-optic imaging of THz beams, *Appl. Phys. Lett.* **69,** 1026–1028 (1996).

56. Z. Jiang and X.-C. Zhang, Electro-optic measurement of THz field pulses with a chirped optical beam, *Appl. Phys. Lett.* **72,** 1945–1947 (1998).

57. F. G. Sun, Z. Jiang, and X.-C. Zhang, Analysis of terahertz pulse measurement with a chirped probe beam, *Appl. Phys. Lett.* **73,** 2233–2235 (1998).

58. J. Shan, A. S. Weling, E. Knoesel, L. Bartels, M. Bartels, A. Nahata, G. A. Reider, and T. F. Heinz, Single-shot measurement of terahertz elctromagnetic pulses by use of electro-optic sampling, *Opt. Lett.* **25,** 426–428 (2000).

59. B. Ferguson, S. Wang, D. Gray, D. Abbot, and X.-C. Zhang, T-ray computed tomography, *Opt. Lett.* **27,** 1312–1314 (2002).

60. E. Abraham, Y. Ohgi, M. Minami, M. Nagai, T. Araki, and T. Yasui, Real-time line projection for fast terahertz spectral computed tomography, *Opt. Lett.* **36,** 2119–2121 (2011).

Chapter 3

Ultrafast Photodiodes

Tadao Ishibashi
NTT Device Innovation Center, NTT Corporation, 3-1, Morinosato-Wakamiya, Atsugi, Kanagawa 243-0198, Japan
ishibashi-tadao@ntt-et.co.jp

Photodiode technology for generating terahertz radiation is described in this chapter. Semiconductor materials for the photodiode considered here are InP and alloy materials lattice-matched to InP, which are well established in fabricating a variety of devices used in long-wavelength fiber optic communication systems. After a brief introduction, basic photodiode structures and their features in frequency response are reviewed. The behavior of electron transport, which is far from the steady state in a photodiode, is also discussed. Then, reported performance of diode-based photomixers is presented.

Handbook of Terahertz Technologies: Devices and Applications
Edited by Ho-Jin Song and Tadao Nagatsuma

ISBN 978-981-4613-08-8 (Hardcover), 978-981-4613-09-5 (eBook)
www.panstanford.com

3.1 Introduction

In many terahertz (THz) wave applications, the use of continuous-wave (CW) radiation, especially that generated by photonic input signals, has indispensable advantages, such as spectral purity, fast spectral scanning, and compactness of the system. Representative devices for such CW generation are photoconductive antennas (PCAs) (described in Chapter 1) and photodiodes (PDs). In a PCA material, the recombination lifetime of photogenerated carriers must be short in order to have a broad frequency bandwidth. Here the anode-to-cathode dimension can be designed wide so that the device capacitance does not limit the bandwidth, which is the basic advantage of PCAs. On the other hand, the bandwidth of a PD is limited by the carrier transit time, even in a submicrometer structure. So, its absorption layer thickness, as well as the anode-to-cathode dimension, must be reasonably small, which tends to increase the diode capacitance. In a typical PD operated in the THz wave range, both the carrier transit time and capacitance effect lower the possible operation frequency.

However, suppose that the carrier transit time of a PD is made equal to a certain recombination lifetime in a PCA material by reducing the diode geometry. When the corresponding reduction in the diode area is done, the resistance–capacitance (RC) time constant becomes the same as that of a PCA. Consequently, there is no essential difference in operation between two kinds of devices. Since the anode-to-cathode spacing is usually much narrower in a PD than in a PCA, a PD produces a much higher output-induced current per one electron–hole pair. This is why diode photomixers generally have better efficiency than PCA photomixers. In general, PCAs have been designed with a smaller RC constant and are more suitable for generating a THz wave in a higher frequency range than PDs. In a frequency range lower than around 1 THz, PDs have been reported to generate much higher CW outputs than PCAs.

Regarding the maximum available power of photomixers, quantitative evaluation is not straightforward. This is because both electronic output saturation mechanisms and thermal limitation depend on many structural factors and operation conditions. It may be a common experience that thermal failure precedes electronic output saturation in a well-designed photomixer. In this sense, a low operation voltage for less self-heating is very important in any photomixer devices.

A variety of PD structures and radiator configurations have been developed to realize efficient THz wave generators. To date, by using a optical heterodyne (or photomixing) input signal, PDs have successfully been operated in a frequency range up to over 1.5 THz and beyond. Figure 3.1 summarizes the reported output of diode-based photomixers. Several review papers also describe the status of diode-based THz photomixers [1–3]. Rectangular waveguide-coupled diode photomixers (indicated by squares) have been fabricated for operation in a frequency range up to the 350 GHz band. Usually, a matching circuit integrated with a diode can help to improve the output by compensating for the effect of the junction capacitance. For the higher frequency range, an antenna-coupled photomixer is more easily packaged since both the diodes and the antenna radiator can be monolithically integrated in the same semiconductor substrate. The output levels of antenna-coupled PDs shown here (indicated by circles) are those with resonant antennas. As discussed later, the outputs could be higher by about 6 dB than those of broadband antennas (such as log-periodic and bow-tie antennas). At 100 GHz, the reported output is as high as 20 dBm, while it lowers to –20 dBm at 1 THz. The steep role-off of the power level with frequency seen here, mostly proportional to f^{-4}, results from the fact that both the intrinsic bandwidth (associated with carrier transit time) and the RC-limited bandwidth strongly dominate the response.

The PD structure used for all of the photomixers shown in Fig. 3.1 is the uni-traveling-carrier photodiode (UTC-PD) [15] or its modified designs. Since UTC-PDs have good linearity of radio frequency (RF) output even at a low operation voltage, they are suitable for photomixer applications. Namely, compared to the classical pin-PD, the UTC-PD design provides a larger bandwidth and wider voltage swing, which can generate a higher output level. Device operation is discussed in the next two sections.

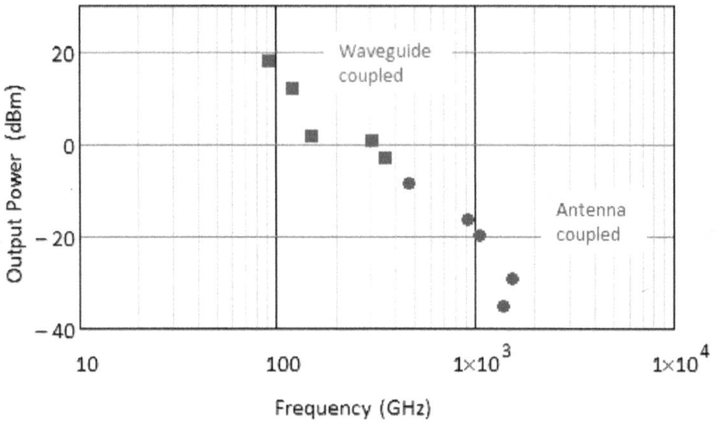

Figure 3.1 Reported output levels of diode-based photomixers operated with a 1.5 μm wavelength band input signal [4–11].

3.2 Basic Intrinsic Photoresponse

The frequency response of any photodetector device depends on the intrinsic response (carrier transit time) and the circuit condition given by parasitics and the effective load. In contrast to PCAs, PDs have a strict trade-off between the intrinsic response time and the capacitance, as described above. The faster intrinsic response time needs a smaller anode-to-cathode dimension, which leads to higher capacitance. Therefore, an important issue in PD design is how to maximize the performance in output power and bandwidth under such a trade-off.

First we examine the intrinsic PD response for two extreme cases, a pin-PD and a UTC-PD with a negligibly thin absorber. In the pin-PD structure, Fig. 3.2A, with a depleted absorber sandwiched by wide-gap contact layers, both electrons and holes contribute equally to the output current. Assuming a uniform carrier generation rate over the depleted absorber (depletion thickness of W_d) and constant saturation velocities v_{se} for electrons and v_{sh} for holes, the short circuit (=intrinsic) current response as a function of frequency $\omega = 2\pi f$ is given as

$$R_{pin}(\omega) = R_{hole}(\omega) + R_{elec}(\omega)$$

$$= \frac{1}{(\omega\tau_h)^2}\left(e^{j\omega\tau_h} - 1\right) - j\frac{1}{\omega\tau_h} + \frac{1}{(\omega\tau_e)^2}\left(e^{j\omega\tau_e} - 1\right) - j\frac{1}{\omega\tau_e} \quad (3.1)$$

where $\tau_h = W_d/v_{sh}$ and $\tau_e = W_d/v_{se}$ are carrier transit times for holes and electrons, respectively [12].

In a UTC-PD with a negligibly thin neutral absorber arranged at the anode end, Fig. 3.2B, the major active carrier is electrons. Therefore, the hole current does not exist in the diode depletion layer. Ignoring the injection delay time for the neutral absorber, we have the current response:

$$\mathrm{R_{UTC}}(\omega) = \frac{1}{j\omega\tau_e}\left(1 - e^{-j\omega\tau_e}\right). \quad (3.2)$$

As easily understood, the two frequency responses, Eqs. 3.1 and 3.2, are represented with $\omega\tau$. Figure 3.3 illustrates how the response (amplitude) changes with $\omega\tau$ (a case for $\tau_h = \tau_e = \tau$) for a pin-PD and a UTC-PD. From this calculation, it is found that a pin-PD has a larger bandwidth than a UTC-PD for the same carrier transit time τ. Here the 10 dB down response is obtained at $\omega\tau = 6.3$ in a pin-PD and 4.6 in a UTC-PD.

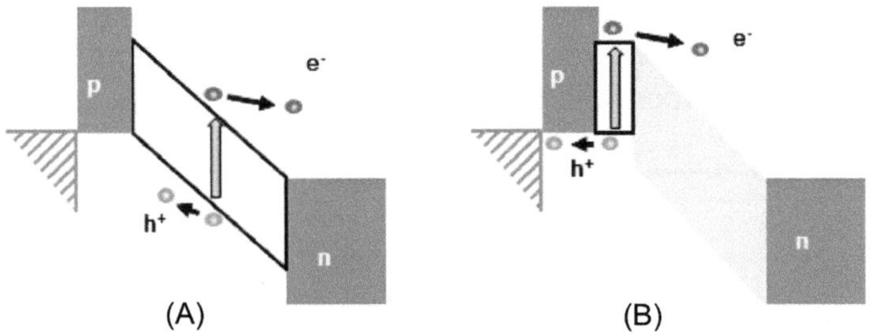

Figure 3.2 Band diagram of a pin-PD (A) and a UTC-PD (B).

In calculating the PD response, a conventional approach uses carrier saturation velocities in a steady-state condition. In reality, however, in a submicrometer dimension or in a several hundred femtosecond time scale, the electron velocity in InGaAs and InP can be much higher than the saturation velocity v_s, and this transient phenomenon is sometimes called the "velocity overshoot effect." In a typical dimension designed for a THz PD, the velocity overshoot significantly contributes to the electron transit time. The velocity overshoot behavior is discussed in more detail in the next section.

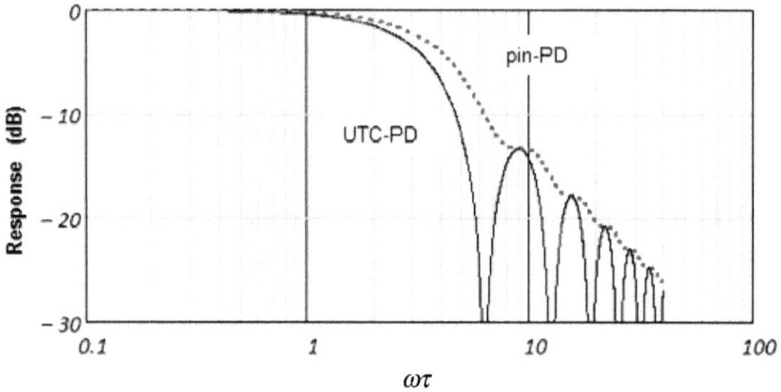

Figure 3.3 Current response of a pin-PD and a UTC-PD as a function of $\omega\tau$.

Here simply assuming the "effective" electron velocity to be v_{se} = 4 × 10^7 cm/s and the hole velocity v_{sh} = 4.5 × 10^6 cm/s, the response for a PD with a depletion thickness W_d = 0.24 mm is calculated as a function of frequency (Fig. 3.4). For the pin-PD, electron and hole current components are shown, as well as the total response. From this comparison, it is clearly seen that the UTC structure provides a higher 3 dB bandwidth f3dB (~740 GHz) than the pin structure (~132 GHz). This inferiority is common to a pin-PD and arises from a situation where the total response is dominated by the hole current, even though the electron current component has a broader f3dB response (~920 GHz) than that of the UTC structure.

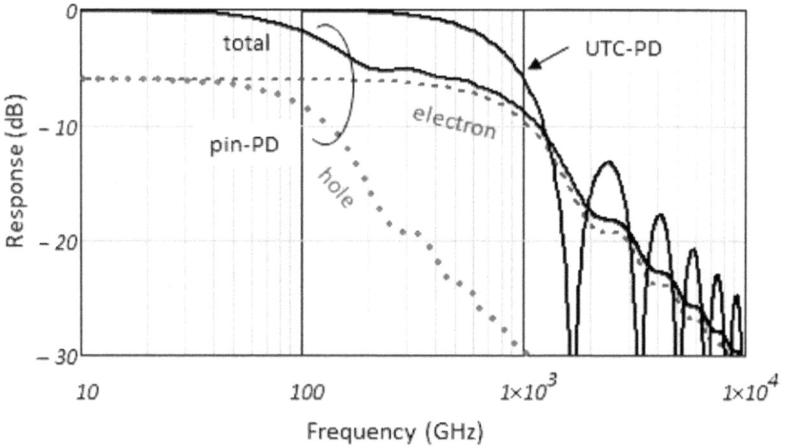

Figure 3.4 Current response of a pin-PD and a UTC-PD with a depletion layer thickness W_d = 0.24 μm.

Due to the quite unequal carrier velocities, the pin structure loses an output of 6 dB over a broad frequency range. Finally, up to a frequency of 1330 GHz ($\omega\tau_e \approx 5$), the UTC structure exhibits a higher intrinsic response than the pin structure. The absorber of the UTC-PD can be made thin with the same capacitance maintained, which is the basic feature of the UTC operation mode. Experimentally, a fabricated UTC-PD (depletion thickness of 0.24 μm) showed an *f*3dB of 310 GHz, evaluated from the pulse photoresponse measured by the electro-optic sampling (EOS) method [13]. Deviation of the measured *f*3dB from the value predicted is due to many factors, including the additional delay time associated with a p-type absorber, an RC time constant, and measurement-related pulse broadening.

3.3 Transient Electron Velocity in a Photodiode

In calculating the PD frequency response above, carrier velocities were assumed constant. However, the electron velocity changes with time and space during the electric field acceleration and scattering process, and the overshoot range depends on field intensity. We will see how such a transient electron velocity depends on the field intensity in a PD structure.

The velocity of each electron v_k is given by the group velocity and approximated to

$$v_k = \frac{p_k}{m^*(1+2\alpha E_k)} \tag{3.3}$$

where k is the wavenumber of electrons, p_k is the electron momentum, m^* is the electron effective mass at the band edge, and E_k is the electron energy. The m^* value is $0.077m_0$ and $0.043m_0$ for InP and InGaAs, respectively. Since the electron energy deviates from parabolic variation with k when k increases, the correction term $(1 + 2\alpha E_k)^{-1}$ must be included using the nonparabolicity parameter α:

$$\alpha = \frac{1}{E_g}\left(1-\frac{m^*}{m_0}\right)^2, \tag{3.4}$$

where E_g is the band-gap energy and m_0 is the electron rest mass. The value of α is 0.63 and 1.22 in InP and InGaAs, respectively. The calculated group velocity is shown in Fig. 3.5. The group velocity can be considerably higher than the saturation velocity on the order of ~1×10^7 cm/s. When comparing materials, InGaAs has a much smaller band-edge effective mass, $m^* = 0.043m_0$ than InP ($0.077m_0$). Nevertheless, InGaAs exhibits only a slightly larger group velocity than InP, and the maximum available group velocities are not very different from each other. This clearly results from the larger band nonparabolicity in InGaAs.

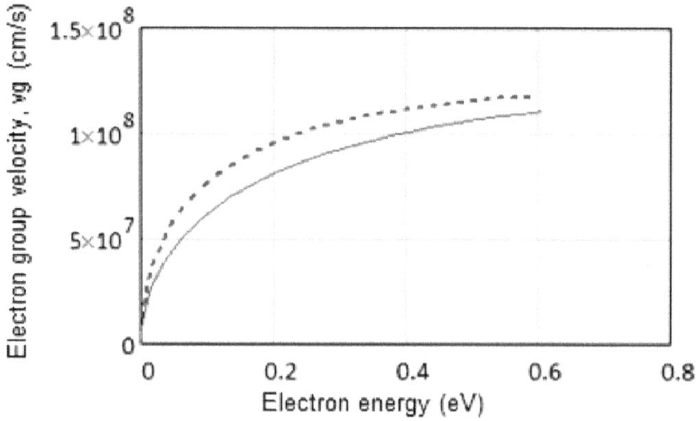

Figure 3.5 Calculated electron group velocity for InP (solid line) and InGaAs (dotted line).

Due to the scattering process, the actual electron drift velocity v_d is reduced from the group velocity (Fig. 3.5). So the scattering rate (or momentum relaxation time) is of key importance for the overshoot. The low-energy momentum relaxation time $\tau_{mLE} = \mu_e m^*/q$ is a basic parameter and experimentally evaluated from mobility μ_e to be 0.20 ps in InP and 0.24 ps in InGaAs, where μ_e of 4,500 cm/Vs for InP and 10,000 cm/s for InGaAs are assumed as values in high-purity material. When electron energy is higher than longitudinal optical (LO)-phonon energy (E_{LO}), the scattering rate becomes significantly high (about a fivefold increase in InP) due to the appearance of the LO-phonon emission process, and the momentum relaxation time is dominated by this process. The high-field momentum relaxation time is a few times smaller than τ_{mLE}. In the phenomenological approach, the temporal variation of the average electron momentum $p(E_e)$ and averaged energy E_e are expressed as

$$\frac{dp(E_e)}{dt} = -qF - \frac{p(E_e)}{\tau_m} \tag{3.5A}$$

$$\frac{dE_e}{dt} = -q\frac{p(E_e)}{m^*(1+2\alpha E_k)}F - \frac{E_{LO}}{\tau_e}. \quad (3.5B)$$

by modifying the relaxation time approximation [22]. Because $p(E_e)$ is an averaged value, it is not directly related to energy through a dispersion relation. However, obtaining the drift velocity v_d as the group velocity, using Eq. 3.3 by taking the average energy E_e into account, is considered to be valid in the present approximation. Instead of energy relaxation, a term of energy loss with a loss rate of $1/\tau_e$ is assumed in Eq. 3.5B.

We calculate the transient electron velocity in InP, while electrons are staying in the Γ-valley ($E_e < \Delta E_{\Gamma\text{-}L}$). Parameters of τ_m and τ_e are determined so that the steady-state solution by Eqs. 3.5A and 3.5B fits the drift velocity and electron energy simulated by Monte Carlo simulation [22]. At a field of 15 kV/cm, τ_m and τ_e are estimated to be $\tau_{mLE}/2.3$ and $\tau_{mLE}/1.59$. The energy loss rate (emission rate - absorption rate) corresponding to $\tau_{mLE}/1.59$ is calculated to be 8×10^{12}/s.

Figure 3.6 shows the transient electron velocity in InP with time and position, numerically simulated by Eqs. 3.5A and 3.5B, where calculations are carried out until the electron energy reaches 0.6 eV (indicated by diamonds). In all curves for 27 kV/cm, 40 kV/cm, and 60 kV/cm, the velocity overshoot feature is seen with peak velocities ranging from 4.1×10^7 cm/s to 6.3×10^7 cm/s. The steep increase of velocity within a short time and traveling distance (<100 fs, <300 A) seen here is possible for a III–V material with a small effective mass. The overshoot ranges while electrons are staying in the Γ-valley are calculated to be 0.32 μm, 0.15 μm, and 0.11 μm at fields of 27 kV/cm, 40 kV/cm, and 60 kV/cm, respectively. According to Eq. 3.5A, the electron momentum $p(E_e)$ tends to saturate during field acceleration. On the other hand, the electron energy continues to be elevated since the field acceleration is relatively quick compared to the energy loss process. In such a condition, the increase in effective mass with electron energy decelerates the electron drift motion, resulting in the velocity overshoot feature.

In a reported THz PD, the optimum field in the diode depletion layer lies around 50 kV/cm. From the calculated velocity profiles, the overshoot in such a field range is found to appear with about

0.15 μm in space and the corresponding time is as short as 0.3 ps. Leitenstorfer et al. characterized the velocity overshoot in an InP pin-diode structure [23]. The evaluated peak velocity was as high as 8×10^7 cm/s at 90 kV/cm. Though the calculated data is not shown here, Eqs. 3.5A and 3.5B predict a peak velocity of 7.5×10^7 cm/s.

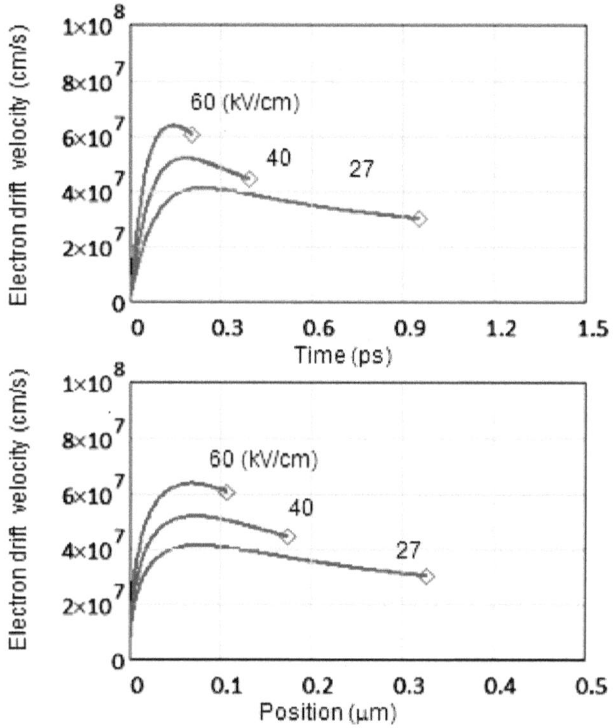

Figure 3.6 Calculated transient electron velocity profiles in InP.

In measurements of the pulse photoresponse of a UTC-PD [13], the highest bandwidth evaluated from the EOS signal (full width, half maximum [FWHM] = 0.97 ps) was 310 GHz at a bias voltage of −0.5 V (field of 50 kV/cm), as shown in Fig. 3.7. At the higher bias voltage beyond the optimum, the FWHM became wide down to 2.4 ps at −5 V, as also seen in Fig. 3.7. Note that the FWHM increases very little, even at 25 kV/cm (close to a bias voltage of 0 V) compared to that for 50 kV/cm, which confirms an existence of a broad overshoot range, as we have discussed above. All these observations clearly demonstrate the velocity overshoot phenomenon in InP.

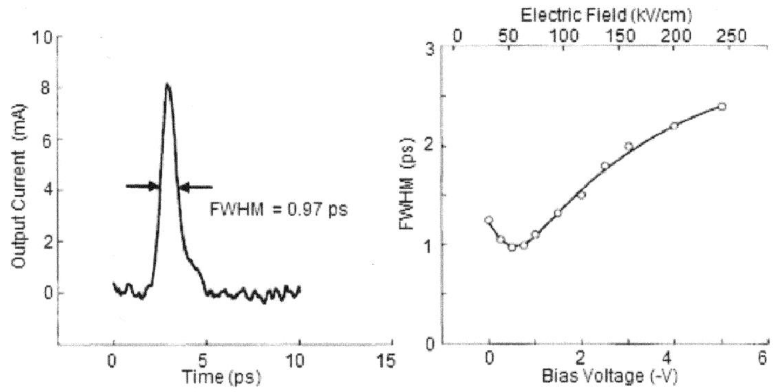

Figure 3.7 Measured pulse photoresponse of an InP/InGaAs UTC-PD.

3.4 Diode Photomixers and Their Performance

The available output power of a diode photomixer depends on many factors: intrinsic bandwidth of a PD used, its junction capacitance, the matching circuit between the PD and load, the effective load condition, and radiator configurations connected to the PD.

In the PD design, the semiconductor layer structure and diode area are optimized by taking the trade-off relation between intrinsic bandwidth and effect of capacitance into account. We briefly discuss the scaling of diode dimensions. To have a k times higher intrinsic bandwidth (here k is the scaling factor), the vertical diode dimension must be reduced by $1/k$. This scaling results in larger capacitance for a unit area by a factor of k. Then to have the same ωC value, the diode area must be scaled by $1/k^2$, leading to an output power proportional to $1/k^4$ for the same current density. This is the inherent constraint in designing THz PDs. When increasing the operation current of a PD, the junction temperature rise limits usually before the electronic output saturation appears, as described in the first section. Therefore, lowering the operation voltage is very important to reduce the device self-heating and to allow a higher operation current level.

3.4.1 Diode Structure

In meeting the basic needs of better bandwidth and low operation voltage, UTC-PDs and their modified structures have played an important role. The dual-depletion-region pin-PD (Fig. 3.8A), in which electron transport dominates in the photoresponse, can also extend the bandwidth, though there has been no report of a photomixer using this structure. When the InGaAs absorption layer occupies 50% of the depletion layer in this structure, for example, the ratio of the average carrier traveling length is 1:3 and the contribution of the hole current to the power response is reduced to only 10% of the total response. The modified UTC-PD structure with a hybrid absorber, Fig. 3.8B [6, 17, 18], is considered to be most suitable for a photomixer device, since the constraint imposed by the bandwidth/efficiency trade-off relation is relaxed. Namely, the efficiency (DC responsivity) can be maximized for a constant required bandwidth by adjusting the p-type neutral and depleted absorber thickness. Typically, about twice the efficiency is achieved by the hybrid absorber design compared to the original UTC-PD. The absorber material is usually InGaAs because it has the highest electron mobility and largest absorption coefficient among material lattice-matched to InP. The electron traveling layer connected to the depleted absorber is usually InGaAsP or InP, in which significant electron velocity overshoot is expected. At a 100 GHz band, the reported output level of UTC-PDs is about 20 dB higher than that of pin-PDs [21].

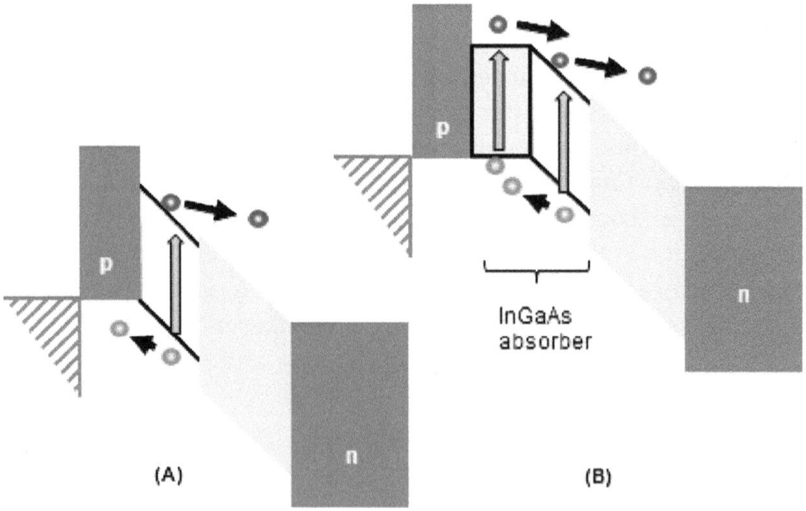

Figure 3.8 Dual-depletion pin-PD (A) and modified UTC-PD with a hybrid absorber.

3.4.2 Optical Coupling

Optical coupling between a fiber and a PD is an important concern, especially when designing antenna-integrated photomixers. A back-illuminated diode geometry is not easy to implement when a diode is packaged with a fiber port. Edge-coupled illumination with a refracting facet is a simple scheme and has been used for photomixers with rectangular waveguide output [5, 6, 9, 10]. Another structure is the evanescently coupled PD with an integrated optical waveguide beneath the diode [24]. By adjusting the matching layer length, good optical coupling can be achieved, even for a small PD length.

To produce output radiation in a quasi-optical manner, an antenna-integrated PD chip (InP substrate) is usually placed on a hemispherical Si lens. The output radiation from the antenna can flow mostly to the semiconductor side because of the difference in refractive index between the semiconductor and air. The radiation is then transmitted through the Si lens surface with a transmittance of 70% without an antireflection coating on Si. The optical coupling to a PD on a Si lens is made also by edge illumination or evanescent coupling [25]. A packaged photomixer using an antenna-integrated PD is shown in Fig. 3.9. The package size is 12.7 (W) × 22.17 (L) × 11.6 (H) mm^3. The coupling of optical input to a PD is made through a refracting facet using two-lens optics.

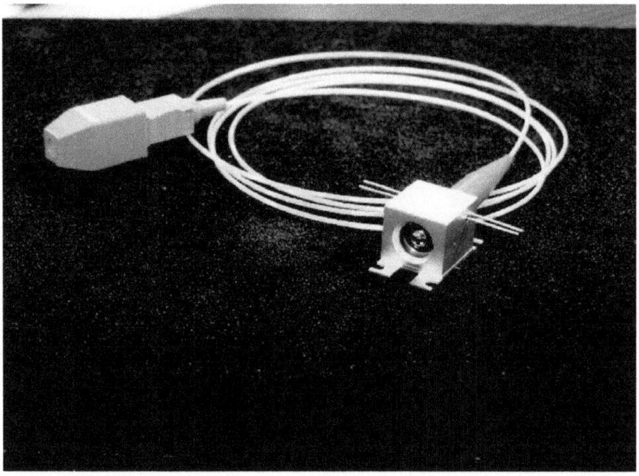

Figure 3.9 Photomixer module packaging an antenna-integrated PD. The diameter of the hemispherical Si lens is 5 mm.

3.4.3 Matching Circuit and Radiator Configurations

In a frequency range up to around 350 GHz, photomixers with a rectangular waveguide output have been successfully fabricated. In these photomixers, a diode is integrated with a resonant matching circuit on the same InP substrate. A typical matching circuit includes a short stub inductor connected to the signal line, which compensates for the reactance of diode capacitance. This allows us to use a larger junction size for a higher operation current. The effective load resistance can also be adjusted to increase the output at the expense of bandwidth. The waveguide coupler line is used to feed the PD's output to the waveguide.

Figure 3.10 shows measured output versus frequency for a J-band photomixer packaging an edge-illuminated UTC-PD. A waveguide coupler made of a quartz line was designed so that the higher-order modes would be well suppressed below –30 dB up to 530 GHz [6]. The bias voltage was fed from the short plane, as seen in Fig. 3.10. The output peaked at 350 GHz with –2.7 dBm, where the DC responsivity was 0.22 A/W. In this case, the effective load was designed to be relatively low, 50 Ω enabling a wider 3 dB down bandwidth of 120 GHz and a 10 dB down bandwidth as wide as about 300 GHz. By combining the output from two monolithically integrated UTC-PDs with a T-junction, a 300 GHz output power as high as 1 mW has been achieved [10].

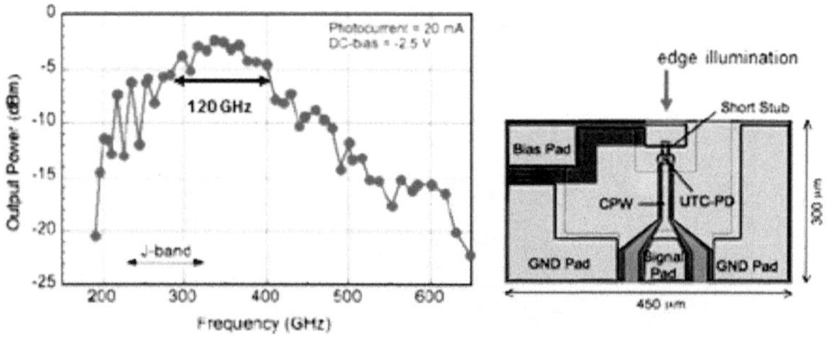

Figure 3.10 Output performance of a J-band photomixer with a WR-3 waveguide output. Reproduced from Ref. [6] by permission of IEEE, © 2008.

The simplest photomixer configuration may be a "chip photomixer" that integrates a PD and an antenna, as the radiator, on a common InP substrate. A photomixer that integrates a UTC-PD and a broadband transverse electromagnetic horn antenna (TEM-HA) [7] has also been fabricated and shown to have excellent performance. The THz radiation obtained ranged in a frequency from 500 GHz to 1800 GHz.

Planar antennas are easily integrated with a PD [1–4] (antenna-integrated PD). Here the whole structure is fabricated with photolithographic preciseness, and there is no need for high-frequency wiring for packaging, which allows us to fabricate/manufacture reproducibly and keep photomixer characteristics. Many types of such antenna-integrated PDs have been investigated. Nakajima et al. reported THz output performance for UTC-PDs integrated with log-periodic antennas and twin-dipole antennas [4], as shown in Fig. 3.11. The PDs integrated with log-periodic antennas had a wide operation frequency range of up to 1.5 GHz. The devices with twin-dipole antennas clearly had a resonant feature with peak frequencies of 0.75 THz, 1.05 THz, and 1.5 THz. At the same DC operation current of 8 mA, the output level of a twin-dipole antenna was 6 dB higher than that of log-periodic antenna (Fig. 3.12). The maximum output could be as high as 11 μW, about five times as large as that obtained by a resonant-type low temperature (LT)-grown GaAs photoconductive switch at around 1 THz [27].

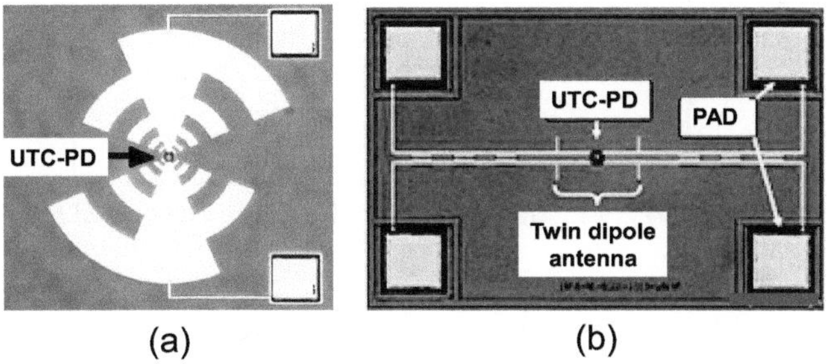

Figure 3.11 Antenna-integrated PD chip. (A) Log-periodic antenna and (B) twin-dipole antenna.

3.5 Summary

Using PDs to generate CW THz radiation has attracted attention over the last decade. Now the available power is approaching the microwatt level in a frequency range of up to 1 THz. A UTC-PD has been widely used to achieve a better frequency response and a higher output. The carrier traveling distance in a UTC-PD is longer than that in a pin-PD for the same depletion thickness. However, due to the electron velocity overshoot, which provides a much higher velocity than the steady-state saturation condition, the UTC-PD can function with a better frequency response. A low operation voltage also contributes to the enhancement of the output. Packaged photomixer modules with waveguide output ports and Si lens ports have been fabricated. Since a PD and a planar antenna can be monolithically integrated on a substrate with photographic preciseness, antenna-integrated PDs are suitable for THz generators operated beyond the THz frequency range.

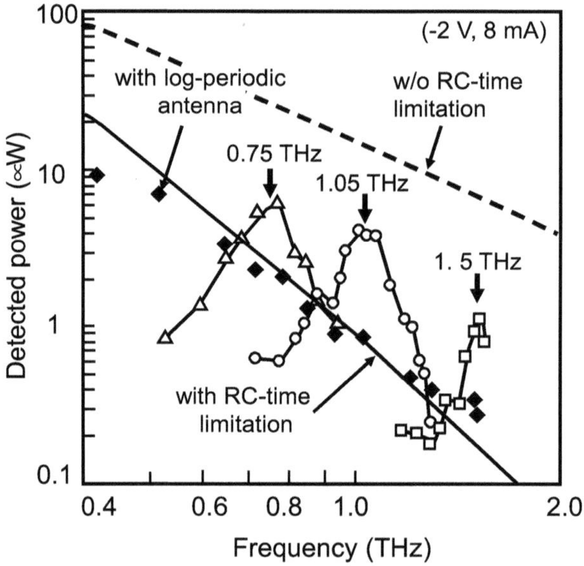

Figure 3.12 THz radiation output of UTC-PDs integrated with log-periodic antennas and twin-dipole antennas.

Acknowledgments

The author acknowledges valuable discussions with people working or those who have worked at NTT Laboratories in the THz science and technology field.

References

1. H. Ito, T. Furuta, F. Nakajima, K. Yoshino, and T. Ishibashi, Photonic generation of continuous THz wave using uni-traveling-carrier photodiode, *J. Lightwave Technol.* **23**(12), 4016–4021 (2005).
2. T. Nagatsuma, Generating millimeter and terahertz waves, *IEEE Microwave Mag.* **June**, 64–74 (2009).
3. M. J. Fice, E. Rouvalis, L. Ponnampalam, C. C. Renaud, and A. J. Seeds, Telecommunications technology-based terahertz sources, *Electron. Lett.* **6**, s28–s31 (2010).
4. F. Nakajima, T. Furuta, and H. Ito, High-power continuous-terahertz-wave generation using resonant-antenna-integrated uni-travelling-carrier photodiode, *Electron. Lett.* **40**(20), 1297–1298 (2004).
5. T. Furuta, T. Ito, Y. Muramoto, H. Ito, M. Tokumitsu, and T. Ishibashiin, D-band rectangular-waveguide-output uni-travelling-carrier photodiode module, *Electron. Lett.* **41**(12), 715–716 (2005).
6. A. Wakatsuki, T. Furuta, Y. Muramoto, T. Yoshimatsu, and H. Ito, High-power and broadband sub-terahertz wave generation using a J-band photomixer module with rectangular-waveguide output port, *33rd Int. Conf. mm THz Waves, 2008. IRMMW-THz 2008*, Pasadena, CA, 1–2 (2008).
7. A. Beck, G. Ducournau, M. Zaknoune, E. Peytavit, T. Akalin, J. F. Lampin, F. Mollot, F. Hindle, C. Yangand, and G. Mouret, High-efficiency uni-travelling-carrier photomixer at 1.55 μm and spectroscopy application up to 1.4 THz, *Electrron. Lett.* **44**(22), 1320–21 (2008).
8. E. Rouvalis, C. C. Renaud, D. G. Moodie, M. J. Robertson, and A. J. Seeds, Traveling-wave uni-traveling carrier photodiodes for continuous wave THz generation, *Opt. Express* **18**(11), 11105–11110 (2012).
9. H. Ito, T. Ito, Y. Muramoto, T. Furuta, and T. Ishibashi, Rectangular waveguide output unitraveling-carrier photodiode module for high-power photonic millimeter-wave generation in the F-band, *J. Lightwave Technol.* **21**(12), 3456–3462 (2003).
10. H.-J. Song, K. Ajito, Y. Muramoto, A. Wakatsuki, T. Nagatsuma, and N. Kukutsu, Uni-traveling-carrier photodiode module generating 300 GHz power greater than 1 mW, *IEEE Microwave Wireless Compon. Lett.* **22**(7), 363–365 (2012).
11. J.-W. Shi, F.-M. Kuo, and M.-Z. Chou, A linear cascade near-ballistic uni-traveling-carrier photodiodes with extremely high saturation-current bandwidth product (6825 mA-GHz, 75 mA/91 GHz) under a 50 Ω load, OSA/OFC/NFOEC 2010, PDPA6 (2010).

12. G. Lucovsky, R. F. Schwarz, and R. B. Emmons, Transit time considerations in p-i-n diodes, *J. Appl. Phys.* **35**(3: Part 1), 622–628 (1964).

13. H. Ito, T. Furuta, S. Kodama, and T. Ishibashi, Photodiode with 310GHz bandwidth, *Electron. Lett.* **36**(21), 1809–1810 (2000).

14. F. J. Effenberger and A. M. Joshi, Ultrafast, dual-depletion region, InGaAs/INP p-i-n detector, *J. Lightwave Technol.* **4**(8), 1859–1863 (1993).

15. T. Ishibashi, N. Shimizu, S. Kodama, H. Ito, T. Nagatsuma, and T. Furuta, Uni-traveling-carrier photodiodes, *Tech. Dig. 1997 Ultrafast Electron. Optoelectron.*, 83–87 (1997).

16. T. Ishibashi and N. Shimizu, Uni-travaling-carrier photodiode as an optoelectronic driver, *Tech. Dig. 1999 Ultrafast Electron. Optoelectron.*, 203–207 (1999).

17. T. Ishibashi, Y. Hirota, and Y. Muramoto, Photodiode, US Patent 6,909,161 B2, June 21, 2005.

18. Y. Muramoto and T. Ishibashi, InP/InGaAs pin photodiode structure maximizing bandwidth and efficiency, *Electron. Lett.* **39**, 1749–1750 (2003).

19. X. Li, S. Demiguel, N. Li, J. C. Campbell, D. L. Tulchinsky, and K. J. Williams, Backside illuminated high saturation current partially depleted absorber, *Electron. Lett.* **39**, 1466–1467 (2003).

20. J.-W. Shi, Y.-S. Wu, C.-Y. Wu, P.-H. Chiu, and C.-C. Hong, High-speed, high-responsivity, and high-power performance of near-ballistic uni-traveling-carrier photodiode at 1.55-μm wavelength, *IEEE Photon. Technol. Lett.* **17**(9), 1929–1931 (2005).

21. H. Ito, S. Kodama, Y. Muramoto, T. Furuta, T. Nagatsuma, and T. Ishibashi, High-speed and high-output InP-InGaAs unitraveling-carrier photodiodes, *IEEE J. Sel. Top. Quantum Electron.* **10**(4), 709–727 (2004).

22. T. L. Maloney and J. Frey, Transient and steady-state electron transport properties of GaAs and InP, *J. Appl. Phys.* **48**(2), 781–787 (1977).

23. A. Leitenstorfer, S. Hunsche, J. Shah, M. C. Nuss, and W. H. Knox, Femtosecond charge transport in polar semiconductors, *Phys. Rev. Lett.* **82**(25), 5140–5143 (1999).

24. A. Beling, H.-G. Bach, G. G. Mekonnen, R. Kunkel, and D. Schmidt, High-speed miniaturized photodiode and parallel-fed traveling-wave photodetectors based on InP, *IEEE J. Sel. Top. Quantum Electron.* **13**(1), 15–21 (2007).

25. M. Weiss, A. G. Steffan, S. Fedderwitz, G. Tsianos, A. Stohr, and D. Jager, Highly-compact fibre-optic package for 30-300GHz wireless transmitter modules, *2nd Electron. Syst. Integrat. Tech. Conf.*, 1111–1114 (2008).

26. T. Ishibashi, A. Wakatsuki, and S. Kodama, Development of photodiode technologies for millimeter/terahertz-wave signal generation, APMP 2012, Kyoto, FA-1 (2012).

27. S. M. Duffy, S. Verghese, K. A. McIntosh, A. Jackson, A. C. Gossard, and S. Matsuura, Accurate modeling of dual dipole and slot elements used with photomixers for coherent terahertz output power, *IEEE Trans. MTT* **49**(6), 1032–1038 (2001).

Chapter 4

Quantum Cascade Lasers as a Terahertz Wave Source

Iwao Hosako and Hiroaki Yasuda
National Institute of Information and Communications Technology, 4-2-1 Nukui-kitamachi, Koganei, Tokyo 184-8795, Japan
hosako@nict.go.jp, yasuda@nict.go.jp

4.1 Device Technology

4.1.1 Quantum Cascade Laser Technology Overview

Quantum cascade lasers (QCLs) are semiconductor lasers operating at midinfrared (mid-IR) frequencies (3–20 μm, i.e., 15–100 THz) and terahertz (THz) (60–200 μm, i.e., 1.5–5.0 THz) bands. The basic concept of these lasers appeared in 1971 [1]; however, a further 20 years elapsed before it reached fruition. Mid-IR QCLs appeared in 1994 [2] and THz QCLs in 2002 [3]. Lead-salt semiconductor lasers (2–30 μm, maximum operating temperature of approximately 240 K in pulse operation, approximately 170 K in continuous operation, normally cooled to 100 K or lower) [4] were used prior to the appearance of mid-IR QCLs. For a number of reasons, including the

Handbook of Terahertz Technologies: Devices and Applications
Edited by Ho-Jin Song and Tadao Nagatsuma

ISBN 978-981-4613-08-8 (Hardcover), 978-981-4613-09-5 (eBook)
www.panstanford.com

need for cooling to 100K or lower, the use of lead-based lasers was limited. In the THz band, light sources, whether cooled or uncooled, were unavailable at small milliwatt-scale outputs, leading to what was referred to as the "THz gap" (see Fig. 4.1). QCLs were seen as a strong candidate to resolve this problem and have been the subject of much research and development since their initial appearance. The number of papers on this subject continues to increase (see Fig. 4.2).

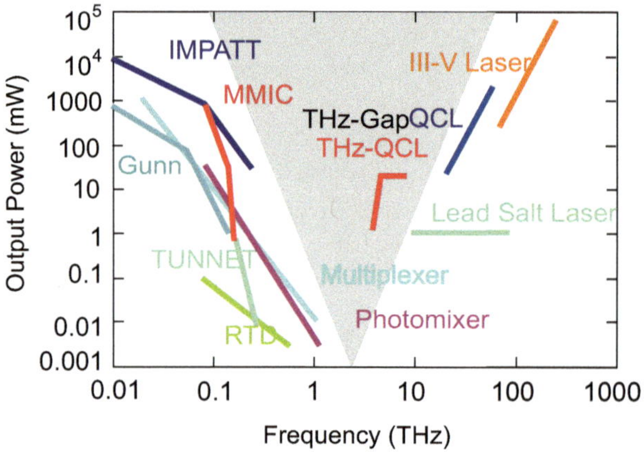

Figure 4.1 Output power continuous wave (CW) of semiconductor devices as a function of operating frequency.

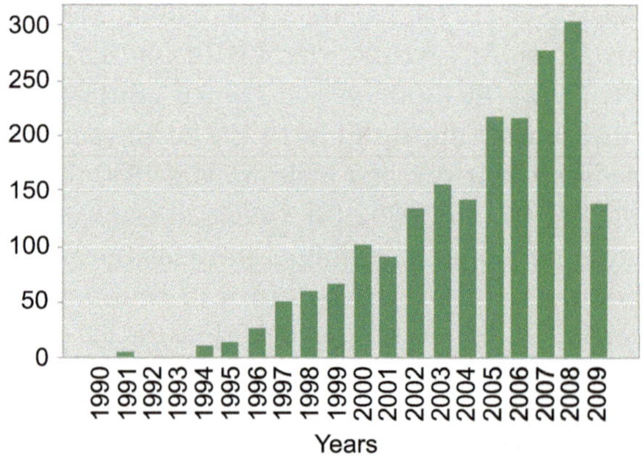

Figure 4.2 Trend in the number of papers, including the keyword "QCL" (including minutes of international conferences). QCLs in the THz band account for approximately 25% of the total (e.g., approximately 75 papers in 2008).

4.1.2 Mid-IR Quantum Cascade Laser Overview

The relationship between the oscillation frequency (wavelength) and the operating temperature of QCLs is shown in Fig. 4.3. Room-temperature continuous oscillators have been implemented at mid-IR in the 3.8–10.6 μm band. High-output QCLs are available at a continuous oscillation output of 3W (293 K, electro-optic conversion efficiency 12.7% [5]) at a wavelength of 4.6 μm and a peak output of 6.3W (300 K, electro-optic conversion efficiency 11% [6]) at a wavelength of 8.6 μm. InGaAs/InAlAs lasers are primarily used at wavelengths of 4 μm or less. In the shorter-wavelength region, strained InGaAs/InAlAs material system is used, for example, in the above-mentioned literature [5], the quantum well/barrier of $In_{0.67}Ga_{0.33}As/Al_{0.64}In_{0.36}As$ containing strain (1% mismatch of lattice constants between well and barrier materials) was used. For a further shorter-wavelength region less than $\lambda = 3.7$ μm, it is important to suppress the intervalley scatterings in the conduction band of the well from a Γ point to an X point, and from a Γ point to an L point, in the case of a InGaAs well and a InAs well, respectively. Moreover, in some cases, to obtain a large conduction band offset, antimony (Sb) has been introduced in the quantum well, for example, the room-temperature pulsed-mode operation at $\lambda = 3.1$ μm with a peak output power of 8 mW (295 K) [7] was realized by introducing the strain-compensated $In_{0.67}Ga_{0.33}As/AlAs_{0.8}Sb_{0.2}/InP$ material system.

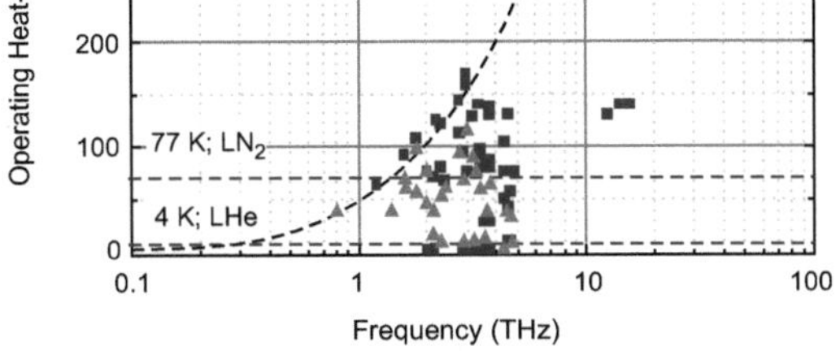

Figure 4.3 Operating temperatures of QCLs.

As an improvement of the active layer structure, an indirect pump (IDP) structure has been proposed and demonstrated [8]. As a result of an introduction of the IDP scheme at the carrier injection barrier of the active layer, the characteristic temperature T_0 of the device reaches up to 303 K. The upper energy level for the laser oscillation (L_{up}) exists at the same energy of the injector in the case of the conventional resonant-longitudinal-optical-phonon (RP)-type active structure (see Fig. 4.3) [13]; however, with the IDP structure, a new laser upper oscillation level ($L_{up\text{-}new}$) is established below this level at a point equivalent to the energy of the longitudinal optical (LO) phonon, and immediately after the carriers (electrons) tunnel the injection barrier, ultrahigh-speed relaxation to the $L_{up\text{-}new}$ occurs at the top energy level in the emission well due to LO phonon resonance scattering. The carriers tunneling the injection barrier in the reverse direction are reduced by reducing the carrier density at the energy level immediately after tunneling the injection barrier in the normal direction. As a result, current density is approximately doubled with the same amount of doping of the injection layer. At the same time, since the carrier density of the injection layer is reduced, Drude absorption of light is controlled by this amount, the threshold current is reduced, and a high T_0 is obtained. As a result of above-mentioned effects, the threshold current becomes lower and T_0 becomes higher. In the actual IDP QCL of λ = 13.5 μm using the lattice-matched InGaAs/InAlAs/InP material system, T_0 reaches 400 K [9] at room temperature. This value is almost equivalent to the T_0 (around 500 K [10]) of the latest 1.3 μm band quantum dot laser operating at room temperature.

Distributed feedback lasers [11] for single-mode operation, and plazmonic antenna loading [12] on the laser end facet of the waveguide to improve the quality of the beam emitted from the end facet, are used to improve the waveguide resonator structure and have the practical effect of improving ease of use.

Gas analysis applications (CO, CO_2, NO, NO_2, SO_2, CH_4, NH_3, C_2H_4, etc.) are proceeding in the mid-IR region. Examples of the development of gas analysis applications include environmental monitoring, waste gas analysis, and diagnoses based on exhaled air. Furthermore, light detection and ranging (LIDAR) in the atmospheric window (3.4–5.5 μm and 8–14 μm in the mid-IR band) is under investigation. These lasers hold promise as high-power (watt-class) light sources for use in plastic fabrication in light of the low transmittance of plastics in the 8–14 μm band. A number of other applications have been proposed, including medical applications and noncontact visualization of matter distribution combining multicolor laser arrays and microbolometer cameras.

4.1.3 Terahertz-Band Quantum Cascade Lasers

In the THz frequency range, the GaAs/AlGaAs material system is primarily used for the THz QCL. GaAs/AlGaAs lasers. The types of active layer structure of the THz QCLs can generally be classified as the RP type [13] described above, the bound-to-continuum (BTC) type [14], and the interminiband type: interminiband transitioning chirped superlattice [15] [(IM-CSL]) (see Fig. 4.4). The IDP structure, which is a kind of improved RP-type active structure, has also been applied to the THz QCL. In both the BTC and IM-CSL, a population inversion for laser action can be realized by using the ultrafast carrier relaxation process from the top of the miniband to the bottom of the miniband. The carrier distribution at the top of the miniband decreased against the upper energy levels of the laser oscillation. In the RP, the resonant LO phonon scattering process is utilized instead of intraminiband scattering. The carriers in the level 2 (L_2) are scattered by the LO phonon to L_1. This process is extremely fast (τ_{32} >> τ_2) and depopulates the L_2. Then, a population inversion arises among levels L_3 and L_2. Finally, the carriers in L_1 are injected into the next active unit.

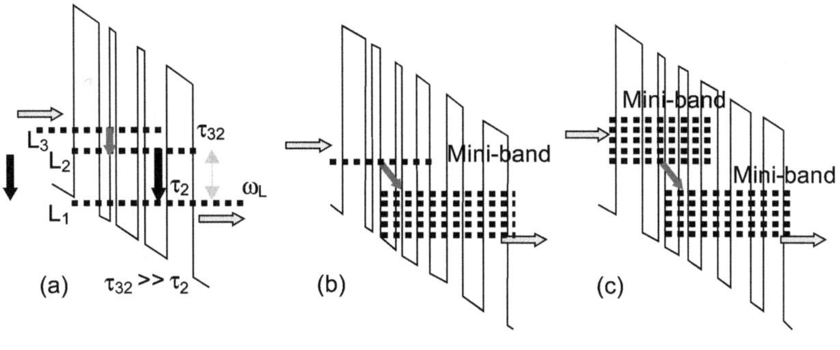

Figure 4.4 Schematic illustrations of the typical active layer structures for THz QCLs. (a) Resonant LO phonon (RP), (b) bound-to-continuum (BTC), and (c) interminiband transition in a chirped superlattice (IM-CSL).

The waveguide structures of THz QCLs differ considerably from those of mid-IR QCLs. While a dielectric waveguide consisting of core (active) and clad (with a lower dielectric constant than the core.) has been usually used in the case of mid-IR QCLs, single-metal [3] or double-metal [16] waveguides have been invented and utilized for THz QCLs. In both cases, the plasmon mode arising from a metal or highly doped semiconductor is utilized to confine the optical mode into the thin active layer (see Fig. 4.5). An end facet of the double-metal-type waveguide (aperture) is rectangular, and the typical size of the short side (crystal growth direction) is 10 μm to 12 μm. This value is smaller than the wavelength of the THz wave (around 30 μm at 3 THz, n = 3.3 is assumed) in the waveguide cavity. The optical (THz) power that has arisen in the resonator cavity can be properly extracted in many cases, especially for the double-metal-type waveguide, due to the aperture size. To improve the efficiency of extraction, to enlarge the output power, and to improve the quality of the output beam, there have been a lot of efforts, such as mounting a hyperhemispherical lens on the edge of the waveguide [17], loading a metal horn antenna on the edge of the waveguide [18], and applying a third-order diffracting grating on the top electrode of the waveguide [19], a ring resonator [20], and a photonic crystal structure [21] for vertical extraction (along with the crystal growth direction) of the laser beam.

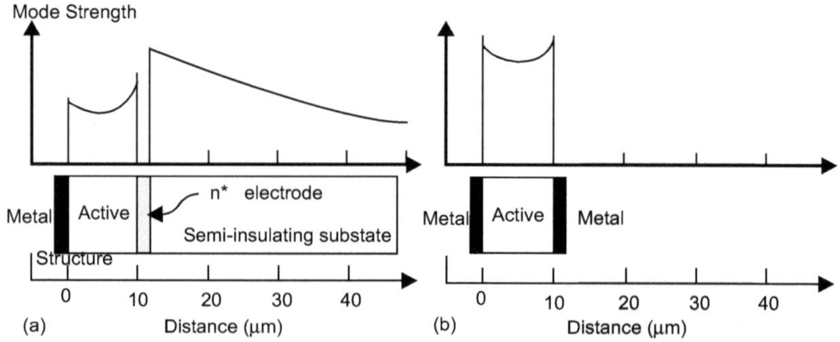

Figure 4.5 Schematic illustrations of cross sections for the waveguides of a THz QCL and optical mode strength for (a) single metal type and (b) double metal type.

The most serious problem encountered when a THz QCL is used for practical applications is the operating temperature. Room-temperature operation is indispensable in practical applications, although a maximum operating temperature of 199.5 K [22] has been achieved in 2012. There is no fundamental reason that limits the room-temperature operation of THz QCLs.

By applying a strong magnetic field (up to 20 tesla) in the direction of crystal growth (i.e., perpendicular to the layers), intrawell conduction in the active layer quantum wells is quantized to the Landau level, permitting injection of carriers at the desired upper level of laser oscillation when the energy difference between upper and lower laser oscillation levels is small and inhibiting thermally activated phonon scattering. Then, laser oscillations at a lower frequency (680 GHz) and higher temperature (225 K) are possible when a magnetic field is applied [23]. While application of a large magnetic field is not practical, a number of experiments on this theme have hinted significantly at the possibility of operation at normal temperatures.

Factors in the inability to operate THz QCLs at room temperature include thermally activated phonon scattering, which does not contribute to emission of light, backfilling of the laser lower oscillation level due to thermal redistribution, and deterioration in injection efficiency in the upper laser oscillation level resulting from thermal distribution. Progress in high-temperature operation of THz QCLs is possible if these problems can be resolved with an active layer structure that does not require application of a magnetic field. An active layer structure that applies the IDP described above to THz QCLs has been proposed as a means of resolving some of these problems. A simulation of this structure based on first principles using the nonequilibrium Green's function (NEGF) has been conducted and indicated gain at 200 K [24].

Research and development focused on THz QCLs operating above room temperature is progressing in two directions. One is based on optimizations on the active layer structure and improvements of heat dissipation of the THz QCL device. Development on this theme, primarily by research groups at the Massachusetts Institute of Technology (MIT), USA, and the National Research Council (NRC), Canada, produced a record operating temperature of 199.5 K [22] in 2012. The THz QCL device in this report is of the RP type, with a multistage unit comprising three quantum wells in one unit of the active layer. The reported laser transition oscillator strength of 0.58 considerably exceeds the value of 0.41 reported by Kumar et al. [25]. The thickness of the tunnel barrier for carrier injection is determined with a simplified density matrix model, and a large tunnel coupling strength and an almost perfect alignment between quantum levels across the barrier were obtained at the designed bias electrical field for the laser oscillation. At temperature of 8 K, a threshold current density of 1 kA/cm^2 and a peak output of 38 mW were recorded. The oscillation frequency changes from 2.6 THz to 2.85 THz by an increase of bias voltage. This change can be explained by the Stark effect. Simulations based on the Monte Carlo method are compared with experimental results. The oscillating frequency is further blue-shifted to 3.22 THz in the vicinity of the maximum operating temperature of 199.5 K. The ratio of oscillating photon energy to environmental thermal energy ($\hbar\omega/kB$) was 1.28. In comparisons using this ratio, the MIT group noted a record of 1.9 (oscillating frequency 1.8 THz, operating temperature 163 K) [26] in 2011. This THz QCL device was designed on a scheme named as scattering-assisted injection (almost same as the IDP scheme proposed by Yamanishi et al. [8, 9]) and exhibited a peak output power exceeding 2 mW at 155 K, attractive for applications in frequency bands up to 2 THz. Furthermore, these results can be seen as the way forward to increasingly high-temperature operation.

A further method focusing on THz-band light sources operating above room temperature is difference frequency generation (DFG) employing mid-IR QCLs, which has a high output power above room temperature. This method operates mid-IR QCLs in two modes using the optically nonlinear characteristics of the active layer to generate THz difference frequencies. Lu et al. [27] have made refinements with THz wave generation and frequencies using Cherenkov phase matching with DFG in which two frequencies are superimposed. Success has been achieved with this method using single-mode oscillation in a range of 1.0–4.6 THz (maximum side-mode suppression ratio 40 dB, maximum output 32 μw).

Examples of device development for room-temperature operation are almost all limited to the above, and it appears that a wall has been reached. While there is not a complete lack of proposals for increasing operating temperatures through the use of GaN materials with their large optical phonon energy [28], results only appear infrequently in the proceedings of international conferences. In the theoretical approach, the laser gain spectrum has been found using the NEGF and compared with experimental data obtained with time-resolved spectroscopy [29], and hypotheses continue to be developed on the nonlinear response of the active layer in order to understand THz QCL operation.

4.2 Applications of Terahertz-Band Quantum Cascade Lasers

4.2.1 Overview

THz QCLs are expected to be used in a wide range of applications such as imaging light sources, local oscillators for heterodyne receivers, and ultrahigh-speed modulation signal sources for very short-distance large-capacity wireless communications.

In the imaging field, label-free biological material detection [30] has been developed using a small, real-time THz imaging device combining a THz-band uncooled microbolometer array camera [31] and a liquid nitrogen–cooled THz QCL light source. Using the characteristics of high brightness and narrow beam, a light source has been developed as a local oscillator for heterodyne receivers. The device is a 1.5 THz QCL using a phase-locked loop (PLL) for frequency stabilization [27].

4.2.2 Wireless Communication

In QCLs with a carrier lifetime in the order of picoseconds, it is thought that the relaxation oscillation frequency is either high (in the order of a few hundred GHz) or nonexistent, and it is therefore considered that ultrahigh-speed modulation operation is possible. Electrical modulation experiments (maximum of 2 GHz [32]) were conducted in the mid-IR band. In the mid-IR band, experiments have not been conducted at greater speeds due to the lack of a high-speed detector. On the other hand, the Schottky barrier diode operates in the THz band as a detector permitting ultrahigh-speed operation, and since it brings the possibility of ultrahigh-speed wireless (up to

100 Gbit/s), modulation experiments on THz-band QCLs are underway. However, the considerable atmospheric absorption in the THz band ensures that wireless communication is at very short distances of less than 10 m indoors. Electrical modulation at up to 24 GHz has been reported [33]. The upper limit of this modulation frequency is constrained by passive components (capacity, resistance, and inductance) parasitic to laser devices and is not a characteristic of the QCL itself. A modulation experiment [34] is underway using photoinjection not subject to the limits due to electrical parasitic components. Demonstrations [35] of the transmission and reception of voice data using THz QCLs (transmitters) and THz-band quantum well–type detectors (receivers) have shown that transmission is possible over a distance of approximately 2 m.

4.2.3 Light Source for Spectroscopic Measurement

Research and development for achieving precision spectroscopic measurement using the THz QCLs is progressing, one example being research and development on a frequency comb for the THz band. The optical frequency comb (or a frequency synthesizer based on an optical frequency comb) permits supply of a high-stability frequency reference grid over a wide band and is considered a groundbreaking method of measurement. Comprehensive efforts to expand the use of this frequency comb in the THz band have commenced. Consolino et al. [36] have succeeded in phase locking of a 2.5 THz QCL to a free-space THz frequency comb generated in the 0.1–6.0 THz frequency range using a $LiNbO_3$ waveguide. The authors report that the beat necessary for phase locking was generated irrespective of the small output (less than 100 nW) of the THz QCL. These results represent technical developments expected to be applied in the manipulation of cooled molecules, astronomical research, and wireless communications. Shiba et al. [37] have used a NbTiN hot electron bolometer mixer (HEB) together with a THz QCL as a local oscillator in development of a 3.1 THz heterodyne receiver. In this receiver, a pair of slot antennas is employed to couple the radio frequency (RF) signal to the mixer. The receiver noise temperature (double sideband [DSB]) is reported to be 5600 K (2100 K if optical losses are considered). This research also shows that when compared to an

NbN HEB mixer normally used in the THz band, performance of the NbTiN HEB mixer is equivalent in practice at 3.1 THz. When taking precise measurements, it is important to know, both theoretically and experimentally, the actual linewidth of the laser. Ravaro et al. [38] measured the noise power spectrum density of a 2.5 THz QCL and obtained a value of an intrinsic linewidth of 230 Hz at a 2 mW output.

Improving single-mode selectivity and expanding the tuning range are both valid for a wide range of applications. Chakraborty et al. [39] introduced a nonperiodic distributed feedback grating into a Fabry–Pérot-type THz QCL resonator and obtained increasing single-mode selectivity and achieved a single-mode tuning range of 160 GHz for a 2.9 THz QCL.

4.2.4 Light Source for Imaging

In the THz imaging field, development of systems combining focal plane array detectors with THz QCL for illumination is demonstrated. Oda et al. [40, 41] have developed a microbolometer-type focal plane array detector with almost flat sensitivity characteristics within a range of 1.5–100 THz. The system is currently employed for real-time imaging. This focal plane array detector employs a 320 × 240 pixel array and is able to acquire 30 frame images per second. A compact THz QCL system employing a Stirling cycle cooler is used for illumination. This system employs a number of THz QCLs and provides 0.5–2 mW illumination at a time-average power over a range of 1.5–5.0 THz. In combination with the focal plane array detector noted above, and with additional lock-in imaging signal processing technology, this system achieves a high signal-to-noise ratio (up to 1000) and permits acquisition of a wide range of real-time images. De Cumis et al. [42] are currently developing a similar system that employs a confocal microscope system using a 2.9 THz QCL. Spatial resolution in horizontal and vertical directions is 70 μm and 140 μm, respectively. In comparison with nonconfocal systems, large contrasts are obtained, and while overlap occurs in the horizontal direction, objects on planes other than the depth direction can be discriminated and detected.

4.3 Simulations for Terahertz Quantum Cascade Lasers

4.3.1 Introduction

Remarkable progress in THz QCLs has been made since their first demonstration in 2002 [3, 44]. Several simulation methods have been proposed for improvement and optimization of THz QCLs. The simulation results serve to understand the fundamental physical processes in THz QCLs and to predict the transport characteristic, the electron distribution, and the gain for design improvements. We will briefly describe the calculation methods developed for THz QCLs, such as the rate equation, the Monte Carlo method, the simplified density matrix model, and the NEGF formalism.

4.3.2 Schrödinger–Poisson Equation

As the QCL utilizes electron transport between subband states in a series of multiquantum well structures, the wavefunction and energy of each sub-band state have to be calculated first. The sub-band wavefunctions can be used as the basis functions of the following simulations. The 1D Schrödinger equation in terms of a single-band effective mass model is given as

$$\left[-\frac{\hbar^2}{2}\frac{d}{dz}\left(\frac{1}{m^*(z)}\frac{d}{dz}\right)+\Delta E_c(z)-e\phi(z)\right]\psi(z)=E\psi(z) \tag{4.1}$$

where z is the direction of the crystal growth, $\Psi(z)$ is the envelope function, m^* is the effective electron mass, ΔE_c is the conduction band profile, e is the electron charge, and ϕ is the electrostatic potential that satisfies the following Poisson equation

$$\frac{d}{dz}\left(\varepsilon(z)\frac{d}{dz}\phi(z)\right)=-e\left(N_D(z)-n(z)\right) \tag{4.2}$$

where $\varepsilon(z)$ is the permittivity and $N_D(z)$ and $n(z)$ are the impurity (dopant) density and the electron density, respectively. The applied bias voltage is used in the boundary condition of Eq. 4.2.

The Schrödinger equation (Eq. 4.1) and the Poisson's equation (Eq. 4.2) are solved self-consistently. The Schrödinger equation can be solved with a transfer matrix method, for example. The Poisson equation can be solved numerically by using the finite difference method. A question is how to set the electron density $n(z)$. The electrons exist mainly in the upper lasing state in the actual THz QCL and do not distribute homogeneously all over the QCL structure. Therefore, another calculation method such as the rate equation is imposed to get the exact electron distribution. To reduce the calculation cost, it is acceptable to assume a homogeneous electron distribution over the one period of the QCL or an electron accumulation in the upper lasing state because the charge density in the normal THz QCL is not so high (around 3×10^{10} cm^{-2} per period) to have the pronounced effect on the potential profile.

4.3.3 Scattering Mechanisms

Intra-subband and inter-subband scatterings take place in THz QCLs, which have an important influence on the performance of the THz QCLs. Polar optical phonon, acoustic phonon, charged impurity, interface roughness, alloy, carrier–carrier, and the other scatterings have been investigated. To describe the phenomena, scattering rates are widely used for many simulations. The scattering rates can be given by using Fermi's golden rule. The important scattering mechanisms for THz QCLs might be LO phonon, interface roughness, and carrier–carrier scatterings. For example, the scattering rate due to the LO phonon scattering, which is utilized for electron depopulation of the lower lasing state in the resonant-phonon depopulation scheme [13], can be given by using the initial and final sub-band wavefunctions Ψ_i, Ψ_j and the phonon interaction Hamiltonian [44]. The lifetime $\tau_{ij}^{(\mathrm{LO\_emi})}(k_i)$ of a carrier in a subband level i with an in-plane wavenumber k_i is expressed as

$$\frac{1}{\tau_{ij}^{(\mathrm{LO\_emi})}(k_i)} = \frac{m^* e^2 \omega_{\mathrm{LO}}}{4\pi\hbar^2}\left(\frac{1}{\varepsilon_\infty} - \frac{1}{\varepsilon_s}\right)(n_{\mathrm{LO}}+1)\Theta\left(k_i^2 - \frac{2m^*\Delta}{\hbar^2}\right) \times \int_{-\infty}^{\infty} \frac{|G_{ij}(q_z)|^2 \, dq_z}{\sqrt{q_z^4 + 2q_z^2(2k_i^2 - 2m^*\Delta/\hbar^2) + (2m^*\Delta/\hbar^2)^2}} \quad (4.3)$$

where ε_s and ε_∞ are the low- and high-frequency dielectric constants, $\Theta(x)$ is the Heaviside function, and q_z is the phonon momentum. The form factor $G_{ij}(q_z)$, the phonon number n_{LO}, and Δ are given by

$$G_{ij}(q_z) = \int_{-\infty}^{\infty} \psi_j^*(z)\exp(-iq_z z)\psi_i(z)dz \tag{4.4}$$

$$n_{\mathrm{LO}} = \left[\exp\left(\hbar\omega_{\mathrm{LO}}/kT\right) - 1\right]^{-1} \tag{4.5}$$

$$\Delta = E_j - E_i + \hbar\omega_{\mathrm{LO}} \tag{4.6}$$

where E_i and E_j are the minima of the 2D subbands *i*, *j*.

The mean scattering rate τ_{ij} can be given by integrating $\tau_{ij}(k_i)$.

$$\frac{1}{\tau_{ij}^{(\mathrm{LO\_emi})}} = \int dE \frac{1}{\tau_{ij}^{(\mathrm{LO\_emi})}(k_i)} f(E) \Big/ \int dE f(E) \tag{4.7}$$

$$E = E_i + \frac{\hbar^2 k_i^2}{2m^*} \tag{4.8}$$

where the Fermi–Dirac distribution *f(E)* uses the electron temperature and the quasi Fermi energy of the subband *i*.

4.3.4 Rate Equation

After getting the wavefunctions of the subband states and the scattering rates, electron distributions and gains of QCLs can be calculated. One of the relatively simple methods is to use a rate equation [46, 47]. The operation principle of a QCL can be explained by using the three-level system model. The sub-band states are labeled 1, 2, and 3 from lower to higher energy. The numbers of electrons in states 1, 2, and 3 are represented by n_1, n_2, and n_3. The sub-band states 3 and 2 are the upper and lower lasing states, respectively. An electron depopulation mechanism is introduced between states 2 and 1. The population inversion is expected to be realized between states 3 and 2. The scattering rate from state *i* to state *j* is denoted as $1/\tau_{ij}$. The rate equations for the three levels are given as

$$\frac{dn_3}{dt} = I - \frac{n_3}{\tau_{32}} - \frac{n_3}{\tau_{31}} + \frac{n_2}{\tau_{23}} + \frac{n_1}{\tau_{13}} \tag{4.9}$$

$$\frac{dn_2}{dt} = -\frac{n_2}{\tau_{21}} - \frac{n_2}{\tau_{23}} + \frac{n_3}{\tau_{32}} + \frac{n_1}{\tau_{12}} \tag{4.10}$$

$$\frac{dn_1}{dt} = \frac{n_3}{\tau_{31}} + \frac{n_2}{\tau_{21}} - \frac{n_1}{\tau_{13}} - \frac{n_1}{\tau_{12}} - I \tag{4.11}$$

where I is the injection rate from state 1 of the previous QCL period to state 3 of the period considered. Some absorption rates can be ignored at low temperatures. In the steady state, the left sides of the equations are set to zero. The rate equation method usually does not take into account the in-plane electron dynamics.

The THz optical gain for the inter-subband transition between the upper lasing state 3 and the lower lasing state 2 is proportional to the product of population inversion and oscillator strength and given as

$$g(\hbar\omega) = \frac{\omega}{n_{\mathrm{r}} c \varepsilon_0} \frac{|\mu_{32}|^2}{(E_3 - E_2 - \hbar\omega)^2 + \gamma^2} (n_3 - n_2) \tag{4.12}$$

$$\mu_{32} = e \int_{-\infty}^{\infty} \psi_3^*(z) z \psi_2(z) dz \tag{4.13}$$

where n_{r} is the refractive index, μ_{32} is the inter-sub-band dipole moment, and γ is the linewidth (the half width at half maximum) [46].

The self-consistent rate equation method does not require much computational power and has been used to improve the QCL performance by optimizing parameters such as the doping density [47]. An optical field can be included in the rate equation. It was used successfully for the analysis of the gain recovery dynamics below and above threshold in mid-IR QCLs [48].

4.3.5 Monte Carlo Method

One of the popular equations to treat the high-field transport in semiconductors is the Boltzmann equation. To calculate the carrier transports based on the Boltzmann equation, the Monte Carlo method was introduced [49, 50]. The ensemble Monte Carlo (EMC) method is widely used for the analysis of QCLs [51]. The EMC method calculates drift and scattering motions of many particles during a short period of time using the scattering rates. X-valley carrier transport can be included in addition to Γ-valley transport [52].

As the Boltzmann equation is semiclassical, the EMC simulation assumes that several quantum mechanical effects such as coherent tunneling are negligible. There has been an argument about whether the physical mechanisms governing charge transport through the

QCL are coherent or incoherent. Pump-probe experiments of mid-IR QCLs demonstrated gain oscillations, giving evidence for a coherent electron motion [53]. As the thickness of the injector barrier has effects on the performance of THz QCLs, attention is paid to coherent tunneling. An inclusion of the quantum effects on carrier transport requires a full quantum mechanical approach such as the density matrix or NEGF method.

4.3.6 Density Matrix Method

To include quantum mechanical effects such as coherent resonant tunneling, a quantum mechanical density matrix approach was incorporated into the Monte Carlo simulation [54, 55]. The simplified density matrix model was developed to reduce the complexity of the full quantum mechanical density matrix model [56, 57].

The density matrix ρ represents the electron populations and coherences in subbands and obeys the following equation of motion known as the Liouville–von Neumann equation:

$$\frac{\partial \boldsymbol{\rho}}{\partial t} = -\frac{i}{\hbar}[\mathbf{H},\boldsymbol{\rho}] \tag{4.14}$$

where the commutator $[\mathbf{H}, \boldsymbol{\rho}] \equiv \mathbf{H}\boldsymbol{\rho}-\boldsymbol{\rho}\mathbf{H}$; the Hamiltonian $\mathbf{H}$ includes perturbations. The equation can be seen as a system of rate equations of the localized states with population and coherent terms concerning electron transport across barriers by resonant tunneling. The phenomenological dephasing terms are also introduced.

Suitable basis states have to be selected to calculate the coherent transport. One period of the THz QCL is usually split into modules. States belonging to different modules are coupled by tunneling.

4.3.7 Nonequilibrium Green's Function Method

The NEGF method is a pure quantum mechanical model first developed in the 1960s by Keldysh [58] and Kadanoff and Baym [59] and applied to the analysis of QCLs by several groups [60, 61]. The conduction band electron Hamiltonian including the in-plane electron dynamics is given by

$$\mathbf{H}_0 = -\frac{\hbar^2}{2}\frac{d}{dz}\left(\frac{1}{m^*(z)}\frac{d}{dz}\right) + \frac{\hbar^2\mathbf{k}_{\parallel}^2}{2m^*(z)} + \Delta E_{\mathrm{c}}(z) - e\phi(z) \tag{4.15}$$

where $\mathbf{k}_{||}$ is the in-plane wavevector. The Dyson equation and the Keldysh equation for the retarded and correlation Green's functions $\mathbf{G}^{\mathrm{R}}$, $\mathbf{G}^{<}$ are, respectively, given by

$$[E\mathbf{I}-\mathbf{H}_0-\Sigma^{\mathrm{R}}]\mathbf{G}^{\mathrm{R}}=\mathbf{I} \tag{4.16}$$

$$\mathbf{G}^{<}=\mathbf{G}^{\mathrm{R}}\Sigma^{<}\mathbf{G}^{\mathrm{R}} \tag{4.17}$$

where Σ represents the self-energies and E is the total energy. The retarded Green's function relates to the density of states. The correlation Green's function determines the electron density $n(z)$ and relates to the Poisson equation. Therefore, the Dyson equation, Keldysh relation, and Poisson equation are solved self-consistently until the Green's functions converge.

The self-energies represent influences of contacts and scattering mechanisms, such as acoustic phonon, optical phonon, charged impurity, and interface roughness scattering. The scattering self-energies are evaluated within the self-consistent Born approximation and are calculated using the Green's functions. Electron–electron interaction is often treated in the Hartree approximation. The real and imaginary parts of the retarded self-energy relate to the energy shift and the energy broadening of the electron state, respectively.

The Green's functions and self-energies can be taken as a function of two spatial coordinates, z and z', the lateral electron momentum $\mathbf{k}_{||}$, and the energy E [61]. The spectral function $A(z, E)$ is given by

$$A(z,E)=i\left[G^{\mathrm{R}}(z,z,\mathbf{0},E)-G^{\mathrm{R}}\ (z,z,\mathbf{0},E)\right] \tag{4.18}$$

The electron density is given by

$$n(z,E)=\frac{2}{(2\pi)^3}\mathrm{Im}\int d\mathbf{k}_{||}G^{<}(z,z,\mathbf{k}_{||},E) \tag{4.19}$$

The optical gain is obtained from the imaginary part of the susceptibility. The susceptibility is calculated from the change in current density in the linear response approximation to the THz optical field [60].

The NEGF calculation demands much computational power. Today's computers cannot calculate the self-energy for the carrier–carrier scattering, including fully the Hartree–Fock exchange term. However, the phase coherence as well as the dynamics in the $\mathbf{k}$-space can be described. One of the most challenging issues for THz QCLs is the increase of the maximum operation temperature. To realize high-temperature operation of THz QCLs, it is important to reduce thermally activated LO phonon scattering [43], where electrons in the upper lasing state acquire sufficient in-plane kinetic energy to emit an LO phonon and relax to the lower lasing state. The thermally activated LO phonon scattering was automatically included in the NEGF formalism [24]. It was found from the NEGF calculations that the gain of GaN-based THz QCL structures is not necessarily sufficient for lasing, though thermal backfilling and thermally activated phonon scattering are well suppressed. The main reason for this is the broadening of sub-band states caused by a strong interaction between electrons and LO phonons in GaN [62].

References

1. R. F. Kazarinov and R. A. Suris, Possibility of amplification of electromagnetic waves in a semiconductor with superlattice, *Sov. Phys. Semicond.* **5**, 707 (1971).
2. J. Faist, F. Capasso, D. L. Sivco, C. Sirtori, A. L. Hutchinson, and A. Y. Cho, Quantum cascade lasers, *Science* **264**, 553 (1994).
3. R. Köhler, A. Tredicucci, F. Beltram, H. Beere, E. Linfield, A. Davies, D. Ritchie, R. Lotti, and F. Rossi, Terahertz semiconductor-heterostructure laser, *Nature* **417**, 156 (2002).
4. D. L. Partin, Lead salt quantum well diode lasers, *Superlattices Microstruct.* **1**(2), 131 (1985).
5. A. Lyakh, R. Maulini, A. Tsekoun, R. Go R, C. Pflugl, L. Diehl, Q. J. Wang, F. Capasso, and C. Kumar N. Patel, 3 W continuous-wave room temperature single-facet emission from quantum cascade lasers based on nonresonant extraction design approach, *Appl. Phys. Lett.* **95**, 141113 (2009).
6. Private communication.
7. S. Y. Zhang, D. G. Revin, J. W. Cockburn, K. Kennedy, A. B. Krysa, and M. Hopkinson, λ~3.1 μm room temperature InGaAs/AlAsSb/InP quantum cascade lasers, *Appl. Phys. Lett.* **94**, 031106 (2009).
8. M. Yamanishi, K. Fujita, T. Edamura, and H. Kan, Indirect pump scheme for quantum cascade lasers: dynamics of electron-transport and very high T_0-values, *Opt. Express* **16**(25), 20748 (2008).
9. M. Yamanishi, K. Fujita, T. Edamura, and H. Kan, Indirect pump scheme for quantum-cascade lasers: electron transport and very high T_0-

values, *Proc. Conf. Lasers Electro-Opt.* CThAA1, May 31–June 5, 2009, Baltimore, MD, USA.

10. K. Takada, Yu Tanaka, T. Matsumoto, M. Ekawa, Y. Nakata, T. Yamamoto, M. Sugawara, and Y. Arakawa, Temperature-stable 10.3-Gb/s operation of 1.3-μm quantum-dot DFB lasers with GaInP/GaAs gratings, *Opt. Fiber Commun. Conf., 2009*, Paper JWA28, http://www.qdlaser.com/news/20090323_PR_eng.pdf (2010.1.7).
11. D. Hofstetter, J. Faist, M. Beck, and A. Müller, Demonstration of high-performance 10.16 μm quantum cascade distributed feedback lasers fabricated without epitaxial regrowth, *Appl. Phys. Lett.* **75**, 665 (1999).
12. N. Yu, R. Blanchard, J. Fan, Q. J. Wang, C. Pflügl, L. Diehl, T. Edamura, M. Yamanishi, H. Kan, and F. Capasso, Quantum cascade lasers with integrated plasmonic antenna-array collimators, *Opt. Express* **16**(24), 19447 (2008).
13. B. Williams, H. Callebaut, S. Kumar, and Q. Hu, 3.4-THz quantum cascade laser based on longitudinal-optical-phonon scattering for depopulation, *Appl. Phys. Lett.* **82**(7), 1015 (2003).
14. G. Scalari, L. Ajili, J. Faist, H. Beere, E. Linfield, D. Ritchie, and G. Davies Far-infrared (λ=87μm) bound-to-continuum quantum-cascade lasers operating up to 90K, *Appl. Phys. Lett.* **82**(19), 3165 (2003).
15. M. Rochat, L. Ajili, H. Willenberg, J. Faist, H. Beere, G. Davies, E. Linfield, and D. Ritchie, Low-threshold terahertz quantum-cascade lasers, *Appl. Phys. Lett.* **81**(8), 1381 (2002).
16. B. Williams, S. Kumar, H. Callebaut, and Q. Hu, Terahertz quantum-cascade laser at λ = 100 μm using metal waveguide for mode confinement, *Appl. Phys. Lett.* **83**(11), 2124 (2003).
17. S. Kumar and A. Lee, Resonant-phonon terahertz quantum-cascade lasers and video-rate terahertz imaging, *IEEE J. Sel. Top. Quantum Electron.* **14**(2), 333 (2008).
18. W. Maineult, P. Gellie, A. Andronico, P. Filloux, G. Leo, C. Sirtori, S. Barbieri, E. Peytavit, T. Akalin, J.-F. Lampin, H.E. Beere, and D. A. Ritchie, Metal-metal terahertz quantum cascade laser with micro-transverse-electromagnetic-horn antenna, *Appl. Phys. Lett.* **93**, 183508 (2008).
19. M. I. Amanti, M. Fischer, G. Scalari, M. Beck, J. Faist, Low-divergence single-mode terahertz quantum cascade laser, *Nat. Photon.*, **3**(10), 586 (2009).
20. L. Mahler, M. I. Amanti, C. Walther, A. Tredicucci, F. Beltram, J. Faist, H. E. Beere, and D. A. Ritchie, Distributed feedback ring resonators

for vertically emitting terahertz quantum cascade lasers, *Opt. Express* **17**(15), 13031 (2009).

21. Y. Chassagneux, R. Colombelli, W. Maineult, S. Barbieri, H. E. Beere, D. A. Ritchie, S. P. Khanna, E. H. Linfield and A. G. Davies, Electrically pumped photonic-crystal terahertz lasers controlled by boundary conditions, *Nature* **457**, 174 (2009).
22. S. Fathololoumi, E. Dupont, C.W.I. Chan, Z.R. Wasilewski, S.R. Laframboise, D. Ban, A. M'aty'as, C. Jirauschek, Q. Hu, and H. C. Liu, Terahertz quantum cascade lasers operating up to ~200 K with optimized oscillator strength and improved injection tunneling, *Opt. Express* **20**(4) 3866–3876 (2012).
23. A. Wade, G. Fedorov, D. Smirnov, S. Kumar, B. S. Williams, Q. Hu, J. Reno, Magnetic-field-assisted terahertz quantum cascade laser operating up to 225 K, *Nat. Photon.* **3**(1), 41 (2009).
24. H. Yasuda, T. Kubis, P. Vogl, N. Sekine, I. Hosako, and K. Hirakawa, Nonequilibrium Green's function calculation for four-level scheme terahertz quantum cascade lasers, *Appl. Phys. Lett.* **94**(15), 151109 (2009).
25. S. Kumar, Q. Hu, and J. Reno, 186 K operation of terahertz quantum-cascade lasers based on a diagonal design, *Appl. Phys. Lett.* **94**, 131105 (2009).
26. S. Kumar, C.-W. I. Chan, Q. Hu, and J. L. Reno, A 1.8-THz quantum cascade laser operating significantly above the temperature of $\hbar$/kB, *Nat. Phys.* **7**, 166-171 (2011).
27. Q. Y. Lu, N. Bandyopadhyay, S. Slivken, Y. Bai, and M. Razeghi, Widely tuned room temperature terahertz quantum cascade laser sources based on difference-frequency generation, *Appl. Phys. Lett.* **101**(25), 251121-1–251121-4 (2012).
28. H. Yasuda, I. Hosako, and K. Hirakawa, Designs of GaN-based terahertz quantum cascade lasers for higher temperature operations, CLEO2012, JW2A91, San Jose, CA, USA.
29. D. O. Winge, M. Lindskog, and A. Wacker, Nonlinear response of quantum cascade structures, *Appl. Phys. Lett.* **101**, 211113 (2012).
30. http://www2.nict.go.jp/pub/whatsnew/press/h21/090618/090618-1.html (2010.1.7 present).
31. N. Oda, H. Yoneyama, T. Sasaki, M. Sano, S. Kurashina, I. Hosako, N. Sekine, T. Sudoh, and T. Irie, Detection of terahertz radiation from quantum cascade laser using vanadium oxide microbolometer focal plane arrays, *Proc. SPIE* **6940**, 69402Y (2008).

32. R. Martini, C. Gmachl, J. Falciglia, F. G. Curti, C. G. Bethea, F. Capasso, E. A. Whittaker, R. Paiella, A. Tredicucci, A. L. Hutchinson, D. L. Sivco, A. Y. Cho, High-speed modulation and free-space optical audio/video transmission using quantum cascade lasers, *Electron. Lett.* **37**(3), 191 (2001).

33. W. Maineult, L. Ding, P. Gellie, L. Lugani, P. Filloux, C. Sirtori, S. Barbieri, S. Guilet, R. Braive, I. Sagnes, H. Beere, and D. Ritchie, High-frequency modulation of bound-to-continuum terahertz quantum cascade lasers up to 24GHz, *Conf. Lasers Electro-Opt./Int. Quantum Electron. Conf., OSA Tech. Dig. (CD)*, Paper CThH, Optical Society of America, 2009.

34. N. Sekine and I. Hosako, Intensity modulation of terahertz quantum cascade lasers under external light injection, *Appl. Phys. Lett.* **95**, 201106 (2009).

35. P. Grant, S. Laframboise, R. Dudek, M. Graf, A. Bezinger, H. Liu, Terahertz free space communications demonstration with quantum cascade laser and quantum well photodetector, *Electron. Lett.* **45**(18), 952 (2009).

36. L. Consolino, A. Taschin, P. Bartolini, S. Bartalini, P. Cancio, A. Tredicucci, H. E. Beere, D. A. Ritchie, R. Torre, M. S. Vitiello, and P. De Natale, Phase-locking to a free-space terahertz comb for metrological-grade terahertz lasers, *Nat. Commun.* **3**, Article #1040 (2012).

37. S. Shiba, Y. Irimajiri, T. Yamakura, H. Maezawa, N. Sekine, I. Hosako, and S. Yamamoto, 3.1-THz heterodyne receiver using an NbTiN hot-electron bolometer mixer and a quantum cascade laser, *IEEE Trans. Terahz. Sci. Technol.* **2**(1), 22–28 (2012).

38. M. Ravaro, S. Barbieri, G. Santarelli, V. Jagtap, C. Manquest, C. Sirtori, S. P. Khanna, and E. H. Linfield, Measurement of the intrinsic linewidth of terahertz quantum cascade lasers using a near-infrared frequency comb, *Opt. Express* **20**(23), 25654–25661 (2012).

39. S. Chakraborty, O. Marshall, C. W. Hsin, Md. Khairuzzaman, H. Beere, and D. Ritchie, Discrete mode tuning in terahertz quantum cascade lasers, *Opt. Express* **20**(26), B306–B314 (2012).

40. N. Oda, A. W. M. Lee, T. Ishi, I. Hosako and Q. Hu, Proposal for real-time terahertz imaging system with palm-size terahertz camera and compact quantum cascade laser, *Proc. SPIE 8363, Terahz. Phys., Devices, Syst. VI: Adv. Appl. Industry Defense*, 83630A (2012).

41. N. Oda, T. Ishi, T. Morimoto, T. Sudou, H. Tabata, S. Kawabe, K. Fukuda, A. W. M. Lee, and Q. Hu, Real-time transmission-type terahertz microscope with palm size terahertz camera and compact quantum

cascade laser, *Proc. SPIE 8496, Terahz. Emitters, Receivers, Appl. III*, 84960Q (2012).

42. Ugo Siciliani de Cumis, J.-H. Xu, L. Masini, R. Degl'Innocenti, P. Pingue, F. Beltram, A. Tredicucci, M. S. Vitiello, P. A. Benedetti, H. E. Beere, and D. A. Ritchie, Terahertz confocal microscopy with a quantum cascade laser source, *Opt. Express* **20**(20) 21924–21931 (2012).
43. B. S. Williams, Terahertz quantum-cascade lasers, *Nat. Photon.* **1**, 517 (2007).
44. P. Harrison, *Quantum Wells, Wires and Dots*, 3rd ed. (Wiley, England, 2010).
45. D. Indjin, P. Harrison, R. W. Kelsall, and Z. Ikonic, Self-consistent scattering theory of transport and output characteristics of quantum cascade lasers, *J. Appl. Phys.* **91**, 9019–9026 (2002).
46. S. L. Chuang, *Physics of Photonic Devices*, 2nd ed. (Wiley, USA, 2009).
47. V. D. Jovanovic, D. Indjin, N. Vukmirovic, Z. Ikonic, P. Harrison, E. H. Linfield, H. Page, X. Marcadet, C. Sirtori, C. Worrall, H. E. Beere, and D. A. Ritchie, Mechanisms of dynamic range limitations in GaAs/AlGaAs quantum-cascade lasers: Influence of injector doping, *Appl. Phys. Lett.* **86**, 211117 (2005).
48. H. Choi, L. Diehl, Z.-K. Wu, M. Giovannini, J. Faist, F. Capasso, and T. B. Norris, Gain recovery dynamics and photon-driven transport in quantum cascade lasers, *Phys. Rev. Lett.* **100**, 167401 (2008).
49. K. Kurosawa, Monte Carlo calculation of hot electron problems, *J. Phys. Soc. Jpn.* **21**Z(Suppl), 424–426 (1966).
50. C. Jacoboni and P. Lugli, The Monte Carlo method for the solution of charge transport in semiconductors with applications to covalent materials, *Rev. Mod. Phys.* **55**, 645–705 (1983).
51. R. Köhler, R. C. Iotti, A. Tredicucci, and F. Rossi, Design and simulation of terahertz quantum cascade lasers, *Appl. Phys. Lett.* **79**, 3920–3922 (2001).
52. X. Gao, D. Botez, and I. Knezevic, X-valley leakage in GaAs-based midinfrared quantum cascade lasers: a Monte Carlo study, *J. Appl. Phys.* **101**, 063101 (2007).
53. F. Eickemeyer, K. Reimann, M. Woerner, T. Elsaesser, S. Barbieri, C. Sirtori, G. Strasser, T. Müller, R. Bratschitsch, and K. Unterrainer, Ultrafast coherent electron transport in semiconductor quantum cascade structures, *Phys. Rev. Lett.* **89**, 047402 (2002).
54. R. C. Iotti and F. Rossi, Nature of charge transport in quantum-cascade lasers, *Phys. Rev. Lett.* **87**, 146603 (2001).

55. H. Callebaut and Q. Hu, Importance of coherence for electron transport in terahertz quantum cascade lasers, *J. Appl. Phys.* **98**, 104505 (2005).
56. S. Kumar and Q. Hu, Coherence of resonant-tunneling transport in terahertz quantum cascade lasers, *Phys. Rev. B* **80**, 245316 (2009).
57. E. Dupont, S. Fathololoumi, and H. C. Liu, Simplified density-matrix model applied to three-well terahertz quantum cascade lasers, *Phys. Rev. B* **81**, 205311 (2010).
58. L. V. Keldysh, Diagram technique for nonequilibrium processes, *Sov. Phys. JETP* **20**, 1018–1026 (1965).
59. L. P. Kadanoff and G. Baym, *Quantum Statistical Mechanics* (Benjamin, New York, 1962).
60. S. C. Lee and A. Wacker, Nonequilibrium Green's function theory for transport and gain properties of quantum cascade structures, *Phys. Rev. B* **66**, 245314 (2002).
61. T. Kubis, C. Yeh, P. Vogl, A. Benz, G. Faschingm, and C. Deutsch, Theory of nonequilibrium quantum transport and energy dissipation in terahertz quantum cascade lasers, *Phys. Rev. B* **79**, 195323 (2009).
62. H. Yasuda, T. Kubis, I. Hosako, and K. Hirakawa, Non-equilibrium Green's function calculation for GaN-based terahertz-quantum cascade laser structures, *J. Appl. Phys.* **111**, 083105 (2012).

Chapter 5

Terahertz Schottky Diode Technology

Jeffrey Hesler
Virginia Diodes Inc., Charlottesville, VA 22902, USA
Department of Electrical and Computer Engineering, University of Virginia, Charlottesville, VA 22904, USA
hesler@vadiodes.com

5.1 Introduction

Schottky diodes have a long history of use in the Terahertz (THz) frequency range for both the detection and the generation of THz power and were a key technology for many of the pioneering measurements in radio astronomy, molecular spectroscopy, and plasma diagnostics that helped open the THz field [1–5].

Schottky diodes are most commonly used for frequency translation or multiplication, rather than for providing gain like active devices. Because of this, Schottky diodes are typically paired in systems with active devices such as Gunn or IMPATT oscillators, backward wave oscillators, and transistor amplifiers and oscillators, among others [6]. Schottky diode–based components are used to extend these active components either up into the THz (the generation of THz power) or to translate THz signals down to the

Handbook of Terahertz Technologies: Devices and Applications
Edited by Ho-Jin Song and Tadao Nagatsuma

ISBN 978-981-4613-08-8 (Hardcover), 978-981-4613-09-5 (eBook)
www.panstanford.com

microwave region where they can be analyzed and processed (the down-conversion of THz signals).

There has been a tremendous amount of development of solid-state hardware at the microwave and millimeter-wave regions, spurred on by the communications, military/radar, and test and measurement industries. One of the key strengths of Schottky diode technology is its ability to leverage this established technological base and extend it into the THz.

5.1.1 Example: Schottky-Based THz Source for ALMA

One recent project that has advanced the state of THz technology is the Atacama Large Millimeter/Submillimeter Array (ALMA), a photograph of which is shown in Figure 5.1. ALMA is an ambitious astronomy project consisting of an array of 66 telescopes covering all the atmospheric windows between 84 GHz and 950 GHz. The astronomical signals are detected using heterodyne mixers based on cryogenic (4 K) superconducting-insulator-superconductor (SIS) junctions that have state-of-the-art detection sensitivity [7, 8]. The frequency range is divided into a series of discrete bands, as shown in Figure 5.1. The local oscillators (LOs) for the mixers were developed at the National Radio Astronomy Observatory (NRAO) and consist of GaAs monolithic microwave integrated circuit (MMIC) amplifiers driving Schottky diode frequency multipliers [9].

Waveguides for ALMA bands

Band	from - to (GHz)
3	84 - 116
4	125 - 163
5	163 - 211
6	211 - 275
7	275 - 370
8	385 - 500
9	602 - 720
10	787 - 950

Figure 5.1 Photograph of antennas at the ALMA site and a table (inset) giving the frequency bands covering 84–950 GHz (Courtesy by ALMA (ESO/NAOJ/NRAO), C. Padilla).

Figure 5.2 shows a schematic of the ALMA Band 6 LO design, which is used to give mixer coverage from 211 GHz to 275 GHz. The LO is driven by a phase-locked YIG-tuned oscillator covering 12.3–14.7 GHz, which feeds a X6 multiplier chain developed at NRAO. A custom-designed GaAs MMIC was used to amplify the 73.6–88.4 GHz signal to a level of about 10 mW. The premultipliers and MMIC amplifier are housed in the warm-cartridge assembly, and a waveguide then delivers this power into the cryogenic dewar. This power drives a Schottky diode frequency tripler developed by Virginia Diodes Inc. (VDI), which has an efficiency of 2%–3% and produces sufficient power to drive the SIS mixers. A total of 132 of these multiplier chains were produced for the project, with 2 chains used for each cartridge to pump the 2 dual-polarized receivers. The ALMA source is fully electronically tunable over the frequency range, an essential requirement for remote operation on the Atacama plateau at 5000 m. ALMA is now operational, and impressive early science results have already appeared (see www.almaobservatory.org).

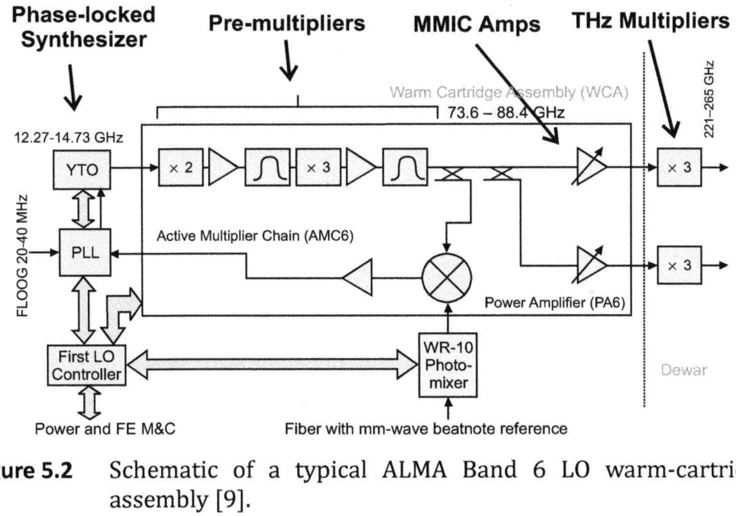

Figure 5.2 Schematic of a typical ALMA Band 6 LO warm-cartridge assembly [9].

5.2 Goals of the Chapter

This chapter will give a brief review of the state of the art of Schottky diode technology at THz frequencies. THz Schottky diodes have a long history, and there are many excellent surveys of the field, and so the reader will be referred to these, when appropriate. The chapter is meant to give a representation of the field and not to be a complete reference.

5.3 Schottky Diodes

The Schottky diode is a two-terminal device consisting of a metal–semiconductor junction that creates a barrier to the flow of electrons. Schottky diodes are a sensitive, flexible, and reliable nonlinear element with proven performance throughout the THz frequency range. Schottky diodes have the advantage of operating at either cryogenic or room temperatures, thus making them suitable for applications requiring compact, rugged, and low-power systems.

5.3.1 Schottky Diode Analysis

Schottky diodes can be analyzed accurately using relatively simple quasi-static approximations for the *I–V* and *C–V* [10–12]. These equations and the diode-equivalent circuit are shown in Figure 5.3. The nonlinear junction resistance r_j is caused by the emission of carriers over the metal–semiconductor barrier. The nonlinear capacitance C_j is caused by the changing depletion depth and is modeled as an abrupt junction capacitor with parallel plate spacing determined by the depletion region. The series resistance R_S is a parasitic element and accounts for ohmic losses in the structure. In addition to this basic model several high-frequency phenomena affect diode operation at THz frequencies. These include phenomena such as the skin effect, diode heating, plasma resonance, anode fringing capacitance, and velocity saturation [13–16].

$$I_d = I_{SAT}\left(e^{\left(\frac{V_j - I_d R_S}{V_0}\right)} - 1\right)$$

$$C_j = \frac{C_{j0}}{\sqrt{1 - V_j/V_{bi}}}$$

(a)

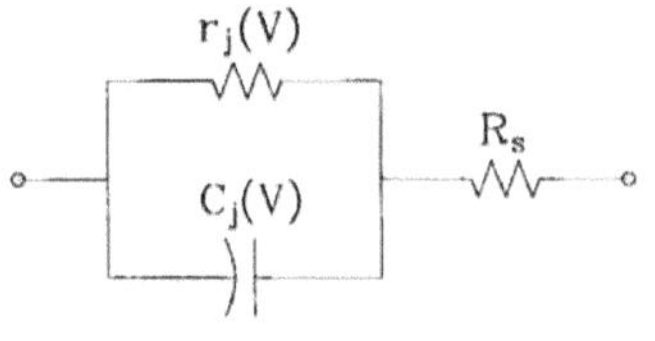

(b)

Figure 5.3 (a) Quasi-static *I–V* and *C–V* equations and (b) the quasi-static equivalent circuit of a Schottky diode.

The diode's nonlinearities enable the translation and multiplication of signals applied to the diode. Components that rely primarily on the *I–V* are termed "varistor based" (i.e., variable resistance r_j), and those that rely primarily on the *C–V* are termed "varactor based" (i.e., variable capacitance C_j). Varactor-based devices tend to be narrowband with high efficiency, while varistor-based devices are wideband but with reduced efficiency. In many cases the same diode can be used as either a varactor or a varistor by changing the diode bias conditions, although this may not yield optimal performance.

The nonlinear diode *I–V* and *C–V* equations can be used along with various analytical methods to determine the radio frequency (RF) performance of the Schottky diode, for example, Penfield & Rafuse for multipliers and Held & Kerr for mixers [17–19]. Numerical nonlinear simulations programs can also be used with the diode equations to determine the RF performance [20].

5.3.2 Schottky Diode Structure and Fabrication

Schottky diode contacts were originally made using a sharpened metal wire pressed onto a semiconductor to make a so-called "whisker" contact. To make the contact more reliable and repeatable modern Schottky diodes are created using a photolithographic process. Figure 5.4 shows a scanning electron micrograph (SEM) of a VDI Schottky diode and a cutaway schematic of the diode chip near the anode. The metal–semiconductor contact is made photolithographically using a narrow metal finger. A surface channel is etched in the GaAs beneath the finger to reduce the parasitic capacitances near the anode. The semi-insulating substrate provides strength to the chip, and the anode contact is quite rugged. The high dielectric material and the planar geometry increase the packaging parasitics when compared to the old "whisker" contact, and these need to be considered during the design.

Figure 5.5 shows the diode fabrication process used in 2001 by the Jet Propulsion Laboratory for its low-frequency diodes [21]. This process is similar to that used by other groups fabricating Schottky diodes and consists of ohmic and mesa formation and Schottky anode and interconnect metal formation followed by device separation [21–23].

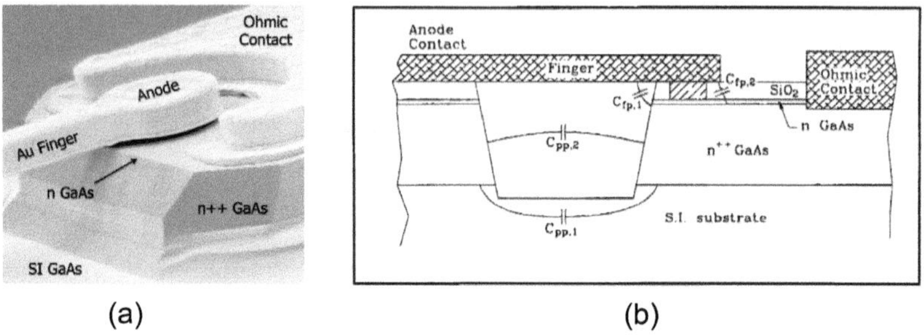

Figure 5.4 (a) Scanning electron micrograph (SEM) of a lithographically defined Schottky diode and (b) a schematic showing the layout and parasitic capacitances of a planar air-bridged Schottky diode.

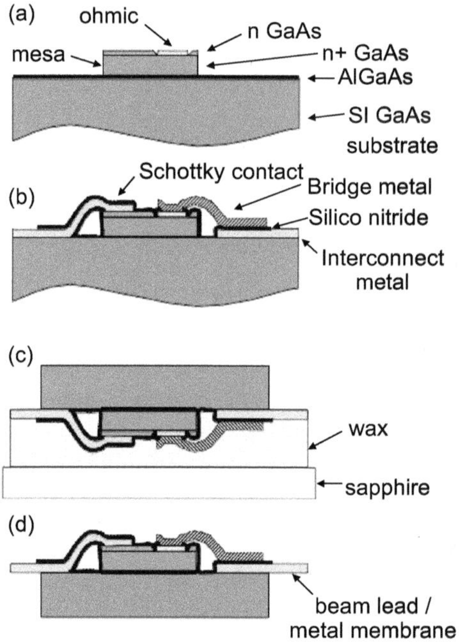

Figure 5.5 Schematic of the low-frequency diode fabrication process for Jet Propulsion Laboratory Schottky diodes.

5.3.3 Schottky Diode Component Design

To design a THz component the packaging parasitics and embedding environment of the Schottky diode should be taken into account.

Early modeling was generally done either using analytical approximations or by scale modeling, excellent examples of both of which are described in the seminal work by Eisenhart [24]. In their experimental validation of the electromagnetic field (EMF) analysis they used an embedded coaxial probe to test the embedding impedance at a particular location inside the structure, as shown in Figure 5.6. By de-embedding to the probe the embedding environment of a device mounted in a post in a waveguide could be measured, showing excellent agreement with the analysis.

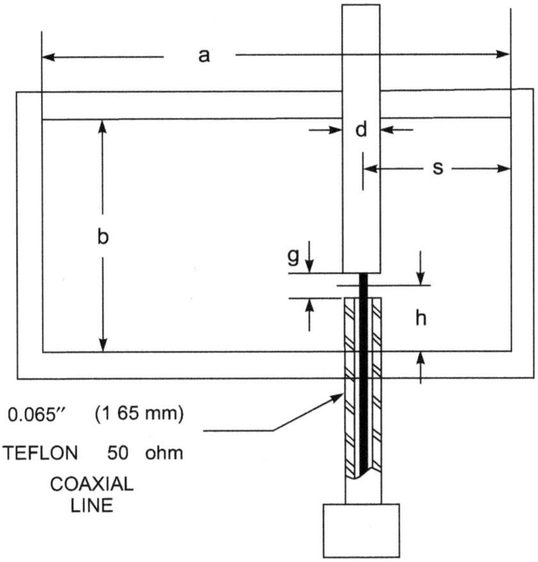

Figure 5.6 Coaxial probe for measurement of the embedding impedance of a post in a waveguide [24].

With the advent of finite-element software such as the High Frequency Structure Simulator (Ansoft/Agilent) and Microwave Studio (CST) this coaxial probe method can be simulated on a computer, thus enabling the rapid analysis of a wide range of embedding networks and topologies. Figure 5.7 shows two methods that have been used to design THz components using computer-aided design. In both cases a coaxial probe was used inside the model to determine the embedding impedance of a Schottky diode mounted in a structure [25, 26].

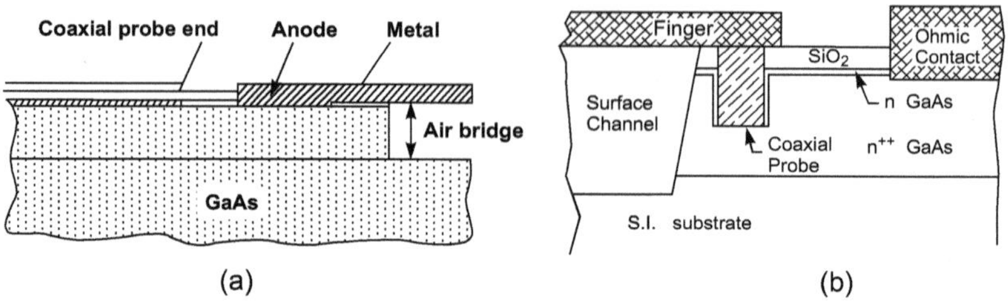

Figure 5.7 Two alternate methods of using computer-aided design software to probe the embedding impedance of a Schottky diode inside an embedding network [25, 26].

5.4 THz Components Using Schottky Diodes

Schottky diodes can be used in a wide range of components, including multipliers, mixers, detectors, nonlinear transmission lines, switches, phase shifters, noise sources, parametric amplifiers, and sideband generators [12, 27, 28]. THz Schottky diode components can be loosely divided between guided-wave structures and quasi-optical structures. Figures 5.8 and 5.9 show typical examples of these two basic component configurations.

5.4.1 Quasi-Optical Schottky Diode Components

Quasi-optical components consist of a Schottky diode coupled to free space using a feedhorn or antenna. For the 250 GHz mixer shown in Figure 5.8a the diode is mounted at the feed of a double-slot antenna located at the focal point of a silicon lens [29]. Other coupling structures can also be used such as log-periodic or log-spiral antennas and conical, diagonal, or corrugated feedhorns. Quasi-optical components become more common above ~1 THz because of challenges with machining and aligning waveguides. Because quasi-optical components are multimode devices they can be designed with very wide bandwidths using, for example, log-periodic antennas.

Coupling into quasi-optical components is done through free space using Gaussian optics [30]. Quasi-optical systems of focusing elements, polarizing grids, attenuators, isolators, interferometers, and beam splitters can be used to build test systems. For example,

Fig. 5.9 shows a 100–1000 GHz quasi-optical detector during characterization of its responsivity. The test setup includes a THz source, collimating and focusing optics and the detector.

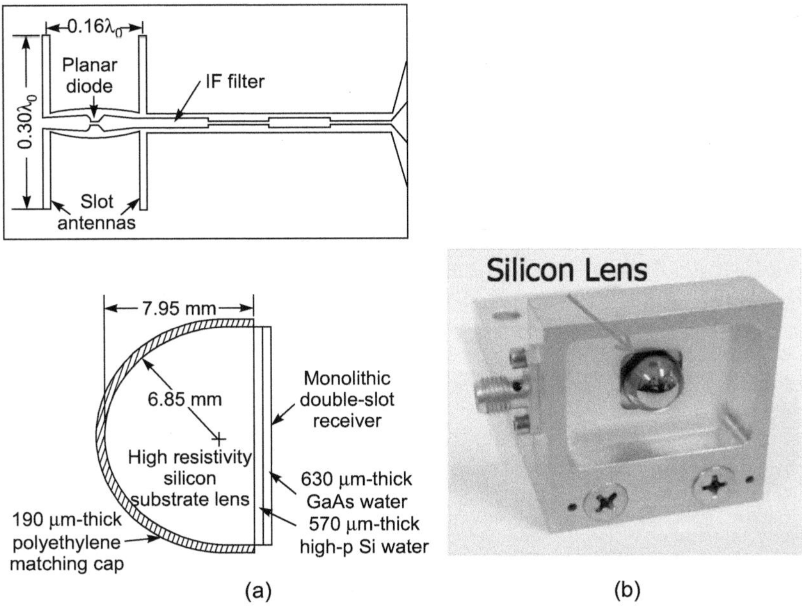

Figure 5.8 (a) Schematic of a Schottky diode mounted in a 250 GHz quasi-optical mixer and (b) photograph of a quasi-optical detector covering 100–1000 GHz (© Virginia Diodes Inc.).

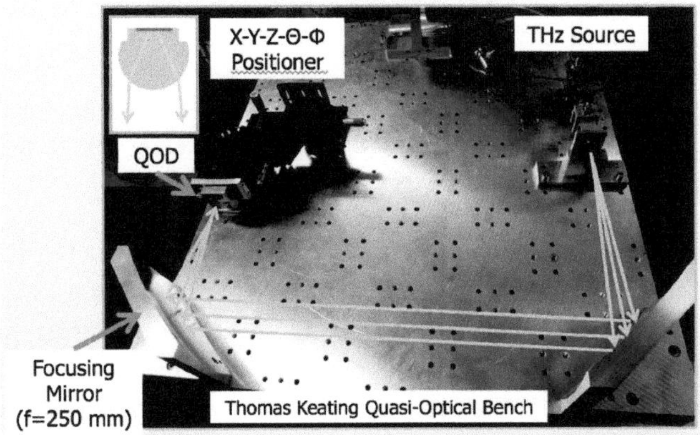

Figure 5.9 Photograph of a quasi-optical detector during responsivity testing.

5.4.2 Waveguide Schottky Diode Components

For guided-wave components a transmission line such as coax, microstrip, or waveguide couples power to the diode. At microwave frequencies the most common guiding and interconnection line is coax, with variants of coax commonly available up to 110 GHz. However, at THz frequencies coaxial lines become prohibitively lossy, for example, at 110 GHz the loss of coax is ~20 dB/m, while a WR-10 waveguide is only ~3 dB/m. So, for THz components it is typical to transition into a waveguide as soon as possible to avoid excessive loss. For example, for the doubler in Fig. 5.10a the microstrip at the output of the diodes is transitioned to a waveguide to carry the signal to the edge of the block.

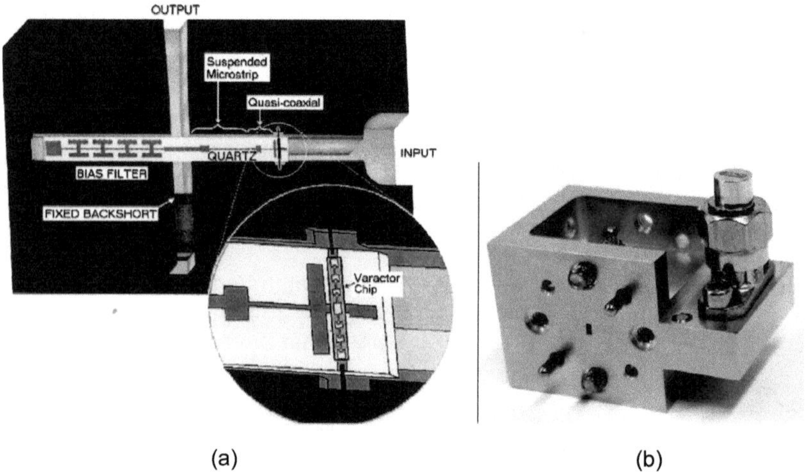

Figure 5.10 (a) Schematic of an 80 GHz Schottky varactor doubler and (b) photograph of a waveguide WR-6.5 detector covering 110–170 GHz (© Virginia Diodes Inc.).

Waveguide interfaces such as the UG-387 variant shown in Fig. 5.10b are used to connect components together [31]. Work is currently underway to standardize this connection for THz interconnections and to develop interfaces with improved performance [32]. The bandwidth of waveguide components is typically limited to the single-mode operation region of the waveguide, with a 40% fractional bandwidth typical.

A wide variety of tests and measurement components are available in waveguides, such as attenuators, isolators, directional couplers, waveguide straights, and bends. For example, Fig. 5.11 shows a typical setup for detector characterization involving a directional coupler, an attenuator, and a power meter, allowing for accurate measurement of the detector's responsivity.

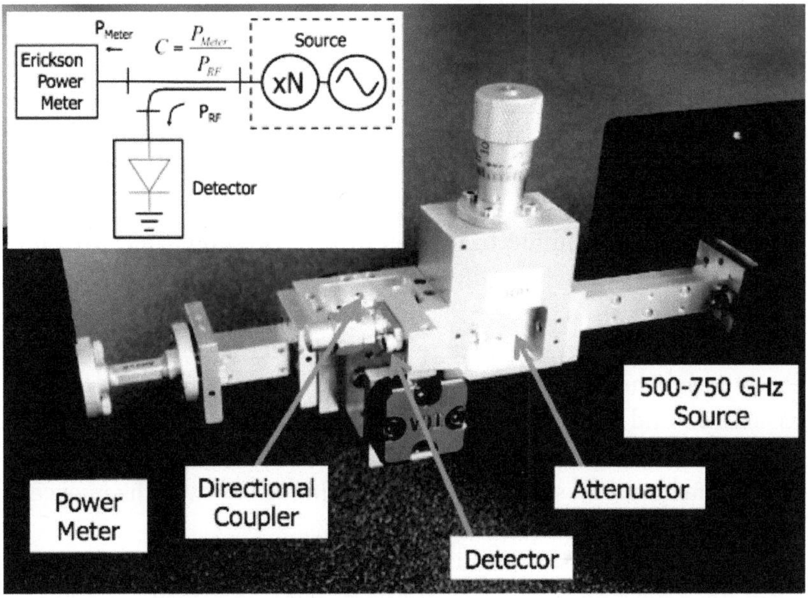

Figure 5.11 Photograph and schematic (inset) of a waveguide test setup to allow calibrated responsivity testing of a Schottky diode detector.

5.4.3 Schottky Diode Multipliers

The nonlinearities of the Schottky diode can be used to produce harmonics of an input signal, thus generating power at integer multiples of the drive frequency. One of the classic treatises on this topic is *Varactor Applications* [17], and the reader is referred to this book for a detailed treatment of the subject. By the use of idlers to reactively terminate intermediate harmonics it is possible to concentrate the spectral content into the desired output harmonic rather than simply generating a comb of output frequencies.

The prime goals of multiplier design are to (a) provide the correct embedding impedances to the multiplier at the various harmonics and (b) separate the input, idler, and output signals. The determination of the correct embedding impedance is accomplished by an analysis using the Schottky diode nonlinearity. The analytical method of Penfield can be used to calculate the embedding impedances and efficiency for varactor doublers. Alternately, a wide variety of nonlinear analysis software is available, such as Advanced Design System (Agilent) and Microwave Office (AWR).

The varactor doubler shown in Fig. 5.10a [33] uses a total of six Schottky diodes in an antiseries configuration to increase power handling and also to improve diode coupling to the waveguide. The optimum embedding impedances at 80 GHz for the individual anodes were determined to be 8+j60 Ω at the input and 16+j30 Ω at the output, for which the predicted efficiency was 72%. The relatively high Q-factor (i.e., ratio of imaginary to real part,) for these impedances is a challenge for achieving good efficiency and bandwidth. The high Q-factor puts constraints on the embedding structure, and the typical fractional bandwidth of a varactor multiplier is 15%–20%. Broadband varistor doublers tend to have lower Q-factors and so can be used to achieve bandwidths in excess of the waveguide bandwidth of 40%–50%.

Multiplier analysis based on quasi-static equations is useful, but it does not take into account effects such as velocity saturation or thermal heating in the diodes. For example, for the doubler in Fig. 5.10a the predicted efficiency was 72%, while the measured efficiency was 48%, a discrepancy that is typical for multiplier simulations using the quasi-static equations. To more accurately predict the performance a number of groups have worked on more sophisticated physics-based simulation methods to provide a more accurate analysis [34–36].

The other key goal of multiplier design is the separation of the various multiplier frequencies. This separation can be accomplished by using a combination of filters (highpass, lowpass, and bandpass) to separate the input and output signals. However, at THz frequencies the filter insertion loss can limit the multiplier efficiency, and so THz multipliers often rely on balanced designs, when possible, to separate harmonics.

Figure 5.12a shows an example of a balanced varactor doubler with an 800 GHz output frequency [37]. This balanced configuration provides a natural separation between the frequencies and can be used either for broadband (varistor) or high-efficiency (varactor) operation. The harmonic separation relies on the junction between a balanced transmission line (waveguide) and an unbalanced line (microstrip), as shown in Fig. 5.12b. The unbalanced input signal in the waveguide drives the two sets of diodes out of phase, while the second harmonic signal currents are in phase with the diodes and thus couple to the unbalanced microstrip line. The third harmonic signal currents are out of phase and so do not couple to the unbalanced microstrip, thus providing natural suppression of the third harmonic at the output. By relying on natural symmetries the losses of the embedding circuit can be kept low, producing efficient devices in the THz region.

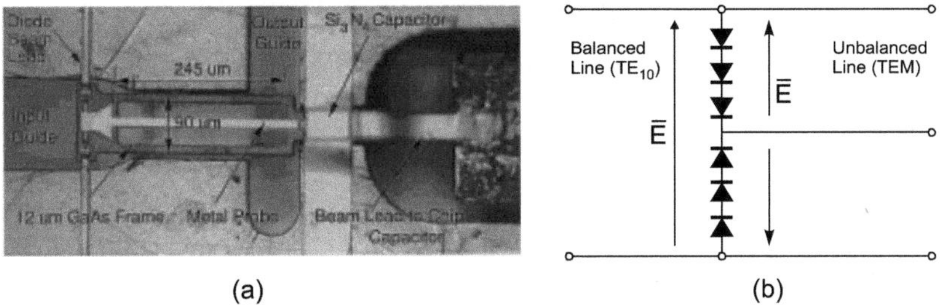

Figure 5.12 (a) Photograph of an 800 GHz balanced varactor doubler (reproduced from Ref. [37] by permission of IEEE, © 2002) and (b) schematic of the field and diode configurations for the balanced doubler topology.

Another way to control the harmonics is to use the intrinsic C–V or I–V of the device to favor or suppress certain harmonics. One successful example of this is the heterostructure barrier varactor (HBV), which has been used to make high-efficiency triplers (Fig. 5.13) and quintuplers. For the HBV the second harmonic is naturally idled by the symmetry of the C–V, and so it is well suited for use in triplers since no external idler circuit is required [38–42].

Figure 5.14 shows the output power for a variety of solid-state sources, ranging from Schottky multipliers, amplifiers, cryogenically cooled quantum cascade lasers, and other sources (Gunn diodes, resonant tunneling diodes, photomixers, etc.). As can be seen in the figure Schottky diode multipliers play a key role in filling the THz gap [33, 37–51].

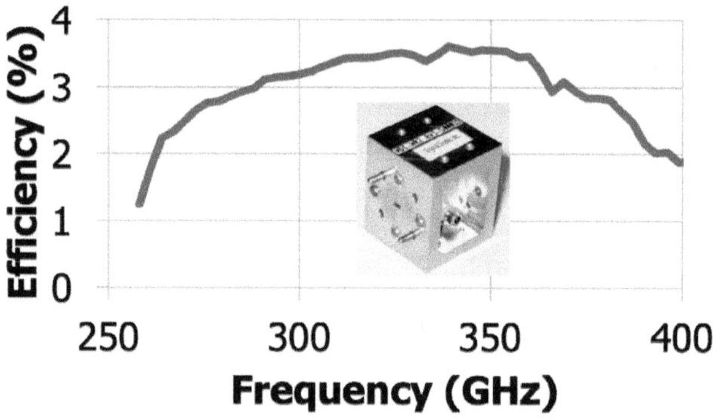

Figure 5.13 Measured performance of a broadband varistor-based tripler to 265–400 GHz (© Virginia Diodes Inc.). This tripler was used as the final stage of the LO chain for ALMA Band 7 [9].

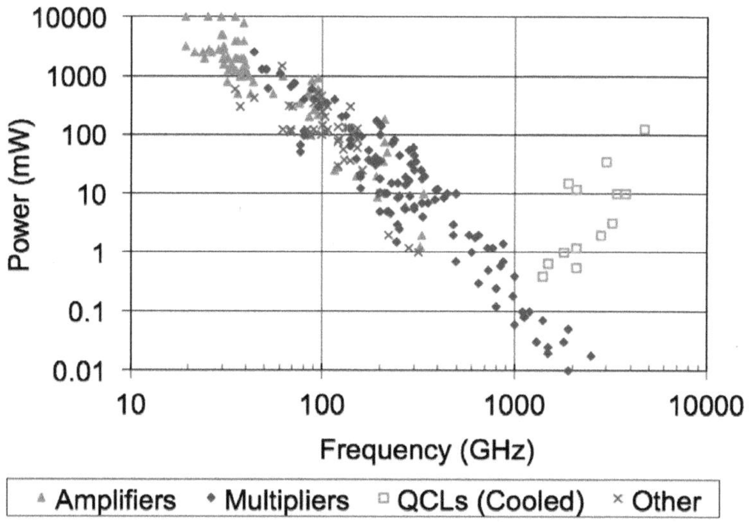

Figure 5.14 State-of-the-art output power for THz solid-state sources.

5.4.4 Schottky Diode Mixers

Schottky diode mixers rely on the resistive nonlinearity of the diode to mix the signal of interest with a large-signal LO [52]. While Schottky diodes are not as sensitive as cryogenic mixers (e.g., SIS or hot-electron bolometer [HEB] mixers) their ability to operate at ambient temperature makes them useful for more general instrumentation. For example, when used as a heterodyne transceiver extremely high sensitivity and dynamic range can be achieved [53], thus enabling a wide range of measurements, as will be discussed in a later section.

The design of Schottky mixers is similar to that described previously for multipliers, with the design goals being to provide optimum embedding impedances and to separate the frequencies. For mixers the frequencies of interest are the RF, the LO and the intermediate frequency (IF). A widely used analysis of mixers has been the harmonic-balance method developed by Kerr [18, 19], which can be used to gain practical intuition into the operation of mixers. Nonlinear simulation software packages such as Advanced Design System or Microwave Office are now generally used to determine the noise and conversion properties for mixers.

As an example of THz mixer design, Table 5.1 gives the predicted performance for a 585 GHz fundamental Schottky diode mixer [26]. The embedding impedance is predominately real, as expected for a varistor-based device, which allows for a wide mixer bandwidth to be achieved.

Table 5.1 Predicted performance for a 585 GHz fundamental planar Schottky diode mixer [26] using the analysis of Held and Kerr [18, 19]

Diode Parameters	
Anode Diameter	1.2 mm
Epilayer Doping	2e17 cm^{-3}
Rs	14 Ohms
I_{SAT}	3e–17 Amps
Ideality	1.17
C_{j0}	2 fF

Predicted Performance	
Loss (dB) SSB)	6.2 dB
T_{MIX} (K) (SSB)	720 K
Z_E (LO & RF)	80 + j40 Ohms
P_{LO}	0.5 mW
Z_E (IF)	200 Ohms

Figure 5.15 shows a schematic of a subharmonic mixer design [54]. A balanced configuration is used to control the embedding impedances and separate the frequencies. In this case an antiparallel pair of Schottky diodes are used, which allows the use of an LO at roughly half the RF and also suppresses amplitude modulation (AM)

noise in the LO and reduces the IF impedance [55–57]. The undesired mixing products and AM noise are trapped in the diode loop and suppressed at the mixer output. The RF and LO waveguides are transitioned to the microstrip before coupling to the diodes. A low-pass filter passes the LO (at ~½ the RF), while reactively terminating the diodes at the RF. The RF waveguide is cut off at the LO and is used to terminate the diodes at the LO by choosing an appropriate length of line between the RF guide and the diodes.

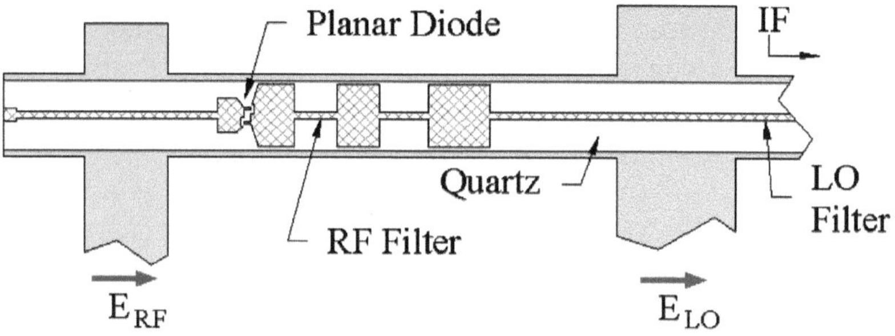

Figure 5.15 Schematic of a waveguide subharmonic mixer design using an antiparallel pair of planar Schottky diodes.

Figure 5.16 shows the measured performance for a WM-470 (WR-1.9 400–600 GHz) subharmonic mixer (VDI). Figure 5.17 gives a summary of the state of the art for room-temperature planar Schottky diode mixers throughout the THz region [26, 54, 58–68].

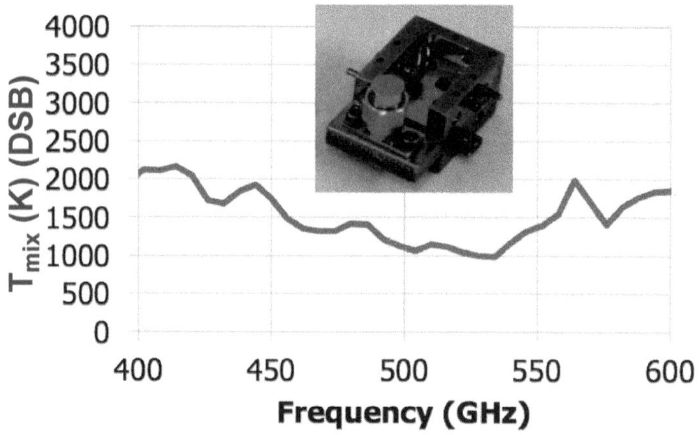

Figure 5.16 Measured performance and photograph (inset) of a WM-470 (WR-1.9) planar Schottky diode subharmonic mixer (VDI).

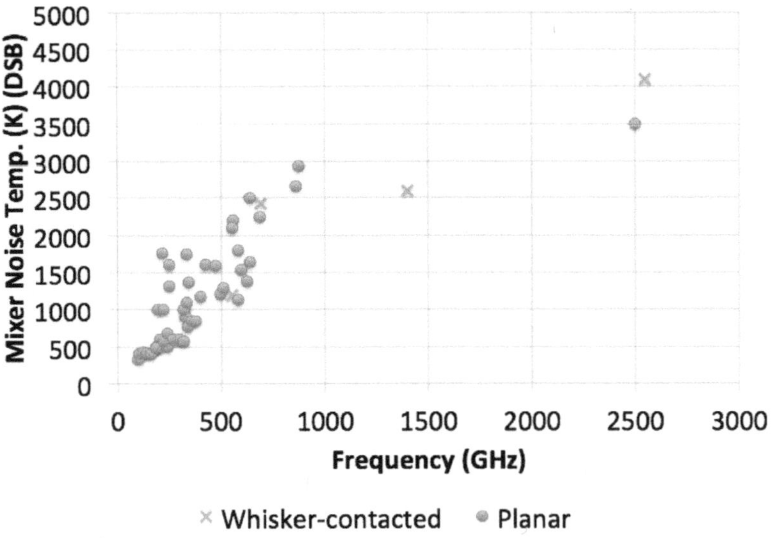

Figure 5.17 Performance summary of ambient-temperature THz Schottky diode mixers.

5.4.5 Schottky Diode Detectors

Another common use for Schottky diodes is as direct detectors, that is, the rectification of a THz signal to DC [69]. There are many room-temperature detectors, for example, Golay cells, pyroelectric detectors, and bolometers [70], but Schottky diode detectors have the benefit of high sensitivity and very fast response time.

Figure 5.18 shows an example of a Schottky diode detector mounted in a rectangular waveguide housing [71]. The diode is mounted in a waveguide channel that is coupled by a waveguide probe to an input THz waveguide. A lowpass filter blocks the THz signal and passes the detected output. The detector embedding circuit is optimized for operation over a full waveguide band. As an example, the voltage responsivity of a WM-570 (WR-2.2) zero-bias detector is shown in Fig. 5.18a.

5.5 Examples of Schottky-Based THz Technology

This section will present examples of a few applications and systems that rely on Schottky diode technology.

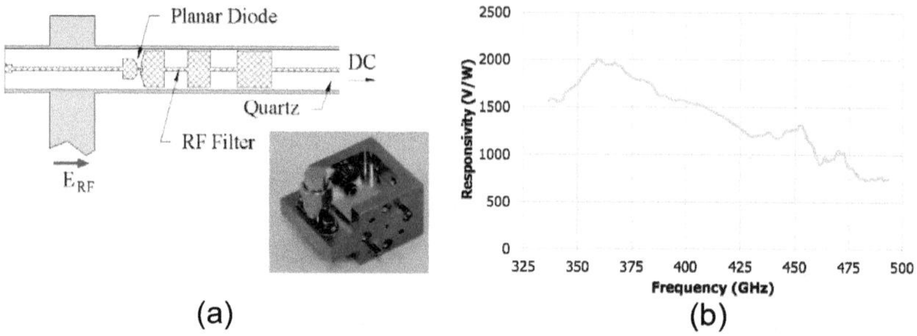

Figure 5.18 (a) Schematic and photograph (inset) of a typical waveguide-based zero-bias detector and (b) measured performance of a WM-570 (WR-2.2 325–500 GHz) Schottky detector (VDI).

5.5.1 Schottky-Based 2.7 THz Source

Figure 5.19 shows a photograph of a 2.7 THz source using Schottky diode technology. A 12.4 GHz synthesizer is used to drive the source and has a typical phase noise of –95 dBc/Hz at 1 kHz offset and a power of 18 dBm. The synthesizer is then fed into a MMIC premultiplier and amplifier with an output frequency of 37 GHz and a saturated output power of 3 W. This MMIC is the final active stage in the change, and after this there are only Schottky diode multiplication stages. For this system there are three high-efficiency Schottky varactor multipliers with outputs at 74 GHz, 148 GHz, and 296 GHz, producing an output power of 50–60 mW at 296 GHz. The final stage is a cascaded pair of broadband Schottky varistor frequency triplers in a single block, which produce an output of 3.5 uW at 2.65 THz.

A multiplier acts as both a frequency and a phase multiplier, and the phase noise at the output is increased by the theoretical 20log(N), where N is the overall multiplication factor from the base oscillator to the output. For this system N is 216, and so the phase noise of the synthesizer will increase by 47 dB to yield a phase noise at 2.67 THz of –48 dBc at 1 kHz offset. This narrow bandwidth allows for narrowband filtering and a high dynamic range.

This source was designed for use as an LO for the GREAT [72] instrument on the Stratospheric Observatory for Infrared Astronomy (SOFIA), a THz telescope mounted in an airplane. The

GREAT instrument was selected for the early science demonstration of SOFIA and has successfully flown on more than a dozen flights.

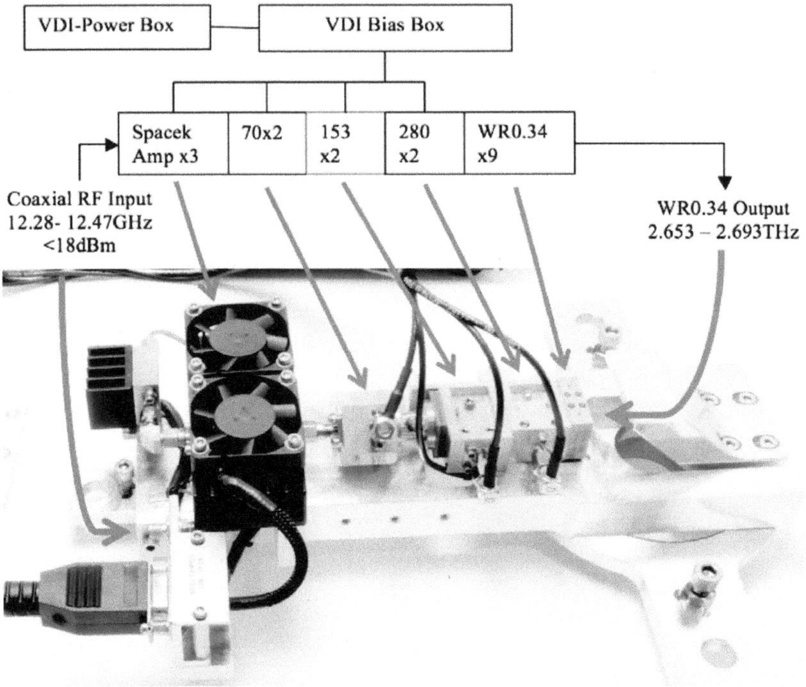

Figure 5.19 Schematic and photograph for a 2.67 THz source using Schottky diode frequency multipliers.

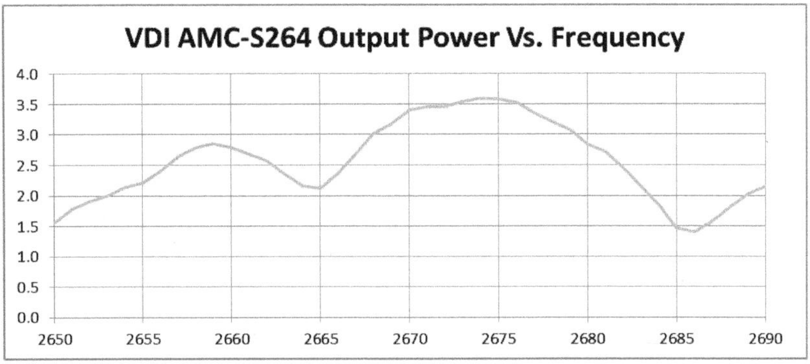

Figure 5.20 Measured output power for a 2.67 THz Schottky diode–based source (VDI).

5.5.2 Schottky-Based Demodulators for Communications

One example of an application for which the fast response time of the Schottky diode detector is useful is communications [73]. The Schottky detector acts as a fast envelope detector, and so the amplitude modulation of a THz signal will be reproduced at the output of the detector. The inset in Fig. 5.21 shows the test setup used for characterizing the demodulation bandwidth of a detector. Two test signals are combined in a waveguide directional coupler that is attached to the input of the detector. The power levels of the two sources are set to approximately –15 dBm at the input of the detector. One of the sources is fixed in frequency, while the other is swept over the desired demodulation bandwidth. The difference signal between the two THz signals is output from the coaxial port of the detector and varies from DC to 40 GHz.

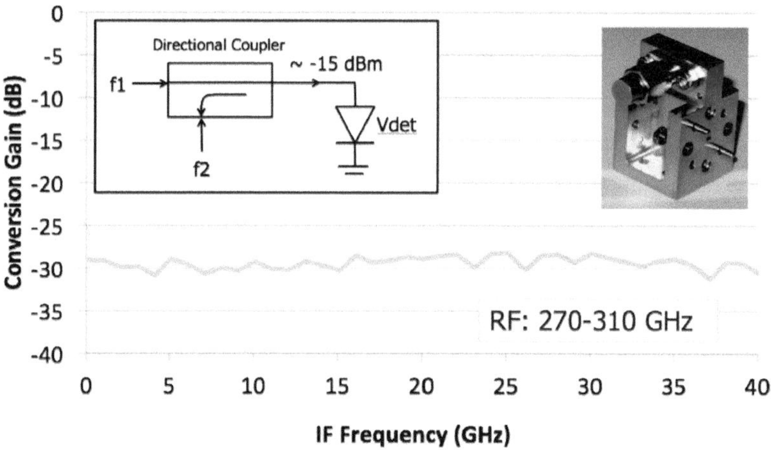

Figure 5.21 Measured demodulation bandwidth performance for a WM-710 (WR-2.8, 265–400 GHz) detector. The inset shows the test setup used to characterize the demodulation bandwidth, and for this case f1 was fixed at 270 GHz and f2 was swept from 270 GHz to 310 GHz, with the total power of the two tones set to roughly –15 dBm.

Measurements on a WM-710 (WR-2.8, 265–400 GHz) detector are shown in Fig. 5.21. For this measurement one source was fixed at 270 GHz and the other ranged from 270 GHz to 310 GHz. The measured demodulation bandwidth extends to 40 GHz. These detectors are thus well suited for use as envelope detectors for wideband THz communications systems.

5.5.3 670 GHz Radar Transceiver

One of the strengths of the heterodyne approach is the ability to build transceiver systems (see Fig. 5.22) that use narrowband filtering to achieve a high dynamic range. The narrow linewidths of microwave sources can be extended to THz, and even with the 20*log(N) increase in phase noise extremely sharp lines can be achieved. These transceivers are widely used in THz systems, including fusion plasma diagnostics, imaging, radar, and spectroscopy.

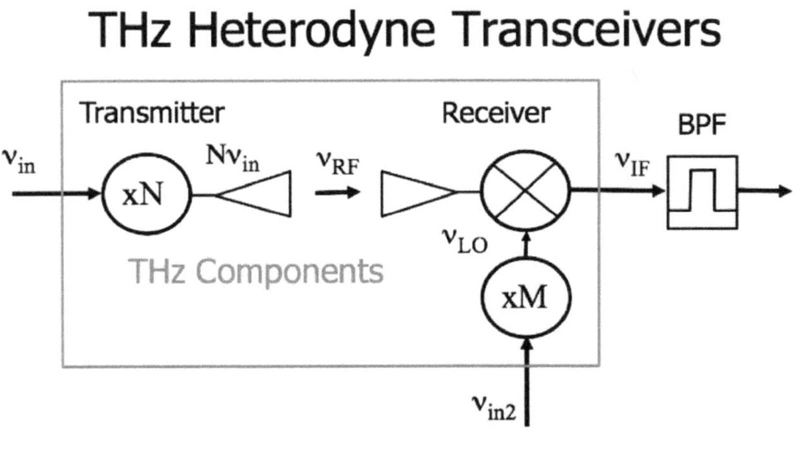

Figure 5.22 Schematic of a THz transceiver using Schottky diode multipliers and mixers. Narrowband filtering (10 Hz or less) can be achieved because of the narrow linewidths of the system, reducing thermal noise and allowing for a high dynamic range.

As an example, Fig. 5.23 shows a 670 GHz frequency-modulated continuous-wave (FMCW) radar developed at the Jet Propulsion Laboratory [74]. This system was designed for 25 m standoff personnel screening (Fig. 5.24), with the goal being to detect objects concealed beneath clothing. The system uses power amplifiers at 110–115 GHz to drive Schottky diode–based multipliers and mixers.

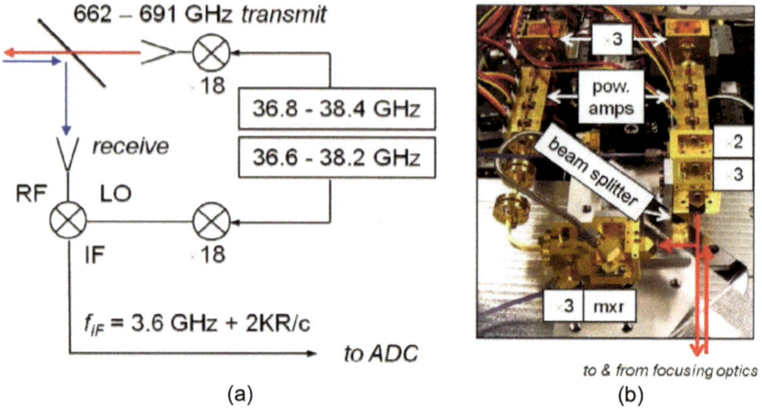

Figure 5.23 (a) Schematic of a frequency-modulated continuous-wave (FMCW) radar system at 670 GHz developed at the Jet Propulsion Laboratory and (b) photograph of the THz transceiver front end. Reproduced from Ref. [74] by permission of IEEE, © 2011.

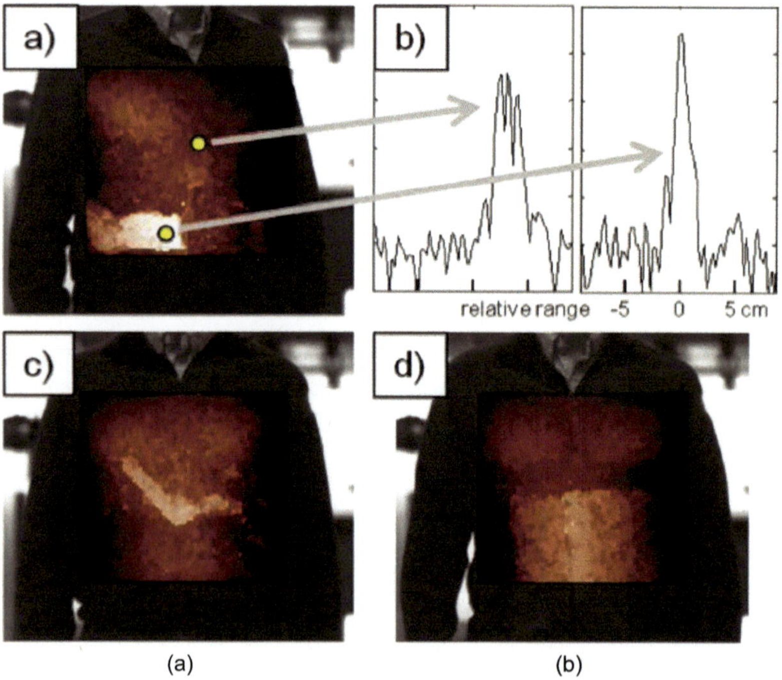

Figure 5.24 Images taken at 25 m standoff using the 670 GHz radar. Reproduced from Ref. [74] by permission of IEEE, © 2011.

5.5.4 Vector Network Analyzer Extension

Vector network analyzers (VNAs) are used to accurately measure the complex scattering parameters for a wide variety of devices in different environments and are a key instrument in the microwave and millimeter-wave frequency ranges [75]. Schottky diode transceivers can be used to extend the frequency range of modern VNAs up to THz frequencies. A schematic of a THz frequency extender configuration is shown in Fig. 5.25. The extender consists of a transmitter followed by back-to-back directional couplers. A receiver is attached to each of the couplers: one to sample the outgoing signal (i.e., the reference mixer) and one for the incoming/returning signal (i.e., the measurement mixer). The two VNA drive signals (one for the transmitter and one for the receiver) are offset by a fixed amount to generate IF signals at a constant frequency, which are then measured by the VNA.

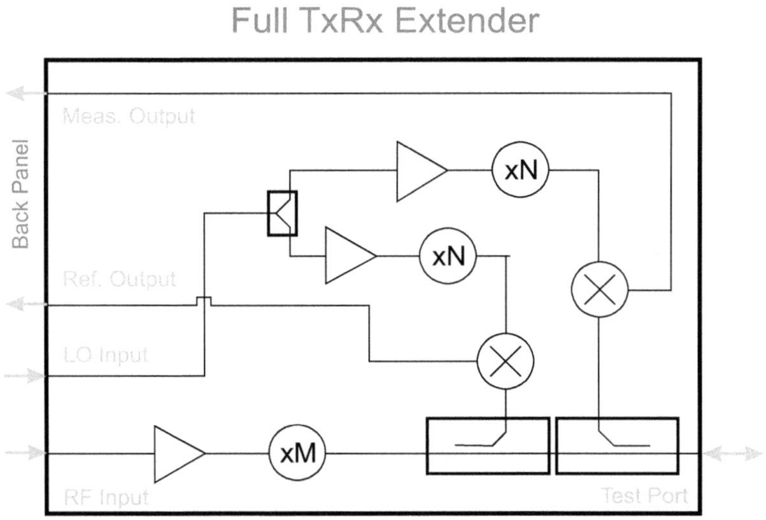

Figure 5.25 Schematic of a THz VNA extender.

A pair of VNA extenders to WM-380 (WR-1.5, 500–750 GHz) are shown in Fig. 5.26. The multipliers and mixers in the extenders rely on full-waveguide-band THz Schottky diode components, for example, doublers, triplers, and mixers. These components produce high output power and low conversion loss, yielding state-of-the-art sensitivity and bandwidth for the extenders. The WM-380 extenders have an average dynamic range at the test port of 100 dB (with 10 Hz bandwidth) and a test port power of –25 dBm typical.

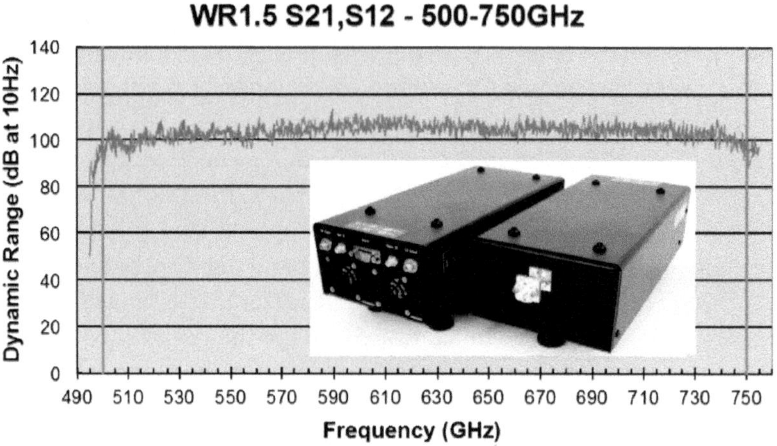

Figure 5.26 Measured dynamic range and photograph (inset) for a WM-380 (WR-1.5) VNA extender covering 500–750 GHz.

Figure 5.27 shows measurements of a 25 mm long section of a straight WM-380 waveguide. The return loss is dominated by reflections from the waveguide interface at each end, which introduces standing waves in the S21 measurement. The measured waveguide loss matches the calculated loss of the smooth-walled gold-plated waveguide, indicating the quality of the machining and gold plating.

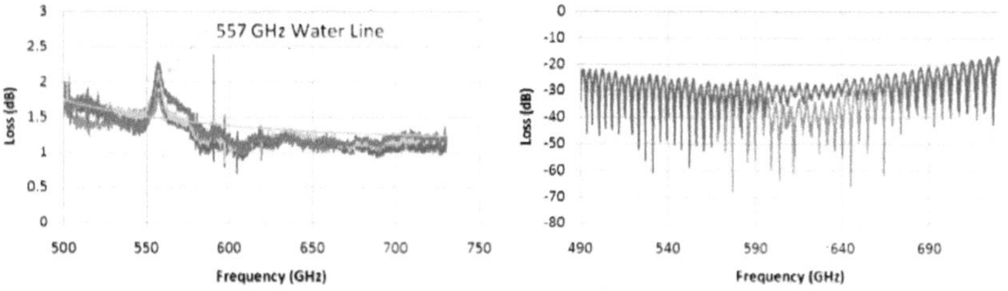

Figure 5.27 Measurement of (a) insertion loss and (b) return loss for a 2.54 cm long section of a straight waveguide at WM-380 (WR-1.5). The measurement disturbance at 557 GHz is caused by the strong water line at that frequency.

These full-waveguide-band VNA extenders are now available up to 1.1 THz and are being used to characterize a variety of components, including on wafer measurements of transistors at 670 GHz and more [76].

5.6 Conclusions

THz Schottky diodes have a wide range of applications at the THz for both scientific measurements but also for more general test and measurement. By enabling the extension of microwave technology into the THz the large technological base developed for radar and communications can be accessed. While transistors are pushing higher and higher in frequency, Schottky diodes will continue to play a key role in THz research.

References

1. J. D. Kraus and M. E. Tiuri, *Radio Astronomy* (Cynus-Quasar, 1980).
2. W. Gordy, *Microwave Spectroscopy* (Wiley, 1953).
3. C. H. Townes and A. L. Schawlow, *Microwave Spectroscopy* (Courier Dover, 1955).
4. F. C. De Lucia, The submillimeter: a spectroscopists view, *J. Mol. Spectrosc.* **261**(1), 1–17 (2010).
5. N. Luhmann, Instrumentation and techniques for plasma diagnostics, in *Infrared and Millimeter Waves, 2* (Academic Press, New York, 1979).
6. G. Chattopadhyay, Technology, capabilities, and performance of low power terahertz sources, *IEEE Trans. THz. Sci. Technol.* **1**(1), 33–53 (2011).
7. A. R. Kerr, S.-K. Pan, S. M. X. Claude, P. Dindo, A. W. Lichtenberger, and E. F. Lauria, Development of the ALMA-North America sideband-separating SIS mixers, arXiv e-print 1306.6085, June 2013.
8. R. Blundell and C. E. Tong, Submillimeter receivers for radio astronomy, *Proc. IEEE* **80**(11), 1702–1720 (1992).
9. E. Bryerton, K. S. Saini, J. Muehlberg, D. Vaselaar, and D. L. Thacker, Low-noise sub-millimeter wave local oscillators for ALMA, in *Microwave Symposium Digest, 2013* (IEEE MTT-S International, 2013).
10. M. V. Schneider, Metal-semiconductor junctions as frequency converters, in *Infrared and Millimeter Waves, 3* (Academic Press, New York, 1980).

11. S. M. Sze and K. K. Ng, *Physics of Semiconductor Devices* (John Wiley & Sons, 2006).
12. S. Yngvesson, *Microwave Semiconductor Devices* (Springer, 1991).
13. L. E. Dickens, Spreading resistance as a function of frequency, *IEEE Trans. Microwave Theory Tech.* **15**(2), 101–109 (1967).
14. K. S. Champlin and G. Eisenstein, Cutoff frequency of submillimeter Schottky-barrier diodes, *IEEE Trans. Microwave Theory Tech.* **26**(1), 31–34 (1978).
15. J. A. Copeland, Diode edge effect on doping-profile measurements, *IEEE Trans. Electron. Devices* **17**(5), 404–407 (1970).
16. E. L. Kolberg, T. J. Tolmunen, M. A. Frerking, and J. R. East, Current saturation in submillimeter-wave varactors, *IEEE Trans. Microwave Theory Tech.* **40**(5), 831–838 (1992).
17. P. Penfield and R. P. Rafuse, *Varactor Applications* (MIT Press, 1962).
18. D. N. Held and A. R. Kerr, Conversion loss and noise of microwave and millimeterwave mixers: part 1; theory, *IEEE Trans. Microwave Theory Tech.* **26**(2), 49–55 (1978).
19. D. N. Held and A. R. Kerr, Conversion loss and noise of microwave and millimeterwave mixers: part 2; experiment, *IEEE Trans. Microwave Theory Tech.* **26**(2), 55–61 (1978).
20. S. A. Maas, *Nonlinear Microwave and RF Circuits* (Artech House, 2003).
21. S. Martin, B. Nakamura, A. Fung, P. Smith, J. Bruston, A. Maestrini, F. Maiwald, P. Siegel, E. Schlecht, and I. Mehdi, Fabrication of 200 to 2700 GHz multiplier devices using GaAs and metal membranes, in *Microwave Symposium Digest, 2001*, Vol. 3 (IEEE MTT-S International, 2001), 1641–1644.
22. D. G. Garfield, R. J. Mattauch, and W. L. Bishop, Design, fabrication, and testing of a novel planar Schottky barrier diode for millimeter and submillimeter wavelengths, in *Southeastcon '88, IEEE Conf. Proc.*, 154–160 (1988).
23. S. M. Marazita, W. L. Bishop, J. L. Hesler, K. Hui, W. E. Bowen, and T. W. Crowe, Integrated GaAs Schottky mixers by spin-on-dielectric wafer bonding, *IEEE Trans. Electron. Devices* **47**(6), 1152–1157 (2000).
24. R. L. Eisenhart and P. J. Khan, Theoretical and experimental analysis of a waveguide mounting structure, *IEEE Trans. Microwave Theory Tech.* **19**(8), 706–719 (1971).
25. J. Tuovinen and N. R. Erickson, Analysis of a 170-GHz frequency doubler with an array of planar diodes, *IEEE Trans. Microwave Theory Tech.* **43**(4), 962–968 (1995).

26. J. L. Hesler, *Planar Schottky Diodes in Submillimeter-Wavelength Waveguide Receivers* (University of Virginia, 1996).

27. H. A. Watson and M. R. Barber, *Microwave Semiconductor Devices and Their Circuit Applications* (McGraw-Hill, 1969).

28. D. M. Pozar, *Microwave Engineering*, 4th ed. (John Wiley & Sons, New Jersey, 2011).

29. S. S. Gearhart and G. M. Rebeiz, A monolithic 250 GHz Schottky-diode receiver, *IEEE Trans. Microwave Theory Tech.* **42**(12), 2504–2511 (1994).

30. P. F. Goldsmith, *Quasioptical Systems* (IEEE Press, New York, 1998).

31. A. R. Kerr, E. Wollack, and N. Horner, *Waveguide Flanges for ALMA Instrumentation*, ALMA Technical Memo Series 278 (1999).

32. N. Ridler and R. Ginley, IEEE P1785: a new standard for waveguide above 110 GHz, *Microwave J. (Cables Connectors Suppl.)* **54**(3), 20–24 (2011).

33. D. W. Porterfield, T. W. Crowe, R. F. Bradley, and N. R. Erickson, A high-power fixed-tuned millimeter-wave balanced frequency doubler, *IEEE Trans. Microwave Theory Tech.* **47**(4), 419–425 (1999).

34. R. E. Lipsey, S. H. Jones, J. R. Jones, T. W. Crowe, L. F. Horvath, U. V. Bhapkar, and R. J. Mattauch, Monte Carlo harmonic-balance and drift-diffusion harmonic-balance analyses of 100-600 GHz Schottky barrier varactor frequency multipliers, *IEEE Trans. Electron. Devices* **44**(11), 1843–1850 (1997).

35. J. V. Siles and J. Grajal, Physics-based design and optimization of Schottky diode frequency multipliers for terahertz applications, *IEEE Trans. Microwave Theory Tech.* **58**(7), 1933–1942 (2010).

36. A. Y. Tang, E. Schlecht, R. Lin, G. Chattopadhyay, C. Lee, J. Gill, I. Mehdi, and J. Stake, Electro-thermal model for multi-anode Schottky diode multipliers, *IEEE Trans. THz. Sci. Technol.* **2**(3), 290–298 (2012).

37. G. Chattopadhyay, E. Schlecht, J. Gill, S. Martin, A. Maestrini, D. Pukala, F. Maiwald, and I. Mehdi, A broadband 800 GHz Schottky balanced doubler, *IEEE Microwave Wireless Compon. Lett.* **12**(4), 117–118 (2002).

38. T. Bryllert, A. O. Olsen, J. Vukusic, T. A. Emadi, M. Ingvarson, J. Stake, and D. Lippens, 11% efficiency 100 GHz InP-based heterostructure barrier varactor quintupler, *Electron. Lett.* **41**(3), 131–132 (2005).

39. X. Melique, A. Maestrini, P. Mounaix, M. Favreau, O. Vanbesien, J. M. Goutoule, G. Beaudin, T. Nahri, and D. Lippens, Record performance of

a 250 GHz InP-based heterostructure barrier varactor tripler, *Electron. Lett.* **35**(11), 938–939 (1999).

40. X. Melique, C. Mann, P. Mounaix, J. Thornton, O. Vanbesien, F. Mollot, and D. Lippens, 5-mW and 5% efficiency 216-GHz InP-based heterostructure barrier varactor tripler, *IEEE Microwave Guided Wave Lett.* **8**(11), 384–386 (1998).
41. J. Stake, L. Dillner, S. H. Jones, C. Mann, J. Thornton, J. R. Jones, W. L. Bishop, and E. Kollberg, Effects of self-heating on planar heterostructure barrier varactor diodes, *IEEE Trans. Electron. Devices* **45**(11), 2298–2303 (1998).
42. Q. Xiao, Y. Duan, J. L. Hesler, T. W. Crowe, and R. M. Weikle, A 5 mW and 5% efficiency 210 GHz InP-based heterostructure barrier varactor quintupler, *IEEE Microwave Wireless Compon. Lett.* **14**(4), 159–161 (2004).
43. G. Chattopadhyay, E. Schlecht, J. S. Ward, J. J. Gill, H. H. S. Javadi, F. Maiwald, and I. Mehdi, An all-solid-state broad-band frequency multiplier chain at 1500 GHz, *IEEE Trans. Microwave Theory Tech.* **52**(5), 1538–1547 (2004).
44. L. E. Dickens, Millimeter wave diodes for harmonic power generation, *IEEE Trans. Microwave Theory Tech.* **15**(1), 32–37 (1967).
45. N. Erickson, High efficiency submillimeter frequency multipliers, in *Microwave Symposium Digest, 1990*, Vol. 3 (IEEE MTT-S International, 1990), 1301–1304.
46. N. R. Erickson, R. P. Smith, S. C. Martin, B. Nakamura, and I. Mehdi, High efficiency MMIC frequency triplers for millimeter and submillimeter wavelengths, in *Microwave Symposium Digest, 2000*, Vol. 2 (IEEE MTT-S International, 2000), 1003–1006.
47. Y. Li, I. Mehdi, A. Maestrini, R. H. Lin, and J. Papapolymerou, A broadband 900-GHz silicon micromachined two-anode frequency tripler, *IEEE Trans. Microwave Theory Tech.* **59**(6), 1673–1681 (2011).
48. A. Maestrini, I. Mehdi, J. V. Siles, J. S. Ward, R. Lin, B. Thomas, C. Lee, J. Gill, G. Chattopadhyay, E. Schlecht, J. Pearson, and P. Siegel, Design and characterization of a room temperature all-solid-state electronic source tunable from 2.48 to 2.75 THz, *IEEE Trans. THz. Sci. Technol.* **2**(2), 177–185 (2012).
49. A. Maestrini, J. Ward, J. Gill, H. Javadi, E. Schlecht, G. Chattopadhyay, F. Maiwald, N. R. Erickson, and I. Mehdi, A 1.7-1.9 THz local oscillator source, *IEEE Microwave Wireless Compon. Lett.* **14**(6), 253–255 (2004).
50. A. Maestrini, J. S. Ward, J. J. Gill, H. S. Javadi, E. Schlecht, C. Tripon-Canseliet, G. Chattopadhyay, and I. Mehdi, A 540-640-GHz high-

efficiency four-anode frequency tripler, *IEEE Trans. Microwave Theory Tech.* **53**(9), 2835–2843 (2005).

51. A. Maestrini, J. S. Ward, J. J. Gill, C. Lee, B. Thomas, R. H. Lin, G. Chattopadhyay, and I. Mehdi, A frequency-multiplied source with more than 1 mW of power across the 840-900-GHz band, *IEEE Trans. Microwave Theory Tech.* **58**(7), 1925–1932 (2010).
52. M. V. Schneider, Metal-semiconductor junctions as frequency converters, in *Infrared and Millimeter Waves*, 3 (Academic Press, New York, 1980).
53. F. C. De Lucia, Noise, detectors, and submillimeter-terahertz system performance in nonambient environments, *J. Opt. Soc. Am. B (Opt. Phys.)* **21**(7), 1273–1279 (2004).
54. J. Hesler, K. Hui, S. He, and T. W. Crowe, A fixed-tuned 400 GHz subharmonic mixer using planar Schottky diodes, in *Proceedings of 1999 ISSTT*, 95–99 (1999).
55. M. Cohn, J. E. Degenford, and B. A. Newman, Harmonic mixing with an antiparallel diode pair, *IEEE Trans. Microwave Theory Tech.* **23**(8), 667–673 (1975).
56. A. R. Kerr, Noise and loss in balanced and subharmonically pumped mixers: part I; theory, *IEEE Trans. Microwave Theory Tech.* **27**(12), 938–943 (1979).
57. A. R. Kerr, Noise and loss in balanced and subharmonically pumped mixers: part II; application, *IEEE Trans. Microwave Theory Tech.* **27**(12), 944–950 (1979).
58. E. R. Carlson, M. Schneider, and T. F. McMaster, Subharmonically pumped millimeter-wave mixers, *IEEE Trans. Microwave Theory Tech.* **26**(10), 706–715 (1978).
59. G. Chattopadhyay, F. Rice, D. Miller, H. G. LeDuc, and J. Zmuidzinas, A 530-GHz balanced mixer, *IEEE Microwave Guided Wave Lett.* **9**(11), 467–469 (1999).
60. N. R. Erickson, Low-noise submillimeter receivers using single-diode harmonic mixers, *Proc. IEEE* **80**(11), 1721–1728 (1992).
61. M. C. Gaidis, H. M. Pickett, C. D. Smith, S. C. Martin, R. P. Smith, and P. H. Siegel, A 2.5-THz receiver front end for spaceborne applications, *IEEE Trans. Microwave Theory Tech.* **48**(4), 733–739 (2000).
62. W. M. Kelley and G. T. Wrixon, Optimization of Shottky-barrier diodes for low-noise low-conversion loss operation at near-millimeter wavelengths, in *Infrared and Millimeter Waves, 3* (Academic Press, New York, 1980).

63. J. E. Oswald, T. Koch, I. Mehdi, A. Pease, R. J. Dengler, T.-H. Lee, D. A. Humphrey, M. Kim, P. H. Siegel, M. A. Frerking, and N. R. Erickson, Planar diode solid-state receiver for 557 GHz with state-of-the-art performance, *IEEE Microwave Guided Wave Lett.* **8**(6), 232–234 (1998).

64. C. R. Predmore, A. V. Raisanen, N. R. Erickson, P. F. Goldsmith, and J. L. R. Marrero, A broad-band, ultra-low-noise Schottky diode receiver from 80 to 115 GHz diode, *IEEE Trans. Microwave Theory Tech.* **32**(5), 498–507 (1984).

65. E. T. Schlecht, J. J. Gill, R. H. Lin, R. J. Dengler, and I. Mehdi, A 520-590 GHz crossbar balanced fundamental Schottky mixer, *IEEE Microwave Wireless Compon. Lett.* **20**(7), 387–389 (2010).

66. P. J. Sobis, N. Wadefalk, A. Emrich, and J. Stake, A broadband, low noise, integrated 340 GHz Schottky diode receiver, *IEEE Microwave Wireless Compon. Lett.* **22**(7), 366–368 (2012).

67. B. Thomas, A. Maestrini, and G. Beaudin, A low-noise fixed-tuned 300-360-GHz sub-harmonic mixer using planar Schottky diodes, *IEEE Microwave Wireless Compon. Lett.* **15**(12), 865–867 (2005).

68. B. Thomas, A. Maestrini, J. Gill, C. Lee, R. Lin, I. Mehdi, and P. de Maagt, A broadband 835-900-GHz fundamental balanced mixer based on monolithic GaAs membrane Schottky diodes, *IEEE Trans. Microwave Theory Tech.* **58**(7), 1917–1924 (2010).

69. A. M. Cowley and H. O. Sorensen, Quantitative comparison of solid-state microwave detectors, *IEEE Trans. Microwave Theory Tech.* **14**(12), 588–602 (1966).

70. P. Griffiths and J. A. D. Haseth, *Fourier Transform Infrared Spectrometry* (John Wiley & Sons, 2007).

71. J. Hesler, K. Hui, and T. Crowe, Ultrafast millimeter-wave and THz envelope detectors for wireless communications, in *2012 International Topical Meeting on Microwave Photonics (MWP)*, 93–94 (2012).

72. S. Heyminck, U. Graf, R. Gusten, J. Stutzki, H.-W. Hubers, and P. Hartogh, GREAT: the SOFIA high-frequency heterodyne instrument, *Astron. Astrophys.* **542**(L1), 1–7 (2012).

73. H.-J. Song and T. Nagatsuma, Present and future of terahertz communications, *IEEE Trans. THz. Sci. Technol.* **1**(1), 256–263 (2011).

74. K. B. Cooper, R. J. Dengler, N. Llombart, B. Thomas, G. Chattopadhyay, and P. H. Siegel, THz imaging radar for standoff personnel screening, *IEEE Trans. THz. Sci. Technol.* **1**(1), 169–182 (2011).

75. J. P. Dunsmore, *Handbook of Microwave Component Measurements: With Advanced VNA Techniques* (John Wiley & Sons, 2012).

76. W. R. Deal, K. Leong, V. Radisic, S. Sarkozy, B. Gorospe, J. Lee, P.-H. Liu, W. Yoshida, J. Zhou, M. Lange, R. Lai, and X. B. Mei, Low noise amplification at 0.67 THz using 30 nm InP HEMTs, *IEEE Microwave Wireless Compon. Lett.* **21**(7), 368–370 (2011).

Chapter 6

III–V High-Electron-Mobility Transistors toward Terahertz Operation

Tae-Woo Kim,[a] Jesus A. del Alamo,[b] Kwang-Seok Seo,[c] Berinder Brar,[d] and Dae-Hyun Kim[a]

[a]*SEMATECH, 257 Fuller Road, Albany, NY 12203, USA*
[b]*Massachusetts Institute of Technology (MIT), 77 Massachusetts Ave., Rm. 39-567A, Cambridge, MA 02139, USA*
[c]*Department of Anesthesiology, Seoul National University (SNU), 1 Gwanak-ro, Gwanak-gu, Seoul, South Korea*
[d]*Teledyne Scientific Company, 1049 Camino Dos Rios, Thousand Oaks, CA 91360, USA*
dae-hyun.kim@sematech.org

It has been more than three decades since the high-electron-mobility transistor (HEMT) was invented [1], revolutionizing the world of high-frequency electronics. First, it was made on GaAs using an n-AlGaAs/GaAs heterostructure, then on InP with an n-InAlAs/InGaAs heterostructure, later on GaAs using an InAs/AlSb heterostructure and a metamorphic buffer, so-called antimonide-based-compound semiconductor (ABCS), and more recently on GaN with an i-AlGaN/GaN heterostructure. Since their inception, HEMTs have steadily been improved to deliver higher levels of performance in terms of operating high-frequency gain, noise, and power. Today,

Handbook of Terahertz Technologies: Devices and Applications
Edited by Ho-Jin Song and Tadao Nagatsuma

ISBN 978-981-4613-08-8 (Hardcover), 978-981-4613-09-5 (eBook)
www.panstanford.com

InAs HEMTs on a GaAs substrate exhibit the best-balanced high-frequency figures of merit (FOMs) in any transistor technology on any material system [2]: high f_T and high f_{max} simultaneously. As a result, InAs HEMTs are uniquely positioned to become the first true THz semiconductor transistor with both f_T and $f_{max} > 1$ THz. This book chapter reviews the evolution of III–V HEMTs for the past three decades and discusses steps to be taken to reach the THz regime.

6.1 Introduction

The high-electron-mobility transistor (HEMT), also known as heterostructure field-effect transistor (HFET) or modulation-doped field-effect transistor (MODFET), was first invented by Mimura et al. at Fujitsu Labs in 1980 [3]. The HEMT was based on the concept of *modulation doping*, first reported by Dingle et al. at Bell Laboratories in 1978 [4]. In principle, the HEMT is an FET that incorporates a junction between two semiconductor materials with different band gaps, forming a heterostructure. Selective doping (also known as modulation doping) of the wide-band-gap semiconductor creates a two-dimensional electron gas (2DEG) on the narrow band-gap material right at the interface, forming a triangular-shaped potential well. The spatial separation between dopants and the 2DEG greatly enhances the mobility of electrons in the potential well that can significantly exceed the bulk mobility value even at relatively high carrier concentrations. This is exactly why this device was named a "high-electron-mobility transistor."

The first demonstration of modulation doping was observed in the AlGaAs/GaAs material system [3]. This was a consequence of molecular beam epitaxy (MBE), which allowed tight control of atomic-layer growth of various compound semiconductors such as GaAs and $Al_xGa_{1-x}As$. With the aid of MBE, heterostructures approaching monolayer-level interface abruptness became possible, heralding a new era of "band-gap engineered" devices. This is at the heart of modulation-doped structures.

This chapter traces the most significant steps of the evolution of the HEMTs from the high-frequency-operation point of view.

6.2 Historical Evolution of High-Frequency HEMTs

Figure 6.1 shows the evolution of the current-gain cutoff frequency (f_T) of GaAs and InP HEMTs as a function of the reported year of each device technology. Soon after their first demonstration, GaAs HEMTs (with the GaAs lattice constant) surpassed GaAs MESFETs in terms of f_T only to be soon overtaken by the newer InP HEMTs (with the InP lattice constant). Since the late 1980s, HEMTs on InP have exhibited the highest f_T of any FET on any material system. The current record is 710 GHz set by a collaboration between National Chiao-Tung University and Tokyo Institute of Technology [5].

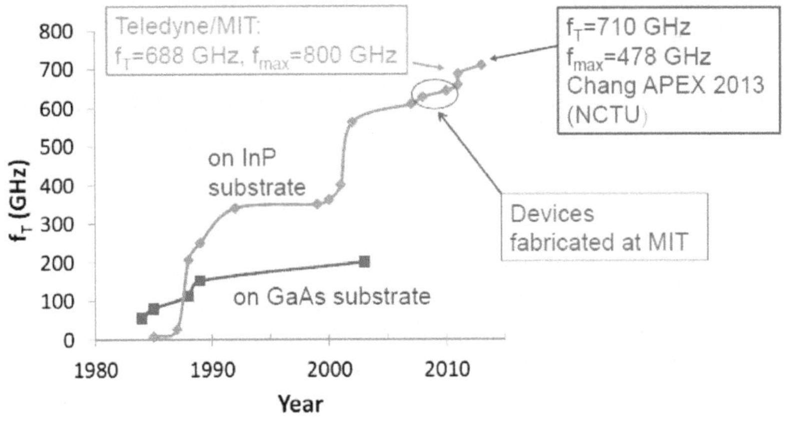

Figure 6.1 Current-gain cutoff frequency (f_T) as a function of the year of demonstration of GaAs HEMTs and InP HEMTs.

More remarkable is the outstanding balance between f_T and the maximum oscillation frequency (f_{max}) that III–V HEMTs have demonstrated. Figure 6.2 summarizes the reported f_T as a function of f_{max} in reported III–V transistors, such as III–V HEMTs and III–V heterojunction bipolar transistors (HBTs). Each locus in this figure is defined as $f_{avg} = (f_T \times f_{max})^{0.5}$, which is a unique figure of merit (FOM) to balance out f_T and f_{max} in a transistor. The record device in Ref. [2] simultaneously demonstrates an f_T of 688 GHz and an f_{max} of 800 GHz. No other transistor comes close to this. Separately, InP HEMTs with f_{max} > 1 THz have been reported [6, 7].

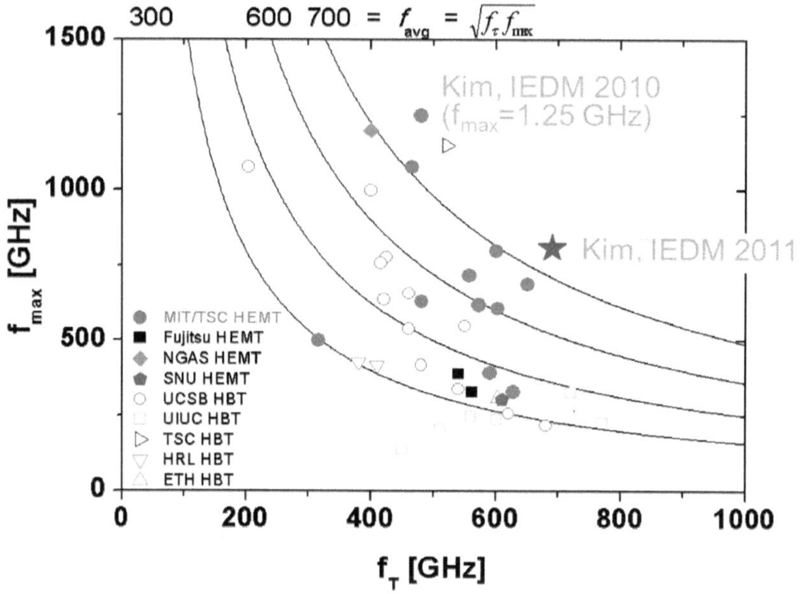

Figure 6.2 Current-gain cutoff frequency (f_T) as a function of the maximum oscillation frequency (f_{max}) for reported III–V HEMTs and HBTs in the literature. Reproduced from Ref. [2] by permission of IEEE, © 2011.

6.3 HEMT versus MESFET and HBT

It is of great importance to compare HEMTs to other types of transistors, such as metal–semiconductor field-effect transistors (MESFETs) and HBTs. The advantages of HBTs are a higher current and power density due to vertical current transport, which offers better utilization of wafer area. This allows compact device architecture with a smaller periphery and, therefore, easy matching with the surrounding circuit. Also, the critical device dimension for HBT speed is vertical, not planar. As a result, high-frequency characteristics of HBTs are independent of the lithography process. However, the HEMT exhibits an excellent high-frequency noise figure (NF) and high-frequency performance. Furthermore, as compared to the MESFET, the HEMT has a lot higher transconductance (g_m), owing to two aspects: First is the close control of channel carriers by the gate, and second is the high mobility of the carriers due to modulation doping. Carrier confinement in the quantum-well channel of III–V HEMTs also improves output behavior, boosting high-frequency gain characteristics.

6.4 III–V HEMTs

6.4.1 Device Overview

The HEMT consists of a multilayer stack of semiconductor materials, which comprises the vertical architecture of the device. These include GaAs/$Al_xGa_{1-x}As$, InGaAs/InAlAs, and InAs/AlSb material systems and typically are grown by MBE or metal–organic chemical vapor deposition (MOCVD). The HEMT relies on the application of a voltage between source and drain to induce current flow through the channel. For a given source–drain voltage, the electron concentration in the channel (typically a narrow-band-gap material, such as GaAs, InGaAs, and InAs), and hence the current flow, is then controlled by the application of a gate voltage, as in Si-MOSFETs. Figure 6.3 shows a generic cross-sectional schematic of a III–V HEMT.

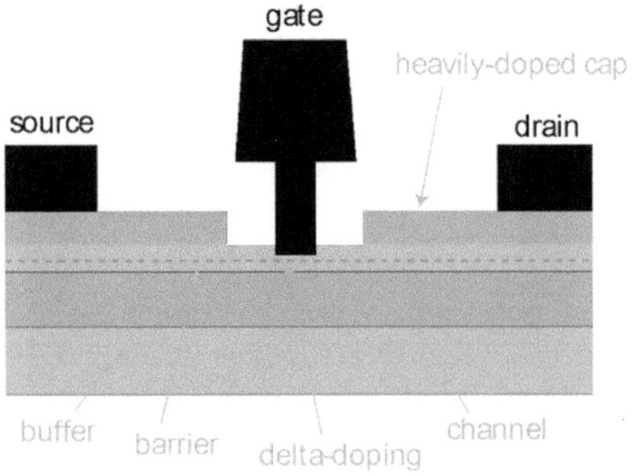

Figure 6.3 Cross-sectional schematic of a III–V HEMT.

The channel of a modern high-frequency HEMT is typically made out of indium-rich $In_xGa_{1-x}As$ and is buried below several other layers, such as spacer, barrier, and cap layers, all of which are grown sequentially in an in situ manner by MBE or MOCVD. In early design, the volume doping in the barrier was used. Later, delta doping (δ-doping) of the barrier was adopted. The cap layer is heavily doped to allow the formation of source and drain ohmic contacts easily. The heavily doped cap is selectively etched away between the source and drain during the gate recess process, exposing the undoped barrier layer. The Schottky gate metal is defined on the barrier layer. The role of the spacer layer is to enhance electron mobility in the channel. The source and drain are either alloyed ohmic contacts using a Ni/Ge/Au metal system or nonalloyed ohmic contacts using Mo. The buffer layer shown in Fig. 6.3 has the dual purpose of providing a high-quality surface with a lattice constant similar to that of the channel for optimal channel growth and to improve electron confinement in the channel, preventing real space transfer and leakage through the buffer. All these materials are grown on semi-insulating GaAs or InP substrates.

6.4.2 State-of-the-Art High-Frequency III–V HEMTs

Figure 6.4 shows a cross-sectional schematic of a state-of-the-art high-frequency InAlAs/InGaAs HEMT, as reported in Ref. [2]. This is referred to as an InGaAs metamorphic HEMT (MHEMT), since all the active layers, such as $In_{0.53}Ga_{0.47}As$ and $In_{0.52}Al_{0.48}As$, are grown on a GaAs substrate using an inverse-step graded InAlAs metamorphic buffer. It features a 70% indium content InGaAs channel with a thickness (t_{ch}) of 10 nm in which the Hall mobility is in excess of 10,000 $cm^2/V{\times}s$ at room temperature. From the device fabrication point of view, it has a T-shaped gate structure with a 40 nm wide gate foot and a 400 nm wide gate head. To mitigate the parasitic capacitance associated with the T-gate, especially the gate stem height was increased to 250 nm. To improve the electrostatic integrity of the device and improve the short-channel effect immunity, a two-step recess based on a 6 nm thick InP etch-stopper was utilized, together with a buried Pt gate [8, 9]. Figure 6.5 shows a transmission electron microscopy (TEM) image of the cross section of the device, reported in Ref. [2]. It features a source-to-drain separation of 2 um. In between the source and drain, there is an L_g = 40 nm T-gate with a stem height of 250 nm. The gate-to-channel distance was 4 nm after the two-step recess and buried Pt gate process.

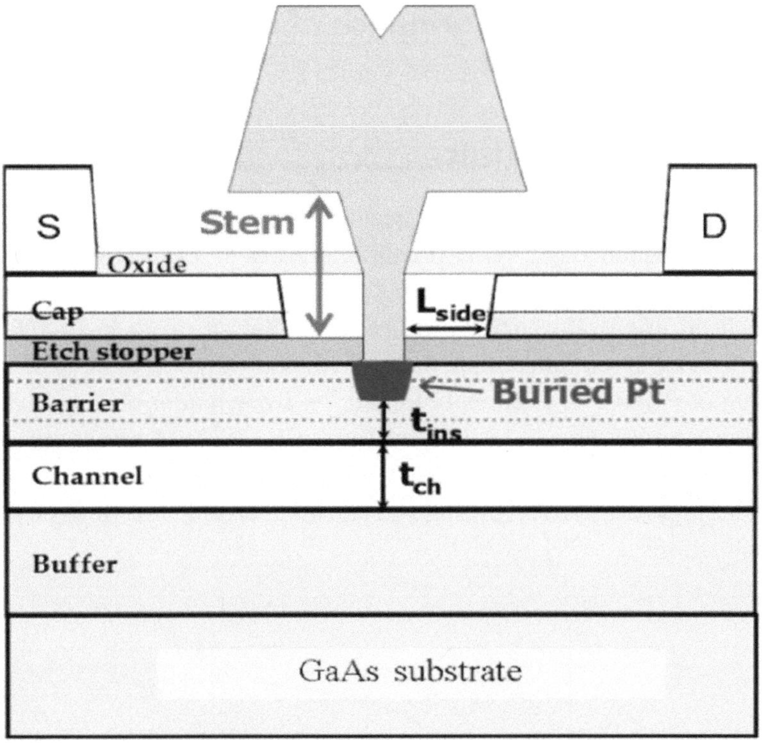

Figure 6.4 Cross-sectional schematic of state-of-the-art InGaAs HEMTs on a GaAs substrate. They are called metamporphic HEMTs (MHEMTs).

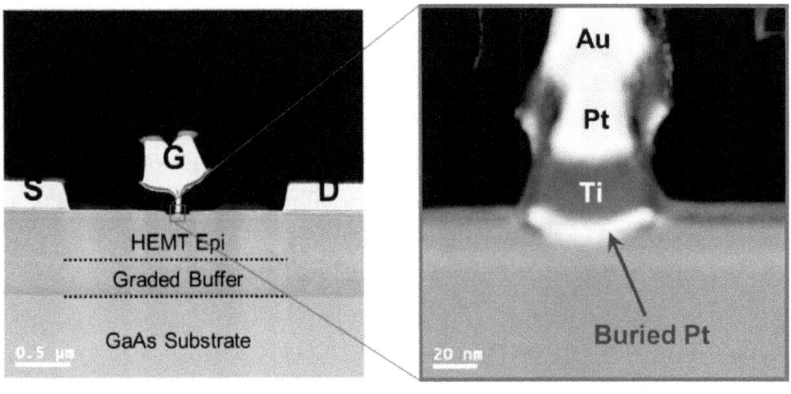

Figure 6.5 Cross-sectional TEM images of state-of-the-art InGaAs MHEMTs. Reproduced from Ref. [2] by permission of IEEE, © 2011.

6.5 Electrical Performance of State-of-the-Art III–V HEMTs

6.5.1 DC Characteristics

Figures 6.6 and 6.7, respectively, show DC output (I_D vs. V_{DS}) and transconductance (g_m vs. V_{GS}) characteristics of an InGaAs MHEMT with L_g = 40 nm, as reported in Ref. [2]. The device exhibits excellent pinch-off and drain current saturation behavior up to V_{DS} = 0.8 V. The device is enhancement mode with V_T = 0.05 V. A very small value of R_{ON} = 280 Ω µm is obtained. As a consequence, the device has outstanding g_{m_max} = 2.75 mS/µm at V_{DS} = 0.8 V and in excess of 2 mS/µm, even at V_{DS} = 0.3 V, both of which make our device technology attractive for high-performance and very-low-power applications.

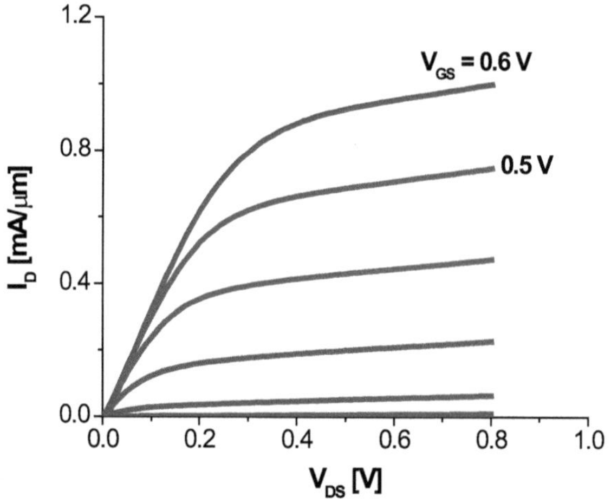

Figure 6.6 Output characteristics of L_g = 40 nm $In_{0.7}Ga_{0.3}As$ MHEMTs on a GaAs substrate. Reproduced from Ref. [2] by permission of IEEE, © 2011.

6.5.2 High-Frequency Characteristics

Perhaps the most important FOM to characterize the high-frequency suitability of a transistor technology is the current-gain cutoff frequency (f_T). This is defined as the frequency at which the short-circuit current gain (h_{21}) in the common-source configuration becomes unity. This can be mathematically defined as follows:

$$h_{21} = \frac{I_2}{I_1} \ @ \ (V_2 = 0), \quad h_{21} = 1 \text{ at } f = f_T \tag{6.1}$$

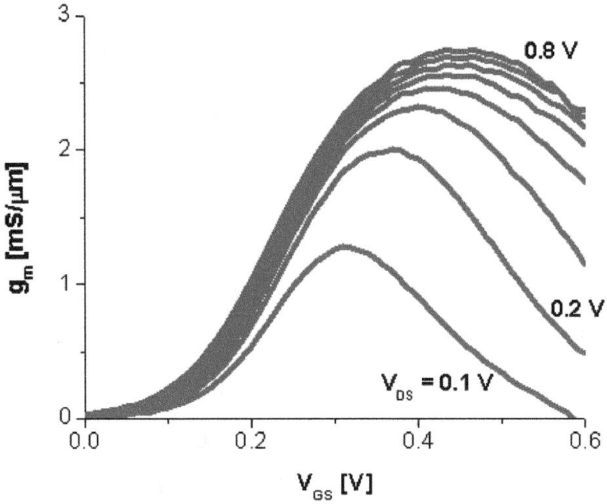

Figure 6.7 Transconductance (g_m) charactersitics of L_g = 40 nm devices. Inset shows subthreshold and gate leakage characteristics. Reproduced from Ref. [2] by permission of IEEE, © 2011.

Generally, $|h_{21}|^2$ in transistors rolls off at –20 dB/decade at high frequencies.

Microwave performance is characterized through a vector network analyzer (VNA) with an off-wafer standard line-reflection-reflection-match (LRRM) calibration. To subtract pad-associated parasitic components, such as pad capacitance and pad inductance, a two-step de-embedding technique is widely utilized in radio frequency (RF) transistor characterization. For example, Fig. 6.8 plots measured short-circuit current gain ($|h_{21}|^2$), the maximum stable gain (MSG), Mason's unilateral gain (U_g), and stability factor (k) against the frequency of a 40 nm MHEMT with $W_g = 2 \times 20$ μm at V_{DS} = 0.6 V and I_D = 0.4 mA/μm [2]. On-wafer open and short patterns are used to subtract pad capacitances and inductances from the measured device S-parameters. Using the de-embedded S-parameters, a small-signal model can also be constructed. A value of f_T = 688 GHz can be obtained by extrapolating $|h_{21}|^2$ with a slope of −20 dB/decade from both measurement systems. The value of f_T in the device is also verified through Gummel's approach [10] (inset) and small-signal modeling, yielding f_T = 690 GHz and 680 GHz, respectively.

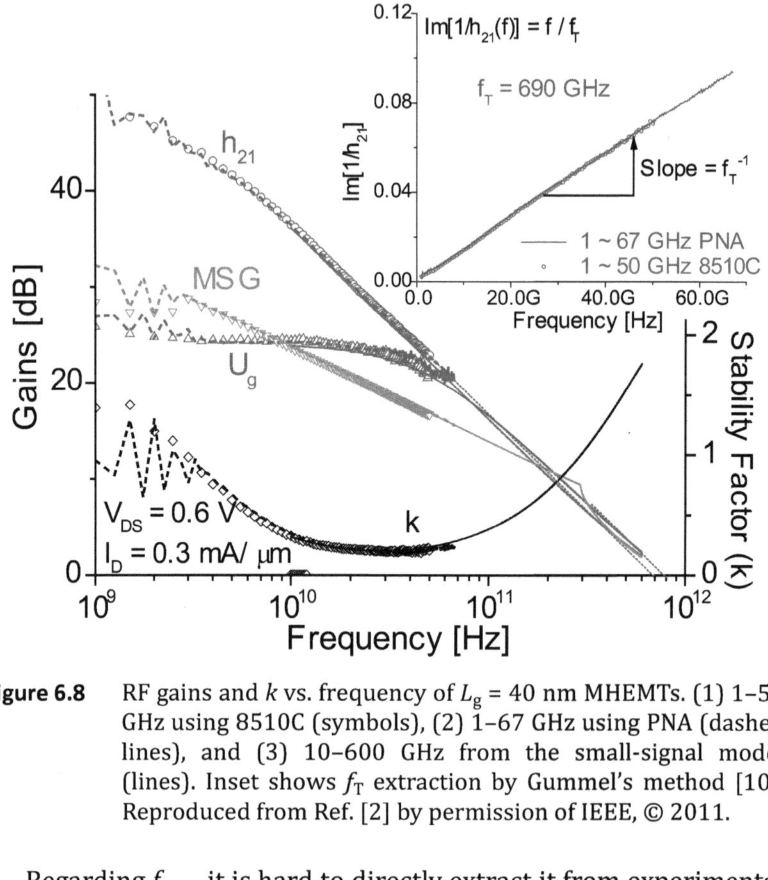

Figure 6.8 RF gains and k vs. frequency of L_g = 40 nm MHEMTs. (1) 1–50 GHz using 8510C (symbols), (2) 1–67 GHz using PNA (dashed lines), and (3) 10–600 GHz from the small-signal model (lines). Inset shows f_T extraction by Gummel's method [10]. Reproduced from Ref. [2] by permission of IEEE, © 2011.

Regarding f_{max}, it is hard to directly extract it from experimental measurements of U_g, especially as V_{DS} increases. This is mostly because the measured U_g does not show a slope of −20 dB/decade in the measured frequency range. Instead, the small-signal model is widely used to estimate f_{max}, as in Refs. [6, 7]. As shown in Fig. 6.8, the modeled U_g nicely predicts the measured one, yielding f_{max} = 800 GHz. The small-signal model indicates a very high value of g_{mi} = 4.4 mS/μm and excellent combinations of g_{mi}/g_{oi} = 5 and C_{gs}/C_{gd} = 10. As shown in Fig. 6.2 before, these results in a 40 nm MHEMT, as reported in Ref. [2], constitute the first demonstration of a transistor in any technology with both f_T and f_{max} above 680 GHz at the same bias point.

6.6 Analytical f_T Model

To understand the origin of the record high-frequency response in our devices, an analytical model for f_T, based on a simple small-signal equivalent circuit, was proposed in Ref. [2]. Figure 6.9 shows a generic schematic of a III–V HEMT, which can be partitioned in intrinsic parameters g_{mi}, g_{oi}, C_{gsi}, and C_{gdi} and extrinsic parameters C_{gs_ext}, C_{gd_ext}, R_S, and R_d.

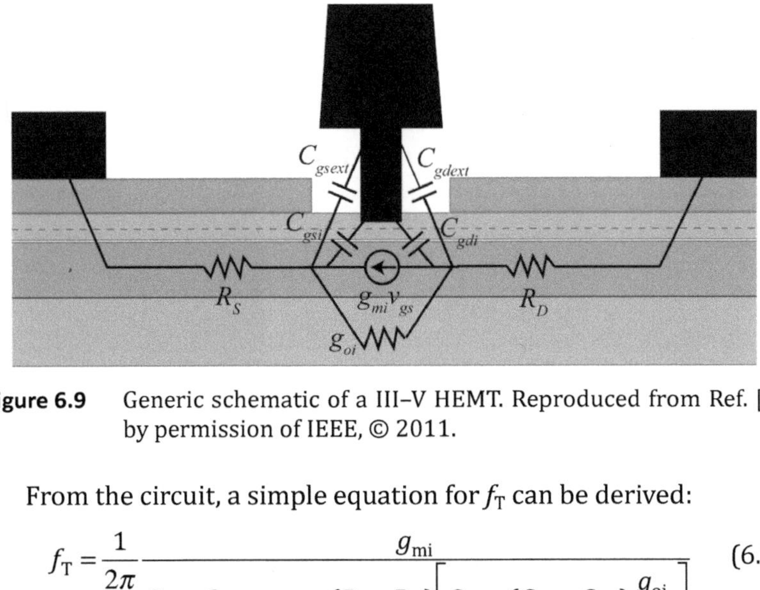

Figure 6.9 Generic schematic of a III–V HEMT. Reproduced from Ref. [2] by permission of IEEE, © 2011.

From the circuit, a simple equation for f_T can be derived:

$$f_T = \frac{1}{2\pi}\frac{g_{mi}}{C_{gs}+C_{gd}+g_{mi}+(R_S+R_D)\left[C_{gd}+(C_{gs}+C_{gd})\frac{g_{oi}}{g_{mi}}\right]} \quad (6.2)$$

f_T can be best understood in terms of the total delay, which can be written as the sum of three components [2]:

$$\tau = \frac{1}{2\pi f_T} = \tau_T + \tau_{ext} + \tau_{par}$$

$$\tau_t = \frac{C_{gsi}+C_{gdi}}{g_{mi}} = \frac{L_g}{<v_e>}$$

$$\tau_{\text{ext}} = \frac{C_{\text{gs\_ext}} + C_{\text{gd\_ext}}}{g_{\text{mi}}}$$

$$\tau_{\text{par}} = (R_S + R_D)\left[C_{\text{gd}} + (C_{\text{gs}} + C_{\text{gd}})\frac{g_{\text{oi}}}{g_{\text{mi}}}\right] \tag{6.3}$$

τ_t is the *transit time* under the gate, which in a simple model is given as L_g/v_e, where v_e is the average electron velocity under the gate. The *extrinsic delay* (τ_{ext}) is the parasitic charging delay through $C_{\text{gs\_ext}}$ and $C_{\text{gd\_ext}}$. Of course, the lower the extrinsic capacitances, the smaller the τ_{ext}. Alternatively, it can be seen that improving carrier transport by increasing g_{mi} is also an effective way to mitigate τ_{ext}. Finally, the *parasitic delay* (τ_{par}) is the delay associated with R_s and R_d [2].

Figure 6.10 illustrates the extracted C_{gs} and C_{gd} against L_g at V_{DS} = 0.6 V. Both C_{gs} and C_{gd} are linearly dependent upon L_g, as expected. Note that $C_{\text{gs\_ext}}$ is larger than $C_{\text{gd\_ext}}$ due to a higher value of reverse bias at the drain side of the gate. Figure 6.11 plots the extracted g_{mi} and g_{oi} against L_g at V_{DS} = 0.6 V. As L_g decreases, g_{mi} initially increases and then saturates at an L_g of around 60 nm, whereas g_{oi} continues to increase.

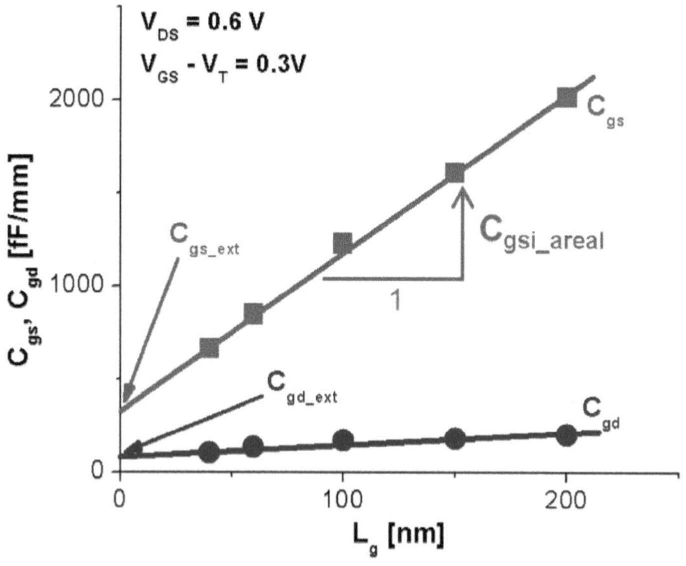

Figure 6.10 C_{gs} and C_{gd} as a function of L_g at V_{DS} = 0.6 V. Both C_{gs} and C_{gd} are linearly dependent upon L_g. Reproduced from Ref. [2] by permission of IEEE, © 2011.

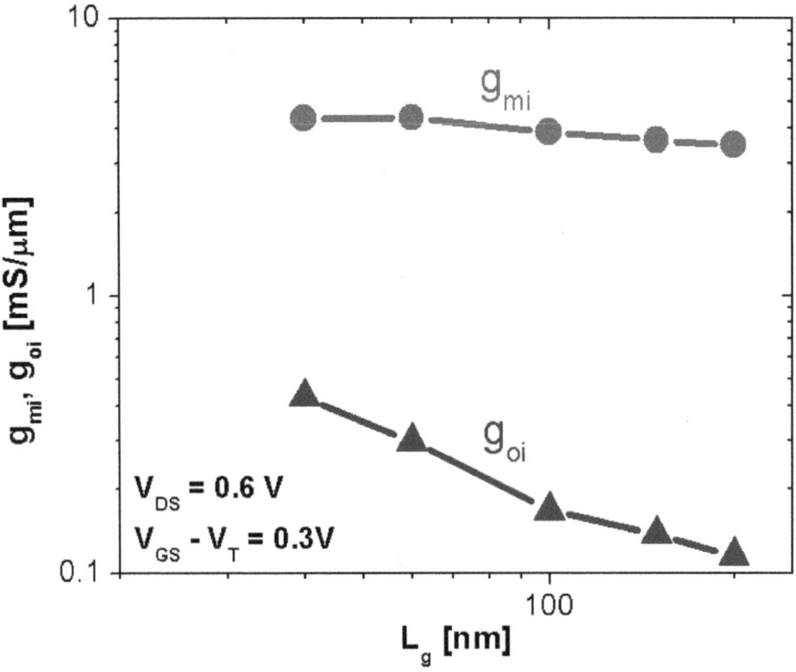

Figure 6.11 g_{mi} and g_{oi} as a function of L_g at $(V_{GS} - V_T) = 0.3$ V and $V_{DS} = 0.6$ V.

Given the values of g_{mi}, g_{oi}, C_{gsi}, C_{gdi}, C_{gs_ext}, and C_{gd_ext}, together with parasitic resistances R_s and R_d, the delay components were computed, as above. Figure 6.12 plots extracted three delay components as a function of L_g [2]. As L_g decreases, τ_t decreases accordingly. However, both τ_{ext} and τ_{par} do not scale with L_g and remain constant. In fact, at L_g = 40 nm, τ_{ext} becomes a dominant factor in the total delay, as shown in Figure 6.13. As a matter of fact, the sum of τ_{ext} and τ_{par} constitutes 65% in the L_g = 40 nm device. This significantly hampers further L_g scaling benefits in f_T. Figure 6.14 plots the electron average velocity (v_e) under the gate, as a function of L_g [2]. As L_g decreases, the extracted v_e increases, thanks to an increase in the electron injection velocity at the top of the potential barrier near the source and an enhanced velocity overshoot effect at the drain side of the gate. At L_g = 40 nm, $v_e = 5 \times 10^7$ cm/s was obtained using our analytical f_T model.

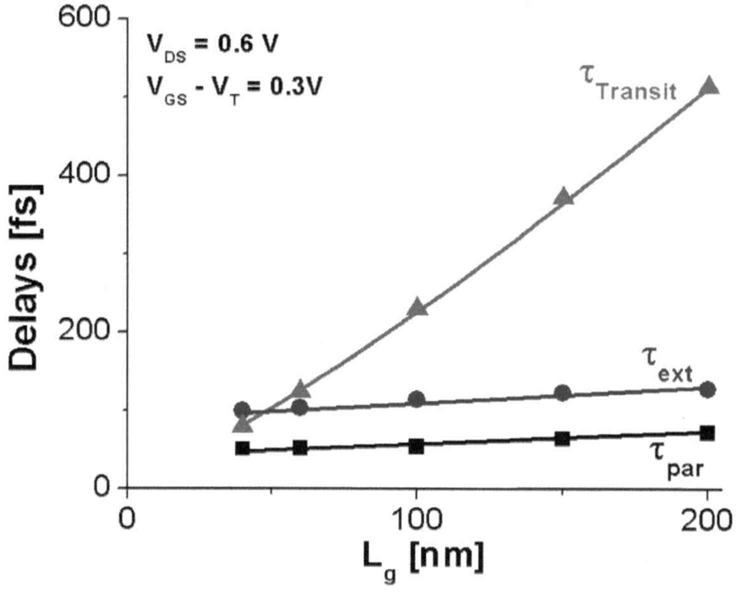

Figure 6.12 Modeled delay components against L_g.

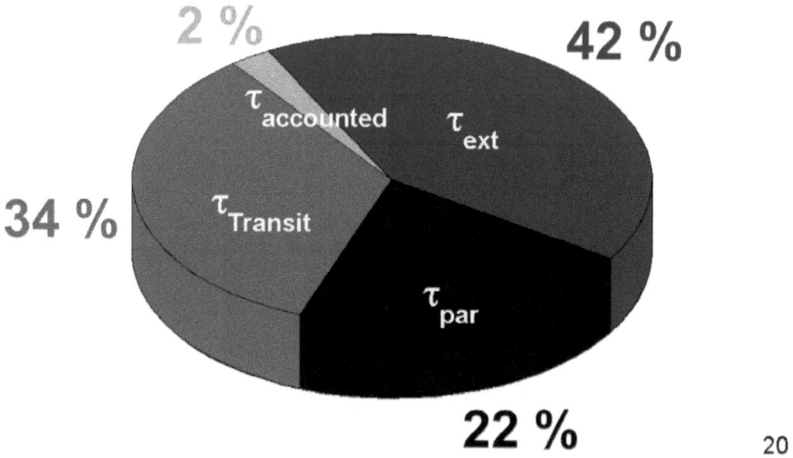

Figure 6.13 Pie diagram of each delay component in L_g = 40 nm InGaAs MHEMTs.

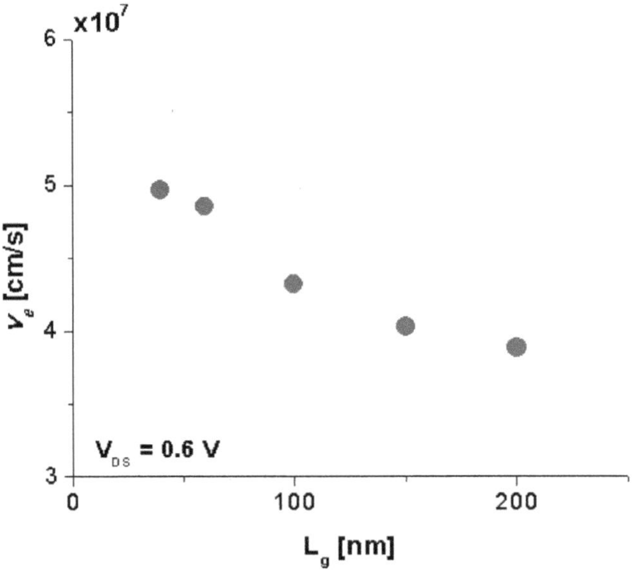

Figure 6.14 Extracted average electron velocity (v_e) under the gate, against L_g. Reproduced from Ref. [2] by permission of IEEE, © 2011.

6.7 Prospects for THz Transistors

The analytical approach to model f_T has been verified by comparing it to the measured one, as reported in Ref. [2]. Figure 6.15 plots both measured and modeled f_T as a function of L_g at V_{DS} = 0.6 V [2]. Excellent agreement is achieved, which increases the credibility of the extraction procedure. The model helps one also to discuss technological options to further improve f_T and reach 1 THz [2].

Starting from the current devices discussed earlier [2], a straightforward reduction of all parasitics by 30% together with gate length scaling to 20 nm should bring us to an f_T of about 1 THz [2]. To accomplish this, series resistance reduction is imperative. Key to this is bringing the metal contacts in very close proximity to the gate. This demands self-aligned designs [11].

Attaining high g_m without degrading anything else is also difficult. Harmonious scaling of all device dimensions is a possible path, but this is not straightforward in conventional designs. A 3D trigate or nanowire FET approach might be required. These designs offer greatly improved electrostatic control of the channel by the gate and therefore potentially higher g_m [12].

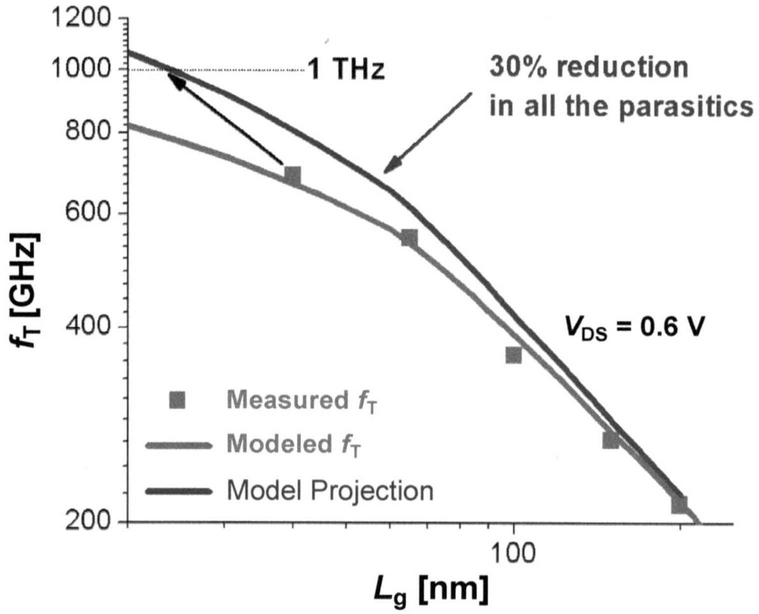

Figure 6.15 Modeled (red line) and measured f_T (rectangle) as a function of L_g at V_{DS} = 0.6 V, together with model projection (blue line) having 30% reduction in all the parasitic components, such as R_s, R_d, C_{gs_ext}, and C_{gd_ext}. Reproduced from Ref. [2] by permission of IEEE, © 2011.

6.8 Conclusions

THz transistors in which both f_T and f_{max} are over 1 THz are just around the corner. InAs-rich InGaAs HEMTs with lattice constants close to that of InP are best positioned to attack this challenge. This book chapter has briefly reviewed the evolution of high-frequency HEMTs and has discussed key paths/innovations to improve the high-frequency response of III–V HEMTs beyond the achievement of the past three decades.

References

1. J. A. del Alamo, The high electron mobility transistor at 30: impressive accomplishments and exciting prospects, 2011 International Conference on Compound Semiconductor Manufacturing Technology, May 16-19, 2011, Indian Wells, California. http://hdl.handle.net/1721.1/87102
2. D.-H. Kim, B. Brar, and J. A. del Alamo, f_T = 688 GHz and f_{max} = 800 GHz in L_g = 40 nm $In_{0.7}Ga_{0.3}As$ MHEMTs with g_{m_max} > 2.7 mS/μm, in *Electron Devices Meeting (IEDM), 2011 IEEE International*, 13.6.1–13.6.4, doi: 10.1109/IEDM.2011.6131548.
3. T. Mimura, S. Hiyamizu, T. Fujii, and K. Nanbu, A new field-effect transistor with selectively doped GaAs/n-$Al_xGa_{1-x}As$ heterojunctions, *Jpn. J. Appl. Phys.* **19**, 225–227 (1980).
4. R. Dingle, H. Stormer, A. Gossard, and W. Wiegmann, Electron mobilities in modulation-doped semiconductor heterojunction superlattices, *Appl. Phys. Lett.* **33**, 665–667 (1978).
5. E.-Y. Chang, C.-I. Kuo, H.-T. Hsu, C.-Y. Chiang, and Y. Miyamoto, InAs thin-channel high-electron-mobility transistors with very high current-gain cutoff frequency for emerging submillimeter-wave applications, *Appl. Phys. Express* **6**, 034001.1–034001.3 (2013).
6. R. Lai, X. B. Mei, W. R. Deal, W. Yoshida, Y. M. Kim, P. H. Liu, J. Lee, J. Uyeda, V. Radisic, M. Lange, T. Gaier, L. Samoska, and A. Fung, Sub 50 nm InP HEMT device with f_{max} greater than 1 THz, in *Electron Devices Meeting (IEDM), 2007 IEEE International,* 609–611, doi: 10.1109/IEDM.2007.4419013.
7. D.-H. Kim, J. A. del Alamo, P. Chen, W. I. Ha, M. Urteaga, and B. Brar, 50-nm E-mode $In_{0.7}Ga_{0.3}As$ PHEMTs on 100-mm InP substrate with f_{max} > 1 THz, in *Electron Devices Meeting (IEDM), 2010 IEEE International,* 30.6.1–30.6.4, doi: 10.1109/IEDM.2010.5703453.
8. T. Suemitsu, H. Yokoyama, Y. Umeda, T. Enoki, and Y. Ishii, High-performance 0.1-μm gate enhancement-mode InAlAs/InGaAs HEMTs using two-step recessed gate technology, in *IEEE Trans. Electron. Devices (TED)* **46**(6), 1074–1080 (1999).
9. A. Mahajan, M. Arafa, P. Fay, C. Caneau, and I. Adesida, Enhancement-mode high electron mobility transistors (E-HEMT's) lattice-matched to InP, *IEEE Trans. Electron. Devices (TED)* **45**, 2422–2429 (1998).
10. H. K. Gummel, On the definition of the cutoff frequency f_T, *Proc. IEEE* **57**, 2159–2159 (1969).

11. T.-W. Kim, D.-H. Kim, and J. A. del Alamo, 60 nm self-aligned-gate InGaAs HEMTs with record high-frequency characteristics, in *Electron Devices Meeting (IEDM), 2010 IEEE International,* 30.7.1–30.7.4, 10.1109/IEDM.2010.5703454.

12. J. A. del Alamo, Nanometre-scale electronics with III-V compound semiconductors, *Nature* **479**, 317–323 (2011).

Chapter 7

Resonant Tunneling Diodes for Terahertz Sources

Masahiro Asada and Safumi Suzuki
Interdisciplinary Graduate School of Science and Engineering, Tokyo Institute of Technology, 2-12-1-S9-3 Ookayama, Meguro-ku, Tokyo 152-8552, Japan
asada@pe.titech.ac.jp

Resonant tunneling diodes (RTDs) have a potential for compact and coherent terahertz (THz) sources operating at room temperature. In this chapter, a basic analysis of oscillation characteristics and recent progress of RTDs for THz sources are described. For high-frequency oscillation, reductions of capacitance, and electron delay time are important. Room-temperature fundamental oscillations up to 1.3–1.4 THz have been achieved until now. By a structure optimization, a fundamental oscillation close to 2 THz is feasible. Although the output power is usually 1–10 μW, an output power of more than 1 mW is possible by impedance matching between the antenna and the RTD. At present, 420 μW has been obtained at 550 GHz in a single oscillator by the offset-fed slot antenna. In a preliminary experiment on two-element arrays, 610 μW has been obtained at 620 GHz

Handbook of Terahertz Technologies: Devices and Applications
Edited by Ho-Jin Song and Tadao Nagatsuma

ISBN 978-981-4613-08-8 (Hardcover), 978-981-4613-09-5 (eBook)
www.panstanford.com

with the mutual injection locking and coherent power combining between the oscillator elements. Spectral characteristics, including the linewidth and frequency change with bias voltage, are shown. A preliminary experiment on wireless data transmission by direct bias modulation is also demonstrated.

7.1 Introduction

For various applications expected in the terahertz (THz) frequency range [1], compact and coherent solid-state sources are important key components. Because the THz range is located between light waves and millimeter waves, both optical and electronic devices are being investigated for THz sources.

Figure 7.1 shows output power versus oscillation frequency for several semiconductor single oscillators at present. From the optical device side, p-type Ge lasers [2] and THz quantum cascade lasers (QCLs) [3–5] are studied. Oscillators with two-terminal devices, such as impact ionization avalanche transit-time (IMPATT) diodes, tunneling transit-time (TUNNETT) diodes, Gunn diodes, and resonant tunneling diodes (RTDs), are being investigated from the electron device side [6–17]. Heterostructure bipolar transistors (HBTs), high-electron-mobility transistors (HEMTs), and Si complementary metal oxide semiconductor (CMOS) transistors are rapidly approaching the THz range [18–26]. Devices based on Bloch oscillation, the plasma effect, and velocity modulation are also being studied [27–30].

RTDs have the highest oscillation frequency among the above electron devices. Room-temperature operation, which has not yet been achieved in a single optical device at present, is also an attractive feature. Research of RTDs began with the theoretical prediction by Tsu and Esaki in 1973 [31], and their behavior of negative differential resistance was experimentally demonstrated at liquid nitrogen temperature in 1974 [32] and at room temperature in 1985 [33]. Oscillation in the microwave range was demonstrated at a low temperature in 1984 [34]. The oscillation frequency was then updated many times to several hundred GHz [35], and a room-temperature fundamental oscillation at 712 GHz was reported in 1991 [9]. A 64-element arrayed oscillator containing RTDs and its oscillation at 650 GHz were reported in 1997 [10]. A harmonic oscillation above 1 THz was also demonstrated [12]. The frequency of the room-temperature fundamental oscillation increased to 831 GHz in 2009 [13] and then exceeded 1 THz in 2010 [14]. The increase in oscillation frequency is still continuing, and oscillations of 1.3–1.4 THz have been obtained until now [15–17]. The output power is also increasing by structure improvement [36, 37].

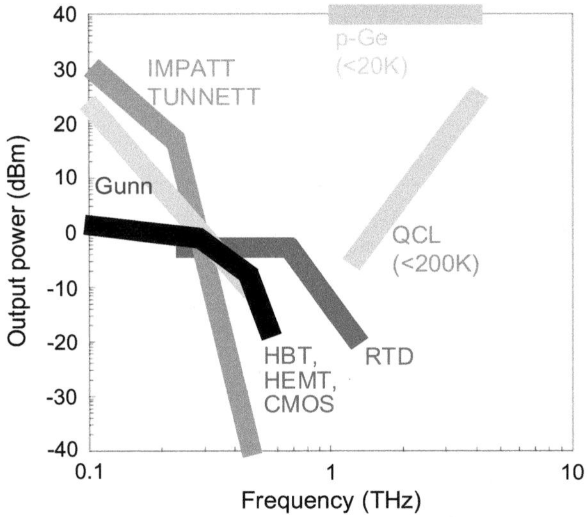

Figure 7.1 Output power as a function of frequency for several semiconductor THz sources at present.

In this chapter, RTDs for THz sources are described. The operation principle and device structures are described in Section 7.2. Oscillation frequency and output power are theoretically discussed in Section 7.3, and experiments for high-frequency oscillation and high output power are shown in Section 7.4. Spectral characteristics and application to wireless communication are described in Section 7.5. The conclusion is given in Section 7.6.

7.2 Oscillation Principle and Device Structure

The RTD is a heterostructure normally composed of two barriers and a quantum well with different semiconductor materials, as shown in Fig. 7.2a. The current–voltage (I–V) curve has a peak, at which the resonance level of electron in the quantum well is close to the conduction band edge of the emitter, as shown schematically in Fig. 7.2b. With increasing voltage over the current peak, the *I–V* curve exhibits a negative differential conductance (NDC) region. A detailed explanation of this behavior is found in Refs. [35, 40]. The oscillator utilizes the NDC region.

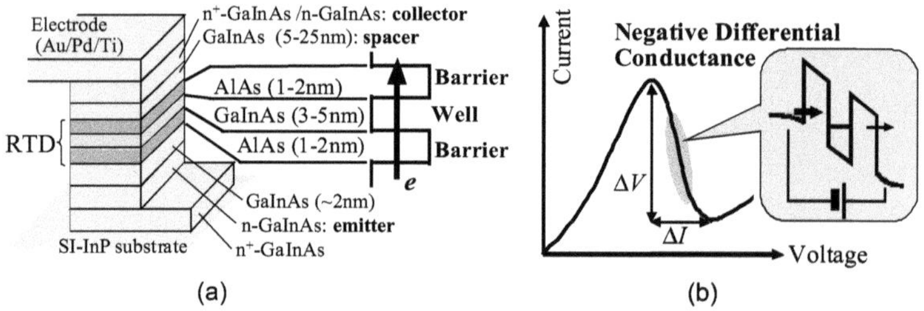

Figure 7.2 Fundamental structure and characteristics of an RTD. (a) Example of the layer structure of a double-barrier RTD and (b) current–voltage characteristics.

Figure 7.3a shows the fundamental structure of the RTD oscillator integrated with a planar slot antenna. The RTD is located at the center of a slot in a metal film. This slot forms a standing wave of the electromagnetic field as a resonator and also acts as an antenna by radiating output power at the same time. The equivalent circuit for this structure is illustrated in Fig. 7.3b. Parasitic elements are neglected in the figure, except the capacitance of the RTD. Oscillation takes place if the absolute value of the NDC exceeds the radiation loss of the slot antenna, that is, $G_{\mathrm{d}} \geq G$. The oscillation frequency f is determined by the parallel resonance of the LC circuit in Fig. 7.3b, as $f = 1/(2\pi\sqrt{LC})$, which corresponds to the formation of the standing wave in Fig. 7.3a.

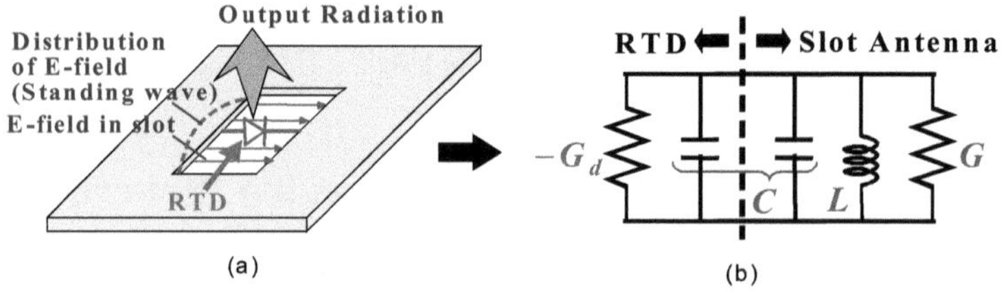

Figure 7.3 Principle of an RTD oscillator. (a) Slot antenna and RTD and (b) its equivalent circuit.

The actual structure of the fabricated device is shown in Fig. 7.4 [11–14]. The electrodes of the RTD are connected to the left and right electrodes of the antenna. At both edges of the antenna, the electrodes are overlapped with a SiO_2 layer between them. Using this structure, reflectors of high-frequency electromagnetic waves are formed, and the separation of DC bias is achieved at the same time. The length of the slot antenna is 10–50 μm, which determines the oscillation frequency through the resonance length for the standing wave. The half wavelength along the slot is approximately given by $\lambda_0/(2\sqrt{\bar{\varepsilon}})$, with $\bar{\varepsilon} = (1+\varepsilon_r)/2$, where λ_0 is the wavelength in the free space and $\bar{\varepsilon}$ is the average of relative dielectric constants between the air (1) and semiconductor substrate (ε_r). However, the resonance length of the oscillator is much shorter than the half wavelength because of the capacitance of RTD, as shown in Fig. 7.3b. The area of the RTD mesa is around 0.5–1 μm^2, which also determines the oscillation frequency through the capacitance.

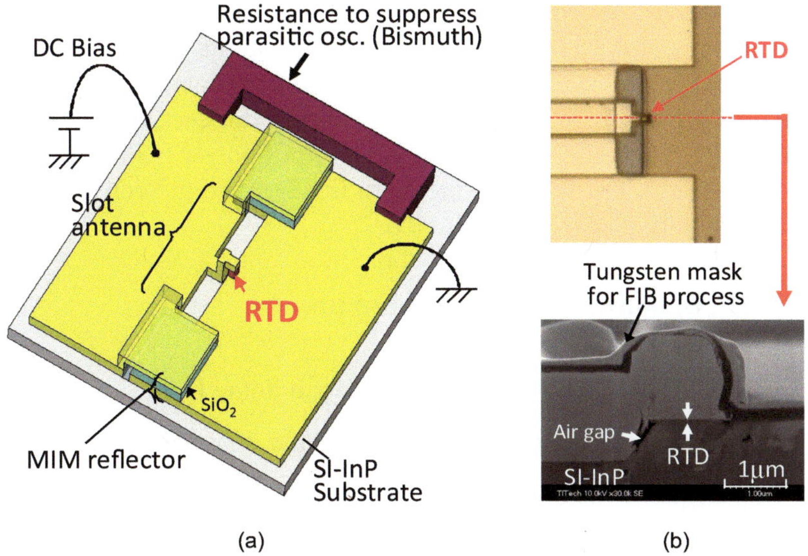

Figure 7.4 Fabricated structure of an RTD oscillator with a slot antenna. (a) Schematic structure and (b) surface micrograph and cross section.

The output power radiated into the substrate is much higher than that into the air (by a factor of $\varepsilon_r^{3/2} \simeq 43$) because of the large dielectric constant [38]. Moreover, the output is reflected at the flat surface of the bottom side due to total internal reflection. The output power was therefore extracted through a hemispherical Si lens under the substrate. The thickness of the substrate is 0.64 mm, and the diameter of the lens is 30 mm in our experiment. Lens-free oscillators were also discussed in Section 7.4.3. A parallel resistance with a bismuth film is connected outside the antenna electrodes, as shown in Fig. 7.4a, to suppress parasitic oscillation of 2–3 GHz due to the resonance formed by external circuits including bias-supplying lines. Measured results of the oscillation characteristics are shown in Section 7.4.

7.3 Theoretical Analysis of Oscillation Characteristics

The oscillation frequency and output power of RTD oscillators are theoretically analyzed in this section. The maximum oscillation frequency is discussed with the electron delay in an RTD and the parasitic elements. Experiments in Section 7.4 are compared with the theoretical calculation based on the results of this section, including possibility of high frequency and high output power with structure improvement. The spectral linewidth is discussed in Section 7.5.1, together with the measurements.

7.3.1 Maximum Oscillation Frequency

7.3.1.1 Frequency limit due to electron delay time

The NDC region of the static I–V curve in Fig. 7.2b is approximated as follows [11] in the present analysis:

$$I_{dc}(V_{dc}) \simeq \frac{\partial I_{dc}}{\partial V_{dc}} V_{dc} + \frac{1}{3!}\frac{\partial^3 I_{dc}}{\partial V_{dc}^3} V_{dc}^3 \equiv -aV_{dc} + bV_{dc}^3 \qquad 7.1$$

where I_{dc} and V_{dc} are the current and voltage at the bias point of the oscillator. The origin of the I–V curve in this expression is defined as a point near the center of the NDC region in Fig. 7.2b. Due to this definition, the terms including even power of V_{dc} are negligible in the analysis shown below. Because the I–V curve in the NDC region is usually unstable and difficult to measure directly, a and b in Eq. 7.1 are approximately expressed with the widths of the current and voltage of the NDC region, that is, ΔI and ΔV in Fig. 7.2b, as

$$a \simeq \frac{3}{2}\frac{\Delta I}{\Delta V} \quad \text{and} \quad b \simeq \frac{2\Delta I}{\Delta V^3} \simeq \frac{4}{3}\frac{a}{\Delta V^2} \tag{7.2}$$

The instantaneous current for the voltage $v_{ac}(t) = V_{ac}\cos\omega t$ vibrating near the center of the NDC region is given by $I_{dc}(v_{ac})$ if the electron delay in the RTD is neglected. Using Eq. 7.1, it is shown that $I_{dc}(v_{ac})$ includes the fundamental component vibrating with ω and the harmonic component with 3ω. The fundamental component is written as $-G_{d0}V_{ac}\cos\omega t$, where G_{d0} is the absolute value of the NDC without the electron delay time, which is written as

$$G_{d0}(V_{ac}) = a - \frac{3b}{4}V_{ac}^2 \tag{7.3}$$

This result has also been derived from the balance of the average power between the NDC and the load conductance [39].

The NDC degrades from that in Eq. 7.3 with increasing ω due to the electron delay time in the RTD. Since electrons stay inside the RTD between the emitter and collector during the delay time, positive charges (i.e., lack of electrons) are induced on the emitter and collector electrodes. As the electrons move from the emitter to the collector, positive charges on the emitter side decrease, while those on the collector side increase. This means that the positive charges transfer from the emitter to the collector through the external circuit connecting the emitter and collector, resulting in external current. It is this current that is measured as the usual RTD current in a circuit. This current has a delay related to the residence of electrons in the RTD. For the voltage $v_{ac}(t)$, this current is given by [11]

$$i(t) = \frac{1}{d_{dep}}\int_0^{d_{dep}} I_{dc}[v_{ac}(t - \tau_{rtd} - z/v_c)]\,dz + \frac{\varepsilon S}{d}\frac{dv_{ac}(t)}{dt} \tag{7.4}$$

where d_{dep} is the thickness of the collector depletion layer and d is the total thickness, including the emitter accumulation layer, barriers and well, and d_{dep}, as shown in Fig. 7.5. ε is the dielectric constant, and S is the mesa area of the RTD perpendicular to the electron flow. The injected electrons are assumed to be resident for the dwell time

τ_{rtd} in the resonant tunneling region (barriers and well), where the movement is neglected, and then to move to the collector across the depletion layer with a velocity of v_c, which is approximated to be constant for position in the depletion region and for vibration with $v_{ac}(t)$. The change of d_{dep} with $v_{ac}(t)$ is also neglected here. In the limit $\tau_{\text{rtd}} \to 0$ and $d_{\text{dep}} \to 0$, that is, if the electron delay time is neglected, the first term of $i(t)$ in Eq. 7.4 coincides with $I_{dc}(v_{ac})$.

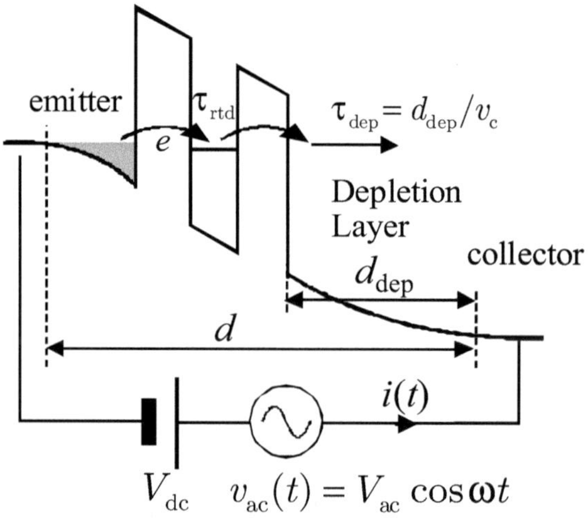

Figure 7.5 Structure of an RTD, indicating parameters used in the analysis of oscillation characteristics.

The amplitude of the current component varying as $\cos\omega t$ included in $i(t)$, which is proportional to $v_{ac}(t)$, is given by

$$I_c = \frac{\omega}{\pi}\int_0^{2\pi/\omega} i(t)\cos\omega t\, dt \tag{7.5}$$

The absolute value of the NDC, including the effect of the electron delay time, is given by $-I_c/V_{ac}$, which is calculated from Eqs. 7.1–7.5 as

$$G_d(V_{ac}) = G_{d0}(V_{ac})\cos\omega\tau \frac{\sin(\omega\tau_{\text{dep}}/2)}{\omega\tau_{\text{dep}}/2} \simeq G_{d0}(V_{ac})\cos\omega\tau \tag{7.6}$$

where $\tau_{\text{dep}} = d_{\text{dep}}/v_c$ is the transit time across the collector depletion region and

$$\tau = \tau_{\text{rtd}} + \frac{\tau_{\text{dep}}}{2} \tag{7.7}$$

is the electron delay time in the RTD. The factor 1/2 for τ_{dep} is also derived in the analysis of transistors [59]. In the low-frequency limit ($\omega\tau << 1$), $G_d = G_{d0}$ in Eq. 7.6. The right-hand side of Eq. 7.6 is a good approximation for $\omega\tau < \pi/2$ (in other words, $f = \omega/2\pi < f_{int}$ with f_{int} shown in Eq. 7.8 below).

G_d decreases with frequency and reaches zero at $f = f_{int}$, where

$$f_{int} = \frac{1}{4\tau} = \frac{1}{2(2\tau_{rtd} + \tau_{dep})} \quad (7.8)$$

G_d is negative when $f > f_{int}$. With a further increase of f, G_d becomes positive again when $f > 3f_{int}$ and repeatedly takes positive and negative values with increasing f, although the peak values of G_d in these regions are smaller than G_{d0}. This property of G_d is equivalent to the transit-time effect utilized in IMPATTs and TUNNETTs.

f_{int} in Eq. 7.8 is a measure of the intrinsic response frequency of RTDs. As discussed in Section 7.4, $\tau = 120$ fs ($\tau_{rtd} = 10$ fs and $\tau_{dep} = 220$ fs) and $f_{int} = 2.1$ THz are obtained from the fitting of theoretical calculation to experimental results for the RTD oscillating up to 1.31 THz. In this example, the collector transit time τ_{dep} is dominant in the electron delay time, and it is important to reduce τ_{dep} with optimized collector structures. The dwell time in the resonant tunneling region τ_{rtd} is discussed in detail in Refs. [35, 40]. τ_{rtd} is reduced by a thin quantum well and barriers in the oscillators shown in Section 7.4. Various effects in double-barrier structures can also be included in τ_{rtd}, such as the Coulomb interaction, by which τ_{rtd} can be longer than the dwell time in the NDC region [41]. The displacement current in the double barrier, which changes the NDC at high frequency [41], may also influence τ_{rtd}.

7.3.1.2 Capacitance due to electron delay time

The current $i(t)$ in Eq. 7.4 has a component proportional to $\sin\omega t$, in addition to the $\cos\omega t$ component in Eq. 7.5. The amplitude of this component is given by

$$I_s = \frac{\omega}{\pi}\int_0^{2\pi/\omega} i(t)\sin\omega t\, dt \quad (7.9)$$

This component is equivalently expressed with a capacitance C_d given by $\omega C_d = I_S/V_{ac}$. From Eqs. 7.4 and 7.9

$$C_d(V_{ac}) = C_0 + C_{rtd}(V_{ac}) \quad (7.10)$$

where C_0 is the geometrical capacitance of the RTD expressed as $\varepsilon S/d$ and $C_{\rm rtd}$ is the additional capacitance due to the delay time given by

$$C_{\rm rtd}(V_{\rm ac}) = \tau G_{\rm d0}(V_{\rm ac}) \frac{\sin\omega\tau}{\omega\tau} \frac{\sin(\omega\tau_{\rm dep}/2)}{\omega\tau_{\rm dep}/2} \simeq \tau G_{\rm d0}(V_{\rm ac}) \quad (7.11)$$

where the right-hand side is the expression in the low-frequency limit ($\omega\tau << 1$). $C_{\rm rtd}$ has a positive value in the frequency region where the RTD works ($G_{\rm d} > 0$), in spite of originated from the phase delay of the current. The capacitance due to the carrier accumulation in the quantum well, which is equivalent to the above capacitance, has been observed and theoretically analyzed [42–44]. The negative inductance in the NDC region introduced as a delay in RTDs [35] is also equivalent to the above capacitance.

The oscillation frequency decreases due to the electron delay time through $C_{\rm d}$ in Eq. 7.11. The dwell time $\tau_{\rm rtd}$ slightly changes with bias voltage, and thus, $C_{\rm rtd}$ also changes through Eqs. 7.7 and 7.11. This results in the oscillation frequency slightly changing with bias voltage, as discussed in Section 7.5.

7.3.1.3 Frequency limit due to parasitic elements

In addition to the intrinsic delay time discussed above, parasitic elements also degrade the NDC. Figure 7.6a shows the equivalent circuit of the RTD oscillator, including parasitic elements and the admittance of the antenna. $-G_{\rm d}$ and $C_{\rm d}$ are, respectively, the NDC and capacitance of the RTD discussed above; $R_{\rm c}$ and $C_{\rm c}$ are the contact resistance and capacitance, respectively; $R_{\rm m}$ is the sum of the RTD mesa resistance and the spreading resistance from the RTD mesa to the bottom electrode; and $L_{\rm m}$ is the inductance of the mesa. $Y_{\rm a}$ is the admittance of the slot antenna.

The RTD connected to the antenna has the admittance of the left-hand side of port 2-2′ in Fig. 7.6a. For the structure in Fig. 7.4a, $L_{\rm m}$ is usually small, and the RTD with the parasitic elements can be simplified, as shown in Fig. 7.6b, with the series resistance $R_{\rm s}$. The real part of the RTD admittance between port 2-2′ is calculated as

$$-G_{\rm d} \cdot \frac{1-\omega^2 C_{\rm d}^2 R_{\rm s}/G_{\rm d}}{1+\omega^2 C_{\rm d}^2 R_{\rm s}^2} \quad (7.12)$$

where $R_{\rm s}G_{\rm d} << 1$ is assumed. From this equation, the NDC decreases with frequency through the parasitic elements as well as through the electron delay time in G_d and C_d given by Eqs. 7.6, 7.10, and 7.11. The maximum oscillation frequency f_{max}, at which the NDC equals zero, is approximately obtained from Eqs. 7.12 and 7.6 as

$$\frac{1}{f_{max}^2} \simeq \frac{1}{f_{int}^2} + \frac{1}{f_{ext}^2} \tag{7.13}$$

where f_{int} is the intrinsic limit given by Eq. 7.8 and f_{ext} is the limit due to the parasitic elements given by

$$f_{ext} \simeq \frac{1}{2\pi C_d}\sqrt{\frac{a}{R_s}} \tag{7.14}$$

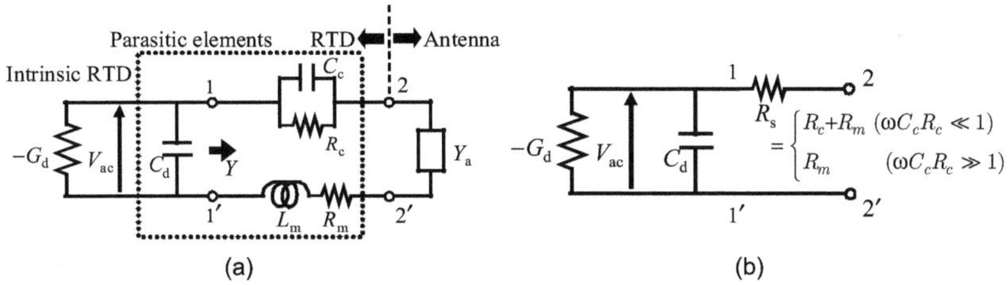

Figure 7.6 (a) Equivalent circuit of an RTD oscillator including parasitic elements and (b) RTD with dominant parasitic elements.

The frequency dependence of C_d is neglected with the approximation in the right-hand side of Eq. 7.11, and $\cos\omega\tau$ in Eq. 7.6 is approximated as $1 - (2\omega\tau/\pi)^2$ in the derivation of Eq. 7.13. As an example, f_{ext} 5.2 THz is obtained with C_d = 2 fF, R_s = 2.3 Ω, and a = 10 mS for the RTD oscillating up to 1.31 THz shown in Section 7.4, and f_{max} is estimated to be 1.9 THz, together with f_{int} = 2.1 THz in the above example. Because of $f_{ext} > f_{int}$, f_{max} is dominated by f_{int} in the present RTD oscillators, and thus, reduction in electron delay time is important for high-frequency oscillation at this stage.

7.3.2 Oscillation Condition

7.3.2.1 Oscillation frequency

Although RTDs can oscillate, in principle, up to f_{max} discussed in the previous subsection, the oscillation frequency in an actual oscillator is limited at the frequency where the NDC can compensate the loss of the antenna, including the radiation loss. The oscillation condition of an actual oscillator within this limit is discussed here in order to estimate the oscillation frequency and output power.

The stationary condition of the oscillation is that the admittance between any two points of the equivalent circuit shown in Fig. 7.6a is equal to zero. For port 1-1′, this condition is written as

$$-G_d(V_{ac}) + j\omega C_d(V_{ac}) + Y = 0 \tag{7.15}$$

where Y is the admittance of the right-hand side of port 1-1′, which is equal to the admittance Y_a of the antenna if the parasitic elements are neglected. Separating Eq. 7.15 into real and imaginary parts and using Eqs. 7.10, 7.11, and 7.3, the oscillation condition is expressed as

$$\mathrm{Im}(Y) + \omega C_0 + \mathrm{Re}(Y)\tan\omega\tau = 0 \tag{7.16}$$

and

$$V_{ac}^2 \simeq \frac{4}{3b}\left[a - \mathrm{Re}(Y)/\cos\omega\tau\right] \tag{7.17}$$

The oscillation frequency is obtained by solving Eq. 7.16 with respect to ω. The third term on the left-hand side of Eq. 7.16 arises from the electron delay. If the electron delay time and parasitic elements are neglected, Eq. 7.16 reduces to $\mathrm{Im}(Y_a) + \omega C_0 = 0$. This equation indicates that the oscillation frequency is determined by the parallel resonance circuit composed of the imaginary part of the antenna admittance and the geometrical capacitance of the RTD. The electron delay time and parasitic elements equivalently result in an increase of the capacitance and a decrease of the oscillation frequency.

7.3.2.2 Output power

The output power is given by the amplitude of oscillation voltage in Eq. 7.17 as

$$P = \frac{1}{2}\left|\frac{Y}{Y_a}\right|^2 G_r \cdot V_{ac}^2 \tag{7.18}$$

where G_r is the radiation conductance included in $\mathrm{Re}(Y_a)$. If the parasitic elements are neglected, Eq. 7.18 reduces to

$$P \simeq \frac{2}{3b}G_r\left(a - G_r/\cos\omega\tau\right) \simeq \frac{1}{2}\frac{G_r}{a}\left(a - G_r/\cos\omega\tau\right)\Delta V^2 \tag{7.19}$$

This equation implies that if the radiation conductance is small, the oscillation easily takes place even for RTDs with a small value of a due to the factor in the brackets but the radiation power is small. The output power in Eq. 7.19 has the maximum at

$$G_r = \frac{a}{2}\cos\omega\tau \quad (7.20)$$

The maximum power is expressed with ΔI and ΔV in Eq. 7.2 as

$$P_{max} = \frac{a^2}{6b}\cos\omega\tau \approx \frac{3}{16}\Delta I \Delta V \cos\omega\tau \quad (7.21)$$

If the electron delay time is neglected, the maximum output power is simply expressed as $(3/16)\Delta I\Delta V$ [39].

7.4 Experimental Results of Oscillation Characteristics

7.4.1 High-Frequency Oscillation

The oscillation frequency is designed as the resonance frequency of the resonator composed of an RTD and an antenna within the frequency region satisfying $V_{ac} > 0$ in Eq. 7.17. Because V_{ac} decreases with increasing frequency, the highest frequency limit of the oscillator is given by the frequency at $V_{ac}= 0$. To increase this frequency limit, it is necessary to enhance a in Eq. 7.2 and to reduce the electron delay time and parasitic elements. The reduction of the latter two factors is equivalent to boosting the maximum frequency in Eq. 7.13.

To increase the oscillation frequency, the geometrical capacitance of the RTD in Eq. 7.10 was reduced by a thick collector spacer in Fig. 7.2b as well as by a small area [13, 49]. To prevent the decrease of a in small-area RTDs, the current density was enhanced with a high doping concentration in the emitter. With this structure, a fundamental oscillation up to 831 GHz was obtained at room temperature [13].

For the reduction of the electron delay time τ in Eq. 7.7, thin barriers and the graded emitter structure were employed [14]. Thin barriers are effective for the reduction of the dwell time τ_{rtd}

in the resonance region. The graded emitter structure reduces the bias voltage required for the NDC, resulting in the reduction of the electric field in the collector depletion region. This may be effective to suppress possible scattering mechanisms of electrons in a high electric field, such as the transition between the Γ and L bands, and to reduce the collector transit time τ_{dep}, although the effect of this structure is not remarkable at present. A spike-doped structure for low bias voltage was also reported [63]. With the thin barriers and graded emitter structure, the oscillation frequency increased up to 1.04 THz [14].

For further reduction of τ_{rtd}, the well width was reduced [16]. Figure 7.7 shows a schematic illustration of the effect of the reduction in well width. With decreasing well width, the resonance width of the energy level increases. Because τ_{rtd} is inversely proportional to the resonance width [35], τ_{rtd} decreases accordingly. The reduction in τ_{rtd} with decreasing well width is also explained with an increase in electron velocity in the well [35, 40]. The current width ΔI in Fig. 7.2b increases with the increase in the resonance width of the energy level. The energy interval between the resonance levels also increases with decreasing well width, resulting in a large voltage widthΔV. The output power increases due to these effects in ΔI and ΔV through Eq. 7.21.

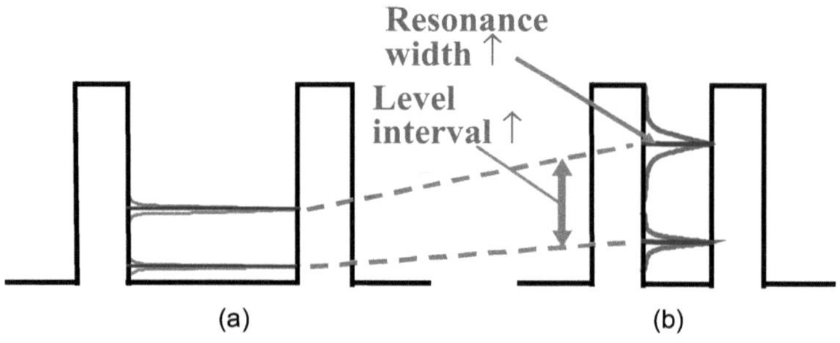

Figure 7.7 Effect of the reduction in well width from (a) a wide well to (b) a narrow well.

Figure 7.8 shows the oscillation frequency as a function of the RTD mesa area measured for two RTDs with the well widths of 4.5 nm and 3.9 nm, respectively, at room temperature. The thicknesses of the barrier and collector spacer are 1 nm and 25 nm, respectively, which are the same for the two RTDs. The total thickness of the collector depletion layer is slightly larger than the spacer thickness (by about 1–2 nm) as the sum of the spacer and the depletion region in the n-collector in Fig. 7.2a. The layer structures in this experiment are described in detail in Ref. [16].

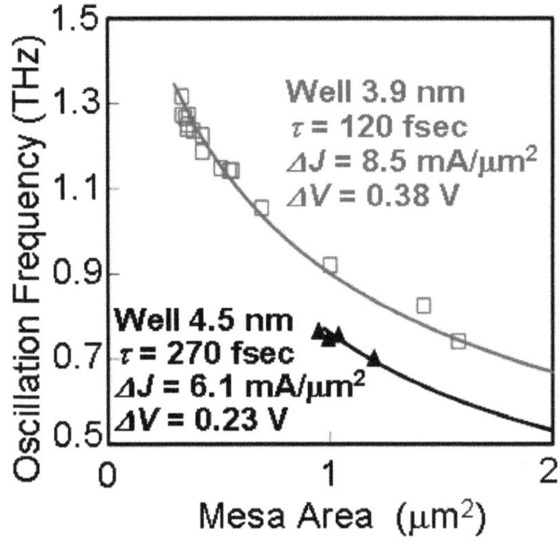

Figure 7.8 Oscillation frequency as a function of the mesa area of an RTD for two different well widths. Reproduced from Ref. [16], © 2012, Japan Society of Applied Physics.

As shown in Fig. 7.8, the oscillation frequency increases with decreasing mesa area due to a reduction in capacitance. The oscillation frequency is higher for a narrower well at the same mesa area because the delay-time capacitance given by Eq. 7.11 is smaller for a narrower well with a shorter dwell time τ_{rtd}. The limit of oscillation frequency is also higher for a narrower well because the NDC in Eq. 7.6, or equivalently the oscillation condition in Eq. 7.17, retains up to higher frequency.

Theoretical curves calculated with the model in the previous section are also shown in Fig. 7.8. For the parasitic elements, the contact resistance was measured in a separate experiment, and the other parameters were estimated from the geometry and doping concentration of the RTD. The widths of the voltage and current density of the NDC region were obtained from the measured *I*–*V* curve. The admittance of the antenna was calculated with a three-dimensional electromagnetic simulation for the actual structure of the slot antenna. The delay time was estimated from the fitting between the measurements and these theoretical curves.

Figure 7.9 shows the output power as a function of frequency measured for the RTD with a 3.9 nm thick well. The output power was 10–30 μW in 0.7–1.3 THz. The highest frequency obtained in this experiment was 1.31 THz with the output power of 10 μW. This frequency was higher than the lowest frequency reported for QCLs [4] and, thus, bridges the THz gap between the electron devices and QCLs.

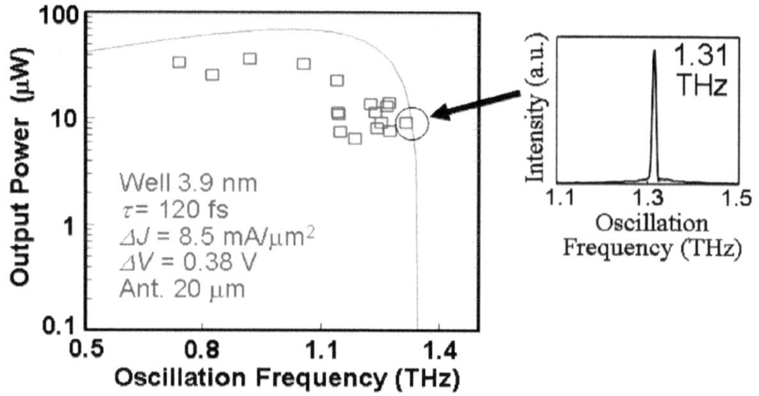

Figure 7.9 Output power as a function of frequency for an RTD with a 3.9 nm thick well and the spectrum for the highest oscillation frequency. Reproduced from Ref. [16], © 2012, Japan Society of Applied Physics.

A more higher frequency (1.4 THz) was reported recently for the RTD oscillator with a patch antenna [17]. In this structure, the radiation conductance seems small, and the output power is low at present. Higher output power keeping the frequency limit will be expected by an optimum design for high radiation conductance of the antenna as well as for a large NDC and short delay time of the RTD. This is the same condition as that in the RTD oscillators with slot antennas described here.

The electron delay time τ was estimated in Fig. 7.8 as 270 fs and 120 fs for 4.5 and 3.9 nm thick wells, respectively. τ_{rtd} and τ_{dep} included in τ in Eq. 7.7 were extracted from the above results, together with that of an RTD having a 4.5 nm thick well and a 2 nm thick spacer [45]. As a result, τ_{rtd} and τ_{dep} are estimated as 30 fs and

480 fs for a 4.5 nm thick well, respectively, and 10 fs and 220 fs for a 3.9 nm thick well, respectively. τ_{rtd} can further be reduced with decreasing well width. However, the resonance level in the well is elevated at the same time, resulting in an increase in bias voltage. Due to this, the electric field approaches the breakdown field. To avoid this situation, a deep quantum well was introduced, narrowing the energy band gap through the change in alloy composition [79]. By this method, the well width was reduced without an increase in bias voltage, and oscillation up to 1.27 THz was obtained [79].

The collector transit time τ_{dep} will become dominant in the electron delay time after the reduction in dwell time τ_{rtd} with thin wells. τ_{dep} can be reduced with a thin collector spacer, but the geometrical capacitance increases at the same time. Therefore, an optimum thickness of the spacer layer exists for high-frequency oscillation. Figure 7.10 shows theoretical output power as a function of oscillation frequency for different thicknesses of the collector spacer. A deep quantum well with a width of 3 nm [79] is assumed. As shown in this figure, by optimizing the spacer thickness (~6 nm), an oscillation frequency close to 2 THz is theoretically expected in the present layer structure of the RTD. Although a current density lower than that in Fig. 7.9 is assumed in Fig. 7.10 because of using a low-biased deep well, higher current densities and higher oscillation frequencies will be possible by an optimum design of the quantum well and emitter doping concentration.

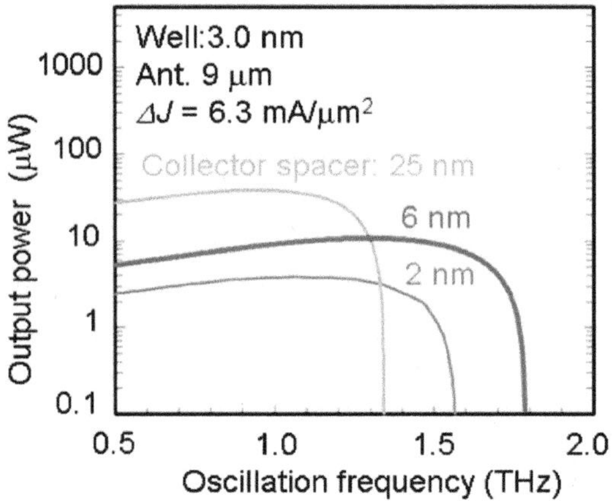

Figure 7.10 Theoretical output power as a function of oscillation frequency for different thicknesses of a collector spacer.

The electron velocity in the collector depletion region was estimated from τ_{dep} obtained above as 1.3×10^7 cm/s. This value is much smaller than that of the ballistic electrons ($\sim 1 \times 10^8$ cm/s). Various scattering processes of electrons in the high electric field are possible reasons for this discrepancy, although it is not clear at present. Selection of materials together with the optimum design of the structure as above may also be useful for the reduction of τ_{dep}.

7.4.2 High-Output-Power Operation

7.4.2.1 Offset and impedance matching

As shown in Eq. 7.21, the maximum output power that can be extracted from the RTD is $(3/16)\Delta I \Delta V$ if the electron delay time is negligible. This value is more than 1 mW in present RTDs. However, the output power of the oscillator in Fig. 7.4 is usually much lower ($\sim$10 μW) than the maximum output power. For example, the RTD with the narrower well in Fig. 7.8 has a value of 1.2 mW for $(3/16)\Delta I \Delta V$ at a mesa area of 2 μm^2, but the output power estimated from Fig. 7.9 is 30 μW. The oscillation frequency is 0.7 THz, and the electron delay time is negligible in this case.

This discrepancy is caused by the small value of the radiation conductance of the antenna. A large radiation conductance can be obtained by the offset slot antenna shown in Fig. 7.11a, in which the position of the RTD is shifted from the center of the slot [46, 47]. Figure 7.11b shows the effect of the offset with the curve of radiation conductance as a function of frequency. As shown in Fig. 7.11b, although the radiation conductance of the slot antenna has the maximum at its own resonance, the oscillation frequency is usually far below the maximum due to the capacitance of the RTD. By shifting the position of the RTD from the center of the antenna, the radiation conductance can be adjusted to the maximum, as shown by the arrowed curve in Fig. 7.11b. Moreover, the maximum value can further be changed with the width of the slot antenna. By this procedure, an impedance-matching condition for the maximum power in Eq. 7.20 can be achieved.

In a preliminary experiment of the offset structure [36], an output power of 420 μW at 550 GHz was obtained with a DC-to-RF conversion efficiency of 1.45% without including the power consumption at the external resistance. The impedance matching for the maximum output power was also demonstrated in RTDs with low current density [48].

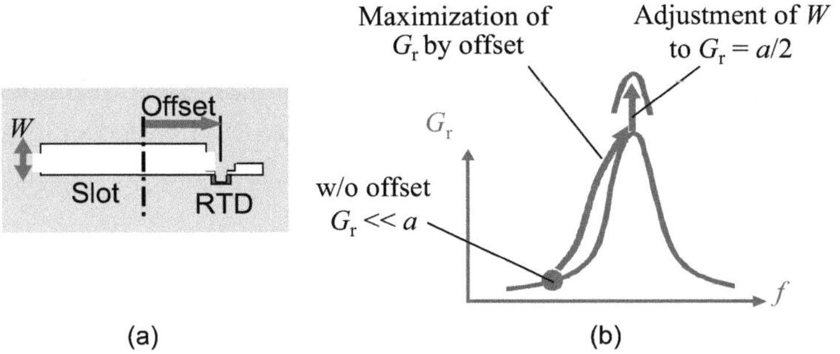

Figure 7.11 (a) Offset slot antenna for high output power and (b) effect of the offset in the curve of radiation conductance of the antenna as a function of frequency.

After achieving impedance matching, the maximum output power can be increased with RTDs having large ΔI and ΔV. A reduction in the well width discussed in Section 7.4.1 is effective for this purpose. Figure 7.12 shows the theoretical output power as a function of slot antenna width for the optimized offset and the RTDs with relatively large ΔI and ΔV, which are available with reduced well widths. The output power has the maximum at the optimum antenna width satisfying impedance matching. An output power of more than 1 mW is theoretically possible, as shown in Fig. 7.12.

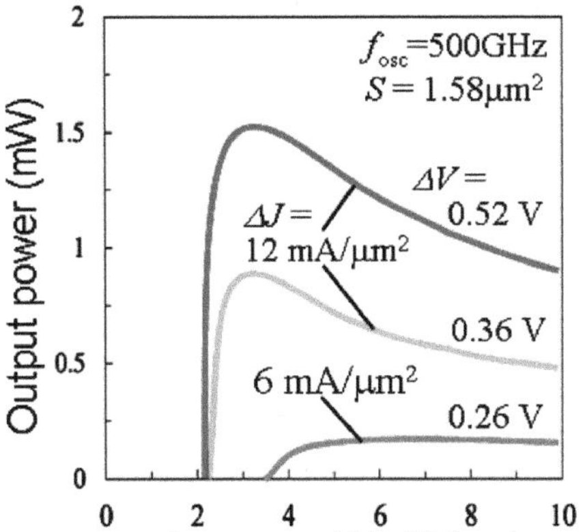

Figure 7.12 Theoretical output power as a function of slot antenna width for RTDs with relatively large ΔI and ΔV.

7.4.2.2 Array configuration

Power combining with array configuration is an effective method for high output power [10, 37, 50–58]. To obtain combined output power in array configuration, all the oscillator elements in the array are required to coherently oscillate with the same phase and frequency. This condition can be satisfied by utilizing mutual injection locking between the elements. Mutual injection locking was observed [37, 50–52] and theoretically analyzed [58].

Figure 7.13 shows an example of a two-element array [37] in which the oscillator elements are coupled with each other through planar circuits composed of angled slot lines and metal-insulator-metal (MIM) layers. Mutual injection locking takes place due to the coupling between the oscillator elements, and the whole device works as a coherent antenna array. Thus, the output power is focused into the substrate direction perpendicular to the surface through improvement of antenna directivity. The power density in this direction is approximately proportional to the square of the element number [58], although the total output power radiated in all directions is proportional to the element number. The radiation conductance and total output power also increase slightly in a strong coupled array.

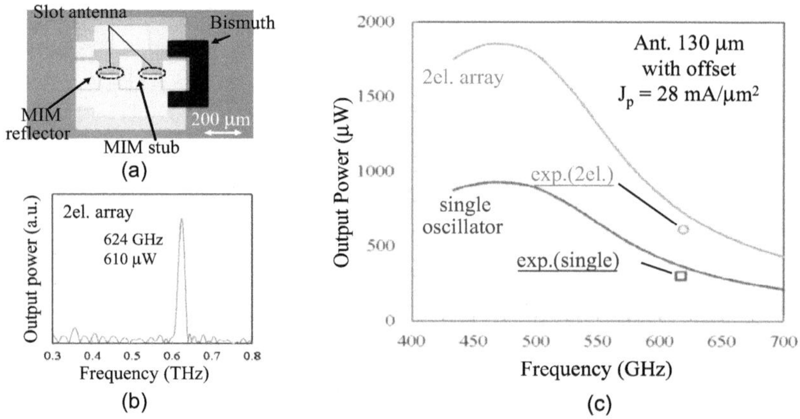

Figure 7.13 Preliminary experiment on power combining in a two-element array of RTD oscillators. (a) Micrograph of a plan view, (b) oscillation spectrum, and (c) output power as a function of frequency for single and array oscillators.

A preliminary result of the array is shown in Fig. 7.13 [37]. A single peak was observed in the oscillation spectrum due to mutual injection locking, and a relatively high total output power, 610 μW, was obtained at 620 GHz, which was about twice that of the single oscillator. The oscillation frequency deviated from that at the maximum output power, as shown in Fig. 7.13, because of the small mesa area of the RTD, and higher output power is expected by optimum design. Relatively high output powers were also obtained at other frequencies: 270 μW and 180 μW at 770 GHz and 810 GHz, respectively [37].

Although a four-element array with the same structure as above was also fabricated, two peaks were observed in the oscillation spectrum [37]. This is because the frequency dispersion between the oscillator elements in free running was too wide for mutual injection locking because of the variation in RTD size. The influence of dispersion in the free-running frequency is theoretically discussed [58]. Dispersion less than a few percent is necessary for stable locking, depending on the coupling strength. Lower dispersion is required for arrays with larger element numbers. The combined output power at the locking also decreases with increasing dispersion of the free-running frequency [58]. To suppress the dispersion, precise control on the size of oscillator elements, in particular in the mesa area of RTDs, is necessary in device fabrication.

7.4.3 Collimating Hemispherical Lens and Lens-Free Structures

Since the oscillators described above have small antennas compared to the oscillation wavelength and are located near the center of a hemispherical Si lens with a large diameter (30 mm) compared to the substrate thickness, the radiation out of the lens is not collimated. Collimation of the output radiation with the collimating Si lens is shown here. In the configuration shown in Fig. 7.14a, the position of the RTD oscillator is shifted in the upward direction from the center of the lens by an amount of approximately $R/(n - 1)$, including the thickness of the oscillator substrate, where R and $n \simeq 3.5$ are the radius and refractive index of the lens, respectively. The output radiation is collimated by this shift, as shown in Fig. 7.14b. Figure 7.15 shows the measured radiation pattern. The measured values of the full width at half maximum (FWHM) are approximately 6 degrees and 8 degrees in H and E planes, respectively. In this experiment, the FWHM is limited by the diffraction, because the size of the lens is not large enough compared to the oscillation wavelength. Misalignment of the lateral position of the RTD may also be included in the results.

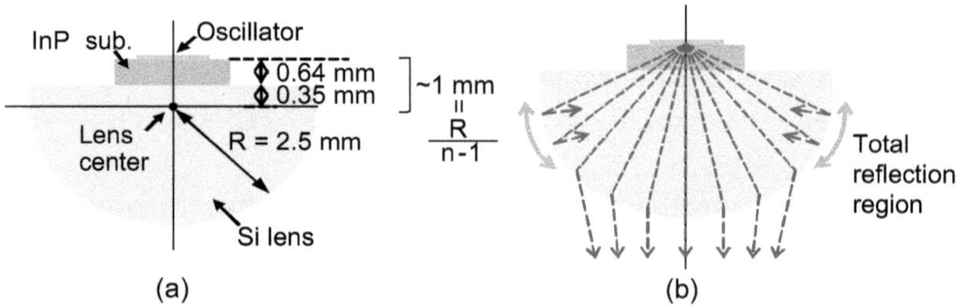

Figure 7.14 RTD oscillator on a collimating hemispherical Si lens. (a) Size and position and (b) ray trace of the radiated output.

Although the Si lens is useful to obtain a collimated beam, it is bulky and requires careful alignment of the oscillator position. In addition, radiation with an angle θ larger than ~46 degrees from the downward direction is not transmitted because of total internal reflection, as shown in Fig. 7.14b. This part is 70% (= cosθ) of the total output power. Moreover, 31% [= $(n - 1)^2/(n + 1)^2$ with n = 3.5] of the power without total internal reflection is reflected at the lens surface. Consequently, only 21% of the total output power is extracted from the collimating hemispherical Si lens if the multiple reflection inside the lens is neglected.

Various structures without a Si lens are reported [15, 17, 60–62], and reasonable values of the directivity are estimated or obtained, although these are smaller than that of the Si lens. Figure 7.16a shows the RTD oscillator with a tapered slot antenna for the horizontal direction [61]. An FWHM of the radiation angle of 40–45 degrees and a directivity of 9 dBi were reported. An RTD oscillator with a patch antenna was also reported [17, 62], which has the highest oscillation frequency (1.4 THz), as mentioned in Section 7.4.1.

Figure 7.16b shows a patch electrode on a dielectric film covering the RTD and the slot antenna. Although the oscillator structure is the same as that in the previous sections, the radiation conductance of a dielectric film with a patch electrode is much higher than that of the substrate side, and thus, high output power in the upward direction is obtained. The directivity of 8 dBi is estimated in this structure. High directivity is expected by the array of patch electrodes connected with strip lines without increasing the oscillator number. The radiation pattern can be flexibly designed with the arrangement of patch electrodes.

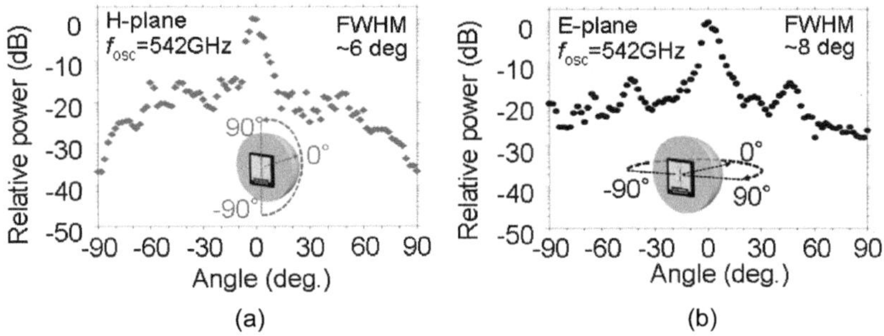

Figure 7.15 Radiation pattern of an RTD oscillator on a collimated hemispherical Si lens in an (a) H plane and (b) E plane.

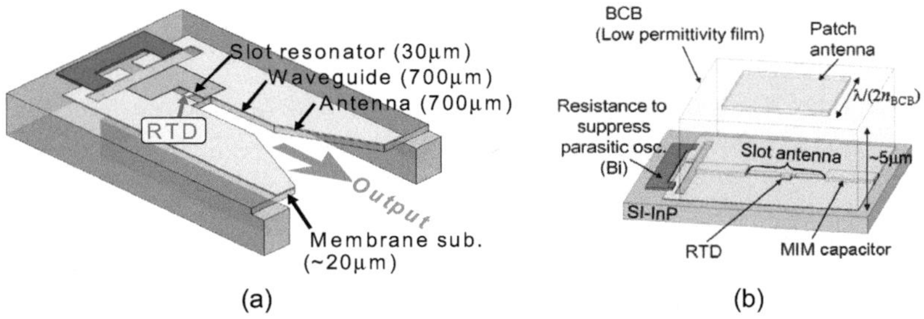

Figure 7.16 RTD oscillators without a Si lens. (a) Tapered slot antenna for horizontal radiation (reproduced from Ref. [61], © 2012, Japan Society of Applied Physics) and (b) patch electrode on a slot antenna for upward radiation.

7.5 Spectral Characteristics, Direct Modulation, and Data Transmission

7.5.1 Spectral Characteristics

The spectral linewidth of RTD oscillators has been measured with harmonic heterodyne detection [64] and the beat frequency between two RTDs [65]. Figure 7.17a shows the measurement setup of harmonic heterodyne detection [64]. The output of the RTD is mixed with the output of the local oscillator (LO) at a Si plate. The oscillation frequencies of the RTD and LO are 550 GHz and 279 GHz, respectively, and the second harmonic heterodyne is employed. The LO is composed of a microwave signal generator and multipliers. The detector is an InP Schottky barrier diode (SBD) integrated with a broad-band bow-tie antenna. Figure 7.17 shows the detected signal. Although the measured spectrum was slowly fluctuated probably due to external noise, the instantaneous linewidth (FWHM) was obtained to be less than 10 MHz.

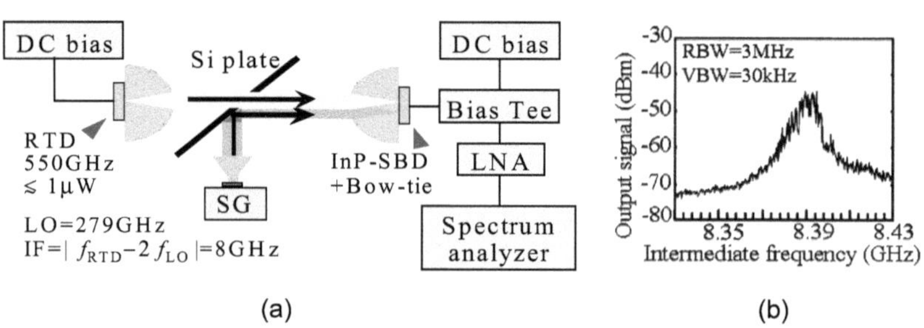

Figure 7.17 Measurement of spectral linewidth with heterodyne detection. (a) Measurement setup and (b) detected signal. Reproduced from Ref. [64], © 2012, Japan Society of Applied Physics.

The theoretical linewidth is given by Ref. [66].

$$\Delta f = \frac{\pi}{4}\frac{p_{\rm nf}}{P}\Delta f_{\rm r}^2 = \frac{\pi}{4}\frac{p_{\rm nf}f^2}{PQ^2} = \frac{1}{4\pi}\frac{e\gamma I_{\rm dc}}{(V_{\rm ac}C)^2} \tag{7.22}$$

where P is the output power, $p_{\rm nf}$ is the noise power density per frequency at the radiation conductance, and $\Delta f_r = f/Q$ is the resonance width (FWHM) of the resonator given by the oscillation frequency f and the Q factor of the resonator. The right-hand side in Eq. 7.22 is expressed with the shot noise in the RTD, where γ is the noise enhancement/suppression factor [69–72], $I_{\rm dc}$ and $V_{\rm ac}$ are the bias current and amplitude of the oscillation voltage, respectively, and C is the capacitance of the resonator composed of the RTD and the antenna. The linewidth in Eq. 7.22 originates from the phase noise of the oscillator, the principle of which is common in optical and electronic oscillators.

The theoretical value calculated with Eq. 7.22 for the RTD in Fig. 7.17b was about 6 MHz, in reasonable agreement with the measurement [64]. A narrow linewidth can be obtained with enhancement of the product of output power and square of the Q factor, as shown in Eq. 7.22. The phase-locked loop (PLL) system making use of the voltage-controlled oscillation (VCO) mentioned below is also effective for a stable and narrow linewidth. Narrowing of the linewidth in array configuration was also analyzed [66].

A frequency change with bias voltage (VCO) has been observed in RTD oscillators [14, 64, 67]. The total change of the frequency was typically 1%–5% of the oscillation frequency. The measured result of the frequency change in the heterodyne detection described above is shown in Fig. 7.18. The mechanism of the frequency change is the change in the dwell time τ_{rtd} included in the electron delay time in Eq. 7.7 with bias voltage [67, 68], which results in bias-dependent capacitance in the resonance circuit through Eq. 7.11. This property is useful for precise control and stabilization of frequency with the PLL system.

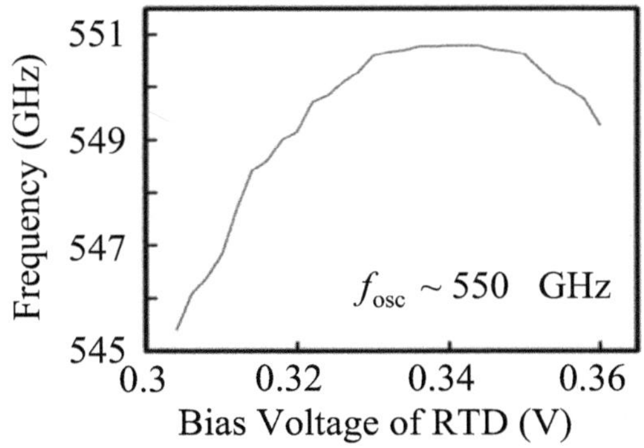

Figure 7.18 Measured result of frequency change with bias voltage. Reproduced from Ref. [64], © 2012, Japan Society of Applied Physics.

7.5.2 Direct Modulation and Data Transmission

High-capacity short-distance wireless communication is one of the valuable applications of THz frequency [73–77]. The direct modulation of RTD oscillators with bias voltage is useful for simple wireless communication. Whether the intensity or frequency modulation is obtained depends on the bias point, because the frequency change is large around the bias point at the peak output power and saturates in the bias region above the output peak [14, 64, 67, 68].

Figure 7.19 shows the measured frequency response of the direct intensity modulation of an RTD oscillator with bias voltage [77]. The 3 dB cutoff frequency of modulation was obtained to be around 4.5 GHz by improvement of the external circuit for the modulation signal, as shown in Fig. 7.19b. Since the intrinsic part of the RTD responds to the THz range, the modulation frequency is limited at present by the large capacitance of the MIM layers at the edges of the slot antenna in Fig. 7.4a and the inductance of the bonding wires in Fig. 7.19a. A higher modulation frequency is possible by reducing these parasitic elements.

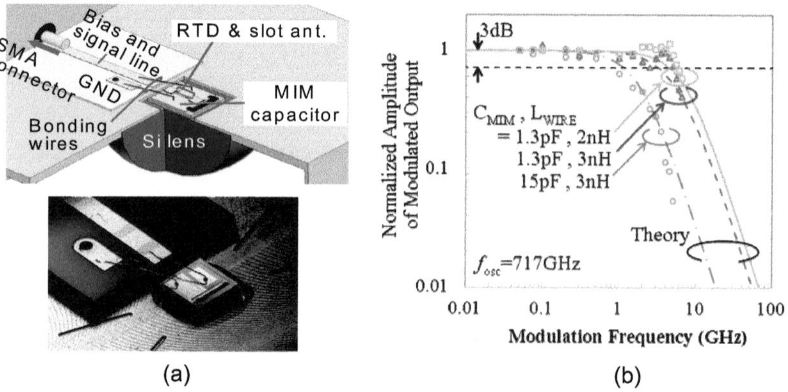

Figure 7.19 Frequency response of direct intensity modulation of an RTD oscillator with bias voltage. (a) Device configuration and (b) measured results.

A preliminary experiment on wireless data transmission by direct ASK modulation of an RTD oscillator was demonstrated [77]. Figure 7.20 shows the experimental setup of the wireless data transmission and the measured bit error rate (BER) as a function of the bit rate. The oscillation frequency and output power of the RTD used in this experiment were 542 GHz and 210 μW, respectively. The distance between the transmitter and the receiver was about 1 cm, and the received power was about 10 μW. As shown in Fig. 7.20b, the BER increases with increasing bit rate. The measured BERs were

2×10^{-8} and 3×10^{-5} for bit rates of 2 Gbps and 3 Gbps, respectively. The BER is limited in this experiment by the modulation bandwidth of the RTD (~1.1 GHz) determined by the external circuit as above and the signal-to-noise ratio composed of the received power and thermal noise in the low-noise amplifier (LNA). The relations between the BER, received power, and distance between the transmitter and the receiver were also measured [78]. The antenna gain of the transmitter was estimated to be 9 dBi from this measurement. This value of the antenna gain may be degraded by misalignment of the RTD position on the lens surface. A higher bit rate can be expected by using an antenna with higher gain and an RTD with a higher modulation bandwidth through improvement of the external circuit.

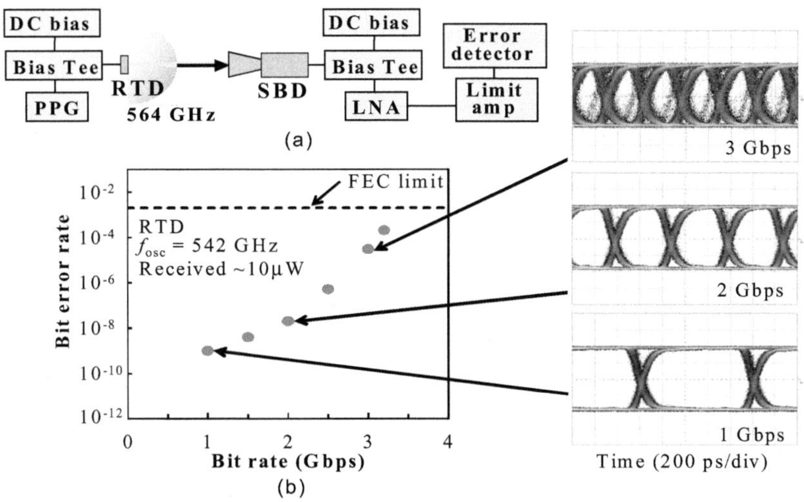

Figure 7.20 Preliminary experiment on wireless data transmission with an RTD oscillator. (a) Measurement setup and (b) BER as a function of bit rate.

Intensity modulation by irradiation of 1.55 μm wavelength laser light into a slot antenna was also demonstrated by using an RTD oscillator with an additional GaInAs layer under the slot antenna [80]. The antenna loss changes with generated free carriers, and the output power is modulated. This modulation method is useful for transformation of an optical-fiber signal to a THz wireless signal.

7.6 Conclusion

Oscillation characteristics, including recent progress, were described for RTD oscillators in the THz range. By the reduction of capacitance and electron delay time, room-temperature fundamental oscillations up to 1.3–1.4 THz have been achieved until now in RTDs with slot antennas and patch antennas. By structure optimization, a fundamental oscillation close to 2 THz is feasible. An output power of more than 1 mW is possible by impedance matching between the antenna and the RTD. At present, 420 μW has been obtained at 550 GHz in a single oscillator by the offset-fed slot antenna. In a preliminary experiment on two-element arrays, 610 μW has been obtained at 620 GHz with coherent power combining between the oscillator elements. Spectral characteristics and a preliminary experiment on wireless data transmission were also demonstrated.

On the basis of these results, we believe that the RTD oscillator is a possible candidate for compact and coherent THz sources. Making use of the advantage of room-temperature oscillation and integratability with other high-speed and functional electron/optical devices, wide applications of RTD oscillators may be expected.

Acknowledgments

The authors thank Emeritus Professors Y. Suematsu and K. Furuya of the Tokyo Institute of Technology for continuous encouragement. The authors also thank Professor S. Arai and Associate Professors Y. Miyamoto, M. Watanabe, and N. Nishiyama of the Tokyo Institute of Technology for their fruitful discussions and encouragement. This work was supported by Scientific Grants-in-Aid from the Ministry of Education, Culture, Sports, Science and Technology, Japan; the Industry–Academia Collaborative R&D Program from the Japan Science and Technology Agency, Japan; and the Strategic Information and Communications R&D Promotion Programme from the Ministry of Internal Affairs and Communications.

References

1. M. Tonouchi, Cutting-edge terahertz technology, *Nat. Photon.* **1**, 97–105 (2007).

2. S. Komiyama, Far-infrared emission from population-inverted hot-carrier system in p-Ge, *Phys. Rev. Lett.* **48**, 271–274 (1982).

3. R. Köhler, A. Tredicucci, F. Beltram, H. E. Beere, E. H. Linfeld, A. G. Davies, D. A. Ritchie, R. C. Iotti, and F. Rossi, Terahertz semiconductor heterostructure laser, *Nature* **417**, 156–159 (2002).

4. B. S. Williams, Terahertz quantum-cascade lasers, *Nat. Photon.* **1**, 517–525 (2007).

5. S. Fathololoumi, E. Dupont, C. W. I. Chan, Z. R. Wasilewski, S. R. Laframboise, D. Ban, A. Mátyás, C. Jirauschek, Q. Hu, and H. C. Liu, Terahertz quantum cascade lasers operating up to ~200 K with optimized oscillator strength and improved injection tunneling, *Opt. Express* **20**, 3866–3876 (2012).

6. J. Nishizawa, P. Plotka, T. Kurabayashi, and H. Makabe, Development of TUNNETT diode as terahertz device and its applications, *Annu. Device Res. Conf.* **VA-3**, Pennsylvania, June 2006.

7. H. Eisele and R. Kamoua, Submillimeter-wave InP Gunn devices, *IEEE Trans. Microwave Theory Tech.* **52**(10), 2371–2378 (2004).

8. H. Eisele, 480 GHz oscillator with an InP Gunn device, *Electron. Lett.* **46**(6), 422–423 (2010).

9. E. R. Brown, J. R. Sönderström, C. D. Parker, L. J. Mahoney, K. M. Molvar, and T. C. McGill, Oscillations up to 712 GHz in InAs/AISb resonant-tunneling diodes, *Appl. Phys. Lett.* **58**, 2291–2293 (1991).

10. M. Reddy, S. C. Martin, A. C. Molnar, R. E. Muller, R. P. Smith, P. H. Siegel, M. J. Mondry, M. J. W. Rodwell, H. Kroemer and S. J. Allen, Monolithic Schottky-collector resonant tunnel diode oscillator arrays to 650 GHz, *IEEE Electron. Device Lett.* **18**, 218–221 (1997).

11. M. Asada, S. Suzuki, and N. Kishimoto, Resonant tunneling diodes for sub-terahertz and terahertz oscillators, *Jpn. J. Appl. Phys.* **47**(6), 4375–4384 (2008).

12. N. Orihashi, S. Suzuki, and M. Asada, One THz harmonic oscillation of resonant tunneling diodes, *Appl. Phys. Lett.* **87**, 233501 (2005).

13. S. Suzuki, A. Teranishi, K. Hinata, M. Asada, H. Sugiyama, and H. Yokoyama, Fundamental oscillation of up to 831 GHz in GaInAs/AlAs resonant tunneling diode, *Appl. Phys. Express* **2**, 054501 (2009).

14. S. Suzuki, M. Asada, A. Teranishi, H. Sugiyama, and H. Yokoyama, Fundamental oscillation of resonant tunneling diodes above 1 THz at room temperature, *Appl. Phys. Lett.* **97**, 242102 (2010).

15. M. Feiginov, C. Sydlo, O. Cojocari, and P. Meissner, Resonant-tunnelling-diode oscillators operating at frequencies above 1.1 THz, *Appl. Phys. Lett.* **99**, 233506 (2011).

16. H. Kanaya, H. Shibayama, R. Sogabe, S. Suzuki, and M. Asada, Fundamental Oscillation up to 1.31 THz in resonant tunneling diodes with thin well and barriers, *Appl. Phys. Express* **5**, 124101 (2012).

17. Y. Koyama, R. Sekiguchi, and T. Ouchi, Oscillations up to 1.40 THz from resonant tunneling diode based oscillators with integrated patch antenna, *Appl. Phys. Express* **6**, 064102 (2013).

18. R. Lai, W. R. Deal, X. B. Mei, W. Yoshida, J. Lee, L. Dang, J. Wang, Y. M. Kim, P. H. Liu, V. Radisic, M. Lange, T. Gaier, L. Samoska, and A. Fung, Fabrication of InP HEMT devices with extremely high Fmax, *Int. Conf. IPRM08* **MoA3.2**, Versailles, May 2008.

19. E.-Y. Chang, C. Kuo, H. Hsu, C. Chiang, and Y. Miyamoto, InAs thin-channel high-electron-mobility transistors with very high current-gain cutoff frequency for emerging submillimeter-wave applications, *Appl. Phys. Express* **6**, 034001 (2013).

20. V. Radisic, L. Samoska, W. R. Deal, X. B. Mei, W. Yoshida, P. H. Liu, J. Uyeda, A.Fung, T. Gaier, and R. Lai, A 330-GHz MMIC oscillator module, *MTT-S Int. Microwave Symp. Dig.* 395–398 (2008).

21. M. Feng and W. Snodgrass, InP pseudomorphic heterojunction bipolar transistor (PHBT) with Ft>750GHz, *Int. Conf. Indium Phosphide Relat. Mater.* **WeA3-1**, Matsue, Japan (2007).

22. M. Seo, M.Urteaga, J. Hacker, A.Young, Z.Griffith, V. Jain, R. Pierson, P. Rowell, A. Skalare, A. Peralta, R. Lin, D. Pukala, and M. Rodwell, InP HBT IC technology for terahertz frequencies: fundamental oscillators up to 0.57 THz, *IEEE J. Solid-State Circuits* **46**(10), 2203–2214 (2011).

23. D. Shim, D. Koukis, D. J. Arenas, D. B. Tanner, and K. K. O, 553-GHz signal generation in CMOS using a quadruple-push oscillator, *Symp. VLSI Circuits, Dig.* 154–155 (2011).

24. O. Momeni, and E. Afshari, High power terahertz and millimeter-wave oscillator design: a systematic approach, *IEEE J. Solid-State Circuits* **46**(3), 583–597 (2011).

25. Y. M. Tousi, O. Momeni, and E. Afshari, A novel CMOS high-power terahertz VCO based on coupled oscillators: theory and implementation, *IEEE J. Solid-State Circuits* **47**(12), 3032–3042 (2012).

26. R. Han, and E. Afshari, A 260GHz broadband source with 1.1mW continuous-wave radiated power and EIRP of 15.7dBm in 65nm CMOS, *IEEE Int. Solid-State Circuits Conf.* **8.2** (2013).

27. L. Esaki and R. Tsu, Superlattice and negative differential conductivity in semiconductors, *IBM J. Res. Dev.* **14**, 61–65 (1970).

28. N. Sekine and K. Hirakawa, Dispersive terahertz gain of a nonclassical oscillator: Bloch oscillation in semiconductor superlattices, *Phys. Rev. Lett.* **94**, 057408 (2005).

29. T. Otsuji, Chapter 9, Plasma-wave devices for terahertz applications, in *Handbook of Terahertz Technologies* (Pan Stanford, Singapore).

30. J. Takeuchi, Y. Iwahashi, and M. Asada, Fabrication and millimeter-wave characterization of semiconductor klystron device using two-dimensional electron gas, *Top. Workshop Heterostruct. Mater.* **ThB-3**, Kisarazu, Japan, Aug. 2007.

31. R. Tsu and L. Esaki, Tunneling in a finite superlattice, *Appl. Phys. Lett.* **22**(11), 562–564 (1973).

32. L. L. Chang, L. Esaki, and R. Tsu, Resonant tunneling in semiconductor double barriers, *Appl. Phys. Lett.* **24**(12), 593–595 (1974).

33. M. Tsuchiya, H. Sakaki, and J. Yoshino, Room temperature observation of differential negative resistance in an AlAs/GaAs/AlAs resonant tunneling diode, *Jpn. J. Appl. Phys.* **24**(6), L466–L468 (1985).

34. T. C. L. G. Sollner, P. E. Tannenwald, D. D. Peck, and W. D. Goodhue, Quantum well oscillators, *Appl. Phys. Lett.* **45**, 1319–1321 (1984).

35. H. C. Liu and T. C. L. G. Sollner, Chapter 6, High-frequency resonant-tunneling devices in *High-Speed Heterostructure Device*, ed. R. A. Kiehl and T. C. L. G. Sollner (Academic Press, San Diego, CA, 1994), 359, and references therein.

36. M. Shiraishi, H. Shibayama, K. Ishigaki, S. Suzuki, M. Asada, H. Sugiyama, and H. Yokoyama, High output power (~400 μW) oscillators at around 550 GHz using resonant tunneling diodes with graded emitters and thin barriers, *Appl. Phys. Express* **4**, 064101 (2011).

37. S. Suzuki, M. Shiraishi, H. Shibayama, and M. Asada, High-power operation of terahertz oscillators with resonant tunneling diodes using impedance-matched antennas and array configuration, *IEEE J. Sel. Top. Quantum Electron.* **19**(1), 8500108 (2013).

38. D. B. Rutledge, D. P. Neikirk, and D. P. Kasilingam, Integrated-circuit antennas, in *Infrared and Millimeter Waves*, ed. K. J. Button (Academic Press, Orland, 1983).

39. C. S. Kim and A. Brändli, High-frequency high-power operation of tunnel diodes, *IRE Trans. Circuit Theory* **8**, 416–425 (1961).

40. H. Mizuta and T. Tanoue, *The Physics and Applications of Resonant Tunneling Diodes* (Cambridge University Press, 1995).

41. M. N. Feiginov, Does the quasibound-state lifetime restrict the high-frequency operation of resonant-tunnelling diodes? *Nanotechnology* **11**, 359–364 (2000).

42. N. Shimizu, T. Waho, and T. Ishibashi, Capacitance anomaly in the negative differential resistance region of resonant tunneling diodes, *Jpn. J. Appl. Phys.* **36**(3B), L330–L333 (1997).

43. T. Wei, S. Stapleton, and E. Berolo, Equivalent circuit and capacitance of double barrier resonant tunneling diode, *J. Appl. Phys.* **73**(2), 829–834 (1993).

44. R. Lake and J. Yang, A physics based model for the RTD quantum capacitance, *IEEE Trans. Electron. Devices* **50**(3), 785–789 (2003).

45. M. Shiraishi, S. Suzuki, A. Teranishi, M. Asada, H. Sugiyama, and H. Yokoyama, Fundamental oscillation of up to 915 GHz in small-area InGaAs/AlAs resonant tunneling diodes with planar slot antennas, *Jpn. J. Appl. Phys.* **49**, 020211 (2010).

46. S. Suzuki and M. Asada, Proposal of resonant tunneling diode oscillators with offset-fed slot antennasin terahertz and sub-terahertz range, *Jpn. J. Appl. Phys.* **46**(1), 119–121 (2007).

47. S. Suzuki, N. Kishimoto, M. Asada, N. Sekine, and I. Hosako, Experimental and theoretical investigation of the dependence of oscillation characteristics on structure of integrated slot antennas in sub-terahertz and terahertz oscillating resonant tunneling diodes, *Jpn. J. Appl. Phys.* **47**(1), 64–67 (2008).

48. H. Shibayama, M. Shiraishi, S. Suzuki, and M. Asada, Dependence of output power on slot antenna width in terahertz oscillating resonant tunneling diodes, *J. Infrared mm THz Waves* **33**, 475–478 (2012).

49. N. Kishimoto, S. Suzuki, A. Teranishi, and M. Asada, Frequency increase of resonant tunneling diode oscillators in sub-THz and THz range using thick spacer layers, *Appl. Phys. Express* **1**, 042003 (2008).

50. S. Suzuki, N. Orihashi, and M. Asada, Mutual injection locking between sub-THz oscillating resonant tunneling diodes, *Jpn. J. Appl. Phys.* **44**(48), L1439–L1441 (2005).

51. S. Suzuki and M. Asada, Coherent power combination in highly integrated resonant tunneling diode oscillators with slot antennas, *Jpn. J. Appl. Phys.* **46**(46), L1108–L1110 (2007).

52. S. Suzuki, K. Urayama, and M. Asada, Coherent power combination in multi-element sub-terahertz resonant tunneling diode oscillators

coupled with metal–insulator–metal stub structure, *Appl. Phys. Express* **1**, 093001 (2008).

53. K. D. Stephan, S.-C. Wang, E. R. Brown, K. M. Molvar, A. R. Calawa, and M. J. Manfra, 5mW parallel-connected renent-tunneling-diode oscillator, *Electron. Lett.* **28**(15), 1411–1412 (1992).

54. D. P. Steenson, J. M. Chamberlain, R. E. Miles, R. D. Pollard, M. Henini, Demonstration of power combining at W-band from GaAs/AlAs resonant tunnelling diodes, *Int. Conf. Superlattices, Microstruct. Microdevices (ICSMM-7)*, Banff, Canada, 98–99, Aug. 1994.

55. M. P. DeLisio, J. F. Davis, S-J. Li, D. B. Rutledge, and J. J. Rosenberg, A 16-element tunnel diode grid oscillator, *IEEE AP-S, Int. Symp.*, 1284, 1995.

56. T. Fujii, H. Mazaki, F. Takei, J. Bae, M. Narihiro, T. Noda, H. Sakaki, and K. Mizuno, Coherent power combining of millimeter wave resonant tunneling diodes in a quasi-optical resonator, *IEEE MTT-S, Int. Microwave Symp.* **WE3F-28**, 919 (1996).

57. H. I. Cantu and W. S. Trunscott, Injection-locking and power combining with double barrier resonant tunnelling diodes, *Electron. Lett.* **37**(20), 1264–1265 (2001).

58. M. Asada and S. Suzuki, Theoretical analysis of coupled oscillator array using resonant tunneling diodes in subterahertz and terahertz range, *J. Appl. Phys.* **103**, 124514 (2008).

59. F. N. Trofimenkoff, Collector depletion region transit time, *Proc. IEEE* **52**, 86–87 (1964).

60. N. Orihashi, S. Hattori, and M. Asada, Millimeter and submillimeter oscillators using resonant tunneling diodes with stacked-layer slot antennas, *Jpn. J. Appl. Phys.* **43**(10A), L1309–L1311 (2004).

61. K. Urayama, S. Aoki, S. Suzuki, M. Asada, H. Sugiyama1, and H. Yokoyama, Sub-terahertz resonant tunneling diode oscillators integrated with tapered slot antennas for horizontal radiation, *Appl. Phys. Express* **2**, 044501 (2009).

62. R. Sekiguchi, Y. Koyama, and T. Ouchi, Subterahertz oscillations from triple-barrier resonant tunneling diodes with integrated patch antennas, *Appl. Phys. Lett.* **96**, 062115 (2010).

63. S. Suzuki, K. Sawada, A. Teranishi, M. Asada, H. Sugiyama, and H. Yokoyama, Fundamental oscillations at ~900 GHz with low bias voltages in RTDs with spike-doped structures, *Electron. Lett.* **46**(14), 1006–1007 (2010).

64. K. Karashima, R. Yokoyama, M. Shiraishi, S. Suzuki, S. Aoki, and M. Asada, Measurement of oscillation frequency and spectral linewidth of sub-terahertz InP-based resonant tunneling diode oscillators using Ni–InP Schottky barrier diode, *Jpn. J. Appl. Phys.* **49**, 020208 (2010).

65. S. Suzuki, K. Karashima, K. Ishigaki, and M. Asada, Heterodyne mixing of sub-terahertz output power from two resonant tunneling diodes using InP Schottky barrier diode, *Jpn. J. Appl. Phys.* **50**, 080211 (2011).

66. M. Asada, Theoretical analysis of spectral linewidth of terahertz oscillators using resonant tunneling diodes and their coupled arrays, *J. Appl. Phys.* **108**, 034504 (2010).

67. N. Orihashi, S. Hattori, S. Suzuki and M. Asada, Voltage-controlled sub-terahertz oscillation of resonant tunnelling diode integrated with slot antenna, *Electron. Lett.* **41**(15), 872–873 (2005).

68. M. Asada, N. Orihashi, and S. Suzuki, Experiment and theoretical analysis of voltage-controlled sub-THz oscillation of resonant tunneling diodes, *IEICE Trans. Electron.* **E89-C**, 968–975 (2006).

69. E. R. Brown, Analytic model of shot noise in double-barrier resonant-tunneling diodes, *IEEE Trans. Electron. Devices* **39**, 2686–2693 (1992).

70. J. H. Davies, P. Hyldgaard, S. Hershfield, and J. W. Wilkins, Classical theory of shot noise in resonant tunneling, *Phys. Rev. B* **46**, 9620–9633 (1992).

71. H. C. Liu, J. Li, G. C. Aers, C. R. Leavens, M. Buchanan, and Z. R. Wasilewski, Shot-noise suppression in resonant tunneling, *Phys. Rev. B* **51**, 5116–5120 (1995).

72. G. Iannaccone, G. Lombardi, M. Macucci, C. Ciofi, and B. Pellegrini, Simulation and measurement of shot noise in resonant tunneling structures, *Analog Integr. Circuits Signal Proc.* **25**, 73–78 (2000).

73. T. Kleine-Ostmann and T. Nagatsuma, A review on terahertz communications research, *J. Infrared mm THz Waves* **32**, 143–171 (2011).

74. H.-J. Song and T. Nagatsuma, Present and future of terahertz communications, *IEEE Trans. THz Sci. Technol.* **1**(1), 256–263 (2011).

75. H.-J. Song, K. Ajito, Y. Muramoto, A. Wakatsuki, T. Nagatsuma and N. Kukutsu, 24 Gbit/s data transmission in 300 GHz band for future terahertz communications, *Electron. Lett.* **48**(15), 953–954 (2012).

76. H.-J. Song and T. Kuerner, Chapter 17, Terahertz wireless communications, in *Handbook of Terahertz Technologies* (Pan Stanford, Singapore).

77. K. Ishigaki, M. Shiraishi, S. Suzuki, M. Asada, N. Nishiyama, and S. Arai, Direct intensity modulation and wireless data transmission characteristics of terahertz-oscillating resonant tunnelling diodes, *Electron. Lett.* **48**(10), 582–583 (2012).

78. S. Suzuki, K. Ishigaki, and M. Asada, Dependence of bit error rate on received power in terahertz wireless communication using resonant-tunneling-diode oscillator, *Int. Symp. Frontiers THz Tech. (FTT 2012)* **Pos1.49**, Nara, Nov. 2012.

79. H. Kanaya, S. Suzuki, and M. Asada, Frequency increase in terahertz oscillating resonant tunneling diodes with keeping bias voltage by deep- and thin-well structure, *Int. Symp. Frontiers THz Tech. (FTT 2012)* **Pos1.14**, Nara, Nov. 2012.

80. S. Kaburaki, S. Suzuki, and M. Asada, Intensity modulation of sub-terahertz oscillating resonant tunneling diode by irradiation of 1.55-μm laser, *IEEE Photon. Conf. (IPC 11)* **ThS-2**, Arlington/VA, Oct. 2011.

Chapter 8

Terahertz Vacuum Electronics

Jung-Il Kim,[a] Seok-Gy Jeon,[a] and Gun-Sik Park[b]
[a]*Korea Electrotechnology Research Institute (KERI), Ansan, 426-170, Korea*
[b]*Center for THz-Bio Application Systems and Department of Physics and Astronomy, Seoul National University, Seoul, 151-742, Korea*
sky@keri.re.kr, gunsik@snu.ac.kr

This chapter introduces compact, high-power terahertz (THz), coherent radiation sources based on vacuum electronic devices (VEDs). The types and issues in THz VEDs are briefly explained. For THz VEDs, the technical issues and developments in the backward wave oscillator (BWO), klystron, traveling wave tube (TWT), gyrotron, spatial-harmonic magnetron (SHM), and free-electron laser (FEL) are presented. The electron beam sources for THz VEDs are also described.

8.1 Introduction

8.1.1 THz Radiation Using Vacuum Electron Beams

There have been considerable demands on compact, high-power, coherent radiation sources up to the THz frequency range due

Handbook of Terahertz Technologies: Devices and Applications
Edited by Ho-Jin Song and Tadao Nagatsuma

ISBN 978-981-4613-08-8 (Hardcover), 978-981-4613-09-5 (eBook)
www.panstanford.com

to their unique physical properties for various applications. THz radiation sources have been developed using electronic and photonic technologies such as vacuum electronic devices (VEDs), solid-state devices, and quantum cascade lasers. However, the development of more compact, low-cost, and high-power THz sources is still required for applications in the practical field. THz VEDs are one of the promising candidates because of their efficient interaction mechanism between an electron beam and an electromagnetic wave in vacuum. VEDs were widely used as a high-power radiation source for communication, various heating systems, diagnostics, and radar systems [1–7]. The origin of energy in coherent electromagnetic wave is from the conversion of kinetic or potential energy of linear or rotating electron beams. In principle, a charged particle that encounters a change of velocity radiates photons. Almost every coherent radiation using free electrons can be categorized into three types by the synchronization condition between an electron beam and an electromagnetic wave. These are Cerenkov radiation, Bremsstrahlung radiation, and scattering, as shown in Fig. 8.1.

Types of radiation	Scheme	Conditions of synchronism	Types of vacuum devices
Cherenkov (and transition)	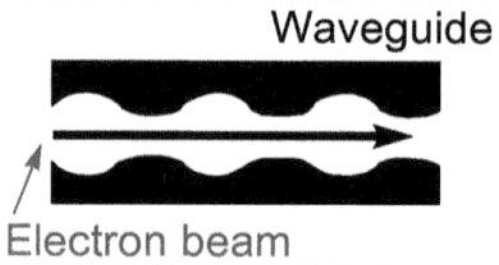	$\omega = k_z v_z$	BWO Klystron TWT Magnetron
Bremsstrahlung	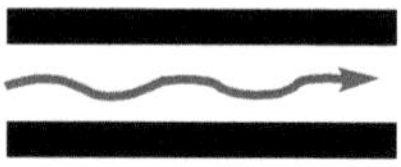	$\omega = k_z v_z + \Omega_{ce}$	Gyrotron gyroklystron FEL
Scattering		$\frac{\omega_s - \omega_i}{k_{zs} - k_{zi}} = v$	Parametric devices

Figure 8.1 Types of vacuum electronic radiation sources.

Among these, Cerenkov radiation occurs by a shock wave, which is activated when the velocity of electrons traveling in a medium is faster than the phase velocity of the electromagnetic wave propagating in the same medium. Cerenkov radiation can be classified into two types by the electromagnetic modes to interact with the electron beam. The electromagnetic modes can be formed as the propagation mode in the linear slow-wave structure (SWS) and the resonant mode in the circularly formed multiresonator in which the resonant modes have a role of a slow wave effectively. In the slow-wave interaction, the circuit and the electron beam modes synchronously encounter each other below the light line. On the basis of the Cerenkov radiation mechanism, the backward wave oscillator (BWO), klystron, traveling wave tube (TWT), and magnetron for THz VEDs have been developed and studied. On the basis of the Bremsstrahlung radiation mechanism resulting from periodically circling or wiggling electrons, the gyrotron, gyroklystron, and free-electron laser (FEL) have been developed and studied.

8.1.2 Issues of THZ VEDs

To develop VEDs at high frequencies beyond 0.1 THz, there are crucial factors. To induce a high-power THz wave radiation, a high-current-density electron beam is generally necessary. A cathode with a high current density up to 100 A/cm^2 is required, and a high field magnetic focusing structure should be prepared for propagation of the dense electron beam through a narrow vacuum channel of circuits. For the cathode, thermionic cathodes like the scandate cathode and cold cathodes like the platinum secondary emission cathode and the carbon nanotube (CNT) cathode are developed. To enhance the output power, sheet electron beams instead of round electron beams were proposed and applied for the planar-type VEDs. Because the structural size of the electronic circuit is proportional to the radiation wavelength, the THz vacuum electronic circuits should be microsized. To satisfy the fabrication accuracy and the surface roughness, the microfabrication method such as the X-ray LIGA, UV LIGA, and microelectromechanical systems (MEMS) deep reactive-ion etching (DRIE) technologies are applied instead of the conventional fabrication methods. The thermal problems are more crucial in the microfabricated THz circuits. Electron impact can produce secondary electrons or discharge and the surface heating problem in the reduction of efficiency and the lifetime [8–10].

8.2 Backward Wave Oscillator

8.2.1 Clinotron

One of the well-known THz VED sources is the BWO. A BWO can be designed to be oscillated and produce a coherent radiation even in the absence of an input signal, provided the electron beam current exceeds a threshold value known as the start oscillation current. The BWO is named so because the SWS mode involved in the resonance has a negative group velocity. Accordingly, the energy transfer into waves is pointing backward of the electron beam direction. To obtain a high output power at the THz frequency range, the BWO must be operated with a high current density of the electron beam, more than 50 A/cm^2. To enhance the interaction between the backward wave and the electron beam, the electron beam must be propagated near the surface of the SWS.

The clinotron is a modified BWO to enhance the output power and efficiency [11–13]. In the clinotron, as shown in Fig. 8.2, the sheet electron beam is propagating near the grating surface with the tilt angle to increase the output power and efficiency. The start oscillation current

$$I_{st} \propto LB \sin^3 \alpha \tag{8.1}$$

depends on the tilt angle expressed by Eq. 8.1, where α is the tilt angle, L is the length of the grating, and B is the width of the grating.

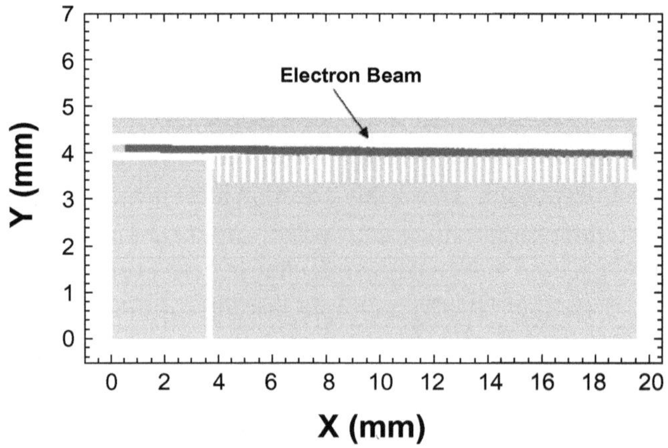

Figure 8.2 Schematic of the 0.1 THz clinotron modeled by 3D particle-in-cell (PIC) code. The sheet electron beam is propagated with the tilt angle.

The grating with the groove period of 310 μm and the groove thickness of 150 μm is fabricated by using the wire–electric discharge machining (wire-EDM) method to be operated at the output frequency of 0.1 THz, as shown in Fig. 8.3. To focus a sheet electron beam with the dimension of 2.5 × 0.14 mm^2, a focusing magnetic field of about 0.4 T is applied. When a beam current of 174 mA and a beam energy of 4.1 keV are used, the continuous-wave (CW) output power of 11 W is observed.

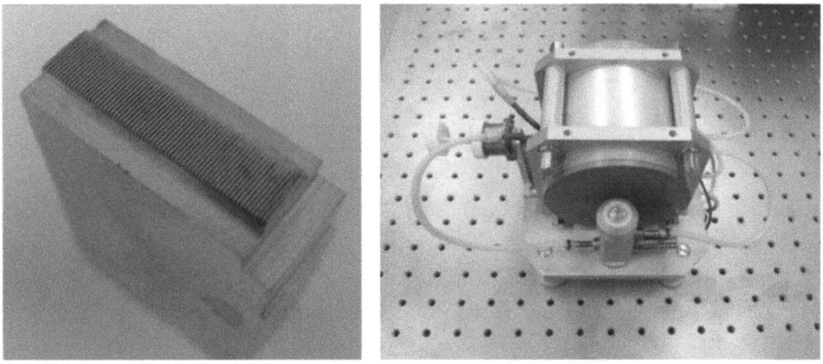

(a) (b)

Figure 8.3 (a) Grating circuit of the 0.1 THz clinotron fabricated by wire-EDM and (b) assembled 10 W 0.1 THz clinotron.

For the microfabrication of a grating to reduce fabrication errors, the MEMS technique based on DRIE is applied, as shown in Fig. 8.4.

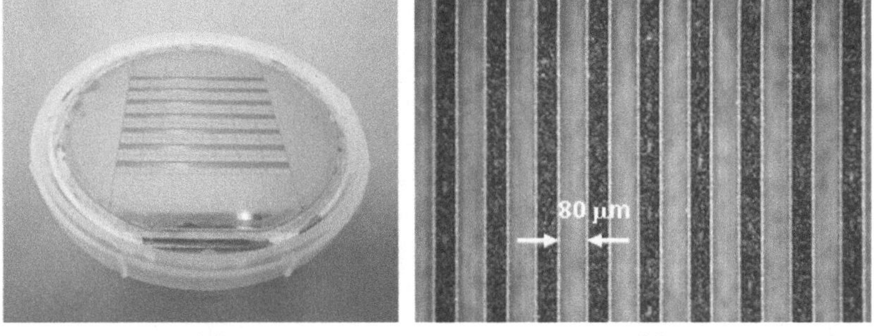

(a) (b)

Figure 8.4 (a) PR patterning on a silicon wafer for a grating circuit of a 0.2 THz clinotron and (b) a 0.2 THz grating fabricated by the DRIE process.

Using photolithography, the pattern of a grating is formed on a silicon wafer, and this pattern is deep-etched into the silicon wafer. As for the 0.2 THz grating circuit, the width of the groove is 80 μm, the height of the groove is 325 μm, and the period is 145 μm. In addition, a pulsed clinotron has demonstrated a 1 kW output power at 0.3 THz. The CW output power of more than 40 mW at 0.39 THz with the electronic frequency tuning of 20% is developed with the help of a 1 T focusing magnetic field [14].

8.2.2 Smith–Purcell BWO

Smith–Purcell (SP) radiation is a coherent radiation from electrons moving along a periodic metallic grating, as shown in Fig. 8.5. SP devices are drawing significant attention as compact and tunable THz radiation sources. SP FEL is based on spontaneous SP radiation. In the SP BWO, the excitation of a surface wave with a high-current electron beam results in power enhancement of SP radiation [15–17].

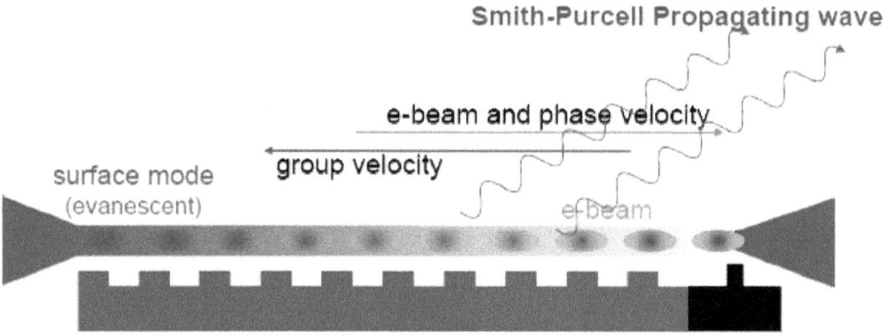

Figure 8.5 Schematic of the SP BWO. Reproduced from Ref. [16] by permission of APS, © 2007.

In an effort to enhance the radiation intensity, various grating structures have been studied with a specific focus on the strong interaction between the electron beam and the evanescent wave that occurs at the sidewall grating and the grating with the resonator. Additionally, SP devices using a train of prebunched electron beams for coherent radiation are investigated. A schematic of the SP BWO with a conventional rectangular grating and the inverse wet-etched grating is shown in Fig. 8.6 [18]. The wet chemical etching of silicon is characterized by anisotropic etching, and the wet-etched profile in the 100-crystal plane is indicated by a sloped sidewall with an angle of 54.7°. The thickness of the copper layer on the grating surface is thicker than the skin depth, and the silicon-based grating can be coated with a conducting material like copper. The backward wave oscillation occurs at the intersection point in the dispersion relation between the 30 keV electron beam and the 0.335 THz evanescent wave. Furthermore, the second harmonics of the evanescent wave with a frequency of 0.67 THz can be propagated from the grating.

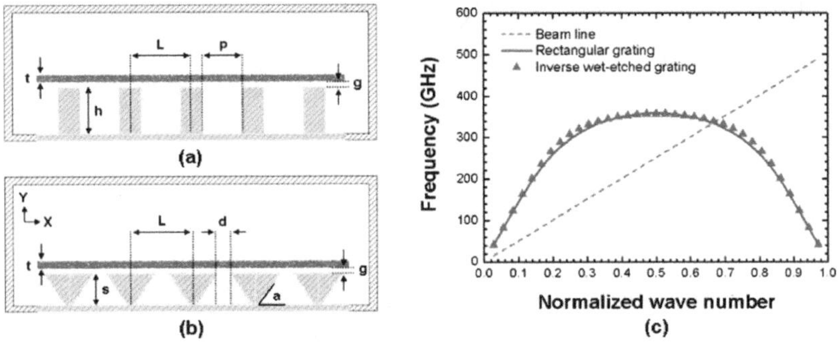

Figure 8.6 Schematic of the 0.6 THz SP BWO with (a) a rectangular grating, (b) an inverse wet-etched grating modeled in the PIC code, and (c) a dispersion relation. The dotted line indicates the electron beam mode at 30 keV. Reproduced from Ref. [18] by permission of APS, © 2011.

When the maximum electric field, E_X, is 1.0 for the inverse wet-etched grating at the central position, E_X is 0.6 for the rectangular grating. Due to the enhancement of the interactions between the electron beam and the evanescent wave by the adjacent surface structure of each pillar in the grating, the radiated power is increased approximately twofold from 7×10^3 W/m to 20×10^3 W/m at a radiated frequency of 0.66 THz.

8.3 Klystron

8.3.1 Extended Interaction Klystron

Klystron amplifies radio frequency (RF) signals by converting kinetic energy in a DC electron beam into RF power. The electron beam passes through an input cavity, which is fed by RF energy with the same frequency of the input cavity to suspend an electric field crossing the cavity gap. The electric field causes the electrons to be bunched periodically. Electron bunches excite a voltage across the gap of the output cavity, and the developed RF energy is coupled out. The extended interaction klystron (EIK) uses a multiple and multigap cavity to strengthen the interaction between the electron beam and the wave. Such a scheme results in the enhancement of cavity impedance expressed by

$$\frac{R}{Q}=\frac{\left(\int |E| ds\right)^2}{2wW_s} \tag{8.2}$$

where R is the shunt resistance, Q is the quality factor, ω is the angular resonant frequency, and W_s is the stored energy per gap [19–21]. In the sub-THz frequency range, the EIK is one of the most powerful and compact sources, and it is a promising device even in the THz range because of the simple circuit structure, which is a suitable structure for modern microfabrication such as wire-EDM, LIGA, and MEMS fabrication.

A CW EIK at 0.218 THz demonstrated an output power up to 9 W when the amplifier gain is 23 dB and the bandwidth is 0.4 GHz. For a 0.67 THz and 43 mW EIK, as shown in Fig. 8.7, a high-current-density electron beam more than 800 A/cm^2 at a beam diameter of 100 μm is investigated [22, 23].

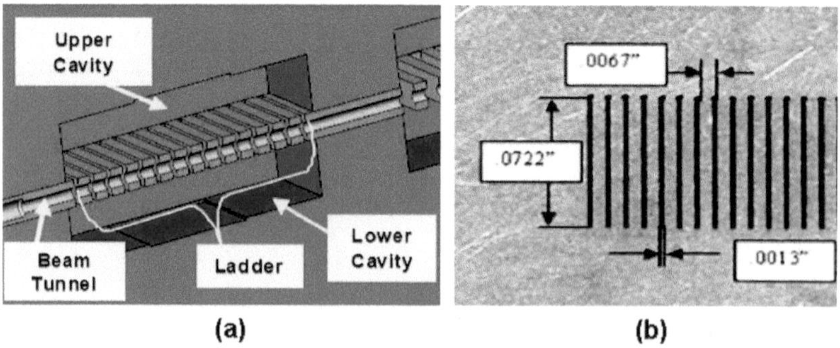

Figure 8.7 (a) Schematic of an EIK cavity with an electron beam tunnel and (b) a 0.67 THz ladder fabricated by using w-EDM.

8.3.2 Reflex Klystron

Modern microfabrication techniques can provide technical solutions for fabrication of a micro- or nanoscale reflex klystron [25, 26], which is suitable for the THz electromagnetic spectrum region. A THz reflex klystron oscillator, called nanoklystron [27], can potentially be used as a local oscillator for high-resolution heterodyne spectroscopy and as a THz wave source for bioimaging, communication, etc. The initial estimation of the operating frequency was 1.2 THz, which is significantly higher than other THz wave sources. For operation of the device, an electron beam having a current density up to a kA/cm^2 is required. Though both cold cathodes and hot cathodes are under development, a cold cathode that emits electrons by field emission is preferred. Multiwalled CNTs synthesized on an electron-beam-patterned catalyst array can generate more than 1.4 A/cm^2 at an electric field lower than 3V/μm. Enhancement of the raw current density at the cathode surface and proper combination of beam focusing can hopefully increase the beam current density to achieve the required current density.

Three-dimensional PIC simulations of 0.3 THz reflex klystrons showed that when an electron beam of 1.0 keV and a net current of 8.9 mA are used, the maximum electronic efficiency can be achieved when the gap transit angle is 0.7 π radians, and the efficiency saturates when the beam current exceeds 10 mA. Figure 8.8 is the schematic of a THz reflex klystron, and Fig. 8.9 is the phase space diagram for normalized axial momentum of electrons. In the simulations, the round-trip transit time in the repeller region is 5.75 times the RF period, which corresponds to the number of modes, five and three quarters. The transit time T_r is given in Eq. 8.3

$$T_r = \frac{2\pi(n-1/4)}{\omega} = \frac{2\pi N}{\omega} \tag{8.3}$$

where ω is the angular resonant frequency of the electromagnetic field, n is any positive integer, and N is the number of modes.

Figure 8.10 shows physical characteristics of a reflex klystron. There are several operation modes, which can be defined by Eq. 8.1. When the mode number n is 6 (or N is 5.75), the device has the maximum efficiency for energy transfer between the beam and the wave. Figure 8.10b shows that the electronic efficiency saturates even though the power of electromagnetic wave continuously increases corresponding to an increase in the beam current.

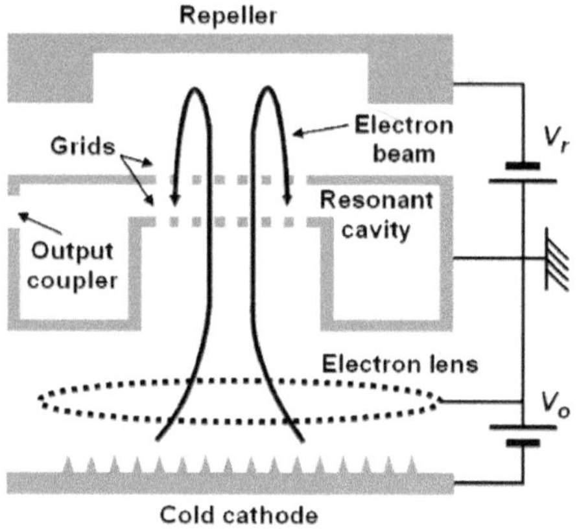

Figure 8.8 In a conventional reflex klystron, a thermionic cathode is located at the position of the cold cathode. Electrons are emitted from the cathode and accelerated and focused by an electron lens. The electron beam is modulated dominantly by the resonant frequency so that a sequence of electron bunches appears by the ballistic process of electrons in the repeller region. Finally the returned electron bunches dump energy into the cavity in the form of an electromagnetic wave. Reproduced from Ref. [24] by permission of JSAP, © 2007.

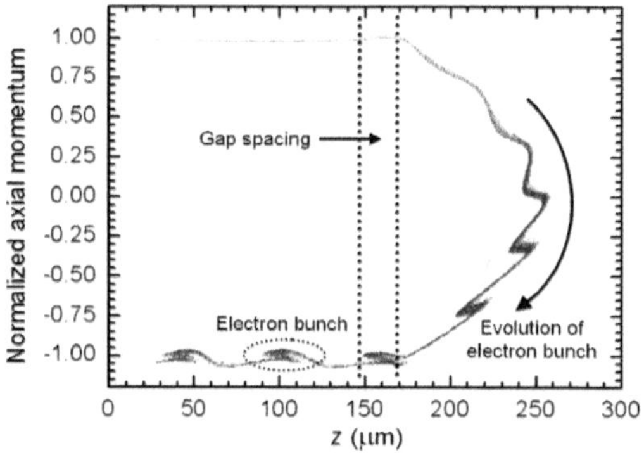

Figure 8.9 Phase space diagram for a normalized axial momentum is shown when electron bunches are fully developed. Injected electrons of an axial momentum of 1.00 undergo ballistic bunching and repelling processes. The simulation had been done by using a commercial code, MAGIC 3D. Reproduced from Ref. [24] by permission of JSAP, © 2007.

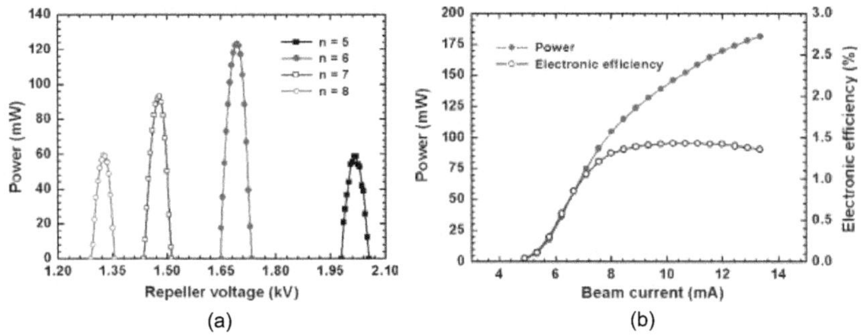

Figure 8.10 Generated power of an electromagnetic wave corresponding to the repeller voltage or the number of modes (a). Generated power of the electromagnetic wave and electronic efficiency corresponding to the beam current (b). Reproduced from Ref. [24] by permission of JSAP, © 2007.

8.3.3 Photonic Crystal Klystron

The reflex klystron is one of the most promising candidates for a microfabricated THz wave source. However, it is categorized as a low-power or low-efficiency device, and it might be more so in the THz wave region.

The main deficiency of a microfabricated THz reflex klystron comes from the size scaling down to match the operation frequency in the THz region. This is an intrinsic property of the device so that a novel concept could be developed to increase the total electromagnetic power. A promising novel concept, which can overcome the problem, had been proposed, and the feasibility was proven by demonstrating its performance in 3D PIC simulations. The concept is simply depicted in Fig. 8.11. The hybrid resonant cavity has a photonic crystal (PC) as one of the sidewalls of the cavity. If a photonic band gap is designed to cover the desired operation frequency of the reflex klystron, a TM_{mn0} standing wave mode can be strongly localized in the cavity region and the resonant mode can be coupled to an external system by inserting a line defect in the 2D PC. Figure 8.12 shows a photonic band gap made by periodically distributed dielectric rods [29]. The novel concept was experimentally demonstrated in the microwave region (around 10 GHz). The type of resonant cavity, which was chosen for the novel reflex klystron, is the hybrid one numbered (3) in Fig. 8.13. In principle, another type of PC cavity numbered (2) in Fig. 8.13 has the same properties with as (3) in Fig. 8.13.

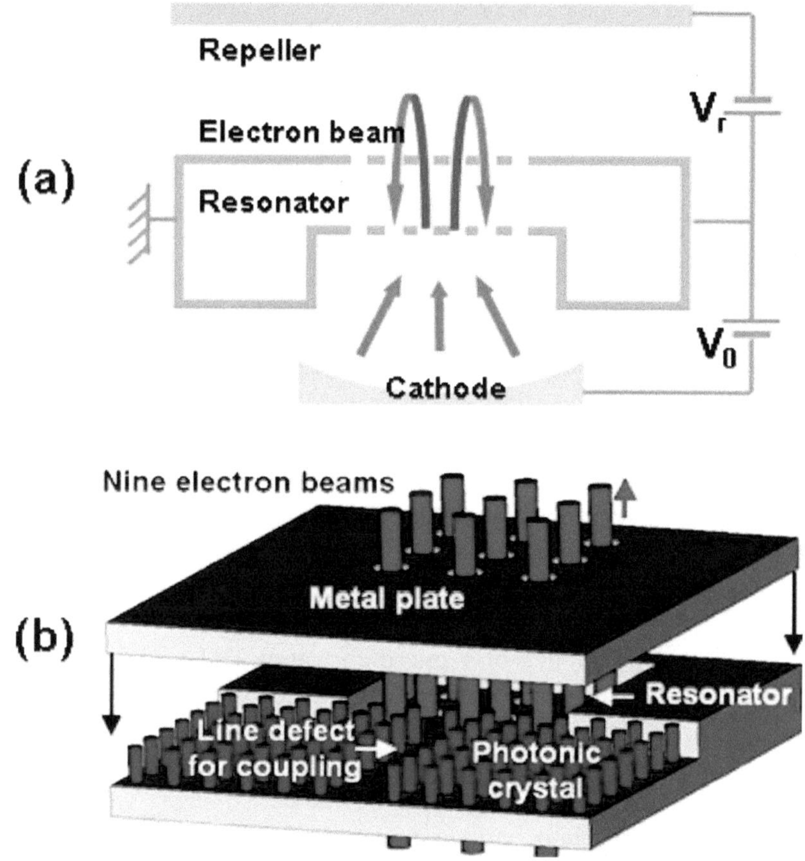

Figure 8.11 Schematic of a reflex klystron is shown in (a) Electrons are returned into the cavity by the repeller voltage, V_r. The hybrid photonic band-gap cavity, which is composed of a 2D photonic crystal and biplanar metal plates sandwiching the cavity space, can be used to increase the total power by adopting multiple electron beams and a high-order transverse magnetic (TM_{mn0}) mode in the cavity. Reproduced from Ref. [28] by permission of AIP, © 2008.

Figure 8.14 shows PIC simulation results. A conventional pillbox-type cavity and a hybrid PC cavity were compared under the same operation conditions, such as beam voltage, current, and operation mode.

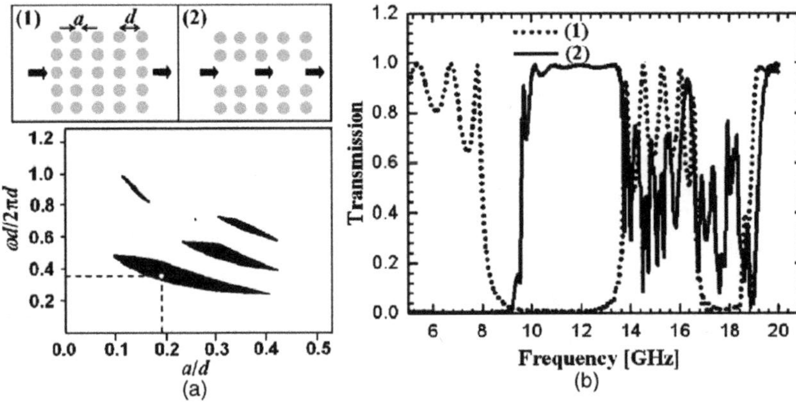

Figure 8.12 Photonic band gap from the designed PC lattice, which has lattice constants of $a = 1.9$ mm and $d = 10$ mm (a). Transmission frequency of the TM mode through the PC lattices are shown in (b) with (dotted line) and without a line defect (solid line). Reproduced from Ref. [29] by permission of AIP, © 2007.

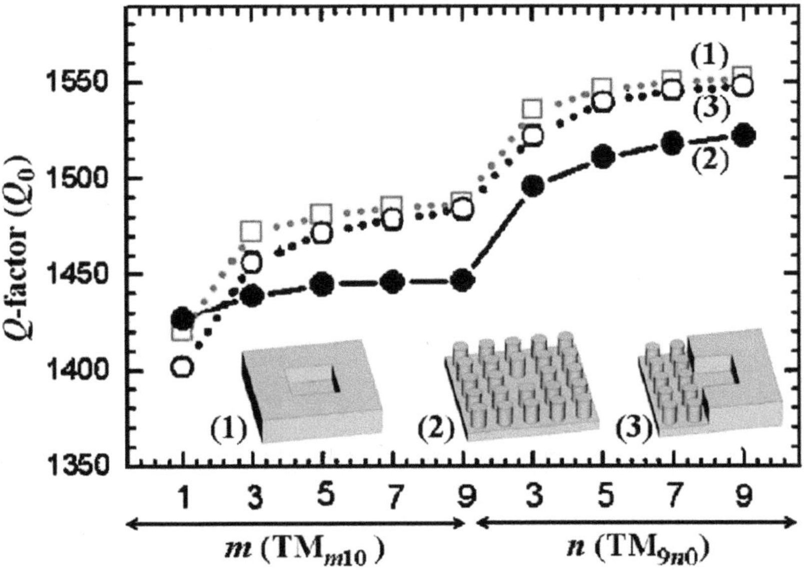

Figure 8.13 Quality factor of a metallic pillbox-type resonant cavity (1), photonic band-gap cavities in different types (2 and 3). Each resonant mode has a different quality factor, and the quality factor increases as the mode number increases. Reproduced from Ref. [29] by permission of AIP, © 2007.

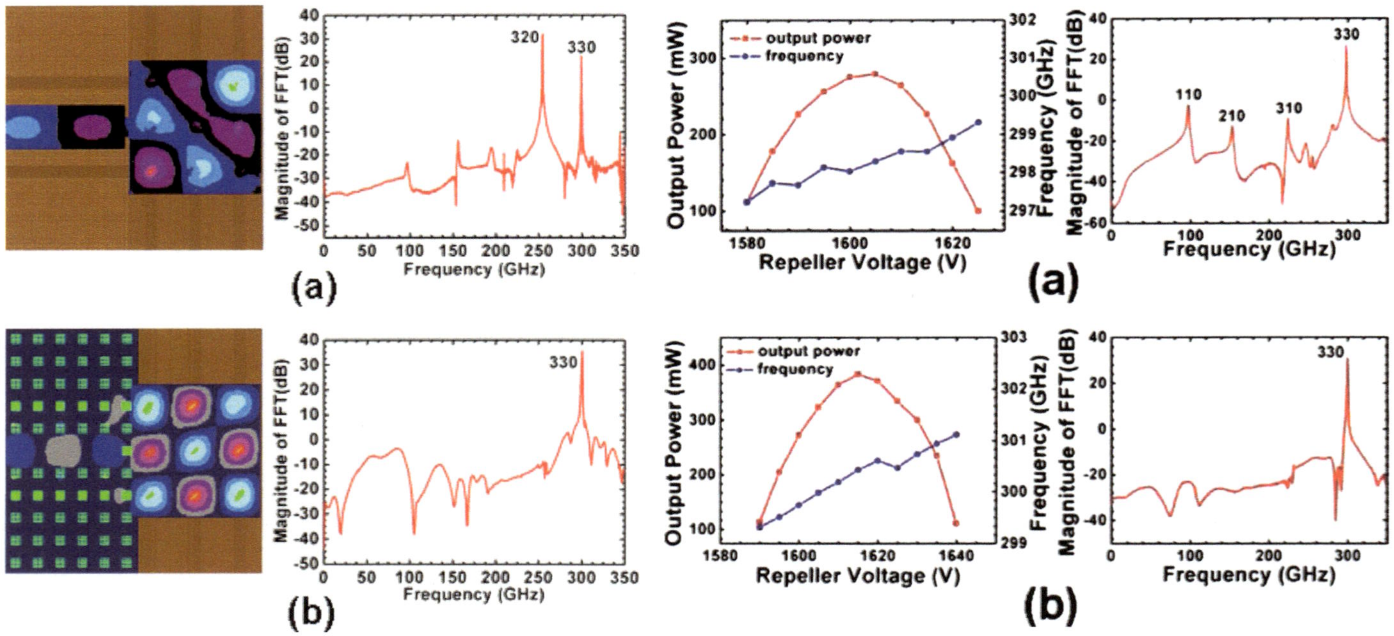

Figure 8.14 Field contours at beam locations and simulated output power and frequency vs. repeller voltage characteristics in (a) conventional pillbox-type resonant cavity applied in a reflex klystron and (b) a hybrid PC resonant cavity in a reflex klystron. Reproduced from Ref. [28] by permission of AIP, © 2008.

Because of the enhancement of the quality factor and high selectivity of the TM_{mn0} mode, a pure TM_{330} mode could be activated in a reflex klystron, adopting a hybrid PC and nine electron beams. Such selectivity of a high-order TM_{mn0} mode makes stable multibeam operation possible, which is a promising technique for increasing the microfabricated THz reflex klystron or other types of VEDs that can be operated with one of the TM_{mn0} modes and a multibeam.

To fabricate 2D silicon PCs that have deeper walls, a simple and effective microfabrication method using wet chemical etching is developed to produce various elements in the THz wave range [30]. The wet chemical etching of Si, based on liquid erosion, is characterized by anisotropic etching. Wet-etched pillars having a sloped sidewall were fabricated to form a PC (16 × 10 square lattices) with a period of 1430 μm and a height of 150 μm, as shown in Fig. 8.15. The formation of a photonic band gap at a frequency of 0.1 THz is demonstrated.

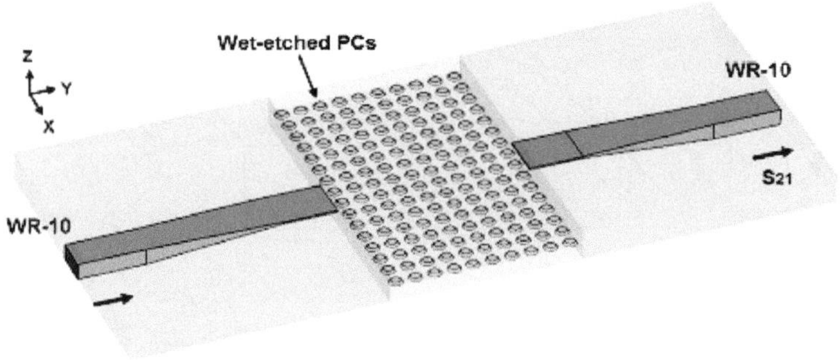

(a)

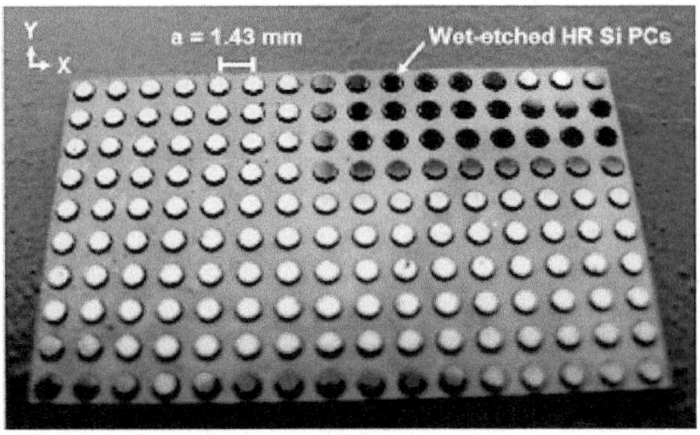

(b)

Figure 8.15 (a) 2D wet-etched PCs coupled to a conventional WR-10 waveguide and (b) picture of 0.1 THz 2D PCs fabricated by the wet etching of HR silicon.

8.4 Traveling Wave Tube

8.4.1 Folded-Waveguide TWT

A TWT consists of an electron beam and an SWS such as the helical-type circuit and the folded-type circuit. In the SWS, the axial velocity of the electromagnetic wave slows down. When the electron beam propagates inside the SWS, physical interaction takes place between the moving axial electric field and the moving electrons. A folded waveguide (FW), as shown in Fig. 8.16, has many advantages as an SWS for a traveling wave interaction due to the wide bandwidth, high power handling capability, and planar structure, which is suitable with the microfabrication method. This draws attention to develop miniaturized radiation sources up to the THz range [31–33].

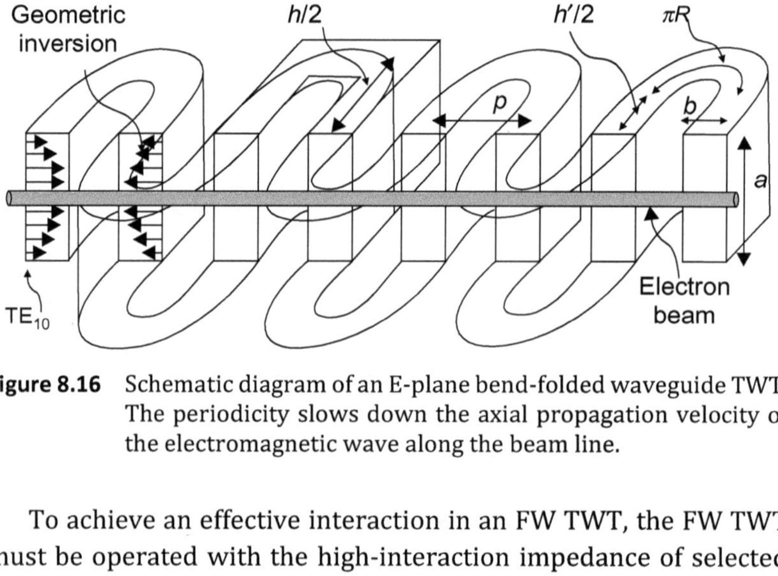

Figure 8.16 Schematic diagram of an E-plane bend-folded waveguide TWT. The periodicity slows down the axial propagation velocity of the electromagnetic wave along the beam line.

To achieve an effective interaction in an FW TWT, the FW TWT must be operated with the high-interaction impedance of selected space harmonics given in Eq. 8.4

$$R = \frac{274b}{c\pi}\frac{w_0 x^5}{\sqrt{1-x^2}}\left(\frac{1}{(2m+1)\pi}\right)^2\left[\frac{\sin(\beta b/2)}{(\beta b/2)}\right]^2 \tag{8.4}$$

where a and b are the broad wall and the narrow wall of the waveguide, respectively, and m is the space harmonics in the FW

SWS. As for the THz FW TWT, microfabrication methods such as LIGA and DRIE are readily applicable, ensuring higher accuracy and tighter tolerance. LIGA can fabricate metal structures with a large aspect ratio and a suitable depth [34].

As shown in Fig. 8.17, the two-step LIGA fabrication is well adopted for the THz FW TWT. The two-step LIGA process consists of repetition of the entire LIGA process, such as X-ray exposure, development, electroplating, and lapping. The first layer is for forming the FW TWT circuit, and the second layer is for forming a beam tunnel. To enhance the low interaction impedance of a normal FW SWS, the FW SWS formed by extending the square beam tunnel to the bends of a normal FW SWS to use the sheet electron beam is presented. An output power more than the CW 100 W in the range of 0.141–0.146 THz is expected, assuming an RF input signal with an average power of 45 mW.

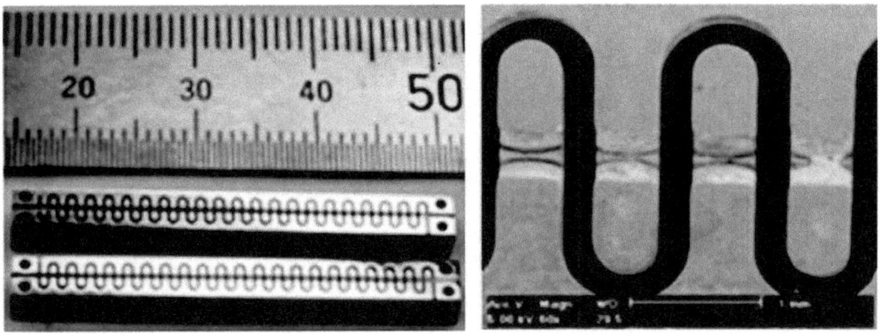

Figure 8.17 Microfabricated 0.1 THz E-plane bend-folded waveguide TWT structure.

8.4.2 Grating TWT

As for the TWT circuit, a grating-based TWT also gets great attention due to the planar circuit structure, which is suitable to match with the sheet electron beam to acquire high power at the THz range [35–37]. To fabricate a THz grating, UV LIGA, SU-8 LIGA, KMPR-LIGA, and Si-DRIE were examined to fabricate the circuit of a high aspect ratio and fine features. SU-8 is a negative-tone photoresist that has a good penetration depth at the i-line ($\lambda = 385$ nm). A KMPR negative-tone photoresist is a chemically enhanced carboxylated epoxy resin. A half-period-staggered double-vane-arrayed TWT was suggested for a ultrawideband THz TWT over a 25% bandwidth at the center frequency of 0.22 THz, and the circuit was fabricated by using the SU-8 LIGA process, as shown in Fig. 8.18. For the copper-based grating circuit at 0.22 THz, an SU-8 process including copper electroforming was developed.

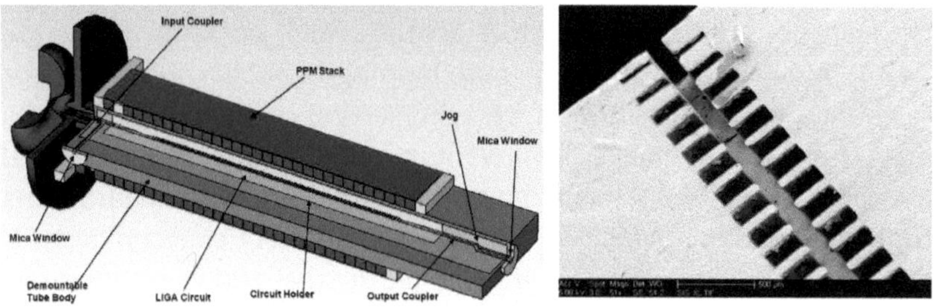

(a) (b)

Figure 8.18 (a) Schematic of a 0.22 THz TWT amplifier using a sheet electron beam and (b) SEM image of a 0.22 THz TWT structure fabricated by the SU-8 LIGA process. Reproduced from Ref. [35], courtesy of Dr. Anisullah Baig, University of California, Davis, USA.

8.5 Gyrotron

A gyrotron is a kind of vacuum electron device that emits powerful millimeter waves (~1 MW) for plasma fusion research or THz waves for many kinds of novel applications. State-of-the-art THz gyrotrons, which generally have their upper frequency limit around 1 THz, can emit moderate or high electromagnetic power from tens of watts to a few kilowatts. Such THz power can be used for dynamic nuclear polarization/nuclear magnetic resonance (DNP/NMR) spectroscopy, imaging for concealed material, and other many interesting applications [38–42]. In the THz wave region, gyrotons can have a several orders larger power level compared to other conventional types of VEDs like THz EIO or THz klystron because the dimension of the gyrotron circuit (cavity) can be much larger than the wavelength. The beam and wave interaction as well as the bunching mechanism is also distinguishable from those in other sources. Electrons in a gyrotron are slightly relativistic so that the bunching process depends on a relativistic effect called the cyclotron resonance maser instability. Figure 8.19 shows the dispersion relation between an electron beam and a wave in a gyrotron cavity. The cyclotron motion caused by the magnetic field elevates the beam line to the wave curve. In the THz frequency region, the harmonic operation denoted by the s-th of cyclotron frequency Ω is typical because such harmonic operation can decrease the requirement of the magnetic field.

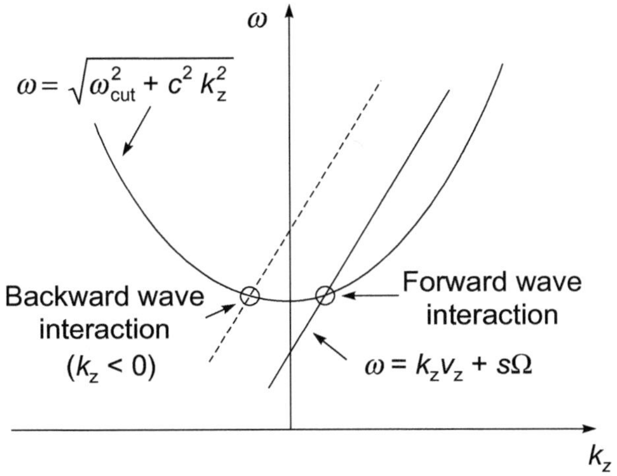

Figure 8.19 Dispersion relation of a beam and a wave is illustrated. The electron beam has both axial and perpendicular momentum in a desired ratio. The perpendicular momentum comes from the cyclotron motion of electrons caused by a strong magnetic field.

A gyrotron is composed of an electron gun, a cavity resonator, a mode convertor, an output window, and a beam collector. Generally a superconducting magnet to obtain a strong magnetic field surrounds the gyrotron and gives a gyrating motion to the electron beam. The most common type of electron gun for gyrotrons is a magnetron injection gun (MIG), which generates a small-orbit electron beam or a magnetic cusp gun, which can generate a large-orbit electron beam [43]. Electrons emitted from the cathode surface are accelerated and start their cyclotron motion. Electrons escaping gun regions move in a gradually increasing magnetic field toward the cavity resonator. The perpendicular momentum of electrons in the form of cyclotron motion increases with the increasing magnetic field until the electrons have a designed ratio transverse to the axial momentum. When electrons get into the cavity resonator, the cyclotron frequency or the *s*-th of that is slightly smaller than the resonant frequency of the electromagnetic wave in the cavity resonator. In the cavity resonator, electrons are bunched by the cyclotron resonance maser instability, and the electron bunches transfer their kinetic energy to electromagnetic energy by being decelerated by the transverse electric field in the cavity resonator. The enhanced electromagnetic field in the cavity resonator is converted to a traveling wave through a mode convertor, which is composed of some mirrors. Finally an electromagnetic wave comes out through the output window. Figure 8.20 shows a simple schematic of a gyrotron oscillator.

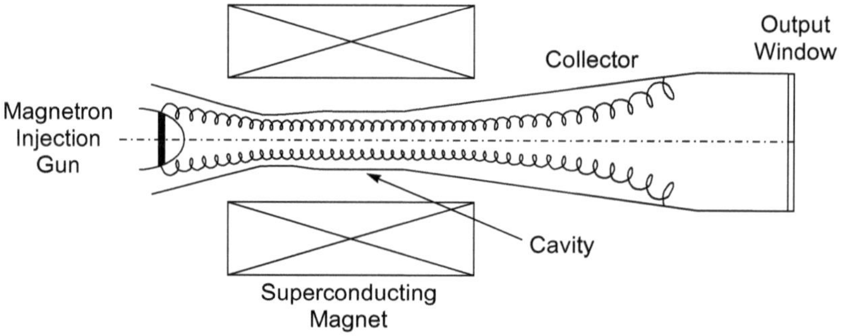

Figure 8.20 Schematic of a typical gyrotron oscillator. An MIG delivers a hollow electron beam, and electric and magnetic fields give axial and perpendicular momentum to the electron beam. The gyrating electron beam loses its transverse momentum inside the cavity resonator, which can be explained by the cyclotron resonance maser instability.

8.6 Spatial Harmonic Magnetron

The magnetron has long been known to generate high-power microwaves with relatively high efficiency and has been widely used for applications such as radars and heating systems. The resonator in the magnetron consists of multislotted resonators, which are formed along the azimuthal direction. This has a number of different modes of oscillation. These correspond to the various resonant frequencies of the slotted resonator, and the desired mode with different angular velocity will interact favorably with

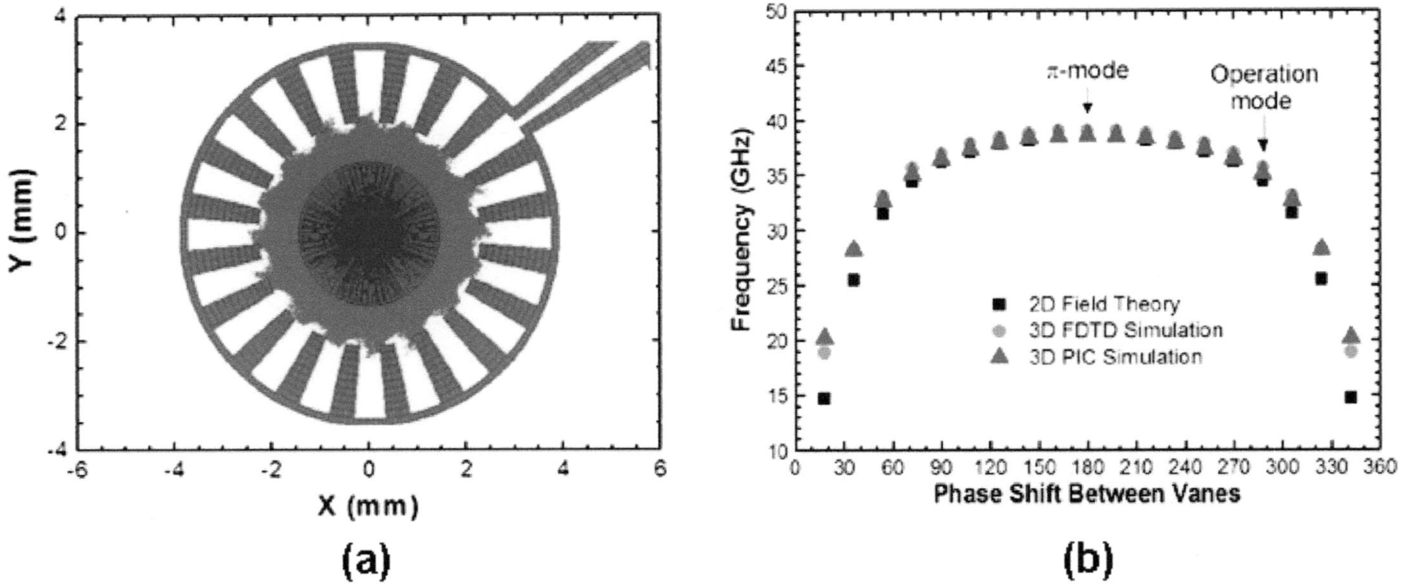

Figure 8.21 (a) Schematic of an SHM molded by the 3D PIC code and (b) the dispersion relation of a 20-vane SHM where the operation mode of $p = 16$. Reproduced from Ref. [45] by permission of IEEE, © 2012.

the electron beam to produce oscillations. For the interaction in a circularly formed slotted resonant mode, electrons are moved along the azimuthal direction by a high DC potential applied between the cathode and the anode, setting up a radial electric field and an axial magnetic field provided by either a permanent magnet or an electromagnet. The spatial harmonic magnetron (SHM) is operated at a spatial harmonic of the RF field of a non-π mode, although the conventional magnetron uses the principal harmonic of the π mode, as shown in Fig. 8.21 [44–47].

Recently, the SHM adopting a cold secondary emission cathode and operating at the spatial harmonic mode has received considerable attention because it can overcome limitations inherent in the conventional π mode operating magnetron, such as the small dimension of the resonator and the high external magnetic field required by a high-frequency operation. Figure 8.22 shows the resonator structure fabricated by wire-EDM and the assembled 35 GHz SHM adopting the secondary electron emission cathode made by platinum. The peak power is 3 kW, and the weight is 2 kg.

(a) (b)

Figure 8.22 (a) Resonator structure fabricated by wire-EDM and (b) the assembled 35 GHz SHM adopting the secondary electron emission cathode.

The resonant condition from the drift-orbital resonance condition for the SHM is expressed as

$$U_a = U_o\left[\left(\frac{B}{B_o}\right)^2 - \left(\frac{1-B/B_o}{1-2n/p}\right)^2\right] \tag{8.5}$$

where $B_o = 2m\omega/ep$, $U_o = (1 - \sigma^2)(m/e)(r_a^2\omega^2/2p^2)$, $\sigma = r_c/r_a$, B is the external magnetic field, ω is the angular oscillation frequency, p is the number of RF field variations along the space interaction perimeter, and n is the number of orbital frequency harmonics.

To overcome the short lifetime of a magnetron, a secondary emission cathode made of platinum is adopted. The maximum secondary emission yield of platinum, as shown in Fig. 8.23, is about 1.8 when the energy of a primary electron is 0.8 keV. Using a platinum cathode, a miniature 1 kW 95 GHz magnetron was experimentally demonstrated that is operated at a 6.5 kV peak anode voltage and the weight with radiators is 0.46 kg. With the help of 3D PIC simulation, the SHM, including the secondary electron emission cathode at the THz frequency range, is studied to understand nonlinear dynamics and output performances.

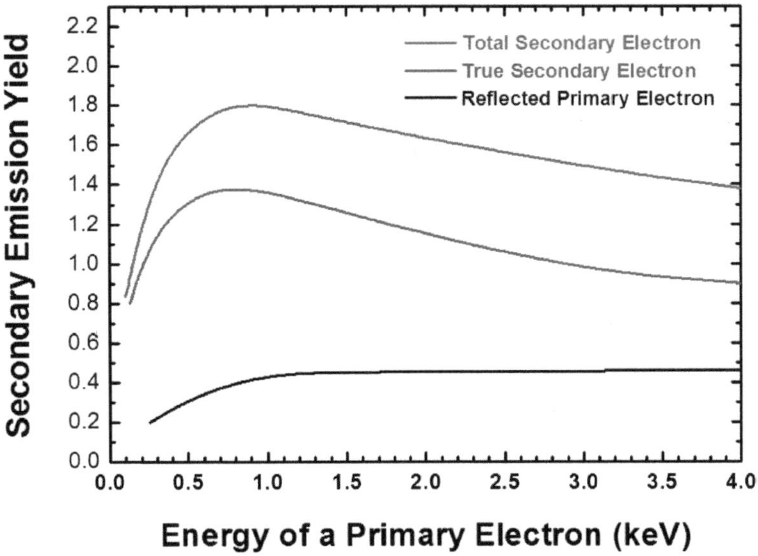

Figure 8.23 Secondary emission yield of platinum when the energy of a primary electron is varied.

The 0.14 THz SHM shows the output power of 11 kW at an anode voltage of 11.5 kV. On the basis of a permanent magnet made of a samarium-cobalt material, 0.21 THz SHMs are produced and tested. The anode diameter is 2.6 mm, and the cathode diameter is 1.55 mm when 40-side resonators are used. When an anode voltage of 15.4 kV and a external magnetic field of 0.95 T are applied, an output power of 1.17 kW is observed at an anode current of 20 A [48–50].

8.7 Free-Electron Laser

The FEL originated from the ubitron. Electron oscillators excited by a spatially periodic magnetic field produce stimulated microwave radiation. Unlike the conventional laser, an FEL uses a relativistic electron beam as a lasing medium, which moves freely through a magnetic structure [51, 52]. To generate FEL radiation, FELs consist of an electron beam emitted from an electron gun, a periodic magnetic structure, which is called "undulator" or "wiggler," and a mirror, as shown in Fig. 8.24. The optical cavity of an FEL is formed by a pair of mirrors. From the microbunched structure of an electron beam resulting from the density modulation of the electron beam, the FEL can emit coherent radiation.

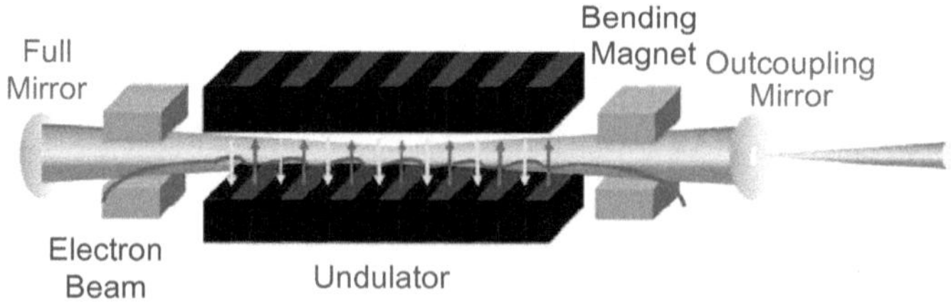

Figure 8.24 Schematic of an oscillator-type FEL. "Undulator" means the periodic magnetic structure.

An FEL demonstrates unprecedented capabilities for various applications due to coherence properties, tunability, intensity, and the operation wavelength, which ranges from the microwave and THz radiation to the visible spectrum and X-rays. Due to superior properties, an FEL at the THz range is investigated. In a quasi-CW mode, the FEL at UCBS provides THz radiation at a frequency from 0.12 THz to 4.8 THz with a peak power of 500 W to 5 kW. NovoFEL in Russia has developed a CW power of a 400 kW FEL from 1 THz to 3 THz. For a compact THz FEL, the FEL operating in the spectral range of 1–3 THz has been developed by KAERI in Korea. The system is fitted in a 2 × 3 m^2 area. For a compact FEL, a microtron is adopted as an electron accelerator [53, 54, 42].

8.8 Electron Beam Source for THz VEDs

8.8.1 Thermionic Electron Beam

Cathodes are used as electron sources in VEDs. Cathode specification is a critical parameter on the operation of VEDs at a higher frequency. Thermionic cathodes emit electrons from the cathode surface by heating the cathode metal to a high temperature. For frequencies ranging from 30 GHz to 1 THz, cathodes with a high current density up to 100 A/cm^2 are required [55–57]. A cathode made by a matrix-type dispenser cathode as a thermionic cathode is one of the candidates. The main distinction of the cathode-manufacturing technique is that compression and sintering, as well as aluminate formation and sponge impregnation, are carried out inside the cathode case. The main sponge body consists of a power mixture of tungsten and rhenium having a grain size smaller than 50 μm.

For the sheet electron beam, as shown in Fig. 8.25, the cylindrical molybdenum case is cup shaped and the tungsten and rhenium sponge is located inside the molybdenum case. In the CW operation mode, the cathode can provide an emission current density up to 50 A/cm^2. To acquire a high-current-density electron beam adopting the Scandia-added tungsten dispenser cathode, a 25:1 aspect-ratio sheet electron beam having a current density of 750 A/cm^2 was also investigated.

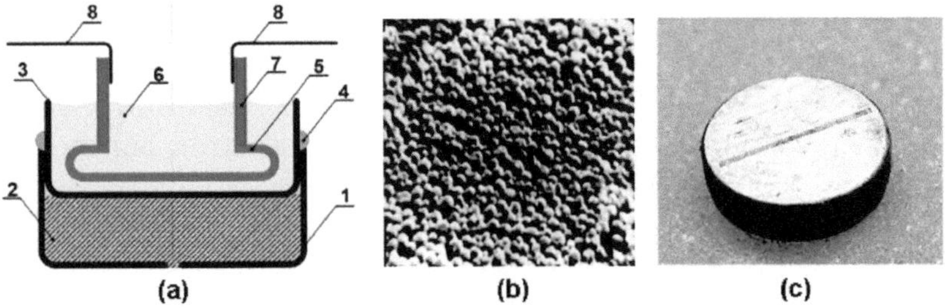

Figure 8.25 (a) Schematic of a cathode heating unit, (b) the emitting surface after compressing and annealing, and (c) a fabricated cathode heating unit for a sheet electron beam.

8.8.2 Secondary Electron Beam

Most vacuum tubes as well as THz vacuum tubes ultimately need a bunched electron beam to generate an electromagnetic wave. Each tube has its own characteristic process to make a continuous flow of electrons into a bunched flow of electrons, which categorizes tube types. One distinguishable technique to make a bunched flow of electrons is called a multipactor electron gun. The gun uses a multipactor discharge, which has been observed on various types of RF devices. When an electron beam irradiates a solid surface, electrons will either be backscattered elastically or inelastically or be absorbed into the material, emitting another electron called the secondary electron. Though the mutipactor has been regarded as a negative effect that has to be eliminated, some novel schemes have been motivated to develop an electron gun [58, 59]. The multipactor electron gun emitting a high-density secondary electron beam as a bunched series is a promising candidate for THz vacuum tubes utilizing a high-order harmonic in electron bunches, because generation of a high-density bunched beam is strictly restricted by many physical principles and technical challenges in the THz wave region. For every type of multipactor phenomena, the fixed-point equation localizing the phase space for secondary electron emission can be found. Equation 8.6 shows the fixed-point equation for a DC-biased single-surface multipactor. The fixed phase θ_o for secondary electron emission is determined by physical properties of a secondary electron, RF parameters, and a positive integer n.

$$\cos\theta_o = \frac{m_e \omega v_o}{eE_a} - \frac{\pi n E_d}{E_a} \tag{8.6}$$

Figure 8.26 shows instantaneous spatial distribution of electrons at a different time frame in a PIC simulation.

Figure 8.27 shows the density profile of electron bunches at each time. The evolution of highly localized electron bunches can be used effectively for high-frequency vacuum tubes like THz vacuum tubes because utilizing a high-order harmonic in a bunched electron beam, which has been regarded as an effective solution to make a THz wave source, needs such highly localized electron bunches.

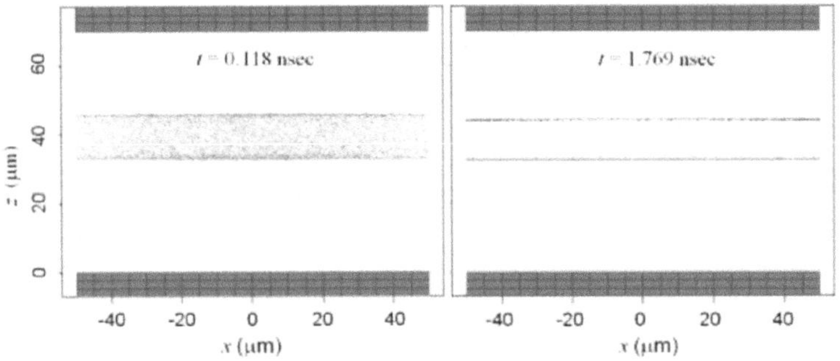

Figure 8.26 At t = 0.118 nsec, an electron cloud is formed in a bounded RF phase. At t = 1.769 nsec, the single-surface multipactor has been evolved to form two highly localized electron bunches. Reproduced from Ref. [58] by permission of AIP, © 2009.

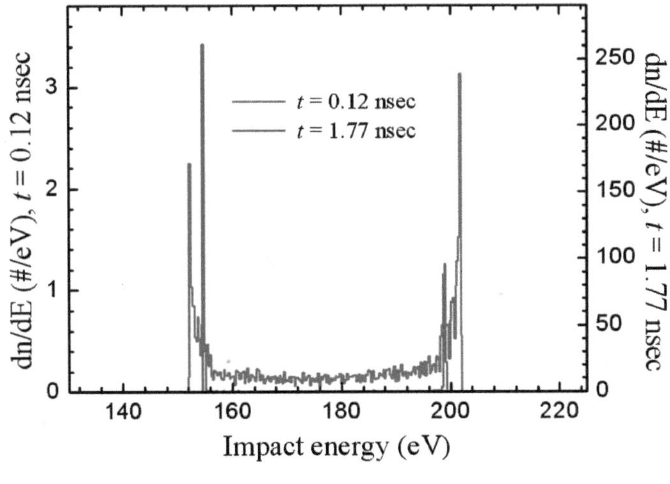

Figure 8.27 Electrons are being gathered at specific points, while secondary emissions and absorption are repeated at the multipactor surface. Reproduced from Ref. [58] by permission of AIP, © 2009.

8.8.3 CNT Electron Beam

In general, dimensional scale-down is essential for developing a THz vacuum tube. Therefore, development of a compact electron beam source, which is compatible with a microfabricated circuit of a THz vacuum tube, is a very important technical issue. One of the most promising technologies is to use cold cathodes, which emit electrons by the field emission process, whose physical principle is illustrated in Fig. 8.28. There are several kinds of cold cathode materials using the field emission mechanism, like moly-tips, nanowire tips, and CNTs [60].

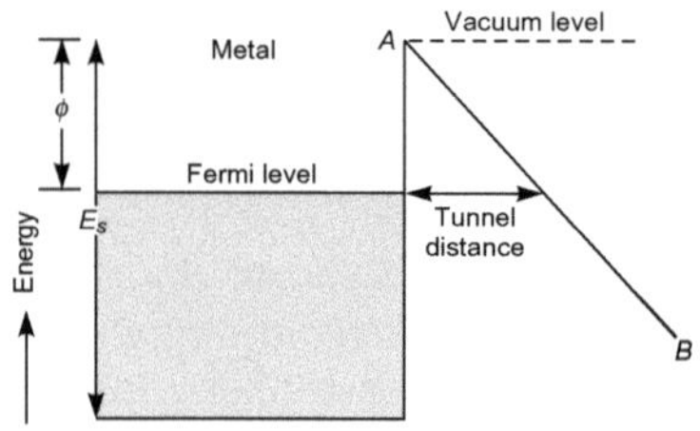

Figure 8.28 Electrons are extracted from a cathode surface to vacuum by quantum tunneling. Highly conductive materials in nanometer size, like CNTs, are generally used because of the high probability of tunneling, which can be explained by the electric field enhancement at the edge of a high-aspect-ratio material.

Though increasing the current density, uniformity, and lifetime has remained as a crucial technical problem, the CNT cold cathode gun has many outstanding merits such as reduction of the heat problem by removing the heaters in conventional vacuum tubes; high compatibility with microfabrication techniques, which have been developed in the semiconductor industry; and the instantaneous and high-frequency switching of beam emission. Figure 8.29 shows an example of a cold cathode from which nine electron beams can be extracted. Accurate microfabrication and alignment for such a cathode are readily available from manufacturing techniques for silicon wafers.

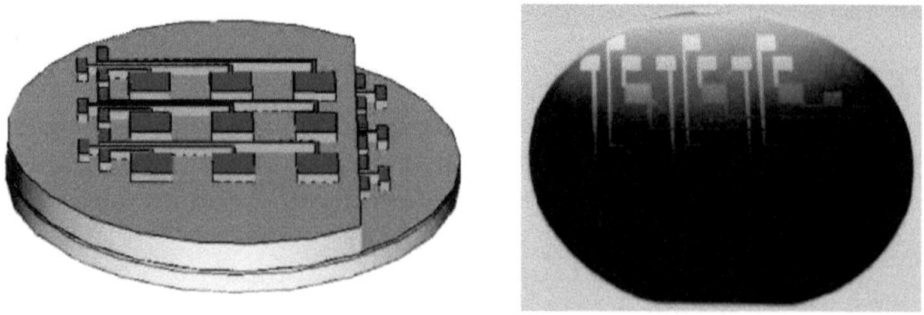

Figure 8.29 Nine patterned cold cathodes by manufacturing a silicon wafer.

8.9 Summary

THz VEDs have received much attention for their efficient interaction mechanism based on bunched electrons in a vacuum. To develop compact, low-cost, high-power THz VEDs for various practical applications, crucial issues such as the microfabrication of THz circuits, the high-current-density electron beam, the focusing structure, breakdown, and thermal problems should be unveiled more clearly. With the help of novel materials for cathodes, realistic computational calculations, and the progress of the microfabrication method, in the near future, many kinds of powerful and practical THz VEDs will be developed and applied for various and innovative applications.

References

1. J. H. Booske, Plasma physics and related challenges of millimeter-wave-to-terahertz and high power microwave generation, *Phys. Plasma* **15**, 055502 (2008).
2. J. H. Booske, R. J. Dobbs, C. D. Joye, C. L. Kory, G. R. Neil, G. S. Park, J. Park, and R. Temkin, Vacuum electronic high power terahertz sources, *IEEE Trans. THz Sci. Technol.* **1**, 54–75 (2011).
3. G. J. Kim, J. I. Kim, S. G. Jeon, J. Kim, K. K. Park, and C. H. Oh, Enhanced continuous-wave terahertz imaging with a horn antenna for food inspection, *J. Infrared mm THz. Waves* **31**, 657–664 (2012).
4. B. Levush, D. K. Abe, J. P. Calame, B. G. Danly, K. T. Nguen, E. J. Dutkowsky, R. H. Abrams, and R. K. Parker, Vacuum electronics: status and trends, *IEEE Aerosp. Electron. Syst. Mag.* **22**, 28–34 (2007).
5. R. K. Parker, R. H. Abrams, B. G. Danly, and B. Levush, Vacuum electronics, *IEEE Trans. Microwave Theory Tech.* **50**, 835–845 (2002).
6. V. Srivastava, THz vacuum microelectronic devices, *J. Phys.: Conf. Ser.* **114**, 1–10 (2008).
7. Y. M. Shin, E. M. Choi, and G. S. Park, Principle of terahertz radiation using electron beams, in *Convergence of Terahertz Sciences in Biomedical Systems, Part 1* (Springer Science, 2012), 3–49.
8. K. Nguyen, J. A. Pasour, T. M. Antonsen, P. B. Larsen, J. J. Petillo, and B. Levush, Intense sheet electron beam transport in a uniform solenoidal magnetic field, *IEEE Trans. Electron. Devices* **56**, 744–752 (2009).

9. A. G. Rozhnev, N. M. Ryskin, D. V. Sokolov, D. I. Trutskov, S. T. Han, J. I. Kim, and G. S. Park, Novel concepts of vacuum microelectronic microwave devices with field emitter cathode arrays, *Phys. Plasma* **9**, 4020–4027 (2002).

10. A. Srivastava, J. K. So, Y. Wang, J. Wang, R. S. Raju, S. T. Han, and G. S. Park, Design of sheet-beam electron gun with planar cathode for terahertz devices, *J. Infrared mm THz. Waves* **30**, 670–678 (2009).

11. J. I. Kim, S. G. Jeon, G. J. Kim, and J. Kim, Numerical and experimental investigation of 0.1 THz clinotron, *IEEE Proc. IRMMW-THz.* (2012).

12. K. Schunemann and D. M. Vavriv, Theory of the clinotron: A grating backward-wave oscillator with inclined electron beam, *IEEE Trans. Electron. Devices* **46**, 2245–2252 (1999).

13. M. Sattorov, E. Khutoryan, K. Lukin, O. Kwon, and G. S. Park, Improved efficiency of backward-wave oscillator with an inclined electron beam, *IEEE Trans. Electron. Devices* **60**, 458–463 (2012).

14. S. Ponomarenko, S. Kishko, S. Khutroryan, E. Tishenko, A. Kuleshov, and B. P. Yefimov, Development of CW clinotron oscillator at 400 GHz, *IEEE Proc. Math. Methods Electromagn. Theory*, 348–352 (2012).

15. V. L. Bratman, A. E. Fedotov, and P. B. Makhalov, Experimental demonstration of smith-purcell radiation enhancement by frequency multiplication in open cavity, *Appl. Phys. Lett.* **98**, 061503 (2011).

16. K. J. Kim and V. Kumar, Electron beam requirements for a three-dimensional smtih-purcell backward-wave oscillator for intense terahertz radiation, *Phys. Rev. ST Accel. Beams* **10**, 080702 (2007).

17. Y. M. Shin, J. K. So, K. H. Jang, J. H. Won, A. Srivastava, and G. S. Park, Superradiant terahertz Smith-Purcell radiation from surface plasmon excited by counterstreaming electron beams, *Appl. Phys. Lett.* **90**, 031502 (2007).

18. J. I. Kim, S. G. Jeon, G. J. Kim, and J. Kim, Enhancement of terahertz radiation in a smith-purcell backward-wave oscillator by an inverse wet-etched grating, *Phys. Lett. A* **375**, 589–592 (2011).

19. Y. M. Shin, G. S. Park, G. P. Scheitrum, and G. Caryotakis, Circuit analysis of an extended interaction klystron, *J. Korean Phys. Soc.* **44**, 1239–1245 (2004).

20. G. Scheitrum, Design and construction of a W-band sheet beam klystron, *7th Workshop on High Energy Density and High Power RF*, Kalamata, Greece (2005).

21. G. Scheitrum, G. Caryotakis, A. Burke, A. Jensen, E. Jongewaard, M. Neubauer, R. Phillips, and R. Steele, W-Band sheet beam klystron research at SLAC, *IEEE Int. Vac. Electron. Conf.*, 481–482 (2006).

22. D. Chernin, R. C. Dobbs, M. Hyttinen, A. Roitman, D. Berry, M. Blank, K. Nguyen, V. Jabotinsky, D. Pershing, E. Wright, J. Calame, B. Levush, J. Neilson, F. Maiwald, T. Gaiser, N. S. Barker, R. Weikle, and J. Booske, Development of a 670 GHz Extended Interaction Klystron Power Amplifier, *GOMACTech-11 Conf.,* 387 (2011).
23. M. Hyttinen, A. Roitman, P. Horoyski, R. Dobbs, E. Sokol, and D. Berry, A compact, high power, submillimeter-wave extended interaction klystron, *IEEE Proc. Int. Vac. Electron. Conf.*, 297 (2008).
24. S. G. Jeon, Y. S. Jin, J. I. Kim, and C. H. Shon, Three-dimensional particle-in-cell simulations of 300 GHz reflex klystrons, *J. Appl. Phys.* **101**, 052419-1–052419-5 (2007).
25. M. J. Smith and G. Philips, *Power Klystrons Today* (Wiley, Taunton-Somerset, 1995).
26. R. H. Varian and S. F. Varian, A high frequency oscillator and amplifier, *J. Appl. Phys.* **10**, 321–327 (1939).
27. H. M. Manohara, P. H. Siegel, M. J. Bronikowski, B. K. Vancil, and K. Hawken, Development of a micromachined THz nanoklystron: A Status Report, *IEEE Int. Conf. Plasma Sci.*, 157 (2004).
28. K. H. Jang, S. G. Jeon, J. I. Kim, J. H. Won, J. K. So, S. H. Bak, A. Srivastava, S. S. Jung, and G. S. Park, High order mode oscillation in a terahertz photonic-band-gap multibeam reflex klystron, *Appl. Phys. Lett.* **93,** 211104-1–211104-3 (2008).
29. S. G. Jeon, Y. M. Shin, K. H. Jang, S. T. Han, J. K. So, Y. D. Joo, and G. S. Park, High order mode formation of externally coupled hybrid photonic-band-gap cavity, *Appl. Phys. Lett.* **90,** 021112-1–021112-3 (2007).
30. J. I. Kim, S. G. Jeon, G. J. Kim, J. Kim, H. H. Lee, and S. H. Park, Two-dimensional terahertz photonic crystals fabricated by wet chemical etching of silicon, *J. Infrared mm THz. Waves* **33**, 206–211 (2012).
31. S. T. Han, J. I. Kim, and G. S. Park, Design of a folded waveguide traveling-wave tube, *Microwave Opt. Technol. Lett.* **38**, 161–165 (2003).
32. J. Tucek, K. Kreischer, D. Gallagher, R. Vogel, and R. Mihailovich, Development and operation of a 650 GHz folded waveguide source, *IEEE Int. Vac. Electron. Conf.* (2007).
33. M. H. Zhang, Y. Y. Wei, G. Guo, L. N. Yue, Y. Hou, S. M. Wang, J. Xu, Y. B. Gong, and W. X. Wang, A novel 140 GHz sheet-beam folded-waveguide traveling-wave tube, *J. Electromagn. Waves Appl.* **26**, 2332–2340 (2012).
34. Y. M. Shin, J. K. So, K. H. Jang, J. H. Won, A. Srivastava, S. T. Han, G. S. Park, J. H. Kim, S. S. Chang, R. K. Sharma, and S. N. Joshi, Experimental investigation of 95 GHz folded waveguide backward wave oscillator

fabricated by two-step LIGA, *IEEE Int. Vac. Electron. Conf.*, 419–420 (2006).

35. A. Baig, Y. M. Shin, L. R. Barnett, D. Gamzina, R. Barchfeld, C. W. Domier, J. Wang, and N. C. Luhmann, Jr., Design, fabrication and RF testing of near-THz sheet beam TWTA, *THz Sci. Technol.* **4**, 181–207 (2011).
36. C. D. Joye, J. P. Calame, M. Garven, and B. Levush, UV-LIGA microfabriation of 220 GHz sheet beam amplifier gratings with SU-8 photoresist, *J. Micromech. Microeng.* **20**, 125016 (2010).
37. Y. M. Shin, L. R. Barnet, N. C. Lhumann, Jr., Strongly confined plasmonic wave propagation through an ultra-wideband staggered double grating waveguide, *Appl. Phys. Lett.* **93**, 221504 (2008).
38. V. S. Bajaj, M. K. Hornstein, K. E. Kreischer, J. R. Sirigiri, P. P. Woskov, M. L. Mak-Jurkauskas, J. Herzfeld, R. J. Temkin, R. G. Griffin, 250 GHz CW gyrotron oscillator for dynamic nuclear polarization in biological solid state NMR, *J. Magn. Reson.* **189**, 251–279 (2007).
39. M. Y. Glyavin, A. G. Luchinin, and G. Y. Golubiatnikov, Generation of 1.5 kW, 1 THz coherent radiation from a gyrotron with a pulsed magnetic field, *Phys. Rev. Lett.* **100**, 015101 (2008).
40. M. K. Hornstein, V. S. Bajaj, R. G. Griffin, and R. J. Temkin, Continuous-wave operation of a 460 GHz second harmonic gyrotron oscillator, *IEEE Trans. Plasma Sci.* **34**, 524–533 (2006).
41. T. Idehara, K. Kosuga, L. Agusu, R. Ikeda, I. Ogawa, T. Saito, Y. Matsuki, K. Ueda, and T. Fujiwara, Continuously frequency tunable high power sub-THz radiation source: gyrotron FU CW VII for 300 MHz and 600 MHz DNP-NMR spectroscopy, *J. Infrared mm THz. Waves* **31**, 775–790 (2010).
42. S. H. Park, K. H. Jang, K. Lee, J. Mun, G. Kazakevich, S. Miginsky, and Y. U. Jeong, Microtron-based beam transfer system for a compact, high power THz FEL, *33rd Int. FEL Conf.*, Shanghai, China (2011).
43. S. G. Jeon, C. W. Baik, D. H. Kim, G. S. Park, N. Sato, and K. Yokoo, Study on velocity spread for axis-encircling electron beams generated by single magnetic cusp, *Appl. Phys. Lett.* **80**, 3703–3705 (2002).
44. J. I. Kim, S. G. Jeon, Y. S. Jin, G. J. Kim, D. H. Kim, and S. S. Jung, Investigation of the 100 GHz reentrant linear magnetron using particle-in-cell simulation, *J. Vac. Sci. Technol.* **27**, 687–691 (2009).
45. J. I. Kim, S. G. Jeon, G. J. Kim, J. Kim, V. D. Yeryomka, O. P. Kulagin, A. S. Tishchenko, and V. Naumenko, Investigation of millimeter-wavelength 20-vane spatial-harmonic magnetron using three-dimensional particle-in-cell simulation, *IEEE Trans. Plasma Sci.* **40**, 1966–1970 (2012).

46. V. D. Naumenko, K. Schunemann, and D. M. Vavriv, Miniature 1 kW 95 GHz magnetrons, *Electron. Lett.* **35**, 1960–1961 (1999).

47. K. Schunemann, S. V. Sosnytskiy, and D. M. Vavriv, Self-consistent simulation of the spatial-harmonic magnetron with cold secondary-emission cathode, *IEEE Trans. Electron. Devices* **48**, 993–998 (2001).

48. N. I. Avtomonov, V. D. Naumenko, D. M. Vavriv, K. Schunemann, A. N. Suvorov, and V. A. Markov, Toward terahertz magnetron: 210 GHz spatial-harmonic magnetron with cold cathode, *IEEE Trans. Electron. Devices* **59**, 3608–3611 (2012).

49. N. N. Esfahani and K. Schunemann, Particle-in-cell simulation of a spatial-harmonic magnetron with a cold secondary emission cathode, *IEEE Trans. Plasma Sci.* **40**, 3512–3519 (2012).

50. N. N. Esfahani, M. Tayarani, and K. Schunemann, Design and 3D particle-in-cell simulation of a 140 GHz spatial-harmonic magnetron, *Prog. Electromagn. Res.* **133**, 443–458 (2013).

51. D. A. G. Deacon, L. R. Elias, J. M. J. Madey, G. J. Ramian, H. A. Schwettman, and T. I. Smith, First operation of a free-electron laser, *Phys. Rev. Lett.* **38**, 892–894 (1977).

52. L. R. Elias, W. M. F. Airbnak, J. M. J. Madey, H. A. Schwettman, and T. I. Smith, Observation of stimulated emission of radiation by relativistic electrons in a spatially periodic transverse magnetic field, *Phys. Rev. Lett.* **36**, 717–720 (1976).

53. Y. U. Jeong, S. H. Park, and K. Lee, Principle of terahertz free electron laser, in *Convergence of Terahertz Sciences in Biomedical Systems, Part 1* (Springer Science, 2012), 51–72.

54. Y. U. Jeong, B. H. Cha, H. K. Cha, Y. Cha, J. S. Jang, K. H. Jang, H. N. Kim, K. N. Kim, B. C. Lee, K. Lee, J. Mun, S. H. Park, J. H. Sunwoo, G. M. Kazakevich, and D. H. Kim, Design status of a KAERI table-top THz free-electron laser, *33rd Int. FEL Conf.*, Shanghai, China (2011).

55. W. Liu, Y. Wang, J. Wang, Y. C. Wang, and B. Vancil, Emission characteristics of nanosized Scandia-doped dispenser cathode, *IEEE Trans. Electron. Devices* **58**, 1241–1246 (2011).

56. Y. Wang, J. Wang, W. Liu, K. Zhang, and J. Li, Development of high current density cathodes with Scandia-doped tungsten powders, *IEEE Trans. Electron. Devices* **54**, 1061–1070 (2007).

57. J. Zhao, N. Li, J. Li, D. Gamzina, A. Baig, R. Barchfeld, L. Barnett, S. Risbud, and N. C. Luhmann, Scandia-added tungsten dispenser cathode fabrication for THz vacuum integrated power amplifier, *THz Sci. Technol.* **4**, 240–252 (2011).

58. S. G. Jeon, J. I. Kim, S. T. Han, S. S. Jung, and J. U. Kim, Theoretical study of dc-biased single-surface multipactors, *Phys. Plasmas* **16**, 073101-1–073101-7 (2009).

59. J. A. Nation, L. Schachter, F. M. Mako, L. K. Len, W. Peter, C-.M. Tang, and T. Srinivasan-Rao, Advances in cold cathode physics and technology, *IEEE Proc.* **87**, 865–889 (1999).

60. K. K. Teo, E. Minoux, L. Hudanski, F. Peauger, J. P. Schnell, L. Gangloff, P. Legagneux, D. Dieumegard, G. A. J. Amaratunga, and W. I. Milne, Carbon nanotubes as cold cathodes, *Nature* **437**, 968 (2005).

Chapter 9

Plasma-Wave Devices for Terahertz Applications

Taiichi Otsuji
Research Institute of Electrical Communication, Tohoku University, 2-1-1 Katahira Aoba-ku, Sendai, Miyagi 980-8577, Japan
otsuji@riec.tohoku.ac.jp

In this chapter recent advances in plasma-wave devices for terahertz applications are described. A plasma wave is defined as a collective charge density wave by the electron dynamics in semiconductors. First, a 2D plasmon is introduced as the quantum of the 2D plasma waves. Nonlinear hydrodynamic properties of 2D plasmons in compound semiconductor heterostructure material systems are presented to demonstrate intense broadband emission and ultrahigh sensitive detection of terahertz radiation. The device structure is based on a high-electron-mobility transistor and incorporates interdigitated dual-grating gates. The second topic focuses on graphene, a monolayer carbon-atomic honeycomb lattice crystal exhibiting peculiar carrier transport and optical properties owing to a massless and gapless energy spectrum. A theoretical

Handbook of Terahertz Technologies: Devices and Applications
Edited by Ho-Jin Song and Tadao Nagatsuma

ISBN 978-981-4613-08-8 (Hardcover), 978-981-4613-09-5 (eBook)
www.panstanford.com

and experimental study on active graphene plasmonics toward the creation of graphene terahertz injection lasers is described.

9.1 Introduction

"Terahertz" is still an unexplored frequency band; there is no commercially available microelectronic device that can generate, detect, or manipulate electromagnetic waves over the entire terahertz (THz) frequency band [1]. The quest for the creation of compact, efficient, and tunable THz sources as well as ultrahigh-sensitive and fast-response THz detectors is currently one of the hottest topics in this research field. Fundamental limits reached by the available sources of electromagnetic radiation based on the classical electronic oscillations radiating at millimeter-wave frequencies, and on electron transitions between quantized energy levels corresponding to infrared (IR) frequencies, give rise to the so-called THz gap [1, 2].

2D plasmons in submicron transistors have attracted considerable attention due to their nature of promoting emission and detection of electromagnetic radiation in the THz regime. The channel of a transistor can act as a resonator for plasma waves, the charge density waves of collectively excited 2D electrons. The plasma frequency depends on the resonator dimensions and the density of 2D electrons; it can reach the sub-THz or even THz range for gate lengths of a micrometer and submicrometer size. Therefore different micro- and nanodevices/structures supporting low-dimensional plasmons were intensively studied as possible candidates for solid-state THz sources [1–18] and detectors [19–30]. Mechanisms of plasma-wave excitation/emission can be divided into two types, (i) incoherent and (ii) coherent. The first is related to thermal excitation of broadband nonresonant plasmons by hot electrons [2, 8, 9, 13, 17, 18]. The second is related to the plasma-wave instability mechanisms like the Dyakonov–Shur (DS) Doppler-shift model [7] and/or the Ryzhii–Satou–Shur (RSS) transit-time model [14, 16], where coherent plasmons can be excited either by hot electrons or by optical phonon (OP) emission under near-ballistic electron motion [31]. On the other hand, hydrodynamic nonlinearities of 2D plasmons in high-electron-mobility transistors (HEMTs) are promising fast and sensitive rectification/detection of THz radiation [19]. When an incoming THz radiation excites plasma waves, the local carrier density as well as the local carrier velocity is modulated by the radiation frequency. This results in generation of a quadratic current component in proportion to the product of the modulated components of the local carrier densities and velocities. The time average of this component is nonzero, leading to rectification. With an asymmetric boundary condition, a rectified component gives rise to a photovoltaic effect. Recently, plasma-wave properties have been successfully used for resonant and nonresonant (broadband) sub-THz and THz detection [20–30]. They can be applied to real-time THz imaging/spectroscopic analysis as well as future THz wireless communications [1].

First a 2D plasmon-resonant microchip emitter featured with an interdigitated dual-grating gate (DGG) structure was proposed [32–35]. The original structure uses a symmetrical dual-grating gate (S-DGG) in which interfinger spaces are all identical, providing room-temperature 0.5–6.5 THz emission with 1 μW order radiation power [34, 35] and rather low detection responsivity of the order of tens of volts/watt [32]. The major causes of broadband emission are considered to be multimode of coherent/incoherent plasmons [36], oblique modes [37], gated and ungated plasmon modes [38], hot plasmons [36], and chirped plasmon modes [34]. Recently an asymmetric dual-grating gate (A-DGG) structure has been proposed and has demonstrated the world's first coherent monochromatic THz emission and ultrahigh-sensitive THz detection with a responsivity of 2.2 kV/W and extremely low noise performance with a noise-equivalent power (NEP) of 15 pW/√Hz at 1 THz [39–43]. These values are lower than those of any commercial room-temperature THz detectors such as Golay cells (200 $pW/Hz^{0.5}$) [44] or Schottky barrier diodes (100 $pW/Hz^{0.5}$) [36]. Very recently a physical model has been developed that explains this photoresponse dependence on gate bias [45].

On the other hand, graphene, a one-atom-thick planar sheet of sp^2-hybridized honeycomb carbon crystals, has attracted considerable attention because of its unique carrier transport and optical properties [46–55]. The groundbreaking discovery of graphene activated the research and development of graphene-based devices in wide aspects among electronic, optoelectronic, and THz photonic devices. Graphene channel transistors increase their cutoff frequencies approaching the THz range [56–59], whereas graphene photodetectors demonstrated high-speed operation [60] in the mid-IR range, which is also expected to operate in the THz range [61]. The conduction band and valence band of graphene have a symmetrical conical shape around the Brillouin zone edges, which are called K and K' points, and contact each other at "Dirac points" at the K and K' points. Electrons and holes in graphene have a linear dispersion relation with a zero band gap, resulting in peculiar features such as massless relativistic fermions with back-scattering-free ultrafast transport [46–49, 50–52, 54, 55, 62–65] as well as negative dynamic conductivity in the THz frequencies under optical or electrical pumping [66–72]. Graphene 2D plasmons hold unique optoelectronic properties and produce extraordinary light–matter interactions. So far graphene 2D plasmons have been intensively studied theoretically [73–95], and very recently electromagnetic responses of graphene surface plasmons have been experimentally observed [96–100]. When graphene is patterned into micrometer to submicrometer ordered structures the plasmon modes fall in the THz range so that it can provide intense THz emission if the cavity boundary conditions allow plasmon instability or the THz dynamic conductivity in the plasmon cavity takes negative values via optical or electrical pumping [76, 86–90].

In this chapter the fundamental basis and recent advances in plasma-wave devices for THz applications are described.

9.2 THz Emission Using 2D Plasmons

9.2.1 Theory

The 2D plasma-wave kinetics can be formulated by the hydrodynamic Euler equation and the continuity equation [4]:

$$m\left(\frac{\partial \mathbf{u}}{\partial t}+\mathbf{u}\frac{\partial \mathbf{u}}{\partial \mathbf{r}}\right)=-e\frac{\partial V}{\partial \mathbf{r}}-\frac{\mathbf{u}}{m\tau}, \tag{9.1}$$

$$\frac{\partial n}{\partial t}+\frac{\partial}{\partial \mathbf{r}}(n\mathbf{u})=0, \tag{9.2}$$

where m is the electron effective mass, $\mathbf{u}(\mathbf{r}, t)$ is the in-plane electron spatiotemporal local velocity, $\mathbf{r}$ is an arbitrary in-plane vector, $V(\mathbf{r}, t)$ is the local potential at $\mathbf{r}$, τ is the electron momentum relaxation time, and $n(\mathbf{r}, t)$ is the spatiotemporal local density of electrons. The first term of the right-hand side in Eq. 9.1 is the Coulomb force, and the second term is the Drude friction. 2D electron channels in HEMTs consist of gated and ungated regions, as shown in Fig. 9.1. The ungated 2D plasmon receives the in-plane longitudinal Coulomb force so that it holds a square-root dispersion relation, which is identical to that for general surface plasmons. The gated 2D plasmon receives the transverse Coulomb force via the gate capacitor, which is far stronger than the in-plane force due to the geometrical situation so that it holds a linear dispersion. In a simple case of gradual channel approximation with infinite channel width (perpendicular to the source–drain direction), the 2D plasma-wave dynamics deduced to 1D systems [4]. The plasma-wave phase velocity s is given by $s = \sqrt{eV_0/m}$, where V_0 is the gate swing voltage [19]. Assuming $V_0 = O[1\ \text{V}]$ and $m = O[0.1\ m_0]$ (m_0 is the electron rest mass in vacuum) for InP-based heterostructure HEMTs, s becomes $O[1 \times 10^6\ \text{m/s}]$, which is at least 2 orders of magnitude higher than the electron drift velocity of any compound semiconductors with superior transport properties. Thus, when we consider a submicrometer gate-length HEMT, the fundamental mode of gated 2D plasmons stays at a frequency in the THz range. This is the main advantage for use in plasmon-resonant modes that can operate in frequencies far beyond the transit frequency limit of transistors [101].

When a single-gate HEMT is situated in source-terminated and drain-opened configuration with the DC potential at the drain terminal with respect to the source terminal, the drain end of the channel becomes depleted so that the drain-side impedance is mainly given by the depletion capacitance and takes a high value at high (THz) frequencies. In such a case, the Doppler-shift effect occurs on the plasma-wave propagation/reflection at the drain boundary, promoting DS instability [4]. Consider the case in which the plasma wave is excited in a HEMT with a constant DC drain bias, causing a background constant DC electron drift flow with velocity v_d, and that the gate length L is shorter than the coherent length of electrons. The plasma-wave-originated local displacement current δj_p is given by the product of the perturbation of the local electron charge density $e\delta n$ and the plasma-wave velocity. The forward (backward) component $\overrightarrow{\delta j_p}$ ($\overleftarrow{\delta j_p}$) traveling to (from) the drain boundary is given by $\overrightarrow{\delta j_p} = e\overrightarrow{\delta n}\cdot(s+v_d)$, $\overleftarrow{\delta j_p} = e\overleftarrow{\delta n}\cdot(s-v_d)$. Since the open-drain boundary conserves the current before and after reflection, $\left|\overrightarrow{\delta j_p}\right| = \left|\overleftarrow{\delta j_p}\right|$, $\overleftarrow{\delta n} = \overrightarrow{\delta n}\cdot(s+v_d)/(s-v_d) > \overrightarrow{\delta n}$. This increment of the electron charge density $\delta n = (\overleftarrow{\delta n} - \overrightarrow{\delta n})$ directly reflects the increment of the gate potential δV_σ via the gate capacitor C: $\delta V_\sigma = e\delta n/C$. Since the source-terminated boundary gives a lossless reflection (reflection coefficient is −1) the gate potential becomes infinite after infinitesimal repetition of plasma-wave reflections, leading to DS instability. When the plasma wave is excited by the incoming THz radiation with angular frequency ω, $e^{-i\omega t}$, the effect of the instability is derived into the imaginary part of ω, ω'', as

$$\omega'' = \frac{s^2 - v_d^2}{Ls}\ln\left|\frac{s+v_d}{s-v_d}\right|. \tag{9.3}$$

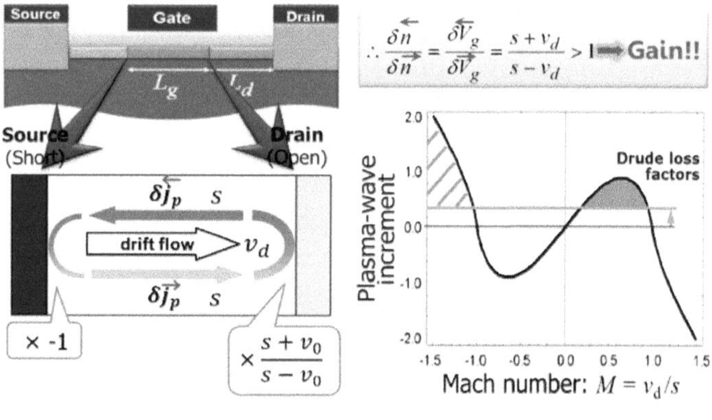

Figure 9.1 DS-type plasma-wave instability in a 2D electron channel under source-terminated and drain-opened boundaries with DC drift velocity v_d [4]. The reciprocal Doppler-shifting plasma waves reflecting at an asymmetric drain-opened boundary promote the increments of their intensity, leading to self-oscillation of instability. The plasma-wave increment is a dimensionless parameter in which the imaginary part of angular frequency is normalized to the fundamental resonant frequency. The positive values of the increment give rise to instability in an idealistic lossless case. In reality with a finite τ value, the Drude loss factor should be considered to obtain an overall gain, which is shown as a threshold level. Reproduced from Ref. [101] by permission of SPIE, © 2013.

When $\omega'' > 0$, the system becomes unstable, giving rise to the condition of DS instability. The plasma-wave increment in units of $s/2L$, a dimensionless plasma-wave increment $2\omega''L/s$ is plotted as a function of the Mach number s/v_{d} [4]. In reality with a finite τ value, the Drude loss factor should be considered to obtain an overall gain, which is shown as a threshold level in Fig. 9.1. So far, DS-instability-driven voltage-tunable millimeter-wave to THz-wave emission has been observed at low and room temperatures from GaAs-, InP-, and GaN-based HEMTs [5, 6, 10, 11].

When the channel pinch-off is insufficient and the drain terminal is not open but yet conductive at THz frequencies, the plasma waves are effectively "absorbed" in the high-field gate-drain region. Hence their reflection is insufficient to promote the necessary positive feedback for the occurrence of DS instability. At large drain–source voltages, the THz conductivity of this region might also be not so small due to the relatively high value of the electron drift velocity v_{gd} in the high-field gate-drain region. In this case v_{gd} becomes much higher than that in the intrinsic channel region v_{d}. Note that the electrons propagating in the high-field gate-drain region induce the AC current in the gated channel and the drain contact. One can find that the electron AC concentration as a function of the coordinate varies as $n_{\omega}(x) = n_{\omega}\,|_{x=L_{\mathrm{g}}}\, e^{-i\omega(x-L_{\mathrm{g}})/v_{\mathrm{gd}}}$ [14, 15]. As a result, the AC current induced in the gated channel is presented. Its frequency dependence is directly reflected by the electron transit time τ_{dd} at the gate-drain region $\tau_{\mathrm{gd}} = L_{\mathrm{d}}/v_{\mathrm{gd}}$ where L_{d} is the length of the gate-drain region, which may contribute to promoting plasma-wave instability [14]. In this case the instability condition is given by the following inequality using the imaginary part of the plasma-wave current:

$$\mathrm{Im}(\omega) = \omega'' \simeq -\frac{\nu}{2} + \gamma_{\mathrm{DS}} + \gamma_{\mathrm{TT}} > 0, \tag{9.4}$$

$$\gamma_{\mathrm{DS}} \simeq \frac{v_{\mathrm{d}}}{L_{\mathrm{g}}}, \tag{9.5}$$

$$\gamma_{\mathrm{RSS}} \simeq -\frac{v_{\mathrm{gd}}}{L_{\mathrm{g}}} r \cos(\omega_n \tau_{\mathrm{gd}}/2) J_0(\omega_n \tau_{\mathrm{gd}}/2), \tag{9.6}$$

where ν is the electron collision frequency, r is a phenomenological parameter ($r \leq 1$), and J_0 is the zeroth Bessel function, and $n = 1, 2, 3$, is the plasmon-mode index [14]. Here, γ_{DS} and γ_{RSS} are the DS instability index and the transit-time-driven RSS instability index, respectively. γ_{RSS} can take both positive and negative values, depending on $\omega_n\tau_{gd}$. The contributions to the plasma-wave instability growth rate of the DS and RSS mechanisms ($\gamma_{DS}/2\pi$ and $\gamma_{RSS}/2\pi$) in an HEMT with typical geometric and material parameters are plotted in Fig. 9.2 as functions of the fundamental plasmon-mode frequency [14]. Due to the nature of the transit-time-driven mechanism RSS instability is sensitive to the plasmon-mode frequency.

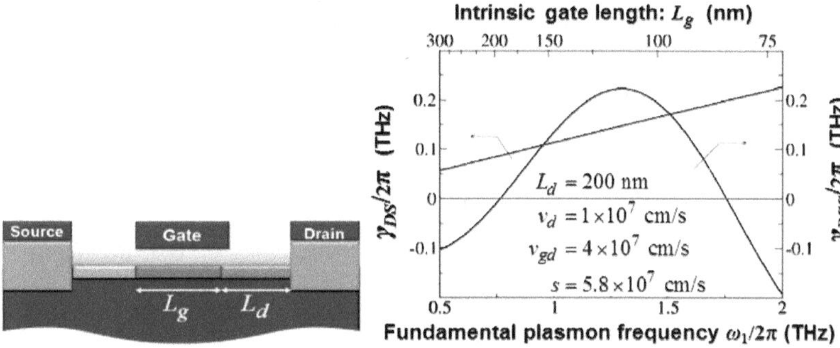

Figure 9.2 Contributions to the plasma-wave instability growth rate of the DS and RSS mechanisms ($\gamma_{DS}/2\pi$ and $\gamma_{RSS}/2\pi$) in an HEMT with typical geometric and material parameters as functions of the fundamental plasmon-mode frequency [14]. Due to the nature of the transit-time-driven mechanism the RSS instability is sensitive to the plasmon-mode frequency. Reproduced from Ref. [101] by permission of SPIE, © 2013.

A 2D plasmon itself is a nonradiative mode so that a metal-wired grating coupler structure is frequently utilized to yield THz electromagnetic-wave emission [7–9, 13, 17, 18]. If the single gate is replaced with a grating-finger-type gate, a plural number of plasmonic cavities (whose electron density is modulated by the bias voltage of the grating-finger gate) are electrostatically coupled (via an interfinger region with less electron density) in a spatially distributed configuration. In this case a constant DC channel current, which is generated by the applied DC drain bias, gives rise to periodic electron velocity modulation over the channel. This may also promote RSS instability [15].

In particular the DGG structure provides an improved confinement of the 2D electrons into the plasmon cavities independent of the tuned gate biases, resulting in more intense resonant plasma excitation with higher quality factors (see Fig. 9.3a) [33–35]. The DGGs can alternately modulate the 2D electron densities to periodically distribute the plasmonic cavities (~100 nm width in micron distance) along the channel by applying a large fraction of gate biases for subgrating G1 and G2 [33–35]. Under pertinent drain–source DC bias conditions, DC electron drift flows may promote plasmon instability, resulting in self-oscillation with characteristic frequencies in the THz regime. Figure 9.3b depicts the potential distribution and equivalent circuit model for RSS instability under a constant DC drain bias condition. The 2D electron channel consists of a periodic series of highly confined 2D plasmon cavity sections underneath the gate finger G1 and a depleted section underneath the gate finger G2. Figure 9.3c shows the Monte Carlo simulation results for spatial distributions of electron potential, density, and velocity for one unit section of the DGG under nominal bias conditions for an InGaAs channel lattice matched to the InP substrate [35]. As seen in Fig. 9.3c, electrons transit through the highly confined cavity section (depleted barrier section) with relatively low (high) velocity.

According to the aforementioned RSS instability mechanism, the depleted barrier section works for the negative conductance gain section, whereas the highly confined section works for the plasmonic resonator modeled by an *L-C-R* circuitry, as shown in Fig. 9.3b [43]. The model accounts for the features of electron transport across the device structure with the two-dimensional electron gas (2DEG) strips separated by the depleted barrier regions and the possibility of plasma oscillations in the 2DEG strips electrically coupled via the barrier regions. We demonstrated that due to the negative dynamic conductance of the barrier regions associated with the electron-transit-time effects, plasma instability leading to self-excitation of plasma oscillations (and the emission of THz radiation) becomes feasible when the DC current exceeds a certain threshold value [34]. Typical simulated results for the plasma-wave growth rate are shown in Fig. 9.3d as a function of the electron transit time through a depleted barrier section τ_t [15]. The parameter j_0/j_m is the background DC channel current j_0 normalized to the thermionic-limited maximum available current j_m.

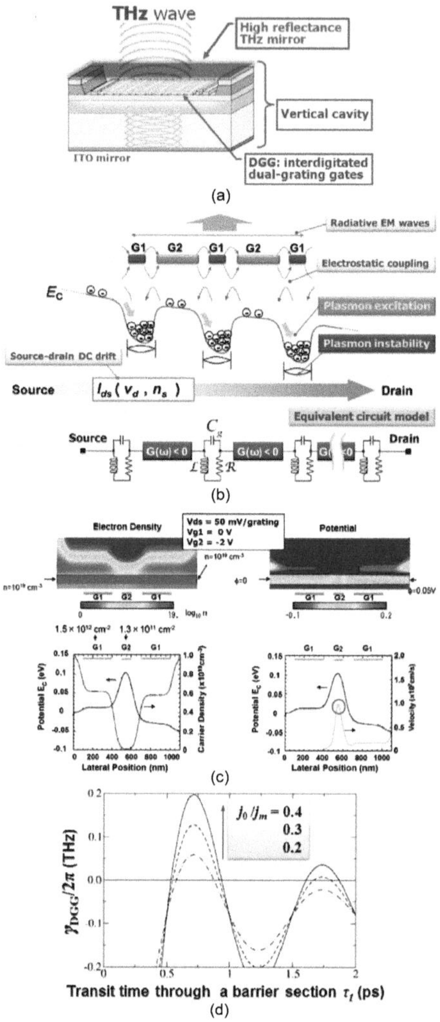

Figure 9.3 A dual-grating gate (DGG) HEMT structure for a THz emitter. (a) Schematic overview, (b) potential distribution and equivalent circuit model for RSS instability under a constant DC drain bias condition, (c) Monte Carlo simulation results for spatial distributions of electron potential, density, and velocity for one unit section of the DGG under the nominal bias conditions for an InGaAs channel lattice matched to the GaAs substrate [34], and (d) typical simulated results for the plasma-wave growth rate as a function of the electron transit time through a depleted barrier section τ_t [15]. The parameter j_0/j_m is the background DC channel current j_0 normalized to the thermionic-limited maximum available current j_m.

9.2.2 Experiments

We have proposed our original DGG-HEMT-type 2D plasmon-resonant microchip emitter as a new THz light source [32–38, 41, 42]. The original structure is a symmetric DGG in which interfinger spaces are all identical. The device was fabricated using InGaP/InGaAs/GaAs and/or InAl/InGaAs/InP material systems [34, 35]. So far a broadband THz emission ranging from 1 to ~6 THz with a maximum output power of ~1 μW at 300 K has been obtained, reflecting a multimode of coherent/incoherent plasmons [36], for which oblique modes [37], gated and ungated plasmon modes [38], hot plasmons [36], and chirped plasmon modes [34] are major causes. The DGG-HEMT THz emitter can work for THz spectroscopic and imaging applications as an incoherent broadband THz microchip source, demonstrating fine identification of water vapor absorptions as well as fingerprints of sugar groups [35].

To realize coherent monochromatic THz emission we introduced our original A-DGG structure (Fig. 9.4a,b) [41]. In the A-DGG structure the DGG is implemented with asymmetric interfinger spaces. THz electric field distribution and the resultant photoresponse were numerically simulated using a self-consistent electromagnetic approach combined with the perturbation theory for the hydrodynamic equations for 2D plasmons in HEMTs under periodic electron density modulation conditions (Fig. 9.4c) [40, 41]. Figure 9.4d shows giant enhancement of the responsivity (by 4 orders of magnitude) in an A-DGG HEMT under drain-unbiased conditions with respect to that for a symmetric DGG HEMT for a DC drain bias current density of 0.1 A/m. When the asymmetric DGG HEMT is DC-drain-biased, the asymmetry of the plasmonic cavity is enhanced enormously, resulting in further enhancement of the responsivity by orders of magnitude. It is reasonable to suggest that similar enhancement should be also exhibited for plasmon instability in the asymmetric DGG HEMT. A chirped DGG structure was also introduced so as to uniform the plasmon frequencies over the entire DGG area under a specific drain–source bias V_{ds} condition. These two features are combined to serve an AC-DGG structure. Moreover, a high-Q vertical photonic cavity structure was installed to stimulate the resonant-enhanced coherent plasmon excitation [18, 40, 102].

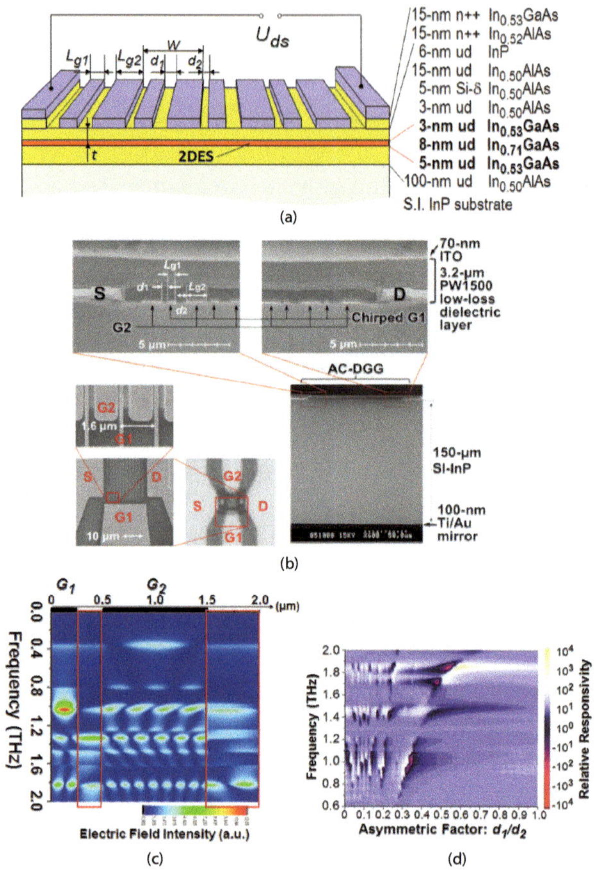

Figure 9.4 (a) Schematic view of an asymmetric dual-grating gate (A-DGG) HEMT THz emitter. (b) Cross-sectional and top scanning electron microscopy (SEM) images for a fabricated AC-DGG HEMT with a resonant-enhanced vertical cavity structure. L_{g1} = 215–430 nm, L_{g2} = 1.6 μm, d_1 = 200 nm, and d_2 = 400 nm. (c) Simulated THz electric field distribution underneath a unit A-DGG cell. Red squares show the asymmetric areas. L_{g1} = 250 nm, L_{g2} = 1.0 μm, $d_1 + d_2$ = 750 nm, and W = 2.0 μm. Electron density under gates G1 and G2 is 2.5×10^{12} cm^{-2} and 2.5×10^{11} cm^{-2}, respectively. A high electric field concentration with the fundamental plasma frequency of 1 THz under G1 can be confirmed. (d) Simulated responsivity enhancement factor for an A-DGG HEMT under drain-unbiased conditions as a function of the asymmetry factor d_1/d_2 with respect to that for a symmetric DGG HEMT for a DC drain bias current density of 0.1 A/m. Reproduced from Ref. [40] by permission of IEEE, © 2013.

AC-DGG HEMTs were designed and fabricated using InAlAs/InGaAs/InP materials (see Fig. 9.4b). Two grating gates G1 and G2 were formed with 70 nm thick Ti/Au/Ti by a standard liftoff process. The asymmetric factor, the ratio of the interfinger spaces, d_1/d_2, was fixed to be 0.5. The grating gates G1 with narrower fingers L_{g1}, serving plasmon cavity gates, were designed to be chirped ranging from 215 nm to 430 nm. The design parameters are summarized in Table 9.1. After processing the AC-DGG HEMT, a high-Q vertical cavity with a high fineness of ~60 was formed with a 4.4 μm thick transparent PW1500 resist as a low-loss buffer layer and a 70 nm thick indium tin oxide (ITO) mirror coat on top and a 100 nm Ti/Au coat on a polished back surface (see Fig. 9.4b). Its fundamental resonant frequency (free spectral range of the Fabry–Pérot modes) was designed to be 65 GHz.

Table 9.1 Design parameters of fabricated devices

Sample no.	**SG1**	**SG2**	**SP1**	**ACP1**
Materials	InGaP/$In_{0.2}$GaAs/GaAs		InAlAs/In_xGaAs/InP[a]	
DGG type[b]	S-DGG			AC-DGG
Cavity[c]	Low Q	High Q	Low Q	High Q
FSR	145 GHz	65 GHz	65 GHz	65 GHz
L_{g1} **(nm)**	100	100	100	215–430
d_1/d_2 **(nm)**	100/100			200/400
L_{g2} **(nm)**	1800	850	800	1600
No. of fingers G1/G2	37/38	30/31	30/31	8/9
Active area (μm²)	75 × 75	30 × 75	30 × 75	20 × 20
Ref.	[34, 35]	–	[34]	[40, 102]

[a] x = 0.53/0.7/0.53 (composite channel)
[b] S: symmetric; AC: asymmetric and chirped
[c] Low Q: Au mirror coat on back surface only

The fabricated HEMTs exhibit normal DC transfer characteristics with good pinch-off and gate modulations for both G1 and G2 with threshold levels of −1.1 V and −0.9 V, respectively. We conducted THz spectroscopic measurements for the fabricated devices using a Fourier transform far-IR spectrometer and a 4.2 K-cooled Si composite bolometer. The gate bias for the plasmon cavities V_{g1} was fixed at 0 V. Material- and structure-dependent emission spectra at 290 K are plotted in Fig. 9.5a in comparison with previously reported GaAs-based and InP-based symmetric DGG HEMTs [40]. The drain bias was set so as to equal the voltage drop along a unit DGG section among all the devices. Superiorities of the A-DGG structure and InP-based materials with higher electron mobility (~11,000 cm^2/V s) reflect intense back-ground-less emission of Fabry–Pérot modes. The highest peak intensity is observed at ~6.2 THz with a sharp line width of ~1.23 cm^{-1} (~37 GHz) corresponding to the fifth plasmon mode. Figure 9.5b plots the dependence of the emission spectra on V_{ds} and V_{g2} under an optimal V_{g1} condition (V_{g1} = −0 V) [40, 102]. By increasing V_{ds} the whole emission intensity of Fabry–Pérot modes raises, reflecting the hot-plasmon-originated broadband background emission. Furthermore, by applying appropriate V_{g2} of −0.55 V to make a strong contrast on the electron densities on plasmonic cavities the peak at ~3.55 THz is enhanced, reflecting the 3rd harmonics of the plasmon resonance driven by the DS instability. When the temperature cools down to 140 K, hot-plasmon-originated residual spurious modes are suppressed, resulting in intense monochromatic emission at the third (at 3.6 THz) plasmon modes whose frequencies match adjacent Fabry–Pérot modes, respectively, as shown in Fig. 9.5b. The emission peak at the fifth (at 6.1 THz) plasmon mode disappears, which is considered to be caused by the temperature dependence of the 2D electron density in the plasmon cavities, reflecting frequency detuning of the plasmon modes. Simulated time evolution of the instantaneous electric field (E_x) distributions under a constant sinusoidal plasmon excitation at 3.4 THz ensures the cooperative oscillation of all the plasmon cavities underneath the subgrating finger G1, as shown in Fig. 9.5c, supporting the superradiant THz emission. The output power of sample ACP1 is estimated to be of the order of 0.1 μW to 1 μW.

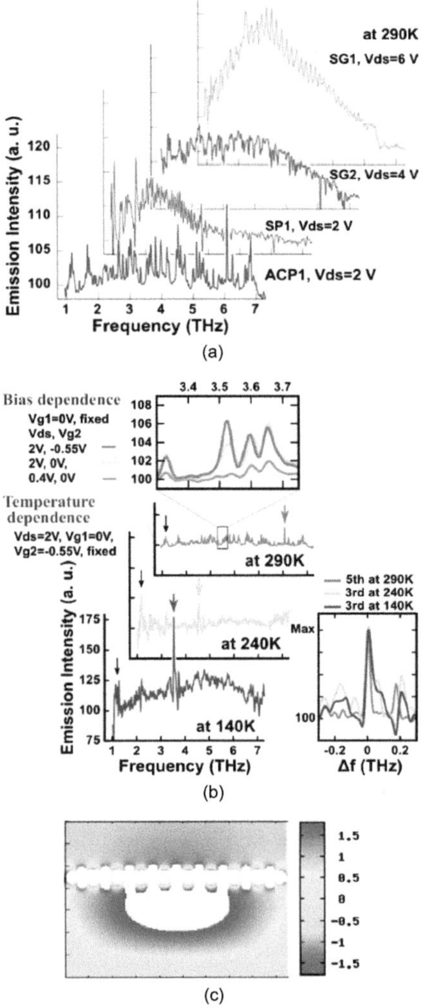

Figure 9.5 (a) Measured emission spectra at 290 K for different material systems and DGG structures. (b) Measured emission spectra at different biases at 290 K (upper) and at different temperatures at optimal bias conditions (V_{ds} = 2 V, V_{g1} = 0 V, V_{g2} = −0.5 V) (lower) of sample ACP1 (InP-based AC-DGG HEMT). Arrows designate the first, second, and third plasmon modes. Inset magnifies the spectral portions of the third and fifth modes. (d) Cross-sectional view of simulated instantaneous electric field (E_x) distributions under a constant sinusoidal plasmon excitation at 3.4 THz. All the plasmon cavities are cooperatively resonated in phase to ensure uniform transverse radiation over the entire DGG area.

9.3 THz Detection Using 2D Plasmons

9.3.1 Theory [101]

Detection of THz radiation by plasmonic nonlinearities in a 2D electron channel of a field-effect transistor (FET) was originally proposed by Dyakonov and Shur [19]. Resonant (frequency-selective) [20, 22, 25, 26, 30] as well as nonresonant (broadband) [21–23] plasmonic detectors have been studied. The frequencies of the plasmon resonances in the FET channel are given by [19]

$$\omega_{\mathrm{p}} = n\frac{\pi}{2L_{\text{g-eff}}}\sqrt{\frac{e^2N^{(0)}d}{m^*\varepsilon\varepsilon_0}}, \quad eN^{(0)} = \varepsilon\varepsilon_0\frac{U_{\mathrm{g}}-U_{\mathrm{th}}}{d}, \tag{9.7}$$

where $N^{(0)}$ is the equilibrium electron density in the channel, d and ε are the thickness and relative dielectric constant of the barrier layer, ε_0 is the dielectric constant in vacuum, $L_{\text{g-eff}}$ is the effective length of the gated 2D electron channel (which is greater than a geometrical length of the gate, L_{g}, due to the electric-field fringing effect), n is an integer, e and m^* are the electron charge and electron effective mass, respectively, U_{g} is the gate-to-source DC bias voltage, and U_{th} is the threshold voltage of the FET. For an asymmetric boundary condition, only the plasmon modes with odd indices n can be excited by the THz radiation normal to the FET channel plane having the electric field component parallel to the channel direction, whereas the plasmon modes with even indices n remain dark. Resonant detection takes place at THz frequencies $\omega = \omega_{\mathrm{p}}$ for high-quality factors of the plasmon resonance, such as $\omega_{\mathrm{p}}\tau > 1$, where τ is the electron momentum relaxation time. The nonresonant plasmonic detection takes place when the relaxation time is low (low mobility case) or the equilibrium electron density in the FET channel decreases so that the inequality $\omega_{\mathrm{p}}\tau \lesssim 1$ becomes valid for $U_{\mathrm{g}} \to U_{\mathrm{th}}$.

The hydrodynamic motion of the plasma-wave kinetics yields a nonlinear plasma-wave current component as the product of the local electron density and local electron velocity, both of which are perturbed by incoming THz radiation, giving rise to the quadratic rectification component in the current time integral, as shown in Fig. 9.6. This is the fundamental principle of THz detection, using plasma-wave dynamics [19]. In the open-circuit condition, the generated (rectified) photocurrent is transformed to a DC voltage component, leading to a photovoltaic signal ∂U. By solving Eqs. 9.1 and 9.2 under pertinent source-terminated and drain-opened boundary conditions, ∂U is derived as follows [19]:

$$\left.\begin{aligned}\partial U &\approx \frac{U_{\mathrm{a}}^2}{4U_0^2}\frac{\omega_0^2}{(\omega^2-\omega_0^2)+\left(\dfrac{1}{2\tau}-\dfrac{v_{\mathrm{d}}}{L}\right)^2}\\ &\approx \frac{U_{\mathrm{a}}^2}{U_0^2}\left(\frac{s\tau}{L}\right)^2\frac{1}{4(\omega-n\omega_0)^2\tau^2+1}, \quad \text{for } \omega\tau \gg 1 \text{ (resonant mode)}\\ &\approx 1+\frac{2\omega\tau}{\sqrt{1+\omega^2\tau^2}}, \qquad \text{for } \omega\tau \ll 1 \text{ (nonresonant mode)}\end{aligned}\right\} \tag{9.8}$$

where s is the plasma-wave velocity, ω_0 is the fundamental plasmon-resonant frequency, n is the harmonic number, and τ is the electron momentum relaxation time.

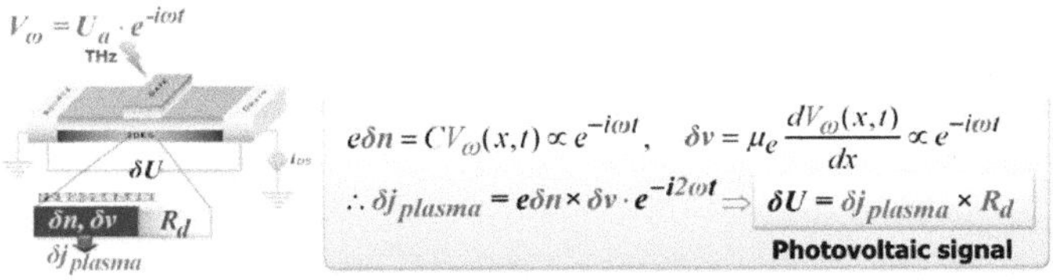

Figure 9.6 Principle of rectification and detection of THz radiation by means of hydrodynamic nonlinearity of 2D plasmons. Reproduced from Ref. [101] by permission of SPIE, © 2013.

As is mentioned in Section 9.2, we propose an A-DGG structure, as shown in Fig. 9.4, which also provides ultrahigh-sensitive THz detection [39–43]. This is because the unit cell of the A-DGG structure can create a strong build-in asymmetric field. The THz photoresponse dramatically increases when the parts of the 2D channel under the fingers of one of the two subgratings are depleted. The device under characterization is identical to the InAlAs/InGaAs/InP A-DGG HEMTs shown in Fig. 9.4, except for uninstalling the vertical cavity.

The A-DGG structure can greatly enhance the asymmetry of the cavity boundaries by applying different gate voltages V_{g1} and V_{g2} to the two different subgratings of the A-DGG. Due to the asymmetry, the impedances at both ends from the 2D plasmon cavity are different. Thus, the photocurrent generated in a 2D plasmon by the THz radiation performs unilateral transport, resulting in a high photovoltaic signal at each cell. These signals at every 2D plasmon cavities are summed up, leading to an effective photovoltaic signal between the drain and the source [42].

9.3.2 Experiments

The device under characterization is identical to the InAlAs/InGaAs/InP A-DGG HEMTs shown in Fig. 9.4, except for uninstalling the vertical cavity. The A-DGG HEMTs have been designed and fabricated using the InAlAs/InGaAs/InP material system [42]. The geometrical parameters of the A-DGG HEMTs are summarized in Table 9.2. The fabricated devices exhibit normal gate modulation operation on their drain–source current–voltage characteristics [42]. For G1 and G2, the threshold voltage V_{th1} and V_{th2} are –1.2 V and –1.0 V, respectively.

Table 9.2 Design parameters of fabricated A-DGG HEMTs

Sample no.	**1-1**	**2-3**	**2-4**
Materials	InAlAs/In$_x$GaAs/InP[a]		
DGG type[b]	AC-DGG		
L_{g1} (nm)	200	215–430	400–705
d_1/d_2 (nm)	200/400	200/400	400/800
No. of fingers G1/G2	8/9	8/9	6/7
Ref.	[39]	[40, 42, 43]	[40, 42, 43]

[a] x = 0.53/0.7/0.53 (composite channel)
[b] AC: asymmetric and chirped
L_{g2} and active area are fixed to be 1600 nm and 20 × 20 μm^2.

Room-temperature THz photovoltaic measurements were carried out with fabricated detectors. Monochromatic THz pulsed waves with frequencies from 1 THz to 3 THz from a ring-cavity THz parametric oscillator source are focused on the detector by a Tsurupica^{TM} lens [39, 40, 42, 43]. The photovoltaic signal was observed as a V_{ds} fluctuation with a lock-in technique. The responsivity was estimated as $R_v = \Delta U \cdot S_t/(P_t \cdot S_d)$, where ΔU is the THz-radiation-induced DC drain voltage, P_t is the total power of the source on the detector plane, S_t is the radiation beam spot area, and S_d is the active area of the detector. Figure 9.7a shows the measured responsivity of detector 2-4 at 1 THz under zero-V_{ds} condition as a function of the gate voltage swing ($V_{g1,2}$–V_{th}); DC voltage of G1: V_{g1} (G2: V_{g2}) is swept, while V_{g2} (V_{g1}) is floated (biased at 0 V) [42]. The best result with R_v = 2.2 kV/W was obtained when sweeping V_{g1} to the threshold V_{th} [42, 43]. It is worth to stress that even at higher frequencies, relatively high responsivities are obtained with detector 2-3 having a shorter L_{g1}. As seen in the inset of Fig. 9.7a, the responsivities monotonically decrease from 1.7 kV/W at 1 THz to 0.52 kV/W at 2 THz. All these values are, to the authors' knowledge, the best ever reported at these frequencies. Figure 9.7b shows the measured responsivity of detector 1-1 at 1.5 THz under V_{ds}-biased conditions. With increasing V_{ds} from 0 V to 0.4 V, the maximum responsivity increases up to 6.4 kV/W, which is, to the best of authors' knowledge, the record responsivity of fast-response detectors at 300 K [39]. For all the photoresponse, the photovoltaic signal increases with decreasing the gate bias and takes a peak value when the gate bias is approaching the threshold level. It is expected to turn down and drop when the gate bias furthermore decreases to the subthreshold region, as was measured for symmetric DGG HEMTs [32] as well as standard single-gate HEMTs [21]. Very recently a physical model has been developed that explains this photoresponse dependence on gate bias [45].

Figure 9.8 shows NEP for detectors (a) 2-4 and (b) 2-3 as a function of $V_{g1,2}$ under a drain-unbiased condition [42]. Detector 2-3 exhibits extremely low NEP, with the minimal value 15 pW/Hz$^{0.5}$ at 1 THz. These values are lower than those of any commercial room-temperature THz detectors such as Golay cells (200 pW/Hz$^{0.5}$) [44] or Schottky barrier diodes (100 pW/Hz$^{0.5}$) [41]. The NEP value is proportional to the square root of the drain resistance $R_d^{0.5}$ and the inverse responsivity R_v^{-1}. On decreasing the gate swing voltage ($V_g - V_{th}$), the responsivity becomes larger in the range ($V_g - V_{th}$) < 0.2 V (see Fig. 9.8), resulting in the decrement of the NEP value. On the other hand, near the threshold condition ($V_g - V_{th}$) ≈ 0, the drain resistance becomes much larger because of the pinch-off, resulting in the increment of the NEP value. As a result, the NEP value becomes minimal around the midpoint $(V_g - V_{th}) \approx -0.1$ V. Due to very similar characteristics of drain resistance and responsivity for both gates at every frequency measured this time, this trend is seen for all the A-DGG HEMT devices and bias conditions.

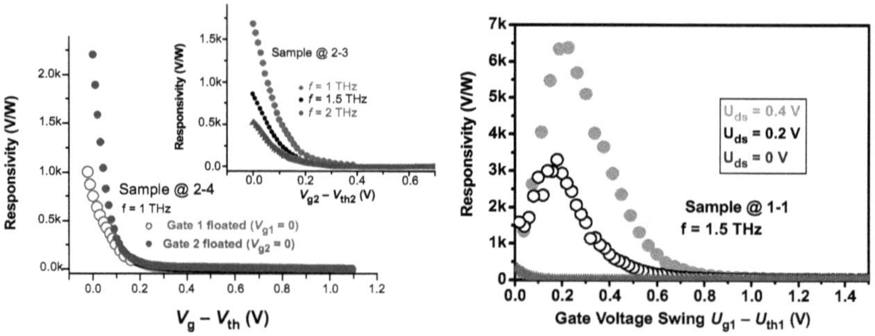

Figure 9.7 (a) Responsivity of detector 2-4 as a function of the gate swing voltage ($V_{g1} - V_{th1}$ or $V_{g2} - V_{th2}$) at 1 THz under a drain-unbiased condition. Inset: Responsivities of detector 2-3 at 1, 1.5, and 2 THz as a function of the gate swing voltage $V_{g2} - V_{th2}$ when G1 is floated. The responsivities become higher with decreasing gate voltages (taking the maximum at the threshold). Reproduced from Ref. [42] by permission of Elsevier, © 2012. (b) Responsivity of detector 1-1 as a function of the gate swing voltage at 1.5 THz radiation under drain-biased conditions. Reproduced from Ref. [39] by permission of IEEE, © 2012.

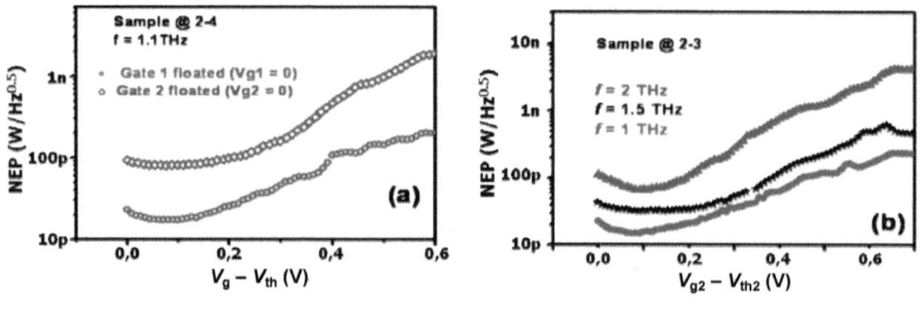

Figure 9.8 Measured NEP for (a) sample 2-4 at 1 THz as a function of the gate swing voltages $V_{g1} - V_{th1}$ and $V_{g2} - V_{th2}$ and (b) sample 2-3 at 1, 1.5, and 2 THz as a function of the gate voltage swing $V_{g2} - V_{th2}$. The NEP becomes minimum at $V_g - V_{th} \approx 0.1$ V. Reproduced from Ref. [42] by permission of Elsevier, © 2012.

The detection performance at relatively low frequency at 292 GHz was also characterized at low temperatures ranging from 50 K to 125 K (see Fig. 9.9) [39]. By decreasing the temperature, one can see a peak appearing as a shoulder and narrowing on the detection curves. This peak is identified as gated plasmon resonance, and its frequency can be estimated as $\delta f = (f/2)*[\delta U_{g1}/ (U_{g1} - U_{th1})]$, where f is the incident frequency and $U_{th1} = -1.2$ V. This width is found to vary from 279 GHz to 100 GHz when decreasing the temperature. The observed shrinking of the resonance peak can be explained by an increase of the plasma waves' damping time τ via an increase of the carrier's mobility. The quality factor of the resonance linewidth is experimentally determined by $\delta f/f = 3$ from which the plasmon decay time is estimated to be about ~2 ps. This quality factor should reflect on the response speed of the plasmon-resonant detection. Therefore, very fast-speed THz detection beyond tens of gigabits per second could be feasible. Its precise characterization is a future subject. The device response speed is characterized indirectly by the resonant-mode detection. Figure 9.9 also shows calculations of channel resistivity R and the difference induced in drain current by the incoming THz radiation ΔI (I with radiation – I without radiation). The $R*\Delta I$ curves (right axis) consistently shows the resonant peak obtained from experiments.

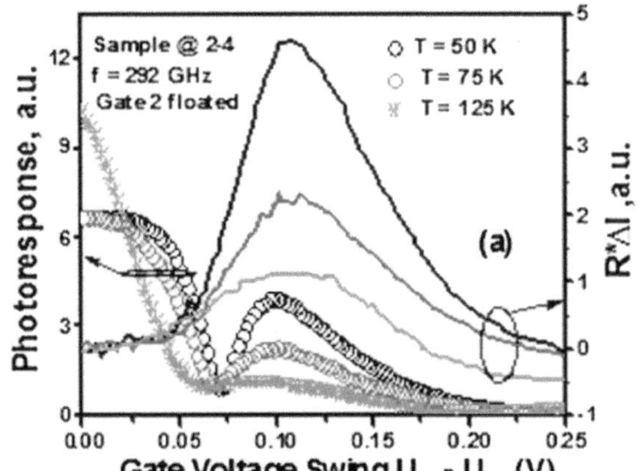

Figure 9.9 Measured photoresponse at 292 GHz for different temperatures as a function of the gate voltage swing (left) and calculated curves $R*\Delta I$ (right), where R is the channel resistivity and ΔI (I with radiation – I without radiation) is the difference induced in drain current by the incoming THz radiation. Reproduced from Ref. [42] by permission of Elsevier, © 2012.

9.4 Applications of Nondestructive THz Spectroscopy

9.4.1 2D Plasmon Emitters for Broadband THz Light Sources in Fourier Transform THz Spectroscopy

The plasmon-resonant symmetric DGG-HEMT emitter "SG-1" shown in Fig. 9.5a was introduced into the Fourier transform infrared (FTIR) system in place of the mercury lamp and examined its applicability as a broadband THz light source, as shown in Fig. 9.10a,b. Atmospheric water vapor and several sugar groups, whose identical absorption/transmission spectra coexist in the emission

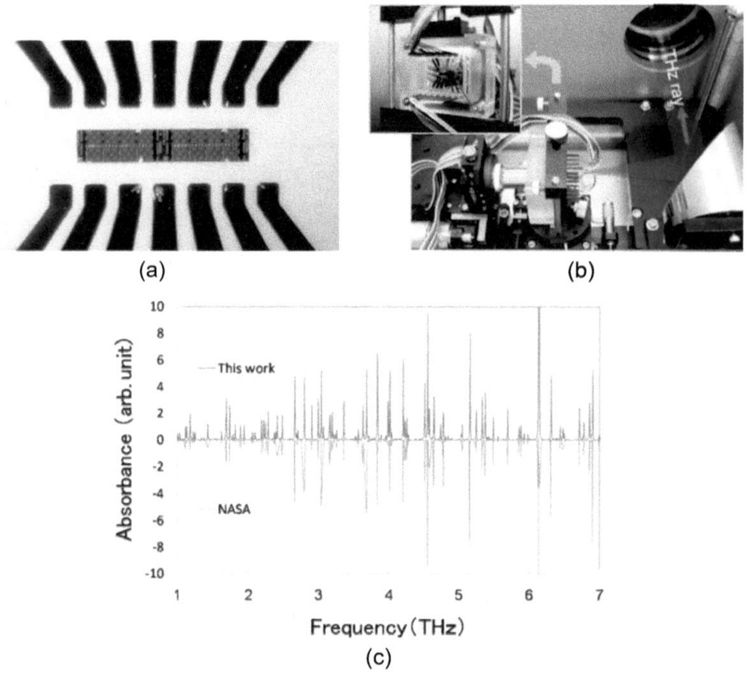

Figure 9.10 (a) 2 × 10 mm die (accommodating 32 independent emitter devices) mounted on a quartz substrate. Two emitters are wire-bonded with metalized patterns on the substrate for an electrical connection. (b) Emitter chip mounted on a sample holder installed in the chamber of the FTIR spectrometer equipment. (c) Measured absorption spectrum of atmospheric water vapor in comparison with the data provided by NASA. Reproduced from Ref. [35] by permission of OSA, © 2009.

spectral range available from the emitter, are chosen as samples under measurement. First, the atmospheric water vapor absorption was measured. The experimental procedure was as follows: first we measured the emission spectra in vacuum inside the spectrometer as the reference. Then atmospheric air was introduced into the spectrometer, and we performed an identical measurement. The obtained result was normalized to the reference data. Figure 9.10c shows the measured absorption spectrum of the water vapor. The result ranging from 1 THz to 6.5 THz well coincides with the spectrum data provided by NASA [35].

Next, we measured the transmission spectra of two different types of sugar groups, honey and maple syrup, both of which contain the featured spectrum in the THz region. Figure 9.11a,b plots the results measured by using the plasmon-resonant emitter (this work) and by using a high-pressure mercury lamp and their main component(s) provided by RIKEN [35]. The measured samples, honey and maple syrup, were in liquid, but the components of sugar groups measured by RIKEN were in dry pellets. Thus, comparison of the absorption peak points among them does not make sense. However, the molecular structures may reflect on the overall spectral shape. They clearly exhibit identical spectral features for both materials. The major bumps for each measured spectrum fairly correspond to those for the main components of each: (a) honey versus glucose and fructose and (b) maple syrup versus sucrose. Other minor parts of the ingredients and/or impurities may also perturb the spectral

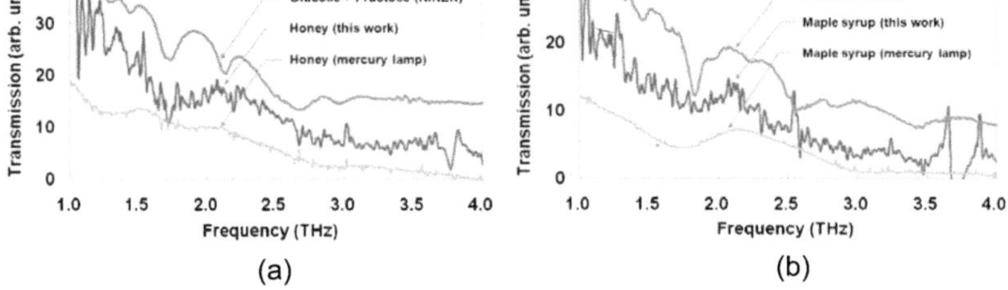

Figure 9.11 Transmission spectra of sugar groups measured by using a plasmon-resonant emitter (this work) and by using a high-pressure mercury lamp and their main ingredient(s) provided by RIKEN. (a) Honey vs. glucose and fructose and (b) maple syrup vs. sucrose. Reproduced from Ref. [35] by permission of OSA, © 2009.

shape. Compared with a high-power mercury lamp, the plasmon-resonant microchip emitter yields higher noise on the spectra due to weak emission intensity. Further improvements in its emission power will be feasible, as is described before, which enables the device to be a promising candidate as a new THz light source.

9.4.2 2D Plasmon Detectors for THz Imaging

The THz imaging experiment was carried out for several DGG HEMT detectors using the experimental setup shown in Fig. 9.12a [43] and Fig. 9.12b [40]. The imaging results are shown in Fig. 9.12c [43] and Fig. 9.12d [40]. The samples under imaging are (c) a tea bag with an aluminum-coated plastic package and (d) an integrated circuit (IC) card and a soap bar. As seen in Fig. 9.12c, THz radiation can transit through the aluminum-coated plastic package so that it can successfully identify the contents of the tea leaves inside of the tea bag. As seen in Fig. 9.12d, measured results clearly demonstrate the imaging of hidden substances of circuit/antenna components inside the IC card and the thickness of the soap bar with a nice contrast in submillimeter resolution.

9.5 2D Plasmons in Graphene and Their THz Device Applications

9.5.1 Fundamental Physics

Collective excitation of 2D plasmons in graphene can provide unique features in their response to THz electromagnetic radiation, such as extremely high absorption as well as giant THz gain [46, 47, 51, 76, 86, 87, 96–100]. The dispersion relations for plasmons have been formulated in various graphene structures: a graphene sheet [78, 80, 81, 83, 91, 93, 95], a gated plane graphene structure [87, 89, 90], gated p-i-n junctions [103], a graphene ribbon array [73–75, 85], and edges [82, 94, 104]. In particular, gated plasmons are a major concern for use in practical frequency-tuned device applications. The dispersion relation and damping rate for gated graphene plasmons are theoretically revealed. A gated plasmon holds a superlinear dispersion in which the gate-to-graphene distance *d*

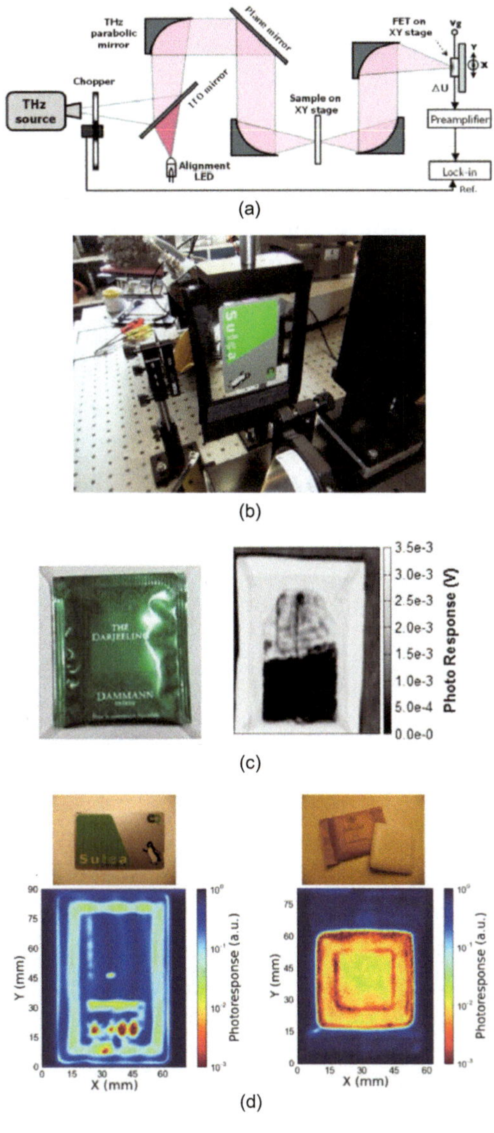

Figure 9.12 Imaging setup and results using an InP-based S-DGG and A-DGG HEMTs. (a) Experimental setup for sub-THz imaging experiments. (b) Photo image of the sample holder of the imaging setup. (c) A tea bag measured by using an S-DGG HEMT. Left: photo image; right: THz image. Reproduced from Ref. [43] by permission of IEEE, © 2013. (d) Photo images and THz imaging measured by using an A-DGG HEMT. Left: an IC card; right: a soap bar. Reproduced from Ref. [40] by permission of IEEE, © 2013.

and the momentum relaxation rate (collision frequency) deform the linear dispersion relations (Fig. 9.13a). The plasmon phase velocity is proportional to the quarter power of the gate bias and of *d*, which is quite different from those for the 2D plasmons in conventional semiconductor quantum wells. Semiclassical Boltzmann equations derive the hydrodynamic kinetics of electron–hole plasma-wave dynamics [92]. Intrinsic graphene and/or photoexcited graphene hold an equidensity of electrons and holes so that bipolar electron–hole modes of plasmons are excited. Due to the freedoms of electron–hole motions, electron–hole plasma waves and charge-neutral sound-like waves are excited (Fig. 9.13b). The electron–hole plasmons are strongly damped, whereas the electron–hole sound-like waves survive. On the other hand, doped graphene holds a large fraction on the majority and minority carriers. The plasma waves originated from the minority carriers are strongly damped, resulting in unipolar modes of plasmons originating from the majority carriers.

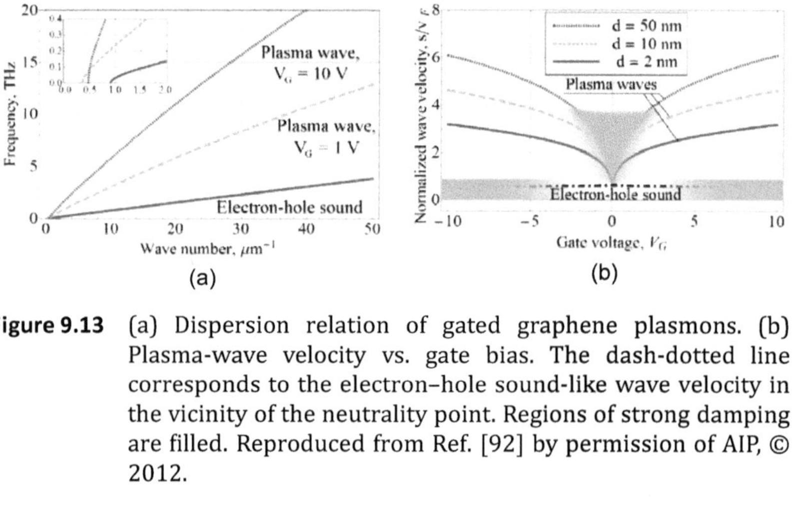

Figure 9.13 (a) Dispersion relation of gated graphene plasmons. (b) Plasma-wave velocity vs. gate bias. The dash-dotted line corresponds to the electron–hole sound-like wave velocity in the vicinity of the neutrality point. Regions of strong damping are filled. Reproduced from Ref. [92] by permission of AIP, © 2012.

THz plasmon modes in a graphene microribbon array structure were experimentally observed to verify the theoretical dispersion relations [96–100]. Plasmon frequencies in graphene microarrays can be tuned through the entire THz range, depending on the direction of plasmon propagation in the array plane by (i) varying the microribbon width, (ii) electrostatic modulation of the carrier density, and (iii) Landau-level splitting under the magnetic field. Plasmons in patterned graphene are strongly coupled to electromagnetic waves, which constitutes a potential for development of tunable graphene-based THz plasmonic metamaterials [84, 99, 105, 106]. Therefore, the prospects of graphene plasmonics for mid-IR and THz applications attract rapidly growing interest [46, 47, 51, 81, 94].

A number of models of the graphene conductivity has been developed so far [100], each being valid in a certain range of frequencies and carrier densities. For a high-frequency (from THz to IR) range, we use the relation for the graphene conductivity derived from the Kubo formula by Falkovsky and Varlamov [77]

$$\sigma_{\mathbf{k}\omega} = \frac{ie^2}{\hbar\pi^2}\sum_{a=1,2}\int\frac{d^2\mathbf{p}v_x^2\{f[\varepsilon_a(\mathbf{p}_-)]-f[\varepsilon_a(\mathbf{p}_+)]\}}{[\varepsilon_a(\mathbf{p}_+)-\varepsilon_a(\mathbf{p}_-)][\hbar\omega-\varepsilon_a(\mathbf{p}_+)+\varepsilon_a(\mathbf{p}_-)]}$$
$$+\frac{2ie^2\hbar\omega}{\hbar\pi^2}\int\frac{d^2\mathbf{p}v_{21}v_{12}\{f[\varepsilon_1(\mathbf{p}_-)]-f[\varepsilon_2(\mathbf{p}_+)]\}}{[\varepsilon_2(\mathbf{p}_+)-\varepsilon_1(\mathbf{p}_-)][(\hbar\omega)^2-[\varepsilon_2(\mathbf{p}_+)-\varepsilon_1(\mathbf{p}_-)]^2]}, \quad (9.9)$$

where the indices 1 and 2 refer to conduction and valence bands, $\varepsilon_1(\mathbf{p}) = |\mathbf{p}|v_F$ and $\varepsilon_2(\mathbf{p}) = -|\mathbf{p}|v_F$, $v_F \simeq 10^6$ m/s, $\mathbf{p}_\pm = \mathbf{p}\pm\hbar\mathbf{k}/2$, $f(\varepsilon)$ is the electron distribution function (the equilibrium Fermi function $f(\varepsilon)=1/[1+e^{(\varepsilon-\varepsilon_F)/T}]$ is assumed), and $v_x = v_F\cos\theta_\mathbf{p}$ and $v_{12} = iv_F\sin\theta_\mathbf{p}$ are the matrix elements of the velocity operator. The first term in Eq. 9.9 corresponds to intraband transitions, whereas the second term corresponds to interband transitions. To allow for momentum relaxation, one should treat the frequency as $\omega\to\omega+i\tau_m^{-1}$. The definite expressions for different scattering processes can be found in Ref. [107]. For example, when electron-phonon scattering is the dominating mechanism, the corresponding relaxation time is given by the following formula [108]:

$$\tau_\mathrm{m}^{-1} = \tau_0^{-1}\frac{\varepsilon_F}{T} \quad (9.10)$$

where the parameter $\tau_0^{-1}\simeq 3\times10^{11}$ s^{-1} at room temperature $T\simeq 26$ meV. The real part in Eq. 9.9 becomes the following simple forms [109–112]:

$$\begin{aligned}\mathrm{Re}\,\sigma_\omega &= \mathrm{Re}\,\sigma_\omega^{\mathrm{intra}}+\mathrm{Re}\,\sigma_\omega^{\mathrm{inter}}\\ &\approx \frac{(\ln 2+\varepsilon_F/2k_BT)e^2}{\pi\hbar}\frac{k_BT\tau_\mathrm{m}}{\hbar(1+\omega^2\tau_\mathrm{m}^2)}+\frac{e^2}{4\hbar}[1-2f(\hbar\omega)]\end{aligned} \quad (9.11)$$

where e is the elementary charge, $\hbar$ is the reduced Planck's constant, k_B is the Boltzmann constant, T is the temperature, and τ is the momentum relaxation time of carriers. The intraband contribution, Re σ_ω^{intra}, corresponds to the Drude absorption and is always positive. The THz frequency domain is situated in the intraband Drude conductivity-dominated region. The plasma-wave kinetic equations (hydrodynamic Euler equation and continuity equation) are derived from semiclassical Boltzmann equations for massless Dirac fermions. As is seen in Fig. 9.14 [112], in the low-frequency region, including the THz range, the intraband Drude conductivity is dominated and decreases with increasing frequency. The interband conductivity vanishes due to the Pauli-state blocking. At the photon energy twice as large as the Fermi level the interband conductivity starts to increase and saturate to a plateau of the quantum conductivity $e^2/4\hbar$.

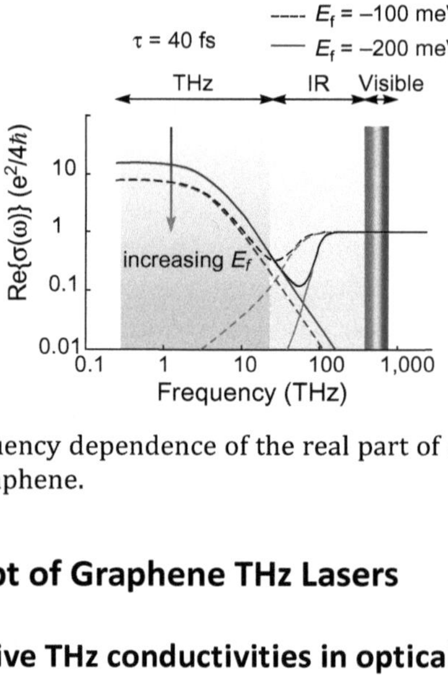

Figure 9.14 Frequency dependence of the real part of optical conductivity in graphene.

9.5.2 Concept of Graphene THz Lasers

9.5.2.1 Negative THz conductivities in optically pumped graphene

Interband population inversion in graphene can be achieved by its optical pumping [69, 72] or carrier injection [68, 71]. At sufficiently strong excitation, the interband stimulated emission of photons can prevail over the intraband (Drude) absorption. In this case, the real part of the dynamic conductivity of graphene, Re[$\sigma(\omega)$], becomes negative at some frequencies ω. Due to the gapless energy spectrum of graphene, Re[$\sigma(\omega)$] can be negative in the THz range. This effect can be used in graphene-based THz lasers with optical or injection pumping, as shown in Fig. 9.15. A graphene photonic laser with the Fabry–Pérot resonators based on dielectric or slot-line waveguides [70, 113] was proposed for lasing THz photons. Stimulated emission of near-IR [114] and THz [66, 67] photons from population-inverted graphene was observed experimentally.

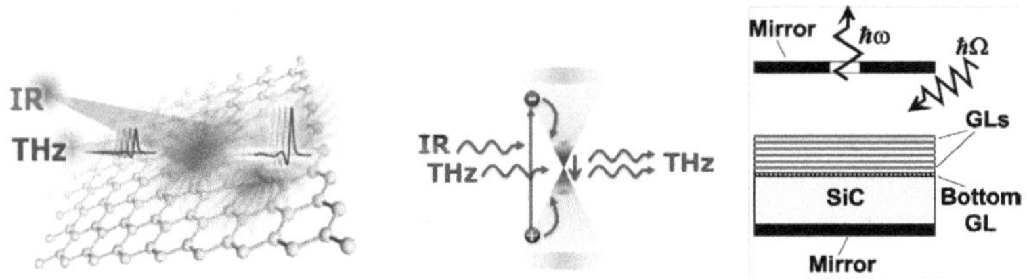

Figure 9.15 Idea for graphene THz lasers. Stimulated emission of THz radiation in IR laser–pumped graphene (left: schematic image; center: band diagram; right: vertical Fabri–Pérot-type THz laser cavity structure in which multiple layers of monolayer graphene are installed.

Carrier relaxation and recombination dynamics in optically pumped graphene are schematically shown in Fig. 9.16. When the photogenerated electrons and holes are heated, collective excitations due to the carrier–carrier (CC) scattering take a dominant roll to perform an ultrafast carrier quasi-equilibration within a 10–100 fs time scale [115–117]. Then carriers at high-energy tails of their distributions emit optical phonons (OPs), cooling themselves and accumulating around the Dirac points [118–120]. For the electron–hole recombination, optical-phonon-assisted or radiative recombination via direct transition may take place, thanks to the gapless symmetrical band structure. Moreover, interband CC scattering like Auger-type and/or impact ionization-type processes, which have been thought to be theoretically forbidden in graphene [121], may take place due to higher-order many-body effects and/or imperfections of crystal qualities, which could also strongly modify the carrier relaxation dynamics [122–125]. Recent experimental and theoretical works demonstrate such aspects on the carrier relaxation dynamics [122–125].

Due to a fast intraband relaxation (picosecond or less) and relatively slow interband recombination (>>1 ps) of photoelectrons/

holes, the population inversion is obtainable, as seen in Fig. 9.16 [67, 117]. When the pumping intensity exceeds a certain threshold level (staying ~10^7 W/cm^2 for 0.8 eV pumping photons), the quasi-Fermi level goes to the positive level, testifying to population inversion. The temporal evolution of the quasi-Fermi level and the carrier temperature in optically pumped graphene with a femtosecond laser pulse are numerically simulated, as shown in Fig. 9.17 [67]. Depending on the graphene quality (reflecting the carrier momentum relaxation time) the gain spectral profile and its temporal profile are different; in high quality of graphene the intraband Drude conductivity loss factor reduces so that a wider gain spectral bandwidth with an increased lower-frequency end is obtainable after few picoseconds from pumping to and last after ~10 ps. Stimulated THz [66, 67] and IR photon emission [114] in optically pumped graphene have recently been observed at room temperature (Figs. 9.18 and 9.19), encouraging us to proceed toward the creation of graphene THz and IR lasers [70, 71, 113, 115, 116].

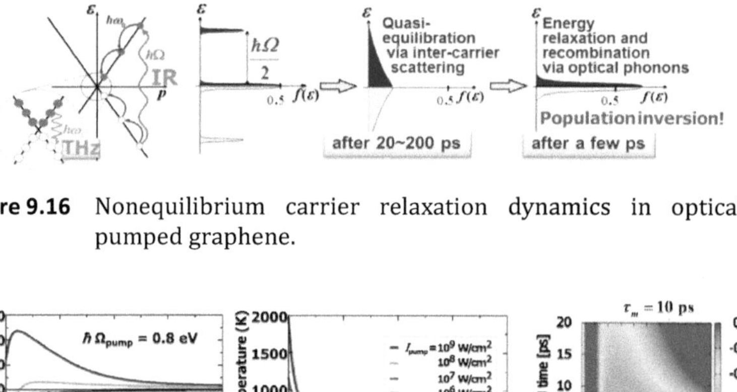

Figure 9.16 Nonequilibrium carrier relaxation dynamics in optically pumped graphene.

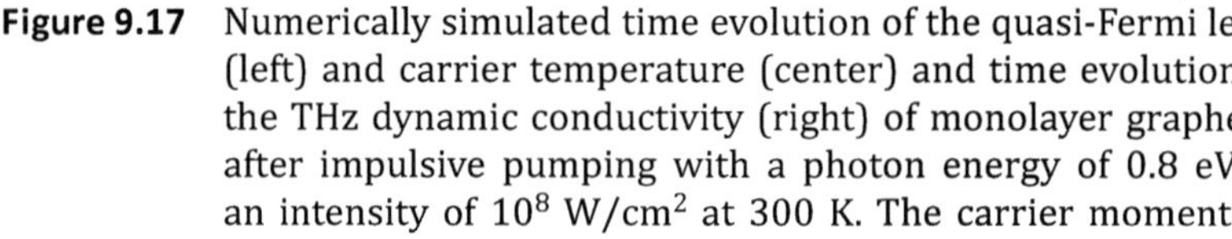

Figure 9.17 Numerically simulated time evolution of the quasi-Fermi level (left) and carrier temperature (center) and time evolution of the THz dynamic conductivity (right) of monolayer graphene after impulsive pumping with a photon energy of 0.8 eV at an intensity of 10^8 W/cm^2 at 300 K. The carrier momentum relaxation time was assumed to be 10 ps for high-quality graphene. Reproduced from Ref. [67] by permission of IEEE, © 2013.

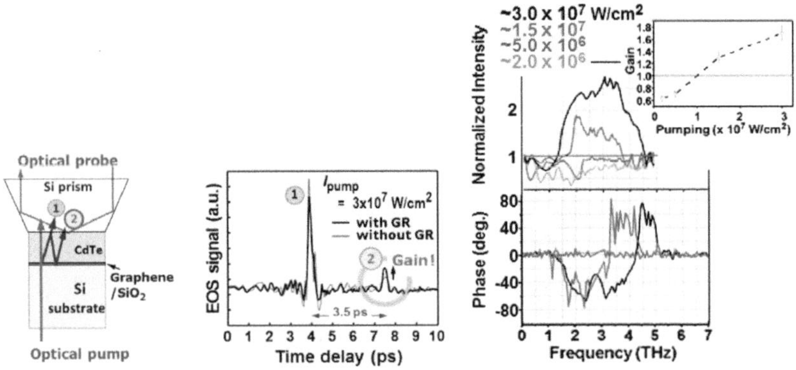

Figure 9.18 Optical pump and THz and optical probe measurement for observation of stimulated THz emission from femtosecond IR laser–pumped exfoliated monolayer graphene. Secondary pulse designated with "2" in the temporal response (center) reflects the interaction of the THz probe pulse and population-inverted graphene, resulting in enhancement of its intensity. The Fourier-transformed gain spectral profiles (right) exhibit a threshold behavior against the pumping intensity, testifying the occurrence of stimulated THz emission in the inverted graphene. Reproduced from Ref. [66] by permission of APS, © 2012.

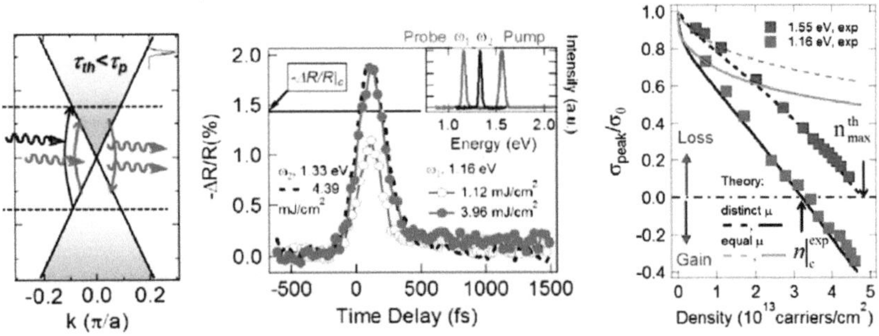

Figure 9.19 Observation of stimulated emission of IR radiation in the femtosecond regime before quasi-equilibration in strongly pumped monolayer graphene. The dependencies of the THz transmission coefficient on the probing photon energy and intensity are plotted (center), showing negative conductivity (THz gain) in the case of a 1.33 eV photon and fluence of 4.30 mJ/cm^2. The dependence of the conductivity values on pumping fluence (photocarrier density) is plotted for two pumping photon energy cases (right). Reproduced from Ref. [114] by permission of APS, © 2012.

9.5.2.2 Ideas for current-injection THz lasers

Optical pumping suffers from significant heating of the electron–hole plasma and the optical phonon system, which suppresses the formation of population inversion, as seen in Fig. 9.17. In the case of optical pumping with sufficiently low photon energy, however, the electron–hole plasma can be cooled down [67]. Another important parameter is the optical phonon decay time, τ_0^{decay}, which stands for the thermal conductivity of the material. A longer τ_0^{decay} (like suspended free-standing graphene) suppresses carrier cooling, preventing population inversion and thus negative conductivity [126]. Current injection pumping is the best solution to substantially reduce the pumping threshold because electrical pumping can serve any pumping energy below the order of "meV" when a p-i-n junction is formed like semiconductor laser diodes. A dual-gate structure can make a p-i-n junction in the graphene channel, as shown in Fig. 9.20 [68, 71]. Gate biasing controls the injection level, whereas the drain bias controls the lasing gain profiles (photon energy and gain). To minimize the undesired tunneling current that lowers the injection efficiency the distance between the dual-gate electrodes must be sufficiently long.

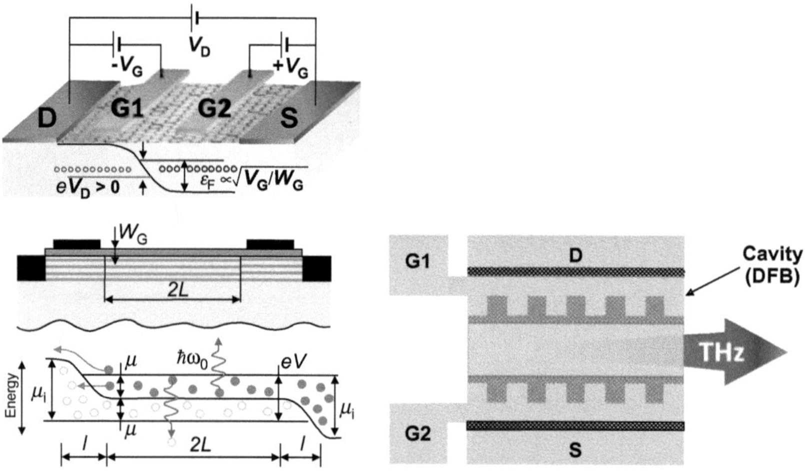

Figure 9.20 Device structure and band diagram of a current-injection-type graphene THz laser based on a dual-gate FET (left) [67, 71] and possible distributed-feedback (DFB) laser cavity structure in a waveguide mode (right).

9.5.3 Giant THz Gain by Excitation of Surface Plasmon Polaritons in Graphene

It is worthy to note that the negative THz conductivity of monolayer graphene is limited to or below the quantum conductivity ($e^2/4\hbar$), as seen in Eq. 9.11. This is because the absorption of THz photons that can contribute to the stimulated emission is only made via the interband transition process whose absorbance is limited by $\pi e^2/\hbar c \sim 2.3\%$ [127]. To overcome this limitation on quantum efficiency, a carrier-recycling process exploited in quantum cascade lasers (QCLs) should be introduced. In this regard, waveguide structures with in-plane THz photon propagation along the graphene sheet are preferable for conforming the laser cavities to maximize the gain overlapping and hence to overcome the quantum mechanical limit in comparison with vertical photon-emitting cavity structures (see Figs. 9.15 and 9.20) [70, 113]. To overcome the 2.3% absorbance limit, introduction of surface plasmon polaritons (SPPs) is a promising, smart, and important idea. There are several factors to exploit graphene plasmons: (i) excitation and propagation of the SPPs along population-inverted graphene [76, 86, 87], (ii) resonant plasmon absorption in structured graphene-like microribbon arrays as well as microdisk arrays [96–100, 112, 128–130], and (iii) superradiant emission mediated by the SPPs [86, 131].

As compared to the lasing associated with the stimulated emission of electromagnetic modes (i.e., photons), the stimulated emission of plasmons by the interband transitions in population-inverted graphene can be a much stronger emission process. The plasmon gain under population inversion in intrinsic graphene has been theoretically studied in Refs. [76, 86]. Nonequilibrium plasmons in graphene can be coupled to the transverse magnetic (TM) modes of electromagnetic waves, resulting in the formation and propagation of SPPs [76]. It is shown in Ref. [76] that the plasmon gain in pumped graphene can be very high due to the small group velocity of the plasmons in graphene and strong confinement of the plasmon field in the vicinity of the graphene layer. Consider the case shown in Fig. 9.21 (left) in which the TM-mode THz photon is irradiated to the xz plane. The propagation index ρ of the graphene SPP along the x coordinate is derived from Maxwell's equations to be

$$\sqrt{n^2-\rho^2}+n^2\sqrt{1-\rho^2}+\frac{4\pi}{c}\sigma_\omega\sqrt{1-\rho^2}\sqrt{n^2-\rho^2}=0 \qquad (9.12)$$

where c is the speed of light in vacuum, n is the refractive index of the substrate, and σ_ω is the conductivity at ω [76]. Thus the absorption coefficient α is obtained as the imaginary part of the wavevector along the x coordinate: $\alpha = \mathrm{Im}(q_z) = 2\mathrm{Im}(\rho\cdot\omega/c)$. Figure 9.21 (right) plots simulated α for monolayer graphene on a substrate for various refraction indices n at 300 K. To set graphene in population inversion and hence negative dynamic conductivity a quasi-Fermi energy ε_F = 20 meV and a carrier momentum relaxation time τ = 10 ps are assumed, respectively. The results demonstrate giant THz gain (negative values of absorption) of the order of 10^4 cm^{-1}. An increase in the substrate refractive index and, consequently, stronger localization of the surface plasmon electric and magnetic fields results in markedly larger gain, that is, a negatively larger absorption coefficient, (compare curves 1 and 2 in Fig. 9.21). As follows from the comparison of curves 2 and 3 in Fig. 9.21, the contribution of the substrate loss to the surface plasmon absorption can be relatively weak at realistic values of the imaginary part of the substrate refractive index (corresponding to curve 3). In particular, in the case of the substrate made of undoped Si (Im $(n) \approx 3 \times 10^{-4}$), the imaginary part of the refractive index can be smaller than those used in the calculations of curves 3–5 in Fig. 9.21. Indeed, waveguiding the THz emitted waves with less attenuation is another key issue to create a graphene THz laser.

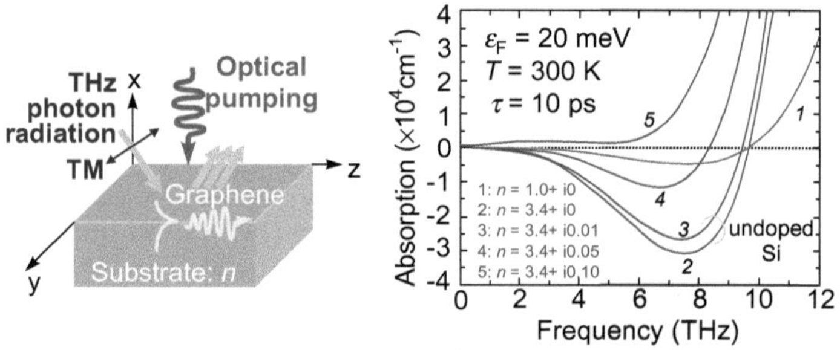

Figure 9.21 Simulated frequency dependences of SPP absorption for monolayer graphene structures with different substrate refraction indices n (1: n = 1.0; 2: n = 3.4; 3: n = 3.4 + i0.01; 4: n = 3.4 + i0.05; 5: n = 3.4 + i0.1) at 300 K, assuming a quasi-Fermi energy ε_F = 20 meV and a carrier momentum relaxation time τ = 10 ps [76]. The results demonstrate giant THz gain (negative values of absorption) of the order of 10^4 cm^{-1}.

We conducted optical pump, THz probe, and optical probe measurement at room temperature for intrinsic monolayer graphene on a SiO_2/Si substrate. The experimental setup is shown in Fig. 9.22, which is identical to the one described in Ref. [66] and is based on a time-resolved near-field reflective electro-optic sampling with a femtosecond-IR laser pulse for optical pumping and a synchronously generated THz pulse for probing the THz dynamics of the sample in a THz photon-echo manner. A 140 μm thick CdTe crystal acting as a THz probe pulse emitter, as well as an electrooptic sensor, was placed on an exfoliated monolayer-graphene/SiO_2/Si sample. A femtosecond-pulsed fiber laser with a full width at half maximum (FWHM) of 80 fs, a repetition rate of 20 MHz, and an average power of ~4 mW was used as the optical pump and probe source. The laser is split into two paths used for pump and probe. The pumping laser beam, being linearly polarized, is mechanically chopped at ~1.2 kHz (for lock-in detection) and focused with a beam diameter of about 120 μm onto the sample and the CdTe from the back side, while the probing beam is cross-polarized with respect to the pump beam and focused from the top side. The CdTe can rectify the optical pump pulse to emit the envelope THz probe pulse. The emitted primary THz beam grows along the Cherenkov angle to be detected at the CdTe top surface as the primary pulse (marked with "①" in Fig. 9.22b) and then reflects being subject to the graphene sample. When the substrate of the sample is conductive the THz probe pulse transmitting through graphene again reflects back to the CdTe top surface, which is electro-optically detected as a THz photon-echo signal (marked with "②" in Fig. 9.22b). Therefore, the original temporal response consists of the first forward-propagating THz pulsation (no interaction with graphene) followed by a photon-echo signal (probing the graphene). The delay between these two pulsations is given by the total round-trip propagation time of the THz probe pulse through the CdTe. The system bandwidth is estimated to be around 6 THz, mainly limited by the Reststrahlen band of the CdTe sensor crystal [66]. The roll-off at a higher frequency of the photon-echo spectrum starts at ~2 THz, continuing monotonically till ~4THz, and lasts at ~6 THz [66].

The sample prepared for this experiment is monolayer graphene, which was exfoliated from graphite and transferred onto a SiO_2/Si substrate, and is identical to the one used in Refs. [66, 67]. The graphene sample was characterized by the surface morphology using atomic force microscopy (AFM), crystallographic properties using Raman spectroscopy, doping effects using Hall-effect measurement, and the microscopic surface potential distribution (fluctuation) using Kelvin force microscopy (KFM), as shown in Refs. [66, 67]. The flake size of the graphene was ~7000 μm^2. The momentum relaxation time was characterized to be 3.5 ps from the G-band-to-D-band peak intensity ratio of the Raman spectra, as also shown in Refs. [66, 67]. Table 9.3 summarizes these properties.

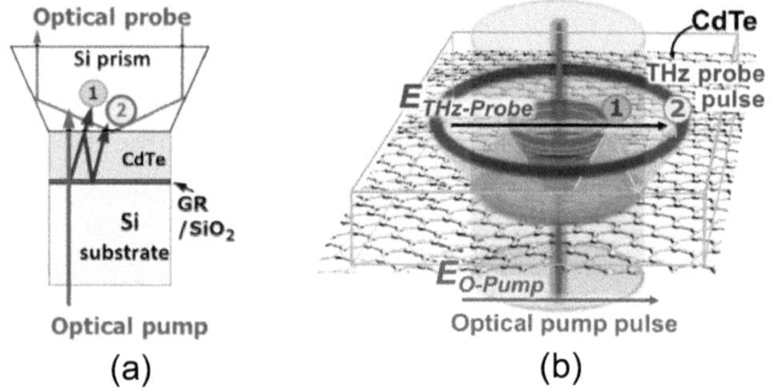

Figure 9.22 Experimental setup of the time-resolved optical pump, THz probe, and optical probe measurement based on near-field reflective electro-optic sampling. (a) Cross-sectional image of the pump/probe geometry and (b) bird's-eye view showing the trajectories of the optical pump and THz probe beams. The polarization of the optical pump and THz probe pulse are depicted with red and dark-blue arrows, respectively. Reproduced from Ref. [132] by permission of IOP Publishing and Deutsche Physikalische Gesellschaft. CC BY-NC-SA, © 2013.

Table 9.3 Graphene sample properties

Synthesis:	Exfoliation from HOPG[a]
Substrate:	300 nm thick SiO_2/560 μm thick Si (100) (resistivity: 0.005 Ω cm)
Flake size:	~>7000 μm^2
Surface height variation as variance:	0.142 nm (in 20 × 20 μm area)
Surface potential variation as variance:	4.02 meV (in 10 × 20 μm area)
Raman G-band peak to D-band peak intensity ratio:	~35
Estimated carrier momentum relaxation time:	3.3 ps at 300 K
Dirac voltage:	~0 V

[a]HOPG = highly ordered pyrolytic graphite

We observe the spatial distribution of the THz probe pulse under the linearly polarized optical pump and THz probe pulse conditions [132]. To measure the in-plane spatial distributions of the THz probe pulse radiation, the optical probe pulse position (at the top surface of the CdTe crystal) was changed step by step by moving the incident point of the optical pump pulse. The pumping intensity I_Ω was fixed at the maximum level, 3×10^7 W/cm^2. We measured 10 times at every point with and without graphene, respectively, and took averages to obtain the electric field intensities and the resultant Fourier transform spectra. The observed field distributions for the primary pulse and the secondary pulse intensity are shown in Fig. 9.23 [132]. The primary pulse field is situated along the circumference with diameter ~50 μm, concentric to the center of the optical pumping position. On the other hand, the secondary pulse (THz photon echo) field is concentrated only at the restricted spot area on and out of the concentric circumference with diameter ~150 μm where the incoming THz probe pulse takes a TM mode being capable of exciting the SPPs in graphene. The distance between the primary pulse position and the secondary pulse position is ~100 μm or longer. The observed field distribution reproduces the reasonable trajectory of the THz echo pulse propagation in the TM modes inside the CdTe crystal, as shown in Fig. 9.22, when we assume the Cherenkov angle of 30 degrees, which was determined by the fraction of the refractive indices between IR and THz frequencies.

How to couple the incoming/outgoing THz pulse photons to the surface plasmons in graphene is the point of discussion, because the defect-less and flat surface of graphene itself has no structural feature that can excite SPPs. One possibility of the excitation of SPPs by the incoming THz probe pulse is the spatial charge density modulation at the area of photoexcitation by optical pumping. The pump beam having a Gaussian profile with diameter ~120 μm may define the continuum SPP modes in a certain THz frequency range, as seen in various SPPs waveguide structures [133, 134]. After a short propagation of the order of ~10 μm the SPPs approach the edge boundary of illuminated and dark areas so that they could mediate the THz electromagnetic emission [135, 136]. The plasmon group velocity in graphene (exceeding the Fermi velocity) and propagation distance give a propagation time of the order of 100 fs. According to the calculated gain spectra shown in Fig. 9.21 (right) the gain enhancement factor could reach or exceed ~10 at the gain peak frequency of 4 THz, which is dominated in the optically probed secondary pulse signals. The obtained gain enhancement factor is ≲50, which is still higher than the calculated results and whose causes should be clarified in further study.

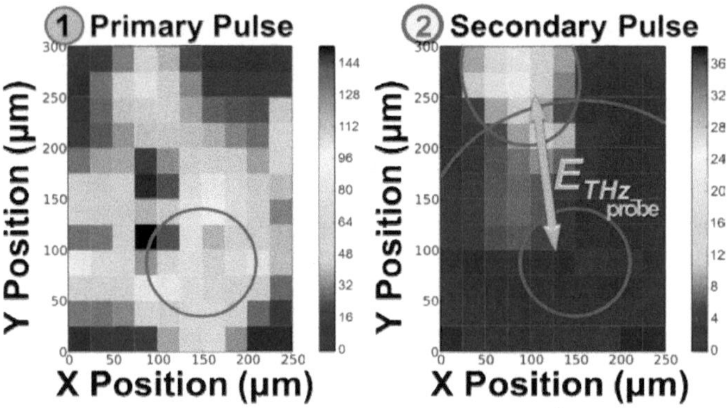

Figure 9.23 Spatial field distribution of THz probe pulse intensities. The primary pulse shows nonpolar distribution, whereas the secondary pulse shows strong localization to the area in which the THz probe pulse is impinged to the graphene surface in TM mode. Reproduced from Ref. [132] by permission of IOP Publishing and Deutsche Physikalische Gesellschaft. CC BY-NC-SA, © 2013.

9.5.4 Active Graphene Plasmonic Metamaterials for Super-Radiant THz Plasmonic Lasing

The amplification of the THz wave by stimulated generation of resonant plasmons in a planar periodic array of graphene plasmonic microcavities has been theoretically derived using a self-consistent electromagnetic approach [86]. Figure 9.24a depicts the graphene device structure. Graphene microcavities are confined between the contacts of the metal grating located on a flat surface of a dielectric (which can be Si or SiC in a practical case) substrate. Suppose that the graphene is pumped either by its optical illumination [66, 67, 69] or by injection of electrons and holes from opposite metal contacts in each graphene microcavity. In the latter case, the opposite ends of each graphene microcavity adjacent to the metal contacts have to be p- and n-doped [68, 71]. One can imagine a biasing scheme for applying DC voltages to successive metal contacts in an interdigital manner [32–36] in order to ensure injection of the carries into each graphene microcavity. The carrier population and hence the dynamic conductivity are characterized by the quasi-Fermi level and carrier temperature, as described in the previous section.

We assume that an external electromagnetic wave (of THz frequency) is incident upon the planar array of graphene microcavities at a normal direction to its plane with the polarization of the electric field across the metal grating contacts (Fig. 9.24a). Then we solve the problem of the amplification of the THz wave by the array of graphene micro-/nanocavities in a semianalytical self-consistent electromagnetic approach similar to that described in Ref. [137]. Figure 9.24b shows the contour map of the calculated absorbance as a function of the quasi-Fermi energy (which corresponds to the pumping strength) and the THz wave frequency for an array of the graphene microcavities with period $L = 4$ μm and the length of each microcavity $a = 2$ μm. The absorbance is defined as the ratio between the absorbed or emitted (which corresponds to negative absorbance) THz power per unit area of the array and the energy flux density in the incoming THz wave. In the amplification regime, the negative value of the absorbance yields the amplification coefficient. The value of $\mathrm{Re}[\sigma_{\mathrm{Gr}}(\omega)]$ is negative above the solid black line in Fig. 9.24b, corresponding to $\mathrm{Re}[\sigma_{\mathrm{Gr}}(\omega)] = 0$ (i.e., to transparent graphene). Above this boundary line, negative absorption (i.e., amplification) takes place at all frequencies and pumping strengths. The plasmon absorption resonances below the $\mathrm{Re}[\sigma_{\mathrm{Gr}}(\omega)] = 0$ line give way to the amplification resonances above this line. Plasmon resonances appear at frequencies $\omega = \omega_{\mathrm{p}}(q)$ determined by the selection rule for the plasmon wavevector $q_{\mathrm{n}} = (2n - 1)\pi/a_{\mathrm{eff}}$, where a_{eff} is the effective length of the graphene micro-/nanocavity. The frequency of the plasmon resonance is determined mainly by the imaginary part of the graphene conductivity, while the real part of the conductivity (Eq. 9.11) is responsible for the energy loss (for $\mathrm{Re}[\sigma_{\mathrm{Gr}}(\omega)] > 0$) or energy gain (for $\mathrm{Re}[\sigma_{\mathrm{Gr}}(\omega)] < 0$).

With increasing ε_{F}, the energy gain can balance the energy loss caused by the electron scattering in graphene so that the net energy loss becomes zero, $\mathrm{Re}[\sigma(\omega)] = 0$, and correspondingly graphene becomes transparent. In this case, the plasmon resonance line exhibits a nonsymmetric Fano-like shape because the real part of graphene conductivity changes its sign across the plasmon resonance. In this case, the plasmon resonance linewidth is given solely by its radiative broadening (because the dissipative damping is close to zero in this case). Above the graphene transparency line $\mathrm{Re}[\sigma_{\mathrm{Gr}}(\omega)] = 0$, the THz wave amplification at the plasmon resonance frequency is several orders of magnitude stronger than away from the resonances (the latter corresponding to the photon amplification in population-inverted graphene [66, 67, 70, 71, 113]). Note that at a certain value of the quasi-Femi energy, the amplification coefficient at the plasmon resonance tends toward infinity with the corresponding amplification linewidth shrinking down to zero. (Of course, the divergence of the amplification coefficient is unphysical; it is a consequence of the linear electromagnetic approach used in this work.) This corresponds to plasmonic lasing in the graphene micro-/nanocavities in the self-excitation regime. The behavior of the amplification coefficient around the self-excitation regime is shown in Fig. 9.24c. The lasing occurs when the plasmon gain balances the electron scattering loss and the radiative loss; see Fig. 9.24c. It means that the plasmon oscillations are highly coherent in this case, with virtually no dephasing at all. The quasi-Fermi energy corresponding to plasmonic lasing in the first plasmon resonance is marked by the red arrow in Fig. 9.24b.

It is worth noting that the plasmons in different graphene microcavities oscillate in phase (even without the incoming electromagnetic wave) because the metal contacts act as synchronizing elements between adjacent graphene microcavities (applying a mechanical analogy, one may think of rigid crossbars connecting oscillating springs arranged in a chain). Therefore, the plasma oscillations in the array of graphene microcavities constitute a single collective plasmon mode distributed over the entire area of the array, which leads to the enhanced *super-radiant* electromagnetic emission from the array [131]. Notice that extraordinary properties of a collective mode in an array of synchronized oscillators are well known in optics: the power of electromagnetic emission from such an array grows as the square of the number of the oscillators in the array [137].

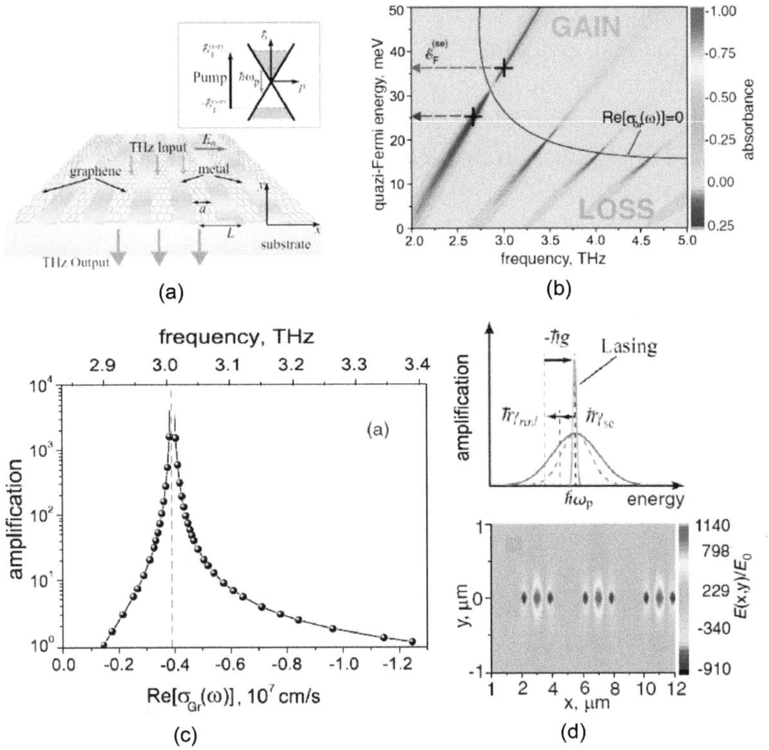

Figure 9.24 (a) Schematic view of the array of graphene micro-/nanocavities. The incoming electromagnetic wave is incident from the top at a normal direction to the structure plane with the polarization of the electric field across the metal grating contacts. The energy band structure of pumped graphene is shown schematically in the inset. (b) Contour map of the absorbance as a function of the quasi-Fermi energy and the frequency of the incoming THz wave for the array of graphene microcavities with period $L = 4$ μm and the length of a graphene microcavity $a = 2$ μm. The electron-scattering time in graphene is $\tau = 10^{-12}$ s. Blue and red arrows mark the quasi-Fermi energies for the maximal absorption and for the plasmonic lasing regime, respectively, at the fundamental plasmon resonance. (c) Variation of the power amplification coefficient along the first-plasmon-resonance lobe. (d) Schematic illustration of the energy rate balance in the plasmon lasing regime (upper) and distribution of the normalized induced in-plane electric field at the moment of time corresponding to the maximal swing of plasma oscillations in the graphene microcavities at the fundamental plasmon amplification resonance (lower). Reproduced from Ref. [86] by permission of APS, © 2012.

It is important to stress that the giant amplification enhancement at the plasmon resonance is ensured by strong confinement of the plasmons in the graphene microcavities; see Fig. 9.24c. As mentioned above, an elevated gain in graphene (approaching the negative of the plasmon radiative damping) is required to meet the self-excitation condition. However, the elevated gain would lead to strong dephasing of a plasma wave over quite a long distance of its propagation (which corresponds to nonresonant stimulated generation of plasmons [76, 87]). Therefore, strong plasmon-mode confinement in a single-mode plasmonic cavity is required to ensure the resonant stimulated generation of plasmons. Plasmon confinement to a single-mode microcavity also enhances the rate of spontaneous electromagnetic emission by the plasmon mode due to the Purcell effect [138]. It is expected that the confinement of plasmons in 2D graphene microcavities (arranged in a chessboard array) could enhance the amplification even more.

9.6 Summary

Recent advances in emission and detection of THz radiation using 2D plasmons in semiconductor nanoheterostructures for nondestructive evaluations were described. The Doppler-shift effect of the plasma-wave velocity under an asymmetric plasmon cavity boundary and/or the spatial modulation of electron transit time in a submicrometer scale 2D plasmon system with a nonuniform 2D electron density distribution can promote the plasmon instability, resulting in self-oscillation of plasmons in the THz regime. The hydrodynamic nonlinearity of 2D plasmons can rectify the incoming electromagnetic radiation, resulting in photovoltaic detection of THz radiation under an asymmetric 2D plasmon cavity boundary. The device structure that can provide practical emission and detection performances were addressed, which are based on an HEMT and incorporates an asymmetrically interdigitated DGG structure. Excellent THz emission and detection performances, including coherent, monochromatic emission beyond the 1 THz range and the record detection responsivity of 2.2 kV/W at 1 THz at 300 K, were experimentally demonstrated by using InAlAs/InGaAs/InP heterostructure material systems. Finally their applications to nondestructive material evaluation based on THz imaging were demonstrated. Obtained results encourage us to proceed to further industrialization of these plasmonic THz devices.

Ultrafast carrier relaxation and recombination dynamics of relativistic Dirac fermions in gapless linear band structures of optically or electrically pumped graphene give rise to negative dynamic conductivity in a wide THz range. This will open a new paradigm to create solid-state THz emitters and lasers using graphene. 2D plasmons in graphene dramatically enhance the light–matter interaction, improving the quantum efficiency and hence leading to higher output power. Graphene ribbon arrays coupled with metallic grating can extensively promote the cooperative excitation of plasmon resonances at the mode frequencies, resulting in superradiant THz emission of a lasing operation. Numerical simulation reveals the new physics in graphene–metal ribbon arrays. It is shown that due to strong confinement of the plasmon modes in graphene microcavities and the super-radiant nature of electromagnetic emission from the array of the plasmonic microcavities, amplification of the THz wave is enhanced by several orders of magnitude at the plasmon resonance frequencies. The simulated gain coefficient exceeds 10^4 cm^{-1} in a wide THz range, which is now under test device fabrication. To this end we have two possibilities of stimulated emission of THz "photons" and "plasmons." These new findings can lead to the creation of new types of room-temperature-operating graphene-based intense THz emitters and lasers.

Nonlinear graphene plasmon dynamics can further extend the functionality from simple THz emission. Mikhailov and Ziegler have theoretically shown that Dirac fermions in graphene with linear dispersion in the presence of an applied AC electric field with an angular frequency ω produce an AC current of not only the original fundamental frequency mode but also odd harmonic modes with angular frequencies $n\omega$ and magnitudes $1/n$ as large as the fundamental one (where n is a series of harmonic odd numbers), even within the linear response without any saturation due to the Pauli blocking [139, 140]. This should be contrasted with conventional semiconductors such as GaAs, which, within the linear response, will only produce an AC current with the original frequency component. Therefore, the dynamics of graphene in an

AC electric field are intrinsically nonlinear, and efficient frequency multiplications for THz generation substantially outperforming any existing materials can be expected for microwave-driven graphene.

References

1. M. Tonouchi, Cutting-edge terahertz technology, *Nat. Photon.* **1**, 97–105 (2007).
2. X.-C. Zhang and J. Xu, *Introduction to THz Wave Photonics* (Springer-Verlag, NY, 2009).
3. P. Bakshi, K. Kempa, A. Scorupsky, C. G. Du, G. Feng, R. Zobl, G. Strasser, C. Rauch, Ch. Pacher, K. Unterrainer, and E. Gornik, Plasmon-based terahertz emission from quantum well structures, *Appl. Phys. Lett.* **75**, 1685 (1999).
4. V. Ryzhii, N. A. Bannov, and V. A. Fedirko, Ballistic and quasiballistic transport in semiconductor structures (review), *Sov. Phys. Semicond.* **18**, 481–491 (1984).
5. N. Dyakonova, F. Teppe, J. Lusakowski, W. Knap, M. Levinshtein, A. P. Dmitriev, M. S. Shur, S. Bollaert, and A. Cappy, Magnetic field effect on the terahertz emission from nanometer InGaAs/AlInAs high electron mobility transistors, *J. Appl. Phys.* **97**, 114313 (2005).
6. A. El Fatimy, N. Dyakonova, Y. Meziani, T. Otsuji, W. Knap, S. Vandenbrouk, K. K. Madjour, D. Théron, C. Gaquiere, M. A. Poisson, S. Delage, P. Pristawko, and C. Skierbiszewski, AlGaN/GaN high electron mobility transistors as a voltage-tunable room temperature terahertz sources, *J. Appl. Phys.* **107**, 024504 (2010).
7. K. Hirakawa, K. Yamanaka, M. Grayson, and D. C. Tsui, Far-infrared emission spectroscopy of hot two-dimensional plasmons in Al0.3Ga0.7As/GaAs heterojunctions, *Appl. Phys. Lett.* **67**, 2326 (1995).
8. R. A. Hopfel, E. Vass, and E. Gornik, Thermal excitation of two-dimensional plasma oscillations, *Phys. Rev. Lett.* **49**, 1667–1671 (1982).
9. R. Hopfel, G. Lindemann, E. Gornik, G. Stangl, A. C. Gossard, and W. Wiegmann, Cyclotron and plasmon emission from two-dimensional electrons in GaAs, *Surf. Sci.* **113**, 118–123 (1982).
10. W. Knap, J. Lusakowski, T. Parenty, S. Bollaert, A. Cappy, V. V. Popov, and M. S. Shur, Terahertz emission by plasma waves in 60 nm gate high electron mobility transistors, *Appl. Phys. Lett.* 84, 2331 (2004).

11. J. Lusakowski, W. Knap, N. Dyakonova, and L. Varani, Voltage tunable terahertz emission from a ballistic nanometer InGaAs/InAlAs transistor, *J. Appl. Phys.* **97**, 064307 (2005).

12. S. A. Mikhailov, Plasma instability and amplification of electromagnetic waves in low-dimensional electron systems, *Phys. Rev. B* **58**, 1517–1532 (1998).

13. N. Okisu, Y. Sambe, and T. Kobayashi, Far-infrared emission from two-dimensional plasmons in AlGaAs/GaAs heterointerfaces, *Appl. Phys. Lett.* **48**, 776 (1986).

14. V. Ryzhii, A. Satou, and M. Shur, Plasma instability and terahertz generation in HEMTs due to electron transit-time effect, *IEICE Trans. Electron.* **E89-C**, 1012 (2006).

15. V. Ryzhii, A. Satou, M. Ryzhii, T. Otsuji, and M. S. Shur, Mechanism of self-excitation of terahertz plasma oscillations in periodically double-gated electron channels, *J. Phys.: Condens. Matter* **20**, 384207 (2008).

16. A. Tredicucci, R. Kohler, L. Mahler, E. HBeere, E. H. Linfield, and D. A. Ritchie, Terahertz quantum cascade lasers-first demonstration and novel concepts, *Semicond. Sci. Technol.* **20**, S222–S227 (2005).

17. D. C. Tsui, E. Gornik, and R. A. Logan, Far infrared emission from plasma oscillations of Si inversion layers, *Solid State Commun.* **35**, 875–877 (1980).

18. R. J. Wilkinson, C. D. Ager, T. Duffield, H. P. Hughes, D. G. Hasko, H. Armed, J. E. F. Frost, D. C. Peacock, D. A. Ritchie, A. C. Jones, C. R. Whitehouse, and N. Apsley, Plasmon excitation and self-coupling in a bi-periodically modulated two-dimensional electron gas, *J. Appl. Phys.* **71**, 6049–6061 (1992).

19. M. Dyakonov and M. Shur, Detection, mixing, and frequency multiplication of terahertz radiation by two-dimensional electronic fluid, *IEEE Trans. Electron. Dev.* **43**, 1640–1645 (1996).

20. A. El Fatimy, J. C. Delagnes, A. Younus, E. Nguema, F. Teppe, W. Knap, E. Abraham, and P. Mounaix, Plasma wave field effect transistor as a resonant detector for 1 terahertz imaging applications, *Opt. Commun.* **282**, 3055–3058 (2009).

21. M. Dyakonov and M. Shur, Shallow water analogy for a ballistic field effect transistor: new mechanism of plasma wave generation by dc current, *Phys. Rev. Lett.* **71**, 2465–2468 (1993).

22. W. Knap, S. Nadar, H. Videlier, S. Boubanga-Tombet, D. Coquillat, N. Dyakonova, F. Teppe, K. Karpierz, J. Lusakowski, M. Sakowicz, I. Kasalynas, D. Seliuta, G. Valusis, T. Otsuji, Y. Meziani, A. El Fatimy, S.

Vandenbrouk, K. Madjour, D. Théron, and C. Gaquière, Field effect transistors for terahertz detection and emission, *J. IR mm THz Waves* **32**, 618–628 (2011).

23. A. Lisauskas, W. von Spiegel, S. Boubanga, A. E. Fatimy, D. Coquillat, F. Teppe, N. Dyakonova, W. Knap, and H. G. Roskos, Terahertz imaging with GaAs field-effect transistors, *Electron. Lett.* **44**, 408–409 (2008).
24. A. Lisauskas, U. Pfeiffer, E. Öjefors, P. Haring Bolivar, D. Glaab, and H. G. Roskos, Rational design of highresponsivity detectors of terahertz radiation based on distributed self-mixing in silicon field-effect transistors, *J. Appl. Phys.* **105**, 114511 (2009).
25. V. M. Muravev, I. V. Kukushkin, J. Smet, and K. von Klitzing, Millimeter/submillimeter mixing based on the nonlinear plasmon response of two-dimensional electron systems, *JETP Lett.* **90**, 197–201 (2009).
26. S. Nadar, D. Coquillat, M. Sakowicz, H. Videlier, F. Teppe, N. Dyakonova, W. Knap, J.-M. Peiris, J. Lyonnet, D. Seliuta, I. Kasalynas, G. Valušis, K. Madjour, D. Théron, C. Gaquière, and M.-A. Poisson, Terahertz imaging using high electron mobility transistors as plasma wave detectors, *Phys. Status Solidi C* **6**, 2855–2857 (2009).
27. F. Schuster, D. Coquillat, H. Videlier, M. Sakowicz, F. Teppe, L. Dussopt, B. Giffard, T. Skotnicki, and W. Knap, Broadband terahertz imaging with highly sensitive silicon CMOS detectors, *Opt. Express* **19**, 7827–7832 (2011).
28. Y. F. Sun, J. D. Sun, Y. Zhou, R. B. Tan, C. H. Zeng, W. Xue, H. Qin, B. S. Zhang, and D. M. Wu, Room temperature GaN/AlGaN self-mixing terahertz detector enhanced by resonant antennas, *Appl. Phys. Lett.* **98**, 252103 (2011).
29. T. Tanigawa, T. Onishi, S. Takigawa, and T. Otsuji, Enhanced responsivity in a novel AlGaN/GaN plasmon-resonant terahertz detector using gate-dipole antenna with parasitic elements, *68th Device Res. Conf. Dig.*, 167–168, Notre Dame, IN (2010).
30. F. Teppe, W. Knap, D. Veksler, M. S. Shur, A. P. Dmitriev, Yu. V. Kachorovskii, and S. Rumyantsev, Room-temperature plasma waves resonant detection of sub-terahertz radiation by nanometer field-effect transistor, *Appl. Phys. Lett.* **87**, 052107 (2005).
31. W. Knap, M. Dyakonov, D. Coquillat, F. Teppe, N. Dyakonova, J. Lusakowski, K. Karpierz, M. Sakowicz, G. Valusis, D. Seliuta, I. Kasalynas, A. El Fatimy, Y. M. Meziani, and T. Otsuji, Field effect transistors for terahertz detection: Physics and first imaging applications, *J. IR mm THz Waves* **30**, 1319–1337 (2009).

32. D. Coquillat, S. Nadar, F. Teppe, N. Dyakonova, S. Boubanga-Tombet, W. Knap, T. Nishimura, T. Otsuji, Y. M. Meziani, G. M. Tsymbalov, and V. V. Popov, Room temperature detection of sub-terahertz radiation in double-grating-gate transistors, *Opt. Express* **18**, 6024–6032 (2010).

33. T. Otsuji, M. Hanabe, T. Nishimura, and E. Sano, A grating-bicoupled plasma-wave photomixer with resonant-cavity enhanced structure, *Opt. Express* **14**, 4815–4825 (2006).

34. T. Otsuji, Y. M. Meziani, T. Nishimura, T. Suemitsu, W. Knap, E. Sano, T. Asano, and V. V. Popov, Emission of terahertz radiation from dual-grating gate plasmon-resonant emitters fabricated with InGaP/ InGaAs/ GaAs material systems, *J. Phys.: Condens. Matter* **20**, 384206 (2008).

35. Y. Tsuda, T. Komori, A. El Fatimy, T. Suemitsu, and T. Otsuji, Application of plasmonic-resonant microchip emitters to broadband terahertz spectroscopic measurement, *J. Opt. Soc. Am. B* **26**, A52–A57 (2009).

36. Y. M. Meziani, Handa, H., W. Knap, T. Otsuji, E. Sano, V. V. Popov, G. M. Tsymbalov, D. Coquillat, and F. Teppe, Room temperature terahertz emission from grating coupled two-dimensional plasmons, *Appl. Phys. Lett.* **92**, 201108 (2008).

37. M. I. Dyakonov, Boundary instability of a two-dimensional electron fluid, *Semiconductors* **42**, 984–988 (2008).

38. M. Dyakonov and M. S. Shur, Current instability and plasma waves generation in ungated two-dimensional electron layers, *Appl. Phys. Lett.* **87**, 111501 (2005).

39. S. Boubanga-Tombet, Y. Tanimoto, T. Watanabe, T. Suemitsu, Y. Wang, H. Minamide, H. Ito, V. Popov, and T. Otsuji, Asymmetric dual-grating gate InGaAs/InAlAs/InP HEMTs for ultrafast and ultrahigh sensitive terahertz detection, *DRC Dig.* **IV.A-10**, 169–170, Pennsylvania State University, PA, USA (2012).

40. T. Otsuji, T. Watanabe, S. Boubanga Tombet, A. Satou, W. Knap, V. Popov, M. Ryzhii, and V. Ryzhii, Emission and detection of terahertz radiation using two-dimensional electrons in III-V semiconductors and graphene, *IEEE Trans. THz Sci. Technol.* **3**, 63–72 (2013).

41. V. V. Popov, D. V. Fateev, T. Otsuji, Y. M. Meziani, D. Coquillat, and W. Knap, Plasmonic terahertz detection by a double-grating-gate field-effect transistor structure with an asymmetric unit cell, *Appl. Phys. Lett.* **99**, 243504 (2011).

42. T. Watanabe, S. Boubanga Tombet, Y. Tanimoto, D. Fateev, V. Popov, D. Coquillat, W. Knap, Y. Meziani, Y. Wang, H. Minamide, H. Ito, and T.

Otsuji, Ultrahigh sensitive plasmonic terahertz detector based on an asymmetric dual-grating gate HEMT structure, *Solid State Electron.* **78**, 109–114 (2012).

43. T. Watanabe, S. Boubanga Tombet, Y. Tanimoto, D. Fateev, V. Popov, D. Coquillat, W. Knap, Y. Meziani, Y. Wang, H. Minamide, H. Ito, and T. Otsuji, InP-and GaAs-based plasmonic high-electron-mobility transistors for room-temperature ultrahigh-sensitive terahertz sensing and imaging, *IEEE Sens. J.* **13**, 89–99 (2013).
44. *OAD-7 Golay Detector Operating Manual* (Cardiff, U.K., 2005).
45. M. Sakowicz, M. B. Lifshits, O. A. Klimenko, F. Schuster, D. Coquillat, F. Teppe, and W. Knap, Terahertz responsivity of field effect transistors versus their static channel conductivity and loading effects, *J. Appl. Phys.* **110**, 054512 (2011).
46. Q. Bao and K. P. Loh, Graphene photonics plasmonics and broadband optoelectronic devices, *ACS Nano* **6**, 3677–3694 (2012).
47. F. Bonaccorso, Z. Sun, T. Hansan, and A. C. Ferrari, Graphene photonics and optoelectronics. *Nat. Photon.* **4**, 611–622 (2010).
48. A. Bostwick, T. Ohta, T. Seyller, K. Horn, and E. Rotenberg, Quasiparticle dynamics in graphene, *Nat. Phys.* **3**, 36–40 (2007).
49. A. H. Castro Neto, F. Guinea, N. M. R. Peres, K. S. Novoselov, and A. K. Geim, The electronic properties of graphene, *Rev. Mod. Phys.* **81**, 109–162 (2009).
50. A. K. Geim and K. S. Novoselov, The rise of graphene, *Nat. Mater.* **6**, 183 (2007).
51. A. N. Grigorenko, M. Polini, and K. S. Novoselov, Graphene plasmonics, *Nat. Photon.* **6**, 749–758 (2012).
52. M. I. Katsnelson, Graphene: carbon in two dimensions, *Mater. Today* **10**, 20 (2007).
53. K. S. Novoselov, A. K. Geim, S. V. Morozov, D. Jiang, Y. Zhang, S. V. Dubonos, I. V. Grigorieva, and A. A. Frisov, Electric field effect in atomically thin carbon films, *Science* **306**, 666 (2004).
54. K. S. Novoselov, A. K. Geim, S. V. Morozov, D. Jiang, M. I. Katsnelson, I. V. Grigorieva, S. V. Dubonos, and A. Firsov, Two-dimensional gas of massless Dirac fermions in graphene, *Nature* **438**, 197 (2005).
55. Y. Zhang, Y.-W. Tan, H. L. Stormer, and P. Kim, Experimental observation of the quantum Hall effect and Berry's phase in graphene, *Nature* **438**, 201 (2005).
56. R. Cheng, J. Bai, L. Liao, H. Zhou, Y. Chen, L. Liu, Y.-C. Lin, S. Jiang, Y. Huang, and X. Duan, High-frequency self-aligned graphene transistors with transferred gate stacks, *PNAS* **109**, 11588–11592 (2012).

57. F. Schwierz, Graphene transistors, *Nat. Nanotech.* **5**, 487–496 (2010).

58. Y. Q. Wu, D. B. Farmer, A. Valdes-Garcia, W. J. Zhu, K. A. Jenkins, C. Dimitrakopoulos, P. Avouris, and Y.-M. Lin, Record high RF performance for epitaxial graphene transistors, *IEDM Tech. Dig. IEEE Ed.* **23.8** (2012).

59. J. Zheng, L. Wang, R. Quhe, Q. Liu, H. Li, D. Yu, W.-N. Mei, J. Shi, Z. Gao, and J. Lu, Sub-10 nm gate length graphene transistors: operating at terahertz frequencies with current saturation, *Sci. Rep.* **3**, 1314 (2013).

60. T. Mueller, F. Xia, and P. Avouris, Graphene photodetectors for high-speed optical communications, *Nat. Photon.* **4**, 297–301 (2010).

61. V. Ryzhii, N. Ryabova, M. Ryzhii, N. V. Varyshnikov, V. E. Karasik, V. Mitin, and T. Otsuji, Terahertz and infrared photodetectors based on multiple graphene layer and nanoribbon structures, *Opto-Electron. Rev.* **20**, 15–25 (2012).

62. T. Ando, T. Nakanishi, and R. Saito, Berry's phase and absence of back scattering in carbon nanotubes, *J. Phys. Soc. Jpn.* **67**, 2857 (1998).

63. M. Fujita, K. Wakabayashi, K. Nakada, and K. Kusakabe, Peculiar localized state at zigzag graphite edge, *J. Phys. Soc. Jpn.* **65**, 1920 (1996).

64. R. Saito, T. Takeya, T. Kimura, G. Dresselhoau, and M. S. Dresselhaus, Raman intensity of single-wall carbon nanotubes, *Phys. Rev. B* **57**, 4145 (1998).

65. K. Wakabayashi, M. Sigrist, and M. Fujita, Spin wave mode of edge-localized magnetic states in nanographite zigzag ribbons, *J. Phys. Soc. Jpn.* **67**, 2089 (1998).

66. S. Boubanga-Tombet, S. Chan, T. Watanabe, A. Satou, V. Ryzhii, and T. Otsuji, Ultrafast carrier dynamics and terahertz emission in optically pumped graphene at room temperature, *Phys. Rev. B* **85**, 035443 (2012).

67. T. Otsuji, S. Boubanga Tombet, A. Satou, M. Ryzhii, and V. Ryzhii, Terahertz-wave generation using graphene-toward new types of terahertz lasers, *IEEE J. Sel. Top. Quantum Electron.* **19**, 8400209 (2013).

68. M. Ryzhii and V. Ryzhii, Injection and population inversion in electrically induced p-n junction in graphene with split gates, *Jpn. J. Appl. Phys.* **46**, L151–L153 (2007).

69. V. Ryzhii, M. Ryzhii, and T. Otsuji, Negative dynamic conductivity in optically pumped graphene, *J. Appl. Phys.* **101**, 083114 (2007).

70. V. Ryzhii, M. Ryzhii, A. Satou, T. Otsuji, A. A. Dubinov, and V. Y. Aleshkin, Feasibility of terahertz lasing in optically pumped epitaxial multiple graphene layer structures, *J. Appl. Phys.* **106**, 084507 (2009).

71. V. Ryzhii, M. Ryzhii, V. Mitin, and T. Otsuji, Toward the creation of terahertz graphene injection laser, *J. Appl. Phys.* **110**, 094503 (2011).

72. A. Satou, F. Vasko, and V. Ryzhii, Nonequilibrium carriers in intrinsic graphene under interband photoexcitation, *Phys. Rev. B* **78**, 115431 (2008).

73. D. R. Andersen and H. Raza, Plasmon dispersion in semimetallic armchair graphene nanoribbons, *Phys. Rev. B* **85**, 075425 (2012).

74. L. Brey and H. A. Fertig, Elementary electronic excitations in graphene nanoribbons, *Phys. Rev. B* 75, 125434 (2007).

75. J. Christensen, A. Manjavacas, S. Thongrattanasiri, F. H. L. Koppens, and F. J. de Abajo, Graphene plasmon waveguiding and hybridization in individual and paired nanoribbons, *ACS Nano* **6**, 431 (2012).

76. A. A. Dubinov, Y. V. Aleshkin, V. Mitin, T. Otsuji, and V. Ryzhii, Terahertz surface plasmons in optically pumped graphene structures, *J. Phys.: Condens. Matter* **23**, 145302 (2011).

77. L. Falkovsky and A. Varlamov, Space-time dispersion of graphene conductivity, *Eur. Phys. J. B* **56**, 281 (2007).

78. G. W. Hanson, Dyadic Green's functions and guided surface waves for a surface conductivity model of graphene, *J. Appl. Phys.* **103**, 064302 (2008).

79. E. H. Hwang and S. Das Sarma, Dielectric function, screening, and plasmons in two-dimensional graphene, *Phys. Rev. B* **75**, 205418 (2007).

80. M. Jablan, H. Buljan, and M. Soljačić, Plasmonics in graphene at infrared frequencies, *Phys. Rev. B* **80**, 245435 (2009).

81. F. H. L. Koppens, D. E. Chang, and F. J. García de Abajo, Graphene plasmonics: a platform for strong light-matter interactions, *Nano Lett.* **11**, 3370 (2011).

82. A. Yu. Nikitin, F. Guinea, F. J. Garcia-Vidal, and L. Martin-Moreno, Edge and waveguide terahertz surface plasmon modes in graphene microribbons, *Phys. Rev. B* **84**, 161407 (2011).

83. A. Yu. Nikitin, F. Guinea, F. J. Garcia-Vidal, and L. Martin-Moreno, Fields radiated by a nanoemitter in a graphene sheet, *Phys. Rev. B* **84**, 195446 (2011).

84. A. Yu. Nikitin, F. Guinea, F. J. Garcia-Vidal, and L. Martin-Moreno, Surface plasmon enhanced absorption and suppressed transmission in periodic arrays of graphene ribbons, *Phys. Rev. B* **85**, 081405(R) (2012).

85. V. V. Popov, T. Yu. Bagaeva, T. Otsuji, and V. Ryzhii, Oblique terahertz plasmons in graphene nanoribbon arrays, *Phys. Rev. B* **81**, 073404 (2010).

86. V. V. Popov, O. V. Polischuk, A. R. Davoyan, V. Ryzhii, T. Otsuji, and M. S. Shur, Plasmonic terahertz lasing in an array of graphene nanocavities, *Phys. Rev. B* **86**, 195437 (2012).

87. F. Rana, Graphene terahertz plasmon oscillators, *IEEE Trans. Nanotechnol.* **7**, 91–99 (2008).

88. F. Rana, J. H. Strait, H. Wang, and C. Manolatou, Ultrafast carrier recombination and generation rates for plasmon emission and absorption in graphene, *Phys. Rev. B* **84**, 045437 (2011).

89. V. Ryzhii, Terahertz plasma waves in gated graphene heterostructures, *Jpn. J. Appl. Phys.* **45**, L923 (2006).

90. V. Ryzhii, A. Satou, and T. Otsuji, Plasma waves in two-dimensional electron-hole system in gated graphene heterostructures, *J. Appl. Phys.* **101**, 024509 (2007).

91. K. W.-K. Shung, Dielecytic function and plasmon structure of stage-1 intercalated graphite, *Phys. Rev. B* **34**, 979–993 (1986).

92. D. Svintsov, V. Vyurkov, S. Yurchenko, T. Otsuji, and V. Ryzhii, Hydrodynamic model for electron-hole plasma in graphene, *J. Appl. Phys.* **111**, 083715 (2012).

93. O. Vafek, Thermoplasma polariton within scaling theory of single-layer graphene, *Phys. Rev. Lett.* **97**, 266406 (2006).

94. A. Vakil, and N. Engheta, Transformation optics using graphene, *Science* **332**, 1291 (2011).

95. B. Wunsch, T. Stauber, F. Sols, and F. Guinea, Dynamical polarization of graphene at finite doping, *New J. Phys.* **8**, 318 (2006).

96. J. Cheng, M. Badioli, P. Alonso-González, S. Thongrattanasiri, F. Huth, J. Osmond, M. Spasenović, A. Centeno, A. Pesquera, P. Godignon, A. Z. Elorza, N. Camara, F. J. García de Abajo, R. Hillenbrand, and F. H. L. Koppens, Optical nano-imaging of gate-tunable graphene plasmons, *Nature* **487**, 77 (2012).

97. Z. Fei, G. O. Andreev, W. Bao, L. M. Zhang, A. S. McLeod, C. Wang, M. K. Stewart, Z. Zhao, G. Dominguez, M. Thiemens, M. M. Fogler, M. J. Tauber, A. H. Castro Neto, C. N. Lau, F. Keilmann, and D. N. Basov, Infrared nanoscopy of Dirac plasmons at the graphene-SiO2 interface, *Nano Lett.* **11**, 4701 (2011).

98. Z. Fei, A. S. Rodin, G. O. Andreev, W. Bao, A. S. McLeod, M. Wagner, L. M. Zhang, Z. Zhao, M. Thiemens, G. Dominguez, M. M. Fogler, A. H. Castro

Neto, C. N. Lau, F. Keilmann, and D. N. Basov, Gate-tuning of graphene plasmons revealed by infrared nano-imaging, *Nature* **487**, 82 (2012).

99. L. Ju, B. Geng, J. Horng, C. Girit, M. Martin, Z. Hao, H. A. Bechtel, X. Liang, A. Zettl, Y. R. Shen, and F. Wang, Graphene plasmonics for tunable terahertz metamaterials, *Nat. Nanotech.* **6**, 630 (2011).

100. H. Yan, X. Li, B. Chandra, G. Tulevski, Y. Wu, M. Freitag, W. Zhu, P. Avouris, and F. Xia, Tunable infrared plasmonic devices using graphene/insulator stacks, *Nat. Nano* **7**, 330–334 (2012).

101. T. Otsuji, T. Watanabe, S. Boubanga Tombet, A. Satou, V. Ryzhii, V. Popov, and W. Knap, Emission and detection of terahertz radiation using two dimensional plasmons in semiconductor nano-heterostructures for nondestructive evaluations, *Opt. Eng.* **53***(3)*, 031206 (2013).

102. T. Watanabe, A. Satou, T. Suemitsu, W. Knap, V. Popov, and T. Otsuji, Plasmonic terahertz monochromatic coherent emission from an asymmetric chirped dual-grating-gate InAlAs/InGaAs/InP HEMT with highly asymmetric resonant cavities, *OTST* **Th3-26**, Kyoto (2013).

103. E. G. Mishchenko, A. V. Shytov, and P. G. Silvestrov, Guided plasmons in graphene p-n junctions, *Phys. Rev. Lett.* **104**, 156806 (2010).

104. W. Wang, P. Apel, and J. Kinaret, Edge plasmons in graphene nanostructures, *Phys. Rev. B* **84**, 085423 (2011).

105. A. E. Nikolaenko, N. Papasimakis, E. Atmatzakis, Z. Luo, Z. Xiang Shen, F. D. Angelis, S. A. Boden, E. D. Fabrizio, and N. I. Zheludev, Nonlinear graphene metamaterial, *Appl. Phys. Lett.* **100**, 181109 (2012).

106. S. Thongrattanasiri, F. H. L. Koppens, and F. J. García de Abajo, Complete optical absorption in periodically patterned graphene, *Phys. Rev. Lett.* **108**, 047401 (2012).

107. S. Das Sarma, S. Adam, E. H. Hwang, and E. Rossi, Electronic transport in two-dimensional graphene, *Rev. Mod. Phys.* **83**, 407–470 (2011).

108. F. Vasko, and V. Ryzhii, Voltage and temperature dependencies of conductivity in gated graphene, *Phys. Rev. B* **76**, 233404 (2007).

109. T. Ando, Y. Zheng, and H. Suzuura, Dynamical conductivity and zero-mode anomaly in honeycomb lattices, *J. Phys. Soc. Jpn.* **71**, 1318–1324 (2002).

110. L. A. Falkovsky and S. S. Pershoguba, Optical far-infrared properties of graphene monolayer and multilayers, *Phys. Rev. B* **76**, 153410 (2007).

111. K. F. Mak, M. Y. Sfeir, Y. Wu, C. H. Lui, J. A. Misewich, and T. F. Heinz, Measurement of the optical conductivity of graphene, *Phys. Rev. Lett.* **101**, 196405 (2008).

112. B. Sensale-Rodriguez, R. Yan, M. Zhu, D. Jena, L. Liu, and H. G. Xing, Efficient terahertz electro-absorption modulation employing graphene plasmonic structures, *Appl. Phys. Lett.* **101**, 261115 (2012).

113. V. Ryzhii, A. Dubinov, T. Otsuji, V. Mitin, and M. S. Shur, Terahertz lasers based on optically pumped multiple graphene structures with slot-line and dielectric waveguides, *J. Appl. Phys.* **107**, 054505 (2010).

114. T. Li, L. Luo, M. Hupalo, J. Zhang, M. C. Tringides, J. Schmalian, and J. Wang, Femtosecond population inversion and stimulated emission of dense Dirac Fermions in graphene, *Phys. Rev. Lett.* **108**, 167401 (2012).

115. T. Otsuji, S. A. Boubanga Tombet, A. Satou, H. Fukidome, M. Suemitsu, E. Sano, V. Popov, M. Ryzhii, and V. Ryzhii, Graphene materials and devices in terahertz science and technology, *MRS Bull.* **37**, 1235–1243 (2012).

116. T. Otsuji, S. A. Boubanga Tombet, A. Satou, H. Fukidome, M. Suemitsu, E. Sano, V. Popov, M. Ryzhii, and V. Ryzhii, Graphene-based devices in terahertz science and technology, *J. Phys. D* **45**, 303001 (2012).

117. A. Satou, T. Otsuji, and V. Ryzhii, Theoretical study of population inversion in graphene under pulse excitation, *Jpn. J. Appl. Phys.* **50**, 070116 (2011).

118. M. Breusing, C. Ropers, and T. Elsaesser, Ultrafast carrier dynamics in graphite, *Phys. Rev. Lett.* **102**, 086809 (2009).

119. J. M. Dawlaty, S. Shivaraman, M. Chandrashekhar, F. Rana, and M. G. Spencer, Measurement of ultrafast carrier dynamics in epitaxial graphene, *Appl. Phys. Lett.* **92**, 042116 (2008).

120. P. A. George, J. Strait, J. Dawlaty, S. Shivaraman, M. Chandrashekhar, and F. R. M. G. Spencer, Ultrafast optical-pump terahertz-probe spectroscopy of the carrier relaxation and recombination dynamics in epitaxial graphene, *Nano Lett.* **8**, 4248–4251 (2008).

121. M. S. Foster and I. L. Aleiner, Slow imbalance relaxation and thermoelectric transport in graphene, *Phys. Rev. B* **79**, 085415 (2009).

122. J. H. Strait, H. Wang, S. Shivaraman, V. Shields, M. Spencer, and F. Rana, Very slow cooling dynamics of photoexcited carriers in graphene observed by optical-pump terahertz-probe spectroscopy, *Nano Lett.* **11**, 4902 (2011).

123. S. Tani, F. Blanchard, and K. Tanaka, Ultrafast carrier dynamics in graphene under a high electric field, *Phys. Rev. Lett.* **109**, 166603 (2012).

124. T. Winzer, A. Knorr, and E. Malic, Carrier multiplication in graphene, *Nano Lett.* **10**, 4839 (2010).

125. T. Winzer and E. Malic, Impact of Auger processes on carrier dynamics in graphene, *Phys. Rev. B* **85**, 241404(R) (2012).

126. V. Ryzhii, M. Ryzhii, V. Mitin, A. Satou, and T. Otsuji, Effect of heating and cooling of photogenerated electron-hole plasma in optically pumped graphene on population inversion, *Jpn. J. Appl. Phys.* **50**, 094001 (2011).

127. R. R. Nair, P. Blake, A. N. Grigorenko, K. S. Novoselov, T. J. Booth, T. Stauber, N. M. R. Peres, and A. K. Geim, Fine structure constant defines visual transparency of graphene, *Science* **320**, 1308 (2008).

128. L. Prechtel, L. Song, D. Schuh, P. Ajayan, W. Wegscheider, and A. W. Holleitner, Time-resolved ultrafast photocurrents and terahertz generation in freely suspended graphene, *Nat. Commun.* **3**, 646 (2012).

129. B. Sensale-Rodriguez, R. Yan, M. M. Kelly, T. Fang, K. Tahy, W. S. Hwang, D. Jena, L. Liu, and H. G. Xing, Broadband graphene terahertz modulators enabled by intraband transitions, *Nat. Commun.* **3**, 780 (2012).

130. B. Sensale-Rodriguez, R. Yan, S. Rafique, M. Zhu, W. Li, X. Liang, D. Gundlach, V. Protasenko, M. M. Kelly, D. Jena, L. Liu, and H. G. Xing, Extraordinary control of terahertz beam reflectance in graphene electro-absorption modulators, *Nano Lett.* **12**, 4518–4522 (2012).

131. P. A. Huidobro, A. Y. Nikitin, C. Gonzalez-Ballestero, L. Martin-Moreno, and F. J. Garcıa-Vidal, Superradiance mediated by graphene surface plasmons, *Phys. Rev. B* **85**, 155438 (2012).

132. T. Watanabe, T. Fukushima, Y. Yabe, S. Boubanga Tombet, A. Satou, A. A. Dubinov, V. Ya Aleshkin, V. Mitin, V. Ryzhii, and T. Otsuji, Gain enhancement effect of surface plasmon polaritons on terahertz stimulated emission in optically pumped monolayer graphene, *New J. Phys.* **15**, 075003 (2013).

133. X.-Y. He, Numerical analysis of the propagation properties of subwavelength semiconductor slit in the terahertz region, *Opt. Express* **17**, 15359–15371 (2009).

134. X.-Y. He, Q.-J. Wang, and S.-F. Yu, Numerical study of gain-assisted terahertz hybrid plasmonic waveguide, *Plasmonics* **7**, 571–577 (2012).

135. B. J. Lawrie, R. F. Haglund, Jr., and R. Mu, Enhancement of ZnO photoluminescence by localized and propagating surface plasmons, *Opt. Express* **17**, 2565 (2009).

136. S. Wedge and W. Barnes, Surface plasmon-polariton mediated light emission through thin metal films, *Opt. Express* **12**, 3672 (2004).

137. M. Benedict, A. Ermolaev, V. Malyshev, I. Sokolov, and E. Trifinov, *Superradiance: Multiatomic Coherent Emission* (IOP, Bristol, 1996).

138. E. M. Purcell, Spontaneous emission probabilities at radio frequencies, *Phys. Rev.* **69**, 681 (1946).

139. S. A. Mikhailov, Non-linear electromagnetic response of graphene, *Europhys. Lett.* **79**, 27002 (2007).

140. S. A. Mikhailov and K. Ziegler, Nonlinear electromagnetic response of graphene: frequency multiplication and the self-consistent-field effects, *J. Phys.: Condens. Matter* **20**, 384204 (2008).

Chapter 10

Terahertz Metamaterials

Seung Hoon Lee,[a] Muhan Choi,[b] and Bumki Min[a]
[a]*Department of Mechanical Engineering, Korea Advanced Institute of Science and Technology (KAIST), 291 Daehak-ro, Yuseong-gu, Daejeon, 305-701, Korea*
[b]*School of Electronics Engineering, Kyungpook National University, Daegu, 702-701, Korea*
troa@kaist.ac.kr, bmin@kaist.ac.kr

Electromagnetic metamaterials are artificially constructed composite media that are composed of subwavelength-scale, periodic metallodielectric structures. They have emerged recently as a very promising platform for the manipulation of electromagnetic waves, such as microwaves and terahertz (THz) waves. To date, various exotic phenomena have been demonstrated experimentally as a result of the benefits derived from well-developed, microfabrication techniques. For example, negative indices of refraction as well as ultrahigh refractive indices, both of which are generally unattainable in nature, have been achieved successfully in the THz frequency regime with so-called "THz metamaterials." Moreover, active control of THz waves, which is regarded as one of the most challenging issues in the field of THz science and technologies, can now be accomplished using THz metamaterials coupled with various active

Handbook of Terahertz Technologies: Devices and Applications
Edited by Ho-Jin Song and Tadao Nagatsuma

ISBN 978-981-4613-08-8 (Hardcover), 978-981-4613-09-5 (eBook)
www.panstanford.com

media, such as conventional semiconductors, two-dimensional electron gas (2DEG) systems, phase transition materials, and carbon nanomaterials, including graphene and nanotubes. In this chapter, we briefly review the historical background, theory, and fabrication methods of THz metamaterials and then discuss recent progress on passive- and active-type THz metamaterials designed for many useful applications at THz frequencies.

10.1 Introduction

Despite the long history of the research on light–matter interactions, similar studies in the terahertz (THz) wave regime have been relatively underdeveloped in fields ranging from the fundamental sciences to practical applications [1]. The main reason for this is the fact that efficient THz sources and detectors have yet to be developed for those kinds of phenomena to be observed reliably. However, in recent years, THz waves have attracted the attention of physicists and material scientists because they offer the possibility of unveiling many important, low-energy physical properties of condensed matters in a noncontact, nondestructive way [2]. For example, the development of THz time domain spectroscopy (THz-TDS) and the quantum cascade laser (QCL) has paved the way for the characterization of a variety of materials as well as the development of many important THz components and devices. However, most naturally existing materials cannot respond efficiently to THz waves, so the development of practical devices to efficiently manipulate THz waves remains still challenging and currently is an important issue in THz research field.

A metamaterial is an artificially constructed composite medium designed specifically to have certain desired electromagnetic responses at any frequency range of interest. Historically, the first practical design of such an artificial material was developed by Sir John Pendry, who modeled a metallic wire array with the Drude model, in which a plasma frequency is located in the microwave frequency range [3]. The plasma frequency can be controlled by modifying the geometrical parameters of the metallic wire array. Similarly, artificial magnetic response (i.e., magnetism) can be derived using nonmagnetic conducting structures as the basic building block [4]. Due to the ability to control both electric and magnetic responses, many interesting phenomena have been observed at microwave frequencies, such as negative indices of refraction. At the same time, many efforts have been made to apply this principle to higher-frequency regimes, such as the THz and visible regimes.

Metamaterials designed to operate in the THz frequency regime are of great importance due to their strong THz responses. The desired electromagnetic properties of THz metamaterials can be obtained by tailoring the electric and/or magnetic responses of so-called "meta-atoms," which are the basic elements of metamaterials. Therefore, in the following sections, the basic concepts of electromagnetism that pertain to metamaterials are reviewed briefly. In addition, the representative fabrication methods for producing various THz metamaterials are addressed. As a specific example, several active-type THz metamaterials that possibly could be used for real-time manipulation of THz waves are discussed briefly, and notable recent works in this area are mentioned.

10.2 Theory of Metamaterials

Meta-atoms generally are made of a conducting material in which the application of an external electromagnetic field causes a surface current to flow. Therefore, the properties of a metamaterial are determined mostly by the geometry of the meta-atom (e.g., arrangement and morphology) rather than the elements that constitute the substance (especially for microwave and THz frequencies). Using innovative designs and arrangements of various resonant meta-atoms, metamaterials can be created that exhibit extraordinary electric and magnetic behaviors that are not likely to occur in nature.

A necessary condition for the design of metamaterials is that the size of a unit cell (or the meta-atom) must be substantially smaller than the wavelength of interest so that the metamaterial can be considered as an effective homogeneous medium. Inside this medium, the meta-atom acts as an individual oscillator driven by an applied time-varying electromagnetic field. Therefore, carefully designed subwavelength-scale metallic structures that satisfy the requirement of effective homogeneity can be used to create new optical properties. When the homogenization condition is satisfied, the electromagnetic properties of metamaterials can be characterized by defining effective optical parameters (e.g., effective permittivity and effective permeability).

In the following two subsections, two representative structures of meta-atoms are reviewed that were designed to provide geometrically controllable electric and magnetic properties.

10.2.1 Tailoring Electric Response

To design a metamaterial to provide specific electrical properties, it is essential to understand the electromagnetic response of the basic substance of the metamaterial. Since the usual requirement of metamaterials is that they consist of conductive materials, it is useful to review the electromagnetic properties of metals.

The response of a noble metal to the electric field of incident electromagnetic waves is described by its permittivity. Using the Drude model, the dielectric function of metals can be expressed as a function of frequency:

$$\varepsilon(\omega) = 1 - \frac{\omega_{\mathrm{p}}^2}{\omega^2 + i\gamma\omega} \tag{10.1}$$

where γ is the collision frequency and ω_{p} is the plasma frequency, which is defined as

$$\omega_{\mathrm{p}} \equiv \sqrt{\frac{ne^2}{\varepsilon_0 m}} \tag{10.2}$$

where n, e, ε_0, and m are the electron density, elementary charge, vacuum permittivity, and mass of the electron, respectively. The plasma frequency is the natural frequency of a collective oscillation of free electrons. In noble metals, the plasma frequency is located in the UV regime.

Among natural materials, it is difficult to identify materials that have plasma frequencies located in the THz or microwave regime. One exceptional example is the outermost part of the earth's atmosphere, known as the ionosphere, which is partially ionized. In this zone, the electron density is very low, so the plasma frequency is approximately 1 MHz.

Materials that have plasma frequencies in the THz frequency regime definitely can be promising for a wide variety of important THz applications, such as telecommunications and THz devices. In the late 1990s, Sir John Pendry first proposed a dilute array of thin metallic wires in order to decrease the material's plasma frequency to the microwave frequency regime [3, 5]. As shown in Fig. 10.1, the lattice constant a of the metallic wire array is much less than the incident wavelength, so the homogenization condition is fulfilled. Thus, this composite material with vacuum and metal wires can be represented by a plasma-like dielectric function, as can be seen later. In this composite medium, the electrons are allowed to move only within the metallic wires, and thus, the effective electron density of the composite n_{eff} can be written as

$$n_{\text{eff}} = \left(\frac{\pi r^2}{a^2}\right).n \tag{10.3}$$

where r, a, and n are the radius of the wire, the lattice constant, and the electron density in the metal, respectively.

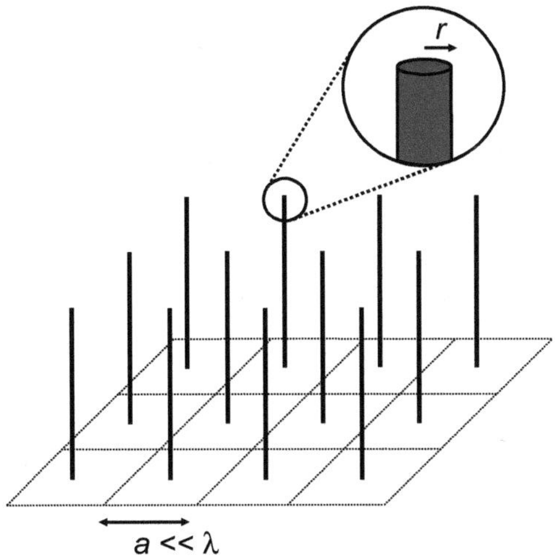

Figure 10.1 An array of infinitely long, parallel, and thin metallic wires of radius r placed on a periodic square lattice with a distant of a, which shows a low plasma frequency for the electric field parallel to the wires [3].

At the same time, the effective mass of electrons also is increased significantly by self-inductance, as follows:

$$m_{\text{eff}} = \left(\frac{\mu \pi r^2 n e^2}{2\pi} \right) . \ln \left(\frac{a}{r} \right) \tag{10.4}$$

Since the two effective quantities presented above are dependent strongly on geometric parameters, the "effective" plasma frequency can be controlled by changing the structural arrangement of the unit cell. For example, the plasma frequency is located near 30 GHz when metallic wires that have radii of 10 μm are arranged periodically with a lattice constant of 10 mm. Therefore, by tailoring the effective plasma frequency with an artificially designed metamaterial, an electric response that is unattainable from natural materials can be obtained in the desired frequency range.

10.2.2 Tailoring Magnetic Response

As mentioned in the previous section, the effective plasma frequency of a metallic wire array can be tuned to the desired frequency range. Meanwhile, the magnetic responses of conventional materials are hardly tunable, and most of them appear in the low-frequency regime. Typically, except for some ferromagnetic materials, it is difficult to achieve magnetic resonance at mid-THz and optical frequencies because the magnetism originates from the alignment of the electron spin or electronic orbital currents [6].

Pendry et al. demonstrated the existence of a strong, artificial magnetic resonance in a desired frequency range by periodically arranging the subwavelength, metallic, split-ring structures, as shown in Fig. 10.2 [4]. The current flowing around the split rings induces magnetic dipoles in the direction of the applied magnetic field. The resonance frequency of this structure is determined by inductance of the current loop and capacitance between the adjacent rings and the gaps. This resonance can be controlled by changing the unit cell's geometry. Using a macroscopic Maxwell's equation, the effective permeability can be stated in the following Lorentzian form:

$$\mu_{\text{eff}} = 1 - \frac{\pi r^2}{a^2} \frac{\omega^2}{\omega^2 - \omega_0^2 + i\Gamma\omega} \tag{10.5}$$

where $\Gamma = (2\rho)/(\mu_0 r)$ is a dissipation parameter that is determined by resistivity, ρ, of the constituting metal. Therefore, the effective magnetic permeability can be controlled by the geometry, shape, and arrangement of the unit cell in the same fashion as the electrical permittivity of the metallic wire array.

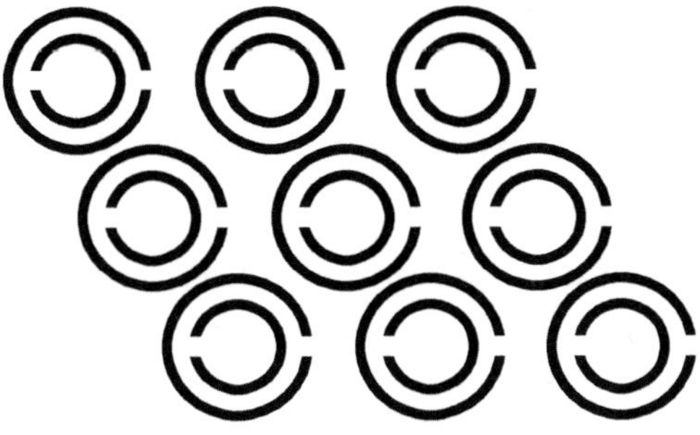

Figure 10.2 An array of split-ring resonators (SRRs) exhibits a rich resonant response inducing large diamagnetism when excited by a magnetic field parallel to the stacking axis [4, 7].

10.2.3 Utilization of Metamaterials at THz Frequencies

THz frequencies, generally defined as the range from 0.1 THz to 10 THz, are of great importance in exploring a wide range of physical phenomena [1, 2]. Various materials show interesting electromagnetic responses at THz frequencies. For example, atomic/molecular vibrations lead to unique absorption spectra in the THz range. With this spectral signature, identification of the substance becomes possible with THz waves, which can be applied to security screening and medical diagnoses [8–11].

Despite the importance of THz waves, naturally occurring materials rarely exhibit strong electric and/or magnetic resonances in the THz regime [6]. Typically, conventional magnetic materials show magnetic resonance at much lower frequencies (below the gigahertz range), because the magnetism arises from unpaired electron spin. The electric resonances of many natural materials induced by lattice vibration or some other mechanism are located in a much higher frequency range. Noble metals, for example, exhibit their electric resonances above infrared (IR) frequencies (typically in the UV regime).

By using a proper meta-atom, the desired electric and/or magnetic responses can be placed selectively in the THz frequency regime. For example, the exotic properties of materials, such as negative indices of refraction, can be accessible by using THz metamaterials [12, 13]. THz metamaterials may be useful in the development of THz imaging systems with resolutions beyond the diffraction limit and the development of perfect THz absorbers, to name a few. In addition, THz metamaterials may be utilized further for active devices to manipulate THz waves, such as switches, modulators, and memory devices.

10.3 Fabrication of THz Metamaterials

Since metamaterials are composed of subwavelength-scale meta-atoms, suitable fabrication methods are required depending on the operating frequencies. In the THz regime, the wavelength is about a few hundred micrometers (e.g., 1 THz ≈ 300 μm in wavelength), and thus, a single unit cell should be in the microscale size to satisfy the homogenization condition. To pattern such micrometer-sized structures, various fabrication methods have been used.

The most common methods used to construct THz metamaterials are the technology used to fabricate conventional integrated circuits (ICs) and microelectromechanical system (MEMS) processes. Typically, meta-atoms are made of metals and dielectrics, and their host medium is chosen selectively from solid-state bulk substrates [14] or flexible polymers [15] according to the intended purpose. These methods are based on photolithography, and they are more appropriate for use in creating 2D, planar, THz metamaterials. Although three-dimensional metamaterials can be made by multistacking [16, 17] or rolling-up 2D layers [18, 19], more complicated 3D structures are difficult to fabricate using conventional photolithography. Recently, nonconventional, microfabrication methods have been used to create more complex 3D THz metamaterials [20]. The representative microfabrication methods for genuine 3D structures are listed below:

- Direct laser writing [21, 22]
- X-ray lithography (e.g., LIGA process) [23, 24]
- Femtosecond laser microlens array (fs-MLA) lithography [17, 25]
- Multilayer electroplating (MLEP) [26–28]
- Laser printing [29]

- Stereolithography (i.e., two-photon polymerization) [30, 31]

10.4 Passive-Type THz Metamaterials

THz metamaterials can be classified as passive- and active-type materials, depending on the controllability of the electromagnetic properties, such as resonance frequency, amplitude/phase in transmission or refection, absorption, and refractive index. In the case of passive-type metamaterials, the electromagnetic properties cannot be altered by additional stimuli once the metamaterials are fabricated. Conversely, the properties of active-type metamaterials, in which various active media generally are embedded, can be tailored by external stimuli, such as optical, electrical, thermal, or mechanical means. In this section, we briefly review two kinds of passive-type metamaterials designed to operate at THz frequencies.

First, we discuss THz metamaterials that have negative refractive indices. Since the first demonstration of metamaterials with negative refractive indices at microwave frequencies by Shelby, Smith, and Schultz [7], the same kind of approach has been used to realize artificial negative refractive indices for higher frequency ranges, such as the THz to IR regime. Realizing such negative refractive indices would show promise for various applications, including super-resolution imaging.

Second, the principles used to design THz metamaterials that have high refractive indices are presented. Such metamaterials have the potential to extend the positive limit of the refractive index in various frequency ranges. For the THz regime, Choi et al. [16] recently demonstrated an artificial material that has an unnaturally high refractive index.

10.4.1 Metamaterials with Negative Refractive Indices

A question for the existence of a material that has a negative refractive index was raised by Victor Veselago in the late 1960s [32]. He introduced a simple plot of the materials' effective parameter called the "epsilon–mu table" to classify the electromagnetic response of materials (Fig. 10.3). According to Maxwell's equations, the refractive index of a material is defined as $\pm\sqrt{\varepsilon\mu}$, the sign of which depends on the signs of ε and μ. Veselago predicted that a negative value of μ would be expected if one could find magnetic monopoles in natural materials, providing the basis for the existence of negative refractive indices. Although he knew that identifying conditions in which materials had negative refractive indices would be physically meaningful work, unfortunately, the magnetic monopoles have yet to be confirmed.

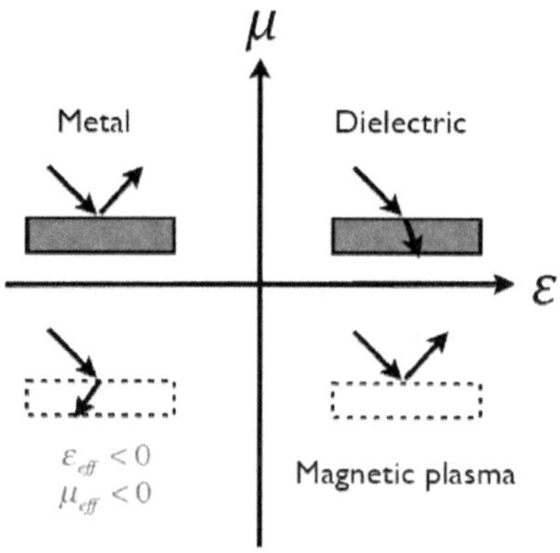

Figure 10.3 Veselago's ε–μ table. Electric permittivity and magnetic permeability determine the propagation of electromagnetic waves in matter. Note that the permittivity of metal can be positive.

In the late 1990s, an artificial material with negative permittivity or permeability was proposed theoretically by Sir John Pendry et al. [3, 4]. By constructing metallic wire arrays with a subwavelength-scale lattice constant, the effective plasma frequency, which determines the range of negative permittivity, could be decreased to microwave frequencies (as was introduced in Section 10.2.1). Similarly, negative permeability was observed near the resonance frequency using split-rings resonators (SRRs), that is, so-called magnetic meta-atoms.

A year later, Smith et al. developed the clever idea of combining two different meta-atoms into a single metamolecule to make a composite medium that simultaneously had negative permittivity and negative permeability [33]. As shown in Fig. 10.4, this composite medium had overlapping frequency regions of negative permittivity and permeability. The first physical manifestation of a negative refractive index was reported in 2001 using the above concept of the composite metamolecule at microwave frequency [7] with copper SRRs and wires on a circuit board. By conducting a Snell's law experiment (i.e., measuring the angle-dependent microwave transmission), it was confirmed that this composite medium had a negative refractive index. This preliminary experiment verified that Veselago's hypothesis could be realized in an artificially constructed material system. Shortly after this demonstration, researchers began to rethink their understanding of the propagation of waves in matter, and the metamaterial with a negative refractive index opened up new frontiers in photonics with many interesting physical concepts. As an example, a perfect lens made of metamaterials that have a negative refractive index enabled the imaging of fine features with subwavelength resolution, overcoming the diffraction limit.

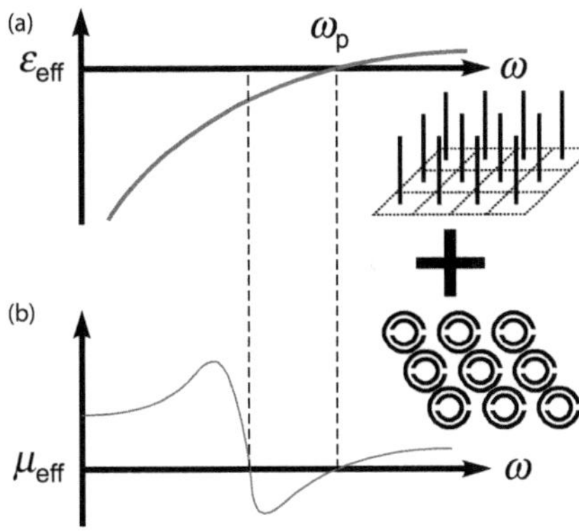

Figure 10.4 Concept for the metamolecules of simultaneously having negative permittivity and permeability. (b) Experimental demonstration of a negative-index metamaterial by combining dilute metallic wires and SRRs [7].

In the early stages of developing THz metamaterials, efforts were devoted to achieving strong magnetic responses [34] in order to extend the frequency ranges for the negative refractive index. Similar to microwave metamaterials, the development of materials with negative refractive indices at THz frequencies was achieved by overlapping the frequency ranges of negative permittivity with those of negative permeability. For this purpose, a combined array of SRRs and thin metallic wires was patterned onto a dielectric, planar substrate using microfabrication methods [12, 22]. Unfortunately, the negative refraction through this structure was valid only in the in-plane propagation direction due to the 2D origin of the SRR patterns from conventional photolithography. More recently, negative refractive indices were demonstrated with THz waves propagating in the direction normal to the substrate plane utilizing different unit cells, such as a fishnet structure [35–39] and a cross pair [13], both of which have been used to develop bulk metamaterials that have negative refractive indices.

10.4.3 Ultrahigh-Refractive-Index Metamaterial

The development of a positive, high refractive index would broaden the research field of metamaterials (Fig. 10.5). For example, a high index would provide better light-bending power and more design flexibility in the manipulation of wave propagation. Typically, naturally existing dielectric materials have small, positive refractive indices, except for a few semiconductors and insulators. Recently, an ultrahigh-index metamaterial was proposed theoretically [40, 41]. However, the proposed structure was difficult to achieve because of its complicated, 3D geometry.

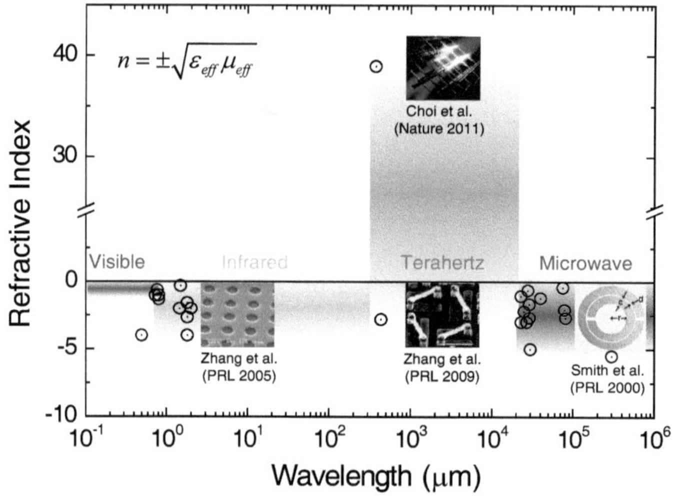

Figure 10.5 Experimentally demonstrated refractive indices of metamaterials at various frequencies. Figures reprinted with permission from Ref. [16], Copyright (2011), Nature Publishing Group, and from Refs. [33], [39], and [42], Copyrights (2000), (2005), and (2009), respectively, the American Physical Society.

By stacking metasurface layers, Choi et al. circumvented this difficulty in fabrication and experimentally proposed a feasible high-index structure for quasi-3D THz metamaterials [16]. Each layer was composed of an array of thin, I-shaped, gold meta-atoms that were patterned within a highly flexible polyimide substrate. The fabricated metamaterials exhibited a very high refractive index for broadband THz frequency ranges. By significantly increasing the effective permittivity through strong capacitive coupling between the meta-atoms, while maintaining effective permeability around/near unity by minimizing the volume fraction of the metal, a peak refractive index of 38.6 was demonstrated experimentally for a single-layer THz metamaterial (Fig. 10.6). As a natural extension of this single-layer metamaterial, quasi-3D, high-refractive-index metamaterials also were fabricated, and a maximum bulk refractive index of 33.2 was obtained.

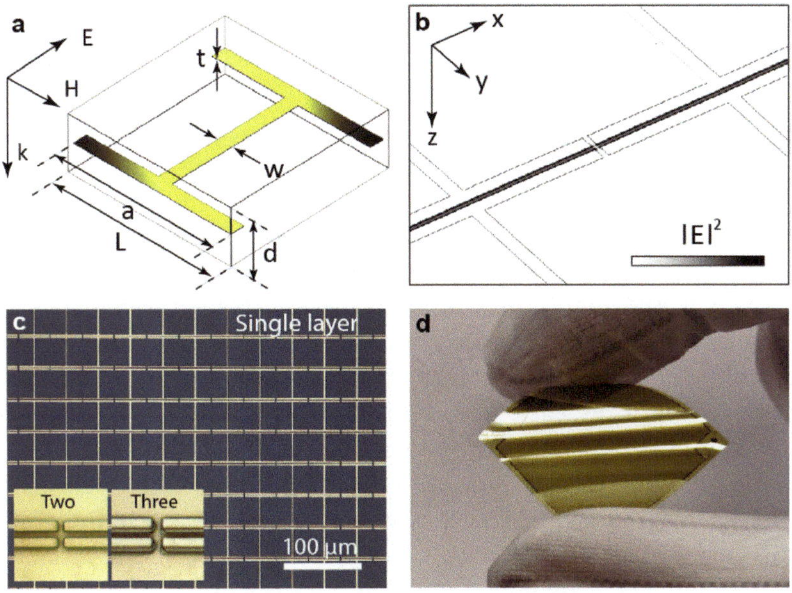

Figure 10.6 A high-index THz metamaterial made of thin I-shaped meta-atoms. (a) Unit cell structure and (b) induced electric in the I-shaped meta-atoms. The charge accumulation at the edge of the meta-atoms greatly increases the polarization. (c) Optical microscope image and (d) photograph of the fabricated metamaterial. Reproduced from Ref. [16], © 2011, Nature Publishing Group.

10.5 Active-Type THz Metamaterials

Active control of electromagnetic waves is at the heart of modern electronics and photonics. During the last few years, active-type THz metamaterials have been studied intensively for the manipulation of THz waves. While most natural materials have a very weak response to THz waves, engineered metamaterials, when combined with active, natural materials, are able to control THz waves more effectively due to their strong electromagnetic resonances. Recently, the importance of active THz metamaterials has been highlighted in the field of THz science and technologies, in which efficient optical components (such as switches, filters, and modulators) are required. Thus, we present a brief overview of recent work on active THz metamaterials. Active metamaterials can be classified into three distinct categories, that is, optically tunable metamaterials, electrically tunable metamaterials, and mechanically tunable metamaterials, according to the methods that are used to tune the electromagnetic properties.

10.5.1 Optically Tunable THz Metamaterials

In 2005, Janke et al. fabricated a 2D, indium antimonide (InSb) grating with square apertures and demonstrated all-optical switching of the transmitted THz waves by controlling photoexcited carriers in the semiconductor [43]. Two years later, Hendry et al. expanded this work to show ultrafast optical switching of the THz response, such as extraordinary transmission using subwavelength metallic apertures patterned onto a silicon substrate [44]. Noting the fact that semiconductors can be used as suitable active-tuning media for THz waves, especially when combined with subwavelength metallic structures, researchers started to integrate specially designed metamolecules with a semiconductor material. In 2006, Padilla et al. demonstrated that the electromagnetic response of THz metamaterials can be controlled dynamically by an optical pump [45]. In their experiment, metallic SRRs were patterned on a semi-insulating gallium arsenide (SI-GaAs) substrate, and optical-pump THz-probe spectroscopy was used to characterize the pump-induced THz responses of the metamaterial. As shown in Fig. 10.7, the real part of permittivity changed noticeably with optical pulse excitation. Without the optical pump, strong resonances of SRRs were clearly observed. However, with increased pump fluence, the resonance strength of SRRs was weakened due to the photoexcited carriers in SI-GaAs.

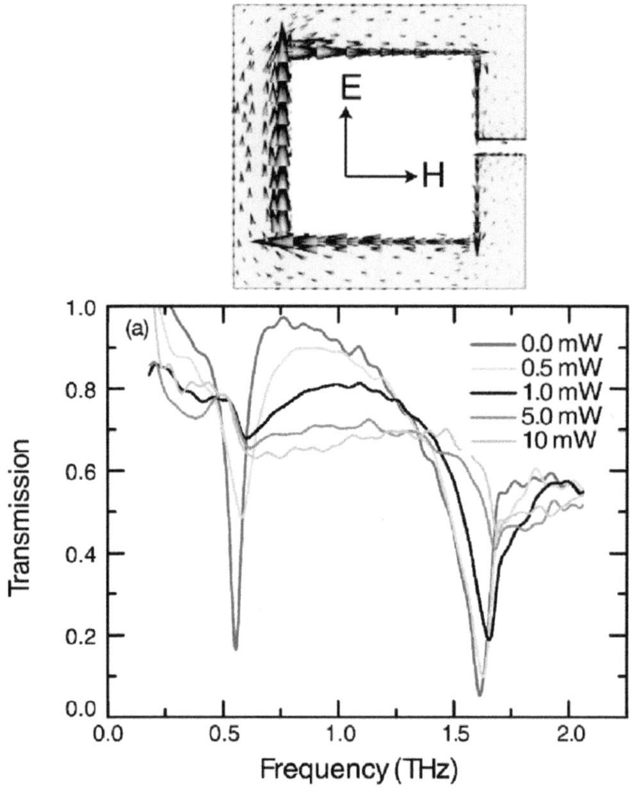

Figure 10.7 Optically controlled THz metamaterial based on SRRs on a SI-GaAs substrate. Reproduced form Ref. [45] by permission of APS, © 2006.

In addition to controlling the resonance strength of metamaterials, shifting of the resonance frequency was demonstrated with a different kind of active metamaterials in which a part of the meta-atom was made of the active semiconductor [46]. Small capacitor plates in the center part of a meta-atom were made of silicon, and these composite meta-atom structures were placed on a quartz substrate. With optical pump excitation, the effective capacitance of the meta-atoms was increased by the photoexcited carriers in the silicon. As a result, the resonance frequency was shifted gradually to lower frequencies as the pump fluence increased.

Utilizing the change in the conductivity of semiconductors under photoexcitation, chirality switching was reported recently [47]. Zhang et al. demonstrated strong rotatory power of the polarization angle and controlled its ellipticity with optical pulse excitation. Similarly, the negative index from the THz chiral metamaterials was controlled by the same method [48]. Optically tunable metamaterials have great potential for use in ultrafast modulation of THz waves when they are combined with semiconductor materials that possess short electron–hole recombination times. An ErAs/GaAs nanoisland superlattice [49] or graphene [50] could be a promising candidate for the semiconducting materials for this purpose because the recombination time of photoexcited carriers for these materials typically is of the order of a few picoseconds.

10.5.2 Electrically Tunable THz Metamaterials

Although optically tunable THz metamaterials have been demonstrated to be a quite versatile platform for the ultrafast control of THz waves, as described in the previous section, a more practical method is required for the realistic implementation of THz devices. Electrical tuning may be a promising alternative to optical control. Series of electrically tunable THz metamaterials were proposed by Chen et al. for efficient real-time control and manipulation of THz waves, such as amplitude [51], phase [52], and spatial modulation [53]. As shown in Fig. 10.8a, these devices are based on the hybrid structure of the *n*-GaAs/metamaterial in which a Schottky junction is formed at the interface. By applying a bias voltage between metallic meta-atoms and the substrate, charge carriers are depleted near the split gap, which causes a change in the resonance strength. Therefore, the amplitude and phase of the transmitted THz waves can be controlled effectively by the bias voltage. In the modulation scheme that utilizes carrier depletion, the speed of modulation generally depends on the mobility of the carrier. Therefore, to increase the modulation speed, high-electron-mobility transistors (HEMTs) with an embedded 2DEG layer were tested. For this metamaterial, modulation speeds up to 10 MHz were achieved [54].

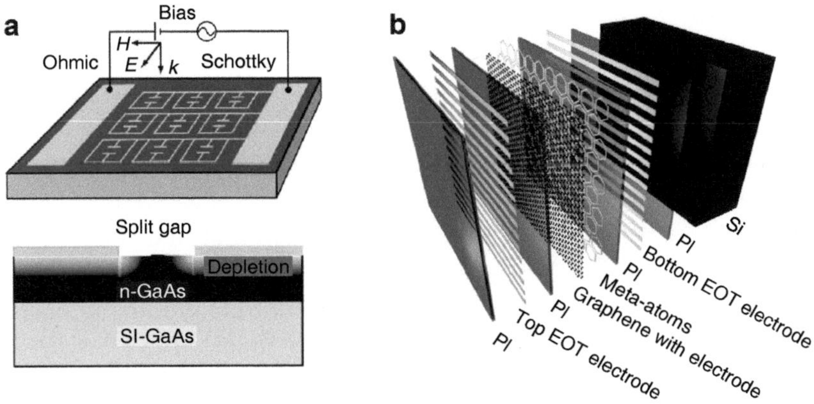

Figure 10.8 Electrically controlled THz metamaterials. (a) *n*-GaAs/metamaterial and (b) graphene/metamaterial hybrid devices for switching THz waves [55].

Another type of electrically tunable THz metamaterials, which used single-layer graphene as the active medium, was reported recently by Lee et al. [55]. A graphene/metamaterial hybrid structure was fabricated within a flexible polyimide substrate and integrated with a "THz" transparent gate electrode (Fig. 10.8b). Unlike near-IR and visible light, THz waves are absorbed in the graphene layer via intraband transition. In their device, the Fermi level of graphene (and the degree of intraband transition) was controlled by electrical gating. To increase the modulation depth, hexagonally shaped meta-atoms, which were strongly resonant at THz frequencies, were placed directly onto the graphene layer. With this approach, amplitude and phase of the transmitted THz waves could be controlled efficiently by electrical gating. More interestingly, the device showed hysteretic behavior in the THz transmission for a cyclic variation of gate voltage, and this hysteretic behavior was utilized for the demonstration of electrically controlled, photonic memory operation.

10.5.3 Mechanically Tunable THz Metamaterials

The electromagnetic response of metamaterials was determined mainly by the structural shapes of the basic building blocks, that is, the meta-atoms. Accordingly, THz wave modulation can be achieved by modifying the structural shape directly or by arranging the meta-atoms. For this reason, mechanically tunable metamaterials are also called reconfigurable metamaterials. The first reconfigurable metamaterials at THz frequencies were fabricated using the MEMS process [56]. In their work, metallic SRRs were attached on a bimaterial microcantilever that can be deflected by a thermal stimulus. As a result, a remarkable transmission change was observed, which was attributed to the structural deformation of the meta-atoms. Another type of reconfigurable THz metamaterials, in which asymmetric SRRs were placed on in-plane moving parts actuated by electrostatic force, was recently reported [57, 58]. One of SRRs in the unit cell was patterned on the moving parts, and the other was placed onto a fixed anchor structure (here, a single unit cell was composed of two SRRs). When a DC bias voltage was applied, the gap between the two SRRs decreased, which modified the inter-meta-atom coupling. As a result, the resonance frequency or the magnetic permeability can be controlled efficiently by changing the structural arrangement of the meta-atoms.

In a very recent study, stretchable metamaterials were used for THz modulation by means of structural deformation of meta-atoms [59, 60]. The elongation of the flexible and stretchable substrate made the variation/change of the morphology or distance between the adjacent meta-atoms different from that of the initial state. As a result, the resonance of the metamaterial was shifted, which could be useful for application in remote strain sensing.

10.6 Conclusion

A metamaterial is an artificial material that is designed to have unnatural electromagnetic properties. Metamaterials are of great importance in THz frequency ranges because many natural materials generally have weak responses in this frequency range. Artificially designed meta-atoms can exhibit exceptionally strong THz responses, and thus, they can be a very useful platform for the manipulation of THz waves. In addition to previously demonstrated passive-type and active-type THz metamaterials, there is a possibility that THz metadevices will become the next enabling technological breakthrough in THz technologies.

References

1. B. Ferguson and X.-C. Zhang, Materials for terahertz science and technology, *Nat. Mater.* **1**(1), 26–33 (2002).
2. M. Tonouchi, Cutting-edge terahertz technology, *Nat. Photon.* **1**(2), 97–105 (2007).
3. J. B. Pendry, A. J. Holden, D. J. Robbins, and W. J. Stewart, Low frequency plasmons in thin-wire structures, *J. Phys.: Condens. Matter* **10**(22), 4785 (1998).
4. J. B. Pendry, A. J. Holden, D. J. Robbins, and W. J. Stewart, Magnetism from conductors and enhanced nonlinear phenomena, *IEEE Trans. Microwave Theory Tech.* **47**(11), 2075–2084 (1999).
5. J. B. Pendry, A. J. Holden, W. J. Stewart, and I. Youngs, Extremely low frequency plasmons in metallic mesostructures, *Phys. Rev. Lett.* **76**(25), 4773–4776 (1996).
6. D. R. Smith, J. B. Pendry, and M. C. K. Wiltshire, Metamaterials and negative refractive index, *Science* **305**(5685), 788–792 (2004).
7. R. A. Shelby, D. R. Smith, and S. Schultz, Experimental verification of a negative index of refraction, *Science* **292**(5514), 77–79 (2001).
8. B. Fischer, M. Hoffmann, H. Helm, G. Modjesch, and P. Uhd. Jepsen, Chemical recognition in terahertz time-domain spectroscopy and imaging, *Semicond. Sci. Technol.* **20**(7), S246 (2005).
9. K. Kawase, Y. Ogawa, Y. Watanabe, and H. Inoue, Non-destructive terahertz imaging of illicit drugs using spectral fingerprints, *Opt. Express* **11**(20), 2549–2554 (2003).
10. D. M. Mittleman, R. H. Jacobsen, and M. C. Nuss, T-ray imaging, *IEEE J. Sel. Top. Quantum Electron.* **2**(3), 679–692 (1996).
11. M. W. Ruth, E. C. Bryan, P. W. Vincent, J. P. Richard, D. A. Donald, H. L. Edmund, and P. Michael, Terahertz pulse imaging in reflection geometry of human skin cancer and skin tissue, *Phys. Med. Biol.* **47**(21), 3853 (2002).
12. M. Gokkavas, K. Guven, I. Bulu, K. Aydin, R. S. Penciu, M. Kafesaki, C. M. Soukoulis, and E. Ozbay, Experimental demonstration of a left-handed metamaterial operating at 100GHz, *Phys. Rev. B* **73**(19), 193103 (2006).
13. O. Paul, C. Imhof, B. Reinhard, R. Zengerle, and R. Beigang, Negative index bulk metamaterial at terahertz frequencies, *Opt. Express* **16**(9), 6736–6744 (2008).

14. T. J. Yen, W. J. Padilla, N. Fang, D. C. Vier, D. R. Smith, J. B. Pendry, D. N. Basov, and X. Zhang, Terahertz magnetic response from artificial materials, *Science* **303**(5663), 1494–1496 (2004).
15. H. Tao, N. I. Landy, K. Fan, A. C. Strikwerda, W. J. Padilla, R. D. Averitt, and Z. Xin, *Flexible Terahertz Metamaterials: Towards a Terahertz Metamaterial Invisible Cloak.* Paper presented at the Electron Devices Meeting, 2008. IEDM 2008. IEEE International.
16. M. Choi, S. H. Lee, Y. Kim, S. B. Kang, J. Shin, M. H. Kwak, K.-Y. Kang, Y.-H. Lee, N. Park, and B. Min, A terahertz metamaterial with unnaturally high refractive index, *Nature* **470**(7334), 369–373 (2011).
17. N. R. Han, Z. C. Chen, C. S. Lim, B. Ng, and M. H. Hong, Broadband multi-layer terahertz metamaterials fabrication and characterization on flexible substrates, *Opt. Express* **19**(8), 6990–6998 (2011).
18. S. Schwaiger, M. Bröll, A. Krohn, A. Stemmann, C. Heyn, Y. Stark, D. Stickler, D. Heitmann, and S. Mendach, Rolled-up three-dimensional metamaterials with a tunable plasma frequency in the visible regime, *Phys. Rev. Lett.* **102**(16), 163903 (2009).
19. C. Zaichun, M. Rahmani, G. Yandong, C. T. Chong, and H. Minghui, Realization of variable three-dimensional terahertz metamaterial tubes for passive resonance tunability, *Adv. Mater.* **24**(23), OP143–OP147 (2012).
20. H. O. Moser and C. Rockstuhl, 3D THz metamaterials from micro/nanomanufacturing, *Laser Photon. Rev.* **6**(2), 219–244 (2012).
21. Y. Cheng, T. Huang, and C. C. Chieng, Thick-film lithography using laser write, *Microsyst. Technol.* **9**(1–2), 17–22 (2002).
22. H. O. Moser, B. D. F. Casse, O. Wilhelmi, and B. T. Saw, Terahertz response of a microfabricated rod–split-ring-resonator electromagnetic metamaterial, *Phys. Rev. Lett.* **94**(6), 063901 (2005).
23. H. O. Moser, L. K. Jian, H. S. Chen, M. Bahou, S. M. P. Kalaiselvi, S. Virasawmy, S. M. Maniam, X. X. Cheng, S. P. Heussler, S. bin Mahmood, and B. I. Wu, All-metal self-supported THz metamaterial: the meta-foil, *Opt. Express* **17**(26), 23914–23919 (2009).
24. H. O. Moser, J. A. Kong, L. K. Jian, H. S. Chen, G. Liu, M. Bahou, S. M. P. Kalaiselvi, S. M. Maniam, X. X. Cheng, B. I. Wu, P. D. Gu, A. Chen, S. P. Heussler, S. bin Mahmood, and L. Wen, Free-standing THz electromagneticmetamaterials, *Opt. Express* **16**(18), 13773–13780 (2008).
25. Z. C. Chen, M. H. Hong, C. S. Lim, N. R. Han, L. P. Shi, and T. C. Chong, Parallel laser microfabrication of large-area asymmetric split ring resonator metamaterials and its structural tuning for terahertz resonance, *Appl. Phys. Lett.* **96**(18), 181101 (2010).

26. K. Fan, A. C. Strikwerda, H. Tao, X. Zhang, and R. D. Averitt, Stand-up magnetic metamaterials at terahertz frequencies, *Opt. Express* **19**(13), 12619–12627 (2011).

27. Y. Jun-Bo, K. Byeong-Il, C. Yun-Seok, and Y. Euisik, 3-D construction of monolithic passive components for RF and microwave ICs using thick-metal surface micromachining technology, *IEEE Trans. Microwave Theory Tech.* **51**(1), 279–288 (2003).

28. F. Kebin, C. S. Andrew, D. A. Richard, and Z. Xin, Three-dimensional magnetic terahertz metamaterials using a multilayer electroplating technique, *J. Micromech. Microeng.* **22**(4), 045011 (2012).

29. H. Kim, J. S. Melinger, A. Khachatrian, N. A. Charipar, R. C. Y. Auyeung, and A. Piqué, Fabrication of terahertz metamaterials by laser printing, *Opt. Lett.* **35**(23), 4039–4041 (2010).

30. S. Kawata, H.-B. Sun, T. Tanaka, and K. Takada, Finer features for functional microdevices, *Nature* **412**(6848), 697–698 (2001).

31. S. H. Park, D. Y. Yang, and K. S. Lee, Two-photon stereolithography for realizing ultraprecise three-dimensional nano/microdevices, *Laser Photon. Rev.* **3**(1–2), 1–11 (2009).

32. G. Veselago Viktor, The electrodynamics of substances with simultaneously negative values of ε and μ, *Soviet Phys. Uspekhi* **10**(4), 509 (1968).

33. D. R. Smith, W. J. Padilla, D. C. Vier, S. C. Nemat-Nasser, and S. Schultz, Composite medium with simultaneously negative permeability and permittivity, *Phys. Rev. Lett.* **84**(18), 4184–4187 (2000).

34. N. Katsarakis, G. Konstantinidis, A. Kostopoulos, R. S. Penciu, T. F. Gundogdu, M. Kafesaki, E. N. Economou, Th. Koschny, and C. M. Soukoulis, Magnetic response of split-ring resonators in the far-infrared frequency regime, *Opt. Lett.* **30**(11), 1348–1350 (2005).

35. K. B. Alici and E. Ozbay, A planar metamaterial: polarization independent fishnet structure, *Photon. Nanostruct.: Fundam. Appl.* **6**(1), 102–107 (2008).

36. G. Dolling, C. Enkrich, M. Wegener, C. M. Soukoulis, and S. Linden, Simultaneous negative phase and group velocity of light in a metamaterial, *Science* **312**(5775), 892–894 (2006).

37. Q.-J. Du, J.-S. Liu, K.-J. Wang, X.-N. Yi, and H.-W. Yang, Dual-band terahertz left-handed metamaterial with fishnet structure, *Chin. Phys. Lett.* **28**(1), 014201 (2011).

38. M. Navarro-Cía, C. García-Meca, M. Beruete, A. Martínez, and M. Sorolla, Dual-band double-negative-index fishnet metamaterial at millimeter-waves, *Opt. Lett.* **36**(21), 4245–4247 (2011).

39. S. Zhang, W. Fan, N. C. Panoiu, K. J. Malloy, R. M. Osgood, and S. R. J. Brueck, Experimental demonstration of near-infrared negative-index metamaterials, *Phys. Rev. Lett.* **95**(13), 137404 (2005).

40. J. T. Shen, P. B. Catrysse, and S. Fan, Mechanism for designing metallic metamaterials with a high index of refraction, *Phys. Rev. Lett.* **94**(19), 197401 (2005).

41. J. Shin, J.-T. Shen, and S. Fan, Three-dimensional metamaterials with an ultrahigh effective refractive index over a broad bandwidth, *Phys. Rev. Lett.* **102**(9), 093903 (2009).

42. S. Zhang, Y.-S. Park, J. Li, X. Lu, W. Zhang, and X. Zhang, Negative refractive index in chiral metamaterials, *Phys. Rev. Lett.* **102**(2), 023901 (2009).

43. C. Janke, J. G. Rivas, P. H. Bolivar, and H. Kurz, All-optical switching of the transmission of electromagnetic radiation through subwavelength apertures, *Opt. Lett.* **30**(18), 2357–2359 (2005).

44. E. Hendry, M. J. Lockyear, J. G. Rivas, L. Kuipers, and M. Bonn, Ultrafast optical switching of the THz transmission through metallic subwavelength hole arrays, *Phys. Rev. B* **75**(23), 235305 (2007).

45. W. J. Padilla, A. J. Taylor, C. Highstrete, M. Lee, and R. D. Averitt, Dynamical electric and magnetic metamaterial response at terahertz frequencies, *Phys. Rev. Lett.* **96**(10), 107401 (2006).

46. H.-T. Chen, J. F. O'Hara, A. K. Azad, A. J. Taylor, R. D. Averitt, D. B. Shrekenhamer, and W. J. Padilla, Experimental demonstration of frequency-agile terahertz metamaterials, *Nat. Photon.* **2**(5), 295–298 (2008).

47. S. Zhang, J. Zhou, Y.-S. Park, J. Rho, R. Singh, S. Nam, A. K. Azad, H.-T. Chen, X. Yin, A. J. Taylor, and X. Zhang, Photoinduced handedness switching in terahertz chiral metamolecules, *Nat. Commun.* **3**, 942 (2012).

48. J. Zhou, D. R. Chowdhury, R. Zhao, A. K. Azad, H.-T. Chen, Costas M. Soukoulis, A. J. Taylor, and J. F. O'Hara, Terahertz chiral metamaterials with giant and dynamically tunable optical activity, *Phys. Rev. B* **86**(3), 035448 (2012).

49. H.-T. Chen, W. J. Padilla, J. M. O. Zide, S. R. Bank, A. C. Gossard, A. J. Taylor, and R. D. Averitt, Ultrafast optical switching of terahertz metamaterials fabricated on ErAs/GaAs nanoisland superlattices, *Opt. Lett.* **32**(12), 1620–1622 (2007).

50. P. A. George, J. Strait, J. Dawlaty, S. Shivaraman, Mvs Chandrashekhar, F. Rana, and M. G. Spencer, Ultrafast optical-pump terahertz-probe spectroscopy of the carrier relaxation and recombination dynamics in epitaxial graphene, *Nano Lett.* **8**(12), 4248–4251 (2008).

51. H. T. Chen, W. J. Padilla, J. M. O. Zide, A. C. Gossard, A. J. Taylor, and R. D. Averitt, Active terahertz metamaterial devices, *Nature* **444**(7119), 597–600 (2006).

52. H.-T. Chen, W. J. Padilla, M. J. Cich, A. K. Azad, R. D. Averitt, and A. J. Taylor, A metamaterial solid-state terahertz phase modulator, *Nat. Photon.* **3**(3), 148–151 (2009).

53. W. L. Chan, H.-T. Chen, A. J. Taylor, I. Brener, M. J. Brener, and D. M. Mittleman, A spatial light modulator for terahertz beams, *Appl. Phys. Lett.* **94**(21), 213511 (2009).

54. D. Shrekenhamer, S. Rout, A. C. Strikwerda, C. Bingham, R. D. Averitt, S. Sonkusale, and W. J. Padilla, High speed terahertz modulation from metamaterials with embedded high electron mobility transistors, *Opt. Express* **19**(10), 9968–9975 (2011).

55. S. H. Lee, M. Choi, T. T. Kim, S. Lee, M. Liu, X. Yin, H. K. Choi, S. S. Lee, C. G. Choi, S. Y. Choi, X. Zhang, and B. Min, Switching terahertz waves with gate-controlled active graphene metamaterials, *Nat. Mater.* **11**(11), 936–941 (2012).

56. H. Tao, A. C. Strikwerda, K. Fan, W. J. Padilla, X. Zhang, and R. D. Averitt, Reconfigurable terahertz metamaterials, *Phys. Rev. Lett.* **103**(14), 147401 (2009).

57. Y. H. Fu, A. Q. Liu, W. M. Zhu, X. M. Zhang, D. P. Tsai, J. B. Zhang, T. Mei, J. F. Tao, H. C. Guo, X. H. Zhang, J. H. Teng, N. I. Zheludev, G. Q. Lo, and D. L. Kwong, A micromachined reconfigurable metamaterial via reconfiguration of asymmetric split-ring resonators, *Adv. Funct. Mater.* **21**(18), 3589–3594 (2011).

58. W. M. Zhu, A. Q. Liu, X. M. Zhang, D. P. Tsai, T. Bourouina, J. H. Teng, X. H. Zhang, H. C. Guo, H. Tanoto, T Mei, G. Q. Lo, and D. L. Kwong, Switchable magnetic metamaterials using micromachining processes, *Adv. Mater.* **23**(15), 1792–1796 (2011).

59. S. Lee, S. Kim, T. T. Kim, Y. Kim, M. Choi, S. H. Lee, J. Y. Kim, and B. Min, Reversibly stretchable and tunable terahertz metamaterials with wrinkled layouts, *Adv. Mater.* **24**(26), 3491–3497 (2012).

60. J. Li, C. M. Shah, W. Withayachumnankul, B. S. Y. Ung, A. Mitchell, S. Sriram, M. Bhaskaran, S. Chang, and D. Abbott, Mechanically tunable terahertz metamaterials, *Appl. Phys. Lett.* **102**(12), 121101–121104 (2013).

Chapter 11

Terahertz-Wave Plastic Fibers and Their Applications

Chih-Hsien Lai[a] and Chi-Kuang Sun[b]

[a]*Department of Electronic Engineering, National Yunlin University of Science and Technology, Yunlin 64002, Taiwan*

[b]*Department of Electrical Engineering and Graduate Institute of Photonics and Optoelectronics, National Taiwan University, Taipei 10617, Taiwan*

chlai@yuntech.edu.tw, sun@ntu.edu.tw

11.1 Introduction

To facilitate terahertz (THz) science and applications, it is essential to develop low-loss and low-cost THz waveguides. Most of the THz systems are large in size because of the usage of free space for THz wave propagation. The major difficulty for THz waveguiding is the lack of low-loss materials in the THz range. High skin-depth loss of metals and strong absorption loss in dielectrics prevent the waveguides, proposed according to conventional microwaves [1, 2] or photonics [3, 4] technologies, from efficient transmission of THz waves. Since the most transparent medium for THz waves is dry air, to minimize the absorption loss, it is feasible to design waveguides with less power within the lossy material and majority of the power

Handbook of Terahertz Technologies: Devices and Applications
Edited by Ho-Jin Song and Tadao Nagatsuma

ISBN 978-981-4613-08-8 (Hardcover), 978-981-4613-09-5 (eBook)
www.panstanford.com

transmitted through air. THz waveguides proposed and based on this concept can be classified, in general, into three types. The first is the subwavelength type [5], which is a thin metal or dielectric wire, with its diameter much smaller than the wavelength and thus high power fraction outside the wire. The second is the hollow type, which is essentially an air-core waveguide surrounded by a structured cladding and confines most of the power within the air-core region [6–11]. The third is the porous type [12, 13], where subwavelength air holes are arranged in the core region to reduce material losses.

In this chapter, two kinds of plastic waveguides will be introduced for THz waveguiding. One is the subwavelength plastic fiber [14], and the other is the pipe waveguide [15]. Working principles and operating characteristics of these THz plastic waveguides will be discussed first. Then, some of their applications, including THz directional couplers and THz fiber-scanning imaging systems, will be described.

11.2 Terahertz Subwavelength Plastic Fiber

A simple plastic wire with a subwavelength diameter can provide low guiding attenuation in a single-mode operation for THz waves [14]. The concept of such low-loss waveguiding was previously demonstrated at both microwave [16, 17] and optical [18] frequencies. In the THz frequency range, we first demonstrated a low-loss plastic fiber for THz waveguiding [14]. In our first study, we adopted a 200 μm diameter polyethylene (PE) wire as the subwavelength THz fiber for a frequency near 300 GHz. With a small diameter-to-wavelength ratio, a large portion of THz waves was guided outside the lossy core such that the attenuation constant is successfully reduced to be of the order of 0.001 cm^{-1}, while the measured free-space coupling coefficient of the fiber can reach up to the order of or much greater than 20%. With direct coupling capability, the subwavelength THz fibers can be easily integrated with current THz systems to improve flexibility and reliability.

11.2.1 Fractional-Material-Loss-Induced Attenuation Constant

To simplify the analysis of a subwavelength THz fiber, we assume that the plastic wire, which has the same geometry as an air-cladding fiber, as shown in Fig. 11.1a, has a circular cross section, a uniform radius a, and a step-index profile with an inner refractive index n_1 and an outer refractive index n_2. Since air serves as the cladding material, we assume that $n_2 = 1$. By solving Maxwell's equations, the mode fields can be analytically expressed. However, because of the lack of a nonabsorptive material in the THz regime, all the guided modes within the dielectric core will be subject to material absorption, and the fiber will be too lossy to compete with a metal-based fiber. A straightforward way to improve the attenuation is to reduce the fractional power inside the dielectric core by using a subwavelength-diameter plastic wire with a small diameter-to-wavelength ratio, which will also ensure single-mode (HE_{11}) operation as long as $(2\pi a/\lambda)\sqrt{n_1^2 - n_2^2} < 2.405$, where λ is the wavelength [19]. To calculate the fractional power flow inside the core area, one needs to solve Poynting vector $\boldsymbol{S}_z$ in the direction of propagation (z direction). Figure 11.1b shows the calculated spatial distributions of the normalized $\boldsymbol{S}_z$ of a 200 μm diameter PE wire for 300, 500, 700, and 900 GHz with $n_1 = 1.5$ [20]. It can be observed that the fractional power flowing outside the PE core decreases as the frequency of the guided waves increases. One can obtain the fractional power η inside the core by

$$\eta = \frac{\int_0^a \mathbf{S}_z(r) r dr d\phi}{\int_0^\infty \mathbf{S}_z(r) r dr d\phi} \tag{11.1}$$

The value of η can be used to estimate roughly the effective fiber attenuation constant influenced by the core absorption. For a small η, which happens when a small-diameter plastic wire is adopted, the attenuation constant will approach that of the air cladding, while for an η close to 1 the attenuation constant will approximate that of bulk PE. To precisely estimate the fiber attenuation caused by material absorption, we adopted a perturbation method [16]. It assumes that the power loss per wavelength in propagation is small compared to the total power flowing along the fiber, which is reasonable for a subwavelength plastic wire in the THz regime. Using the method simplified with the Poynting theorem [16], we can find the fiber attenuation constant α, given by

$$\alpha = \left|\frac{1}{P}\frac{dP}{dz}\right| = \frac{\sigma \int |E|^2 \, d\tau}{\left|\int \mathbf{S}_z d\tau\right|} \tag{11.2}$$

where τ is an infinite plane perpendicular to the fiber axis and σ is the conductivity. For the plastic core, conductivity is a function of the refractive index n_1 and the absorption constant of the plastic α_m, which can be expressed as $\sigma = n_1 c \alpha_m / 4\pi$, where c is the velocity of light in free space. For air cladding, the conductivity approximates zero because of its negligible absorption of air. The calculated α as a function of electromagnetic frequency is shown as a dashed curve in Fig. 11.2, for which it is assumed that $\alpha_m = 1\ \text{cm}^{-1}$.

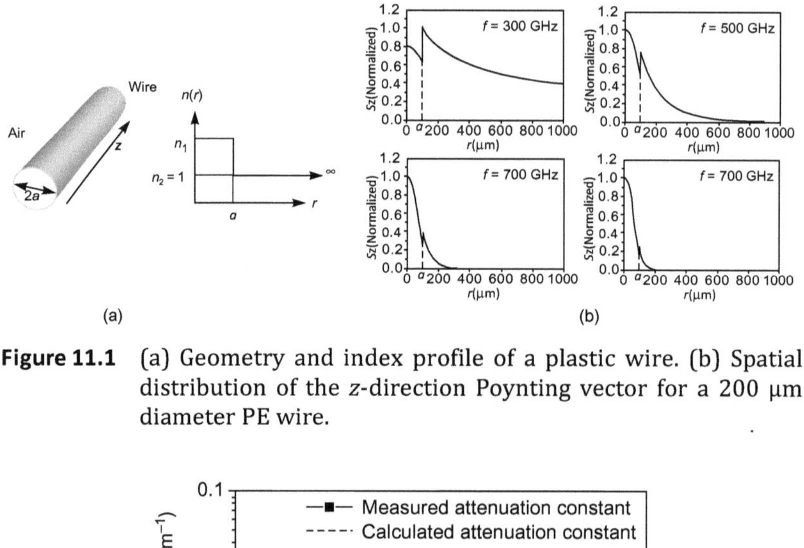

Figure 11.1 (a) Geometry and index profile of a plastic wire. (b) Spatial distribution of the z-direction Poynting vector for a 200 μm diameter PE wire.

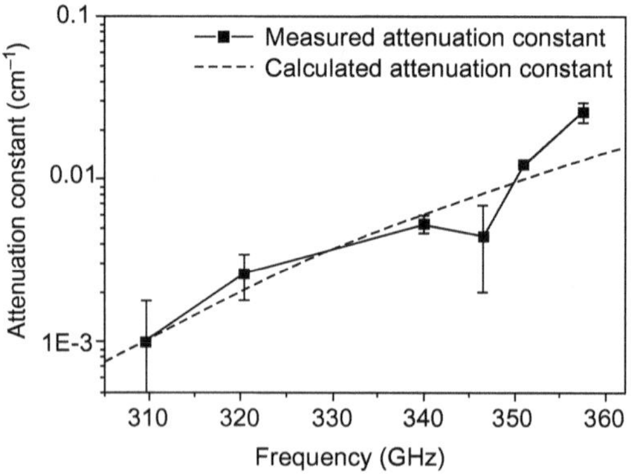

Figure 11.2 Measured and calculated attenuation constants of a 200 μm diameter PE wire.

We used PE wires for the subwavelength THz fibers because PE has a relatively low α_m, with no significant absorption lines in the THz window. One of the experimental arrangements employed for the fiber attenuation measurement is shown in Fig. 11.3. Since in this first study our collected PE wires (fishing wires) came from the fishing industry and had a relatively poor material quality and many bends that seriously degrade their guiding performance, the fiber to be measured was first heated and straightened with steam, which made α_m higher than expected. The ends of the fiber were polished with sandpaper to enhance coupling efficiency. The fiber attenuation constant was measured by the cut-back method. For consistency, the measurement was performed with the same PE fiber. Three fiber lengths, 17.5, 13.5, and 6.0 cm, were measured. We began with the measurement of output THz power for the longest length by using a bolometer and a lock-in amplifier at different radiation frequencies. Then we cut a small piece of fiber off the output end and repeated the measurements. By analyzing the transmitted THz power for different fiber lengths, we obtained the fiber attenuation constant.

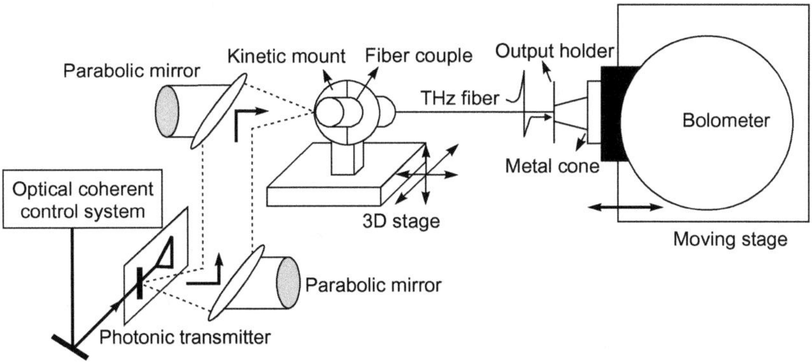

Figure 11.3 Experimental setup of the THz subwavelength fiber measurement.

For a sufficiently small core-diameter-to-wavelength ratio ($2a/\lambda$), the frequency of interest is ~300 GHz. In Fig. 11.2a, we show the experimentally obtained attenuation constant as a function of frequency from 310 GHz to 360 GHz, corresponding to the $2a/\lambda$ ratio from 0.20 to 0.24. Similar to the theoretical expectation, with a higher THz frequency and thus a larger η, the measured α increases with frequency for a fixed-diameter PE wire. Tuning the

THz frequency from 310 GHz to 360 GHz results in a tenfold increase in the measured α value, from the order of 10^{-3} cm^{-1} to the order of 10^{-2} cm^{-1}, as expected from the theoretical calculation. It is important to note that this is to our knowledge the first time that an α value of the order of or less than 10^{-2} cm^{-1} was achieved for a THz fiber.

11.2.2 Diameter-Variation-Induced Radiation Loss

From Fig. 11.2, a consequent question may naturally arise. What will the attenuation constant be if we kept measuring with frequencies lower than 310 GHz? They are not provided in Fig. 11.2 because we could not obtain consistent results at such low frequencies, due to the short fiber length (<20 cm) available in commercial fishing wires, which are all twisted and bended for easy storage. For example, the measured attenuation constant for our first study at 300 GHz was actually negative, which was attributed to measurement limitation.

To investigate the lower limit of the fractional-material-loss-induced attenuation constant for the subwavelength plastic fibers, we then made three different diameter PE wires by a melt-spinning method for an early study [21]. The longest available lengths for 210, 155, and 550 μm diameter fibers were 105.0 cm, 42.5 cm, and 29.0 cm, respectively. Attenuation constants measured by the cut-back method are shown in Fig. 11.4 as a function of wavelength. Interestingly, we found that melt-spun PE fibers of different diameters have minimum attenuation constants, typically of the order of 10^{-2}–10^{-3} cm^{-1} at distinct wavelength ranges, and a thinner fiber delivers THz waves with minimal attenuation at a shorter wavelength. It should be noted that the length difference between two measured 210 μm fibers was much longer than the other two fibers such that the measured interferogram suffered less noise. Therefore, the measured attenuation constant of the 210 μm fiber can reach a smaller value than those of 155 μm and 550 μm ones. As mentioned previously, the dominant loss mechanism for THz waveguiding is the fractional material absorption, which strongly depends on the fractional power inside the waveguide. For a subwavelength fiber, the shorter the wavelength, the larger portion of the power confined in the core, leading to a higher attenuation constant. On the short-wavelength (left) side of the minimum attenuation point, the attenuation constant, as expected, shows an increasing trend toward smaller wavelength and higher frequency. However, on its long-wavelength (right) side the attenuation constant increases with wavelength in contrast to the expected decreasing trend owing to the declined fractional power inside the core. This result indicated that the fractional material absorption is not the only primary contributor to the loss curve and that there must be another loss mechanism existing in the subwavelength fiber: nonuniformity-induced radiation loss.

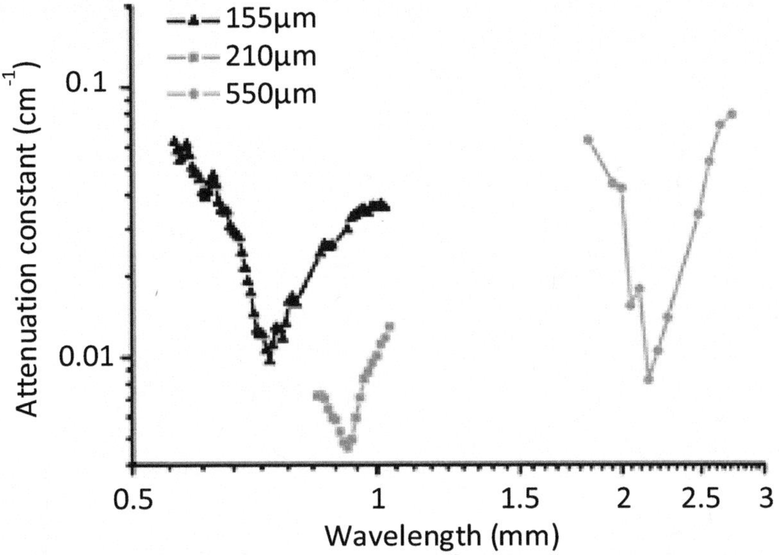

Figure 11.4 Attenuation constants of three different subwavelength THz fibers with diameters of 155, 210, and 550 μm, as a function of wavelength. Reproduced from Ref. [21] by permission of OSA, © 2007.

Previous optical nanotaper fiber experiments with a core diameter much smaller than the wavelength had found a significant loss increase with reduced fiber radius [18, 22], which was attributed to bending loss [22]. On the basis of the theory of nonadiabatic transitions developed in quantum mechanics, Sumetsky proposed the theory that minute nonuniformity of optical subwavelength fiber diameter could induce strong radiation loss when the diameter of the fiber is much smaller than the wavelength [23]. With a large portion of energy guided outside the subwavelength THz fiber, this proposed loss mechanism could thus pose a lower limit on the guidable fraction power in the core.

The PE fibers in the study shown in Fig. 11.4 were fabricated by a melt-spinning method, which is generally with a coefficient of variation (CV) in diameter smaller than 3%. The value of CV is defined as the ratio of the standard deviation to the mean. A subwavelength fiber with a large portion of power outside the core could thus suffer serious loss from diameter-variation-induced radiation. Slight changes of the core diameter cause the modes of the fiber to couple with one another. If we excite a pure guided mode at the input end of the fiber, diameter variation could make the mode field couple to the radiation modes, resulting in fiber losses, since the power is carried away from the core into the infinite cladding [23]. In Sumetsky's theory, the fiber diameter variation is considered smooth and can be adiabatically shown as a function of the fiber length. The shape of the optical taper fiber is expressed as

$$d(z) = d_\infty - (d_\infty - d_0)\,\{1 + [(z - z_0)/L]\}^{-1}$$

where d_0 and d_∞ are the taper's diameters at its center and ends, respectively. For a silica microtaper with a small waist diameter of $d \ll \lambda$, $\Delta\beta_{12}(z)$, the difference between the propagation constant of the HE_{11} mode, $\beta_1(z)$, and that of the radiation threshold, $\beta_2(z) = 2\pi/\lambda$, is approximated as

$$\Delta\beta_{12}(z) \approx 0.57[\lambda/d^2(z)]\exp[-0.27\lambda^2/d^2(z)] \qquad (11.3)$$

Substituting $\Delta\beta_{12}(z)$ into the Landau–Dykhne formula, the fiber transmission loss can thus be roughly estimated.

To apply Sumetsky's theory in our case, the expression of $\Delta\beta_{12}$ is modified according to our numerical results. We combine the calculated loss spectrum from fractional material absorption with that from diameter-variation-induced radiation to compare with our experimental results. The parameters are defined as follows: d_0 is the fiber diameter, and $d_v = d_\infty - d_0$ is the fiber diameter variation. The experimental results and calculated results with d_v = 5 μm and d_v = 50 nm for a fixed L = 1 m are shown in Fig. 11.5. Curve 2 (calculated fractional material absorption) intersects both curves 3 and 4 (calculated diameter-variation-induced radiation loss) at a frequency near the one at which the experimental minimum attenuation constant occurs, indicating that the diameter-variation-induced radiation is a dominant loss mechanism in subwavelength THz fiber when the core fraction power is low.

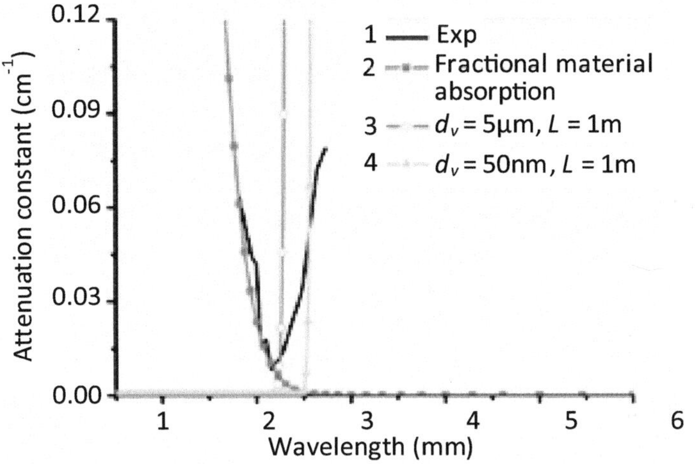

Figure 11.5 THz spectral loss characteristic of a 550 μm diameter PE fiber. Curve 1: measured result; curve 2: calculated loss due to fractional material absorption; curves 3 and 4: calculated losses due to diameter-variation-induced radiation with d_v = 5 μm and d_v = 50 nm, respectively. Characteristic length L is assumed to be 1 m. Reproduced from Ref. [21] by permission of OSA, © 2007.

From the above discussion, it is shown that there are two dominant loss mechanisms in subwavelength THz PE fibers operating in the regime of low fractional power in the core (<10%). Combining these two loss mechanisms, fractional material absorption and diameter-variation-induced radiation, the minimum attenuation frequency can be predicted. In fact, the minimum attenuation wavelength (the best operating wavelength) is found to be between four to five times of the PE fiber diameter.

11.2.3 Free-Space Coupling Efficiency

In fiber optics, coupling efficiency is always an important issue. This efficiency describes the power transfer from the input source, through free space, to the fiber and can be evaluated by an overlap integral of the input beam and the mode field, which is defined as follows:

$$\eta_{\text{coupling}} = \frac{\left|\int E_{\text{input}} E_{\text{mode}}^{*} ds\right|^2}{\int \left|E_{\text{input}}\right|^2 ds \int \left|E_{\text{mode}}\right|^2 ds} \tag{11.4}$$

where E_{mode} is the guided mode (the HE_{11} mode) of the subwavelength fiber and E_{input} is the field distribution of the input beam, which is usually Gaussian-like. Since the Bessel-function-based HE_{11} mode has a field distribution similar to the Gaussian, theoretically, the free-space coupling efficiency for the subwavelength plastic fiber could be very high (near 100%) as long as the spot size of the input beam is properly adjusted to match the mode size of the subwavelength fiber. In experiments, we determined the coupling efficiency by measuring the ratio of the output-to-input power and then calibrating it with the loss caused by the attenuation of the fiber. Figure 11.6 shows the measured values for a PE wire with a 600 μm diameter. As theoretically predicted, a direct coupling efficiency as high as 95% was experimentally achieved for the THz subwavelength plastic fibers.

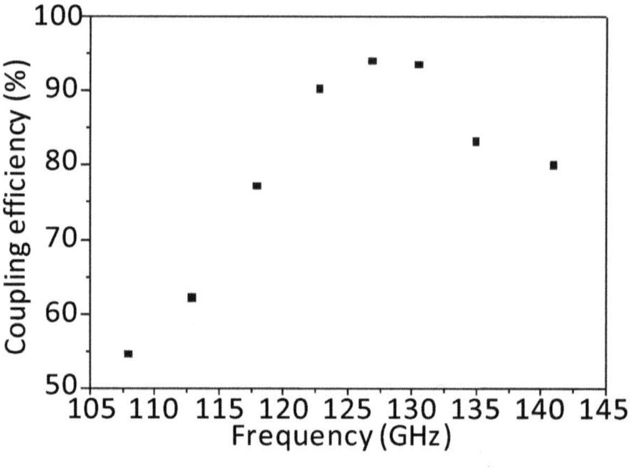

Figure 11.6 Measured coupling efficiency of a PE wire with a 600 μm diameter.

11.2.4 Subwavelength Hollow-Core Fiber

Due to its low attenuation constant, the subwavelength single-mode fiber shows a promising prospect in an evanescent-wave-based THz fiber-sensing application. It is because the large portion of the evanescent wave outside the fiber can provide great sensitivity. However, in some aspects, a hollow waveguide may be more convenient than a solid wire, such as gas or liquid molecular sensing. Therefore, here we also show a subwavelength hollow-core PE fiber with an outer diameter of 600 μm and an inner diameter of 280 μm [24]. In our experiment, three different fiber lengths, 8.6 cm, 11.5 cm, and 14 cm, were chosen. We measured the frequency dependence of power transmission for the three lengths. The fiber attenuation constants calculated from the measured data are shown in Fig. 11.7 with a similar but higher value compared to that of the solid-core fiber shown in Fig. 11.2.

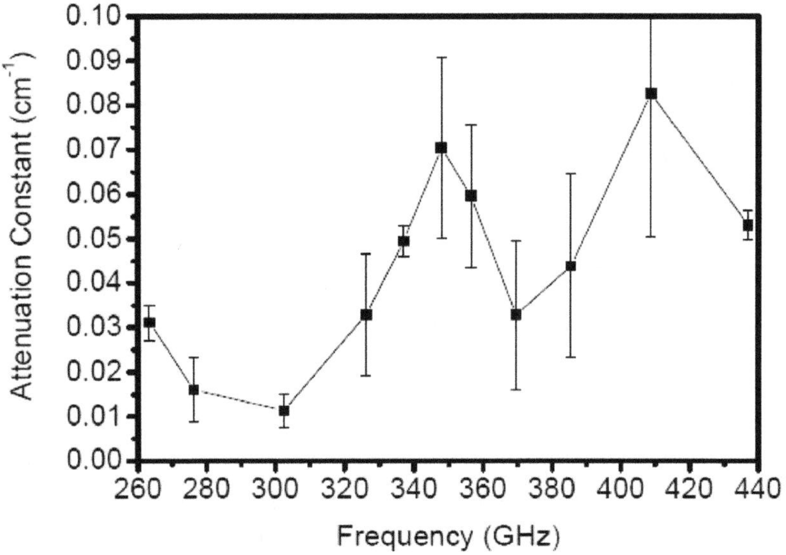

Figure 11.7 Measured attenuation constants of a subwavelength hollow-core PE fiber. Reproduced from Ref. [24] by permission of SPIE, © 2006.

In the preliminary experiment for subwavelength hollow-core fibers, it was demonstrated that a plastic tube can successfully guide the THz wave with an attenuation constant of the order of 0.01 cm^{-1} for a frequency around 300 GHz. Note that the guided mode of this subwavelength hollow-core fiber is similar to the solid-core case where the subwavelength core yields an extended evanescent field, and the hollow-core fiber has a subwavelength cladding, which causes the THz wave to extend well outside the lossy cladding and thus a lower material-induced loss. However, slightly different from the above-mentioned subwavelength solid-core fiber mode, the field confinement in the central region (air core and PE cladding) is relatively stronger in the subwavelength hollow-core fiber case [25]. A detailed comparison between the subwavelength solid-core and hollow-core fibers can be found in a later publication [26].

Note that there is a second type of hollow-core THz fiber, where a subwavelength air hole was also adopted in the core region [27] but with thicker cladding. Owing to the effect of low-index discontinuity, THz power is tightly confined in the air hole to reduce the material loss. The operation principle of porous fibers [12, 13] with more subwavelength air holes arranged in the dielectric core is similar to that of this second type one-subwavelength-air-hole THz fiber [27].

11.3 Terahertz Pipe Waveguides

In this section, we introduce another plastic waveguide, a pipe waveguide, for THz wave transmission [15, 28]. Much different from all previously reported air-core THz waveguides, which bear high reflection materials [6–9] or periodic structures [10, 11] to provide high cladding reflection, the structure of a THz pipe waveguide is very simple, and it is only a thin pipe consisting of a large air core (with a diameter greater than the wavelength) and a thin dielectric layer (with a thickness much smaller than the core diameter) with a uniform but low index. The guiding mechanism of a pipe waveguide is similar to that of an antiresonant reflecting optical waveguide (ARROW) [29], which is mostly in planar and solid-core form with multiple cladding layers, but is realized here with a single dielectric layer. Compared to other ARROW-like circular air-core THz waveguides [7, 11], our study indicates that a single dielectric layer can achieve excellent THz mode confinement without the need of multiple-layered periodic structures [11] or high-reflection metallic coatings [7]. Using commercially available Teflon air pipes up to 3 m long, it was experimentally confirmed that THz waves can be successfully guided in the central air core with excellent mode qualities, high coupling efficiency (as high as 80%), and controllable bandwidth, while the measured attenuation constants can be as low as 0.0008 cm^{-1} [15].

11.3.1 Guiding Mechanism of the Core Modes

A transverse cross section of a THz pipe waveguide is shown in Fig. 11.8a. It has a large air core ($n_1 = 1$) with diameter D and a thin layer of cladding with thickness t. The medium outside the cladding is also air. The material of the cladding is assumed to be a dielectric with a uniform low refractive index n_2.

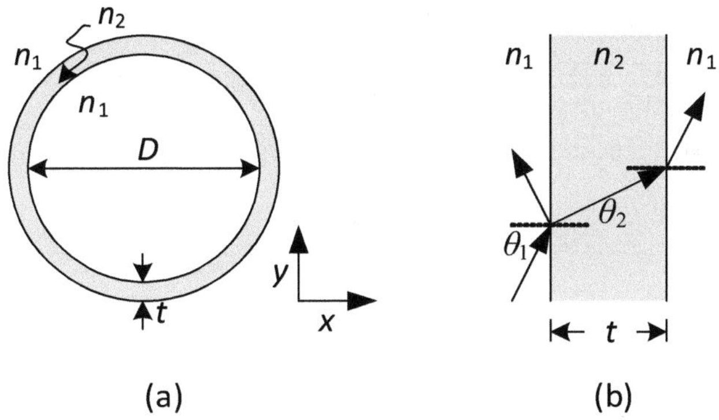

Figure 11.8 (a) Cross section of a THz pipe waveguide, where $n_1 = 1$ (air). (b) The cladding can be viewed as a Fabry–Pérot etalon.

The pipe waveguide supports cladding modes for which fields are confined within the cladding region. These modes are guided on the basis of the total internal reflection (TIR) owing to the higher refractive index of the cladding. However, the cladding modes attenuate rapidly since high material absorption losses are encountered. Alternatively, the pipe waveguide also supports core modes with fields confined in the air-core region. As the refractive index of the core is less than that of the cladding, fields of the core modes will oscillate and radiate through the cladding, making the core modes leaky. However, as the core modes suffer less material absorption losses than the cladding modes, they are the dominant modes. Here, only the core modes are investigated.

The guiding mechanism of the core modes in a THz pipe waveguide is similar to that of ARROWs [29]. The mechanism can be briefly described by viewing the cladding of the pipe waveguide as a Fabry–

Pérot etalon, as shown in Fig. 11.8b, where θ_1 and θ_2 are angles of inclined wave vectors with respect to the interface normal in media 1 and 2, respectively. It is well known that a Fabry–Pérot etalon behaves like a resonator [30]. Its transmission spectrum is periodic, and narrow peaks corresponding to transmission maximums occur at the frequencies at which the etalon resonates. That is to say, at or near the frequencies at which the cladding of a THz pipe waveguide is resonant, nearly no reflection takes place at the core-cladding interface and thus fields could hardly exist inside the core region. On the other hand, under the antiresonant conditions of the cladding, that is, at frequencies away from the resonant ones, considerable reflections occurring at the core-cladding interface cause the core modes to appear, resulting from the fact that waves bounce back and forth inside the core region.

Resonant frequencies of the cladding are given by [30]

$$f_m = \frac{mc}{2n_2 t\cos\theta_2}, \quad m = 1,2,3,... \tag{11.5}$$

where m is an integer. In the regime where the core modes are with glancing reflections, θ_1 is nearly 90°. Applying Snell's law and after some manipulations, Eq. 11.5 becomes

$$f_m = \frac{mc}{2t\sqrt{n_2^2 - n_1^2}}, \quad m = 1,2,3,... \tag{11.6}$$

Equation 11.6 can be used to predict the frequencies at or near which the core modes of the pipe waveguide do not exist.

11.3.2 Attenuation Constant of the Fundamental Mode

In the following section, modal characteristics of the pipe waveguide are first examined numerically over the frequency range from 200 GHz to 900 GHz. Numerical simulations are performed by a full-vectorial finite-difference frequency domain (FDFD) mode solver [31]. If not particularly specified, we assume the parameters of the pipe waveguide as core diameter D = 9 mm, cladding thickness t = 0.5 mm, and cladding refractive index n_2 = 1.4 (Teflon).

Figure 11.9 shows the simulated attenuations of the fundamental mode. To investigate the dependences of the attenuation constant on D, t, and n_2, simulations of respective variations are shown in

Fig. 11.9a–c. We first examine the effect of the core diameter, and the calculated results are shown in Fig. 11.9a. Like the Fabry–Pérot etalon, periodic passbands are clearly shown. It is observed that there are discontinuities in Fig. 11.9a. At these frequencies, no core modes can be found by the mode solver. Substituting m = 1 and 2 in Eq. 11.6, the obtained resonant frequencies are 306 GHz and 612 GHz, respectively, which coincide with the discontinuity frequencies in Fig. 11.9a. In fact, the discontinuities correspond to the resonance conditions of the cladding under which core modes of the pipe waveguide do not exist. From Fig. 11.9a, obviously, the attenuation constant decreases as D increases. That is to say, pipe waveguides with a larger core diameter suffer less waveguide losses. We further plot the attenuation constant as a function of D in Fig. 11.9d, and the log–log plots shown in the inset reveal that the attenuation constant is proportional to $1/D^4$, different from the $1/D^3$ dependence for the hollow dielectric waveguides with an infinite cladding [32].

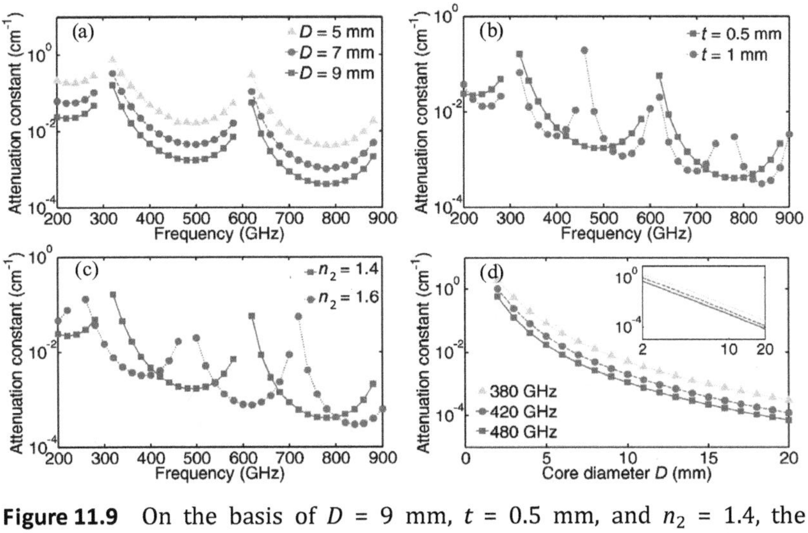

Figure 11.9 On the basis of D = 9 mm, t = 0.5 mm, and n_2 = 1.4, the attenuation constants of THz pipe waveguides are calculated for variations in the following parameters: (a) core diameter, D; (b) cladding thickness, t; and (c) refractive index of the cladding, n_2. (d) Attenuation constant as a function of D. Inset: log–log plot of the attenuation constant.

Next, we increase the cladding thickness from t = 0.5 mm to t = 1.0 mm. It is found the discontinuity frequencies, which are relevant to the cladding resonant frequencies, depend significantly on the thickness of the cladding. As can be observed in Fig. 11.9b, when the thickness doubles, the period of the discontinuity frequency becomes one-half. This phenomenon complies with what can be deduced from Eq. 11.6. In addition, the cladding resonant frequencies for t = 1.0 mm derived from Eq. 11.6 with m = 2, 3, 4, and 5 are 306 GHz, 459 GHz, 612 GHz, and 765 GHz, respectively, which also coincide with the discontinuity frequencies shown in Fig. 11.9b. If the bandwidth of the passband is approximated as the period of the resonant frequencies of the cladding, from Fig. 11.9b, it is clear that when t doubles, the passband width reduces to one-half, indicating that the passband width is proportional to $1/t$. Moreover, from Fig. 11.9c, the passband width decreases if n_2 increases.

From Fig. 11.9, it is found that discontinuity frequencies of the pipe waveguide are independent of the core diameter and are inversely proportional to the cladding thickness. These characteristics agree with those of the cladding resonant frequencies, as described by Eq. 11.6. Moreover, values of the discontinuity frequencies coincide with those obtained from Eq. 11.6. Therefore, we can conclude that the discontinuity frequencies are exactly the same as the resonant frequencies of the cladding, and hence confirm the guiding mechanism of the pipe waveguide to be antiresonant reflecting guiding.

Generally speaking, for a THz pipe waveguide to be low loss, larger air core is preferred, while to possess a high passband width, a thin cladding layer and a low-index material are more favorable [15, 28, 33].

11.3.3 Characteristics of Higher-Order Modes

A THz pipe waveguide is a multimode waveguide. The first four lowest-order modes (HE_{11}, TM_{01}, HE_{21}, and TE_{01}) of a pipe waveguide are solved at a frequency of 380 GHz. Modal intensity distributions and electric field vector distributions of these modes are shown in Fig. 11.10a and Fig. 11.10b, respectively. Calculated attenuation constants of the fundamental mode (HE_{11}) and the three higher-order modes (TM_{01}, HE_{21}, and TE_{01}) are shown in Fig. 11.10c. It is clear that basic characteristics of the higher-order modes are the same as those of the fundamental mode, including the same discontinuity

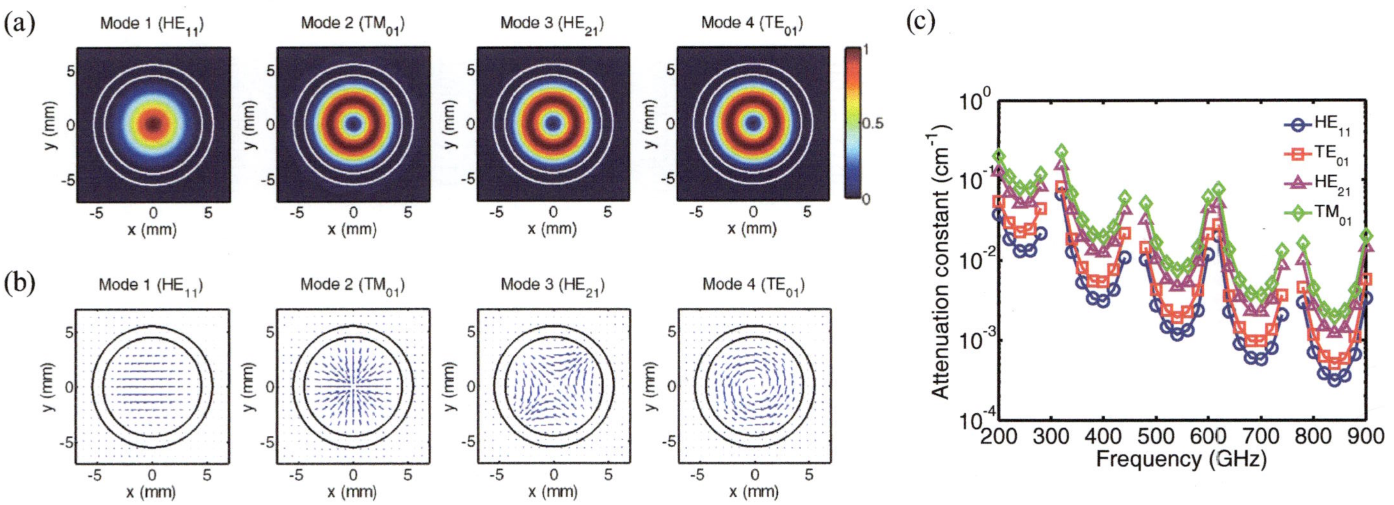

Figure 11.10 (a) Modal intensity distributions, (b) electric field vector distributions, and (c) attenuation constants of the first four lowest-order modes of a THz pipe waveguide.

frequencies. Moreover, compared with the higher-order modes, the HE_{11} mode has the smallest attenuation constant and thus is the dominant mode for the pipe waveguide investigated here.

11.3.4 Performances of THz Pipe Waveguides

Previous discussions assume that the refractive index of the cladding is real. In other words, material absorption is not considered. Here, the effect of material absorption is examined by assuming the refractive index of the cladding is $n_2 = 1.4 - 0.0015j$. From the calculated attenuation constants shown in Fig. 11.11a, it is clear that the effect of material absorption is not obvious in the lower frequency region. However, as the frequency increases, the attenuation constant is getting affected owing to the gradual dominance of the material loss over the waveguide loss. Several commercially available thin pipes made of various materials had been experimentally tested for THz waveguiding. It was found that pipes of acrylic glass (polymethylmethacrylate, PMMA), Teflon (water and air pipes), polypropylene (straw pipes used for drinking), and glass are all capable of delivering terahertz waves, and the measured attenuation loss for the Teflon pipe was the lowest owing to its lowest material absorption.

The effect of environmental variation is examined by changing the refractive index of the outside region surrounding the cladding. It is taken to be 1 in all previous discussions and denoted here as n_{1o}. Note that the refractive index of the inside core is still assumed to be 1. Calculated results shown in Fig. 11.11b indicate that the attenuation constant is highly sensitive to the variation of the outside environment. Even only a 1% increase of n_{1o} (from 1 to 1.01) would yield a significant increment to the attenuation constant. Such sensitivity would make the pipe waveguide ideal for sensing applications.

For practical applications, THz power emitted from the source is usually butt-coupled into the pipe waveguide directly. By assuming that the field distribution of the source is Gaussian-like, Fig. 11.11c shows coupling efficiencies calculated according to Eq. 11.4 as a function of the beam waist $2w_0$ for different frequencies, where D is assumed to be 9 mm. It is clear the coupling efficiency can be easily over 90% if a proper beam waist is applied.

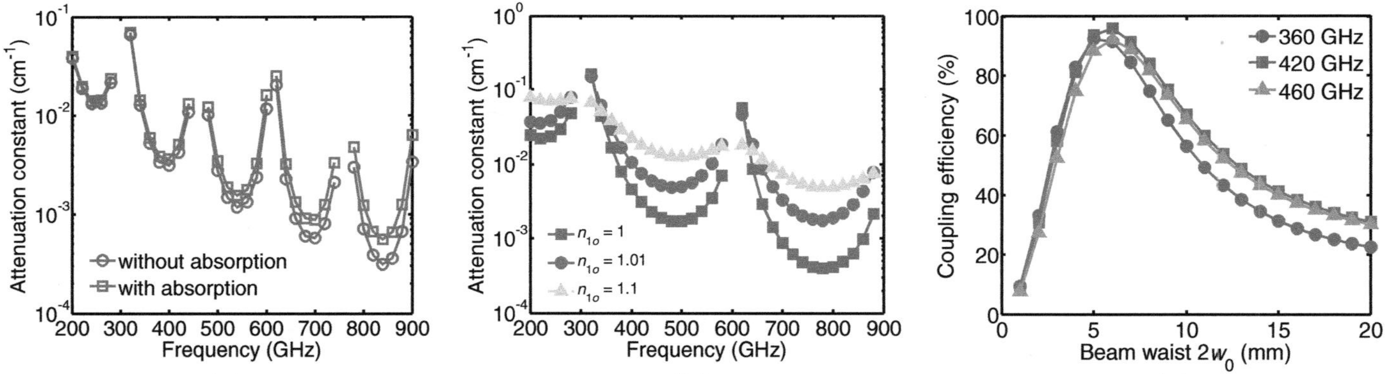

Figure 11.11 Some calculated results for a THz pipe waveguide. (a) Effect of material absorption, (b) effect of environment variation, and (c) coupling efficiency.

To demonstrate the structure of THz waveguiding, commercially available Teflon air pipes were used for THz wave transmission. The measured refractive index of Teflon is dispersive with an average value of 1.4. The experimental setup is shown in Fig. 11.12a. The intensity distributions of the output waves were measured using a metallic pinhole after a 3 m long Teflon pipe with D = 9 mm and t = 0.5 mm. Measured distributions at 356 GHz and 455 GHz, shown in Fig.11.12b, indicated that the THz waves were indeed guided inside the air-core region with an excellent mode quality over a wide frequency range. The intensity distributions resemble that of the fundamental mode (the HE_{11} mode) shown in Fig. 11.10, which also implies that higher-order modes were absent after such a propagation distance. The attenuation constants of Teflon pipes were measured for pipes with the same diameter, D = 9 mm, but different thicknesses, t = 0.5 and t = 1 mm. Experimental results shown in Fig. 11.12c display the same trend as the simulated ones shown in Fig. 11.9b. For example, the measured passband-width for t = 0.5 mm is about twice that of t = 1 mm. Note that for t = 0.5 mm, the passband-width is relatively broad with at least 200 GHz. Moreover, low attenuation constants were obtained of the order of or lower than 0.005 cm^{-1}, and the lowest value measured was (7.85 ± 0.05) × 10^{-4} cm^{-1}. Coupling efficiencies from the source to the pipe waveguide were also measured, and a maximum of 84% was obtained for t = 0.5 mm. It has been experimentally demonstrated that high butt-coupling efficiency (near 100%) between pipe waveguides with different shapes of cross section, for example, from a circular pipe waveguide to a square one, can be achieved as well if the two waveguides are with the same air-core sizes [34].

Compared with the simulated spectra, a slight shift of the cladding-resonant frequencies can be observed for the measured results. This is unavoidable due to cladding-thickness variations in the commercial Teflon pipes. Cladding-thickness was found to be 0.49–0.56 mm for t = 0.5 mm and 0.93–1.13 mm for t = 1 mm. Hence, the frequency mismatch could be partially attributed to the non-uniform cladding thickness, as shown in Fig. 11.12d, in which both t = 0.5 and t = 0.55 mm are assumed for simulation. Other discrepancy source includes the frequency-dependent cladding refractive index.

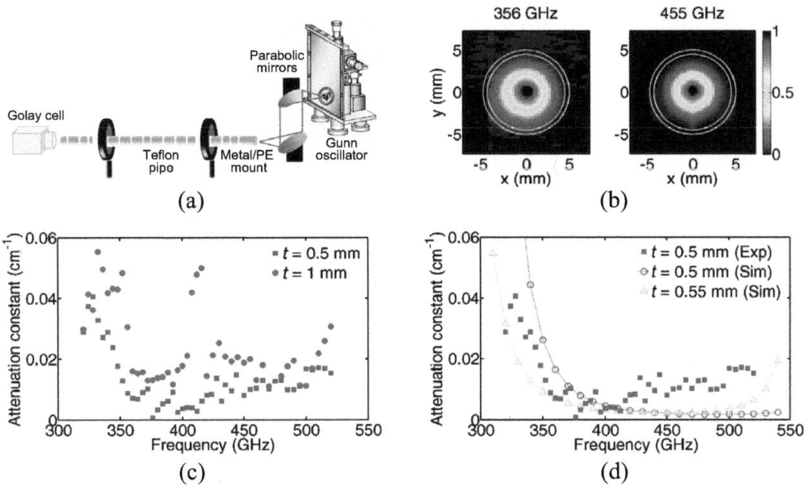

Figure 11.12 (a) Experimental setup for measuring the intensity distributions and attenuation constants of Teflon pipes. (b) Measured intensity distributions of commercially available Teflon pipes 3 m long. (c) Measured attenuation constants of Teflon pipes for t = 0.5 mm and t = 1 mm. (d) Comparison of the attenuation constants obtained from experiment and simulation for t = 0.5 mm.

Since the resonant frequency is an important feature of the pipe waveguides, it is further verified using PMMA pipes with better cladding uniformity by measuring the transmission spectra through the standard THz time domain spectroscopy [35]. Normalized transmission spectra of PMMA pipes with two different sizes, one with D = 8 mm and t = 1 mm and the other with D = 4 mm and t = 2 mm, are shown in Fig. 11.13a and Fig. 11.13b, respectively. Resonant frequencies of the cladding obtained from Eq. 11.6 are also shown for comparison. Theoretically, transmissions at the resonant frequencies are expected to drop because there are no core modes existing to deliver THz powers. It is clear in Fig. 11.13 that the transmission minima and resonant frequencies match very well, indicating that the guiding mechanism of the pipe waveguides is indeed that of antiresonant reflecting guiding.

Before leaving this section, modal profiles of the HE_{11} modes for the Teflon pipe waveguide and the subwavelength PE fiber are numerically compared at 200 GHz and 400 GHz. The results with x polarization are shown in Fig. 11.14. For the pipe waveguide, the parameters are fixed to D = 7 mm and t = 0.5 mm. For the subwavelength fiber, the diameters are 300 μm and 150 μm, respectively, which are chosen according to Section 11.2.2 to have minimum loss ($2a/\lambda = 0.2$). As can be seen, although there is no outer boundary for the subwavelength fiber to confine the THz wave, the modal size of a subwavelength fiber is much smaller than its pipe waveguide counterpart. The modal size of a subwavelength fiber is actually smaller than its guiding wavelength, while the modal size of the pipe waveguide is much greater than its guiding wavelength.

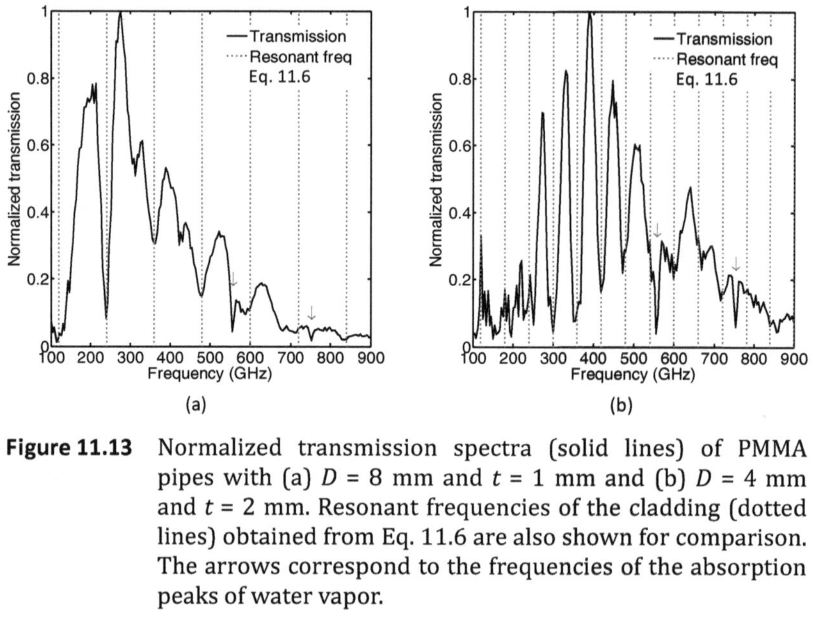

Figure 11.13 Normalized transmission spectra (solid lines) of PMMA pipes with (a) D = 8 mm and t = 1 mm and (b) D = 4 mm and t = 2 mm. Resonant frequencies of the cladding (dotted lines) obtained from Eq. 11.6 are also shown for comparison. The arrows correspond to the frequencies of the absorption peaks of water vapor.

11.4 Applications of Terahertz Plastic Fibers

11.4.1 THz Directional Couplers

A directional coupler is a power-dividing device, which is primarily used for power splitting, power switching, frequency selecting, and polarization filtering. It is usually formed by placing two single-mode waveguides in close proximity, with the total field being composed of two system modes, one even (symmetric) and one odd (antisymmetric). Since the propagation constants of the two system modes are different, the power can be totally transferred from one fiber to the other at the coupling length [30]. For communication, sensing, and imaging systems in the optical regimes, many kinds of directional couplers are applied; thus, multiple streams of information can be sent and received at the same time, while the phase difference of waves in two paths could be obtained to achieve high-resolution and high-sensitivity interferometric detection [36, 37].

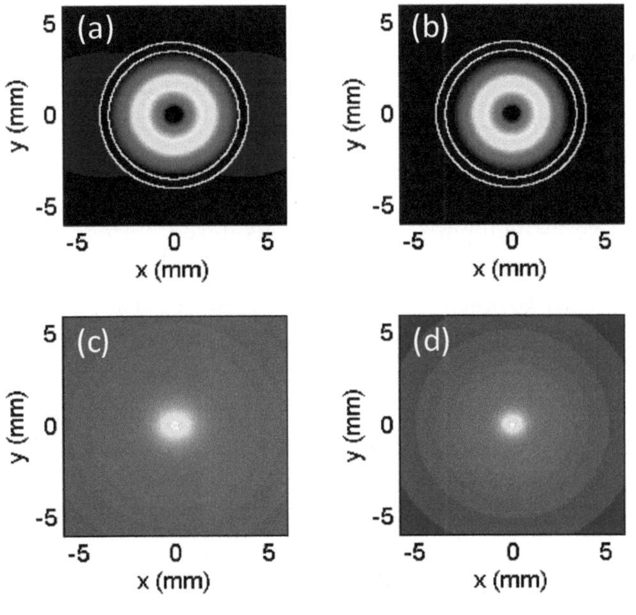

Figure 11.14 Modal profiles of the HE_{11} mode with *x* polarization. (a) Pipe waveguide at 200 GHz, (b) pipe waveguide at 400 GHz, (c) subwavelength fiber with 300 μm diameter at 200 GHz, and (d) subwavelength fiber with 150 μm diameter at 400 GHz.

To increase the coupling, most fiber-optic directional couplers are composed of two fibers put very close or touched together, often by complicated fabrication techniques, such as fusing, polishing, and etching, due to the highly confined fields within the core [38, 39]. Alternatively, for a THz subwavelength plastic fiber, since most of the guided fields are not confined in the solid core but in the air cladding, it provides a unique opportunity to constitute a strong coupling directional coupler without any fabrications that would distort the original fiber structure [40]. While for a directional coupler composed of two touching THz pipe waveguides [41, 42], unlike conventional directional couplers where one even system mode and one odd system mode are excited, three (one even and two odd) system modes will participate in the power transfer process at the antiresonant frequencies. As a result, the coupling length can be significantly reduced.

In this section, two types of directional couplers are discussed for THz operation. The first type is composed of two subwavelength plastic fibers with a separation between the fibers [40], while in the second type, two touching pipe waveguides are adopted, and a three-mode-beating behavior will be shown at the antiresonant frequencies [41, 42].

11.4.1.1 THz couplers with subwavelength plastic fibers

The THz subwavelength coupler considered here contains two uniform and identical subwavelength dielectric fibers, which are placed parallel to each other with a core separation s [40]. The schematic diagrams are shown in Fig. 11.15, where P_{in} represents the power of the input HE_{11} mode of the input fiber and P_{out} represents the power of the output HE_{11} mode of the coupled fiber. The total field in the coupling region of the directional coupler is composed of one even system mode and one odd system mode. At the input plane, the input HE_{11} mode would excite the even mode and the odd mode simultaneously. Since the even and odd modes propagate through the coupler region with different propagation constants β_e and β_o, respectively, total power transfer occurs at the distance where the two modes are out of phase. This distance is often called the coupling length L_c [30] of the directional coupler and is given by $\pi/(\beta_e - \beta_o)$. The coupling behavior of directional couplers can also be described by the coupled-mode theory [30], in which the mutual coupling factor κ is expressed as the overlap integral between the two HE_{11} electric fields, that is, E_i and E_c of the input fiber and the coupled fiber

$$\kappa = \frac{-j\omega\varepsilon_0}{4} \int_{\substack{\text{core area of the}\\ \text{coupled fiber}}} (n_{core}^2 - 1)(E_i \bullet E_c^*)da \tag{11.7}$$

where * denotes the operation of the complex conjugate. According to the coupling of modes in space solutions, the ratio of P_{out} to P_{in} can be expressed as [30]

$$\frac{P_{out}}{P_{in}} = \sin^2(|\kappa|\ell)\times e^{-\alpha\ell} \tag{11.8}$$

where ℓ is the length of the coupler region and α is the attenuation constant of the HE_{11} mode of the single fiber. Note that the exact exponential term in Eq. 11.8 should be $\exp[-(\alpha_e + \alpha_o)\ell/2]$, where α_e and α_o are the attenuation constants of the even and odd system modes, respectively. However, since α_e and α_o are both close to α, the exponential term that represents the propagation loss through the coupler is replaced by $e^{-\alpha\ell}$. From Eq. 11.8, the coupling length L_c at which the maximum P_{out} occurs is $\pi/2|\kappa|$. Generally, from the device point of view, it is desirable to have stronger field coupling, which means more field overlap and thus a larger $|\kappa|$, so that the coupling length L_c would be shorter for power switching from the input fiber to the coupled one, and there would be less propagation loss through the coupler region. Without the need of etching, fusing, or polishing to touch the two fibers together to increase the field coupling, as usually observed in conventional fiber-optic directional couplers, the subwavelength directional coupler described here can easily achieve strong coupling with a core separation between the two fibers. This is contributed from the wide field distribution outside the subwavelength PE core, which leads to a considerable field overlap between the propagation modes.

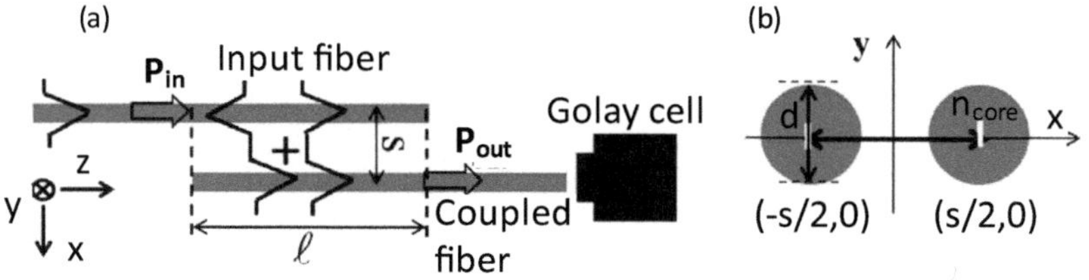

Figure 11.15 (a) Schematic diagram of a THz directional coupler with a fiber separation s and (b) index distribution of the directional coupler in the xy plane. The core coordinates of the two fibers are $(-s/2, 0)$ and $(s/2, 0)$, respectively.

However, it is notable that as one makes the core separation s smaller and smaller to enhance the coupling, the modal excitation loss from the HE_{11} mode field to the field of the linear combination of the two system modes would become more and more significant [43]. As a result, the modal excitation loss may exceed the propagation loss and become the dominant loss factor for the subwavelength directional coupler. Furthermore, when the core separation is eventually too small, for example, the fibers touch each other ($s = d$), the odd mode may encounter cutoff and thus cannot be guided in the coupler [44]. Under the cutoff condition, the THz power is unable to switch back and forth.

In one experiment, the core diameter d of the PE fibers is 300 μm. The measured attenuation constant α for a single PE fiber at 324 GHz was 0.008 cm^{-1}. We first conducted the length-of-the-coupler-region (ℓ)-dependent measurement with a fixed core separation s to investigate the power switching between the two subwavelength fibers. Numerical simulations are also presented, which were performed by the FDFD mode solver. Figure 11.16 shows both measured and simulated results of $(P_{out}/P_{in})e^{\alpha\ell}$ as a function of ℓ with the core separation s = 2 mm. The oscillatory behavior of power transfer between the two fibers can be clearly observed. The maximum $(P_{out}/P_{in})e^{\alpha\ell}$ is close to 100%, which indicates that the modal excitation loss is negligible. We can observe that the coupling length L_c is about 9 cm when s is 2 mm.

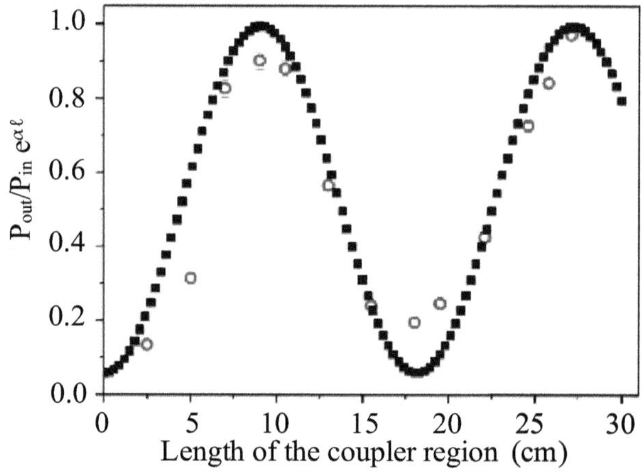

Figure 11.16 Measured (red open circles) and simulated (black solid squares) results of $(P_{out}/P_{in})\ e^{\alpha\ell}$ as a function of ℓ, the length of the coupler region. Reproduced from Ref. [40] by permission of IEEE, © 2012.

The core separation s was then varied from 1.17 mm to 2.55 mm (odd-mode cutoff occurs when $s < 1.1$ mm). Figure 11.17 shows both measured and simulated coupling lengths L_c as a function of s, and the results are close to each other. With a smaller s, L_c is shorter, owing to more field overlap between the two fibers. The measured shortest coupling length L_c is 2.8 cm (30λ) when s is 1.17 mm. Note that such a short coupling length, in terms of wavelength, is hard to achieve in the optical domain.

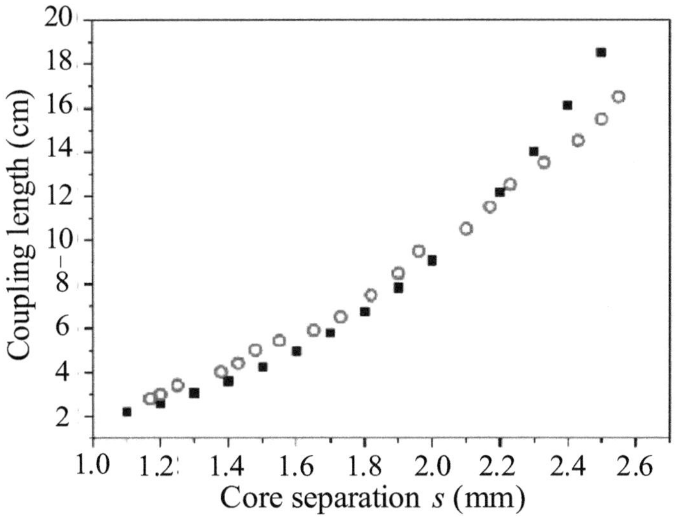

Figure 11.17 Measured (red open circles) and simulated (black solid squares) results of the coupling length L as a function of core separation s. Reproduced from Ref. [40] by permission of IEEE, © 2012.

We then investigated the insertion loss of this subwavelength directional coupler. The values of P_{out}/P_{in} as a function of core separation are shown as red open circles in Fig. 11.18a, where P_{out} was measured at the corresponding coupling length L_c. According to Fig. 11.14, L_c varies with s. From Fig. 11.18a, we can observe that the maximum P_{out}/P_{in} value is between s = 1.4 mm and s = 1.8 mm. We then use the term $\exp(-\alpha L_c)$ to estimate the propagation loss of the coupler, where α = 0.008 cm^{-1} is the measured attenuation constant of a single subwavelength PE fiber. To compare the values of the measured P_{out}/P_{in} and the propagation loss, that is, $\exp(-\alpha L_c)$, we plot the results together in Fig. 11.18a, and the propagation loss is shown as blue open triangles. It can be found that when 1.7 mm < s < 2.55 mm, the P_{out}/P_{in} and the propagation loss agree with each other, meaning that the propagation loss is the dominant loss factor and the modal excitation loss can be neglected. The trend of P_{out}/P_{in} decreases with increased core separation. With 1.17 mm < s < 1.7 mm, P_{out}/P_{in} becomes less than the propagation loss, a clear indication for the existence of the modal excitation loss [43]. We then calculate the insertion loss with the definition of 10 $\log_{10}(P_{in}/P_{out})$. The measured and simulated insertion losses as a function of core separation are shown in Fig. 11.18b. Experimentally, an insertion loss of less than 0.3 dB can be achieved.

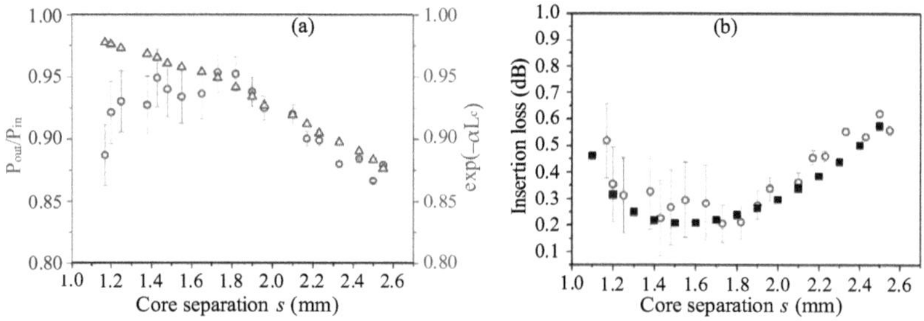

Figure 11.18 (a) Red open circles show the measured P_{out}/P_{in} as a function of core separation s. The blue open triangles show the estimated propagation losses at the coupling lengths. (b) Measured (red open circles) and simulated (black solid squares) results of the insertion loss as a function of core separation s. Reproduced from Ref. [40] by permission of IEEE, © 2012.

As stated previously, when the core separation is too small, the odd mode may encounter cutoff [44]. Previously we demonstrated this phenomenon by placing two identical PE fibers touching together. The core diameter of the PE fibers was 210 μm, and the working frequency was 304 GHz. Numerical simulation indicates that the odd mode is cut off below 715 GHz. Under the odd-mode cutoff, only the even mode exists. Hence it is expected that the power in each fiber will be equal and the power transfer ratio, defined as $P_2/(P_1 + P_2)$ where P_1 and P_2 are the powers of the input and coupled fibers, will be 50% and independent of the length of the coupler region ℓ. Figure 11.19 shows the measured power transfer ratio as a function of ℓ. It is clear that the measured result agrees with the theoretical expectation and the phenomenon of odd-mode cutoff is thus evidenced.

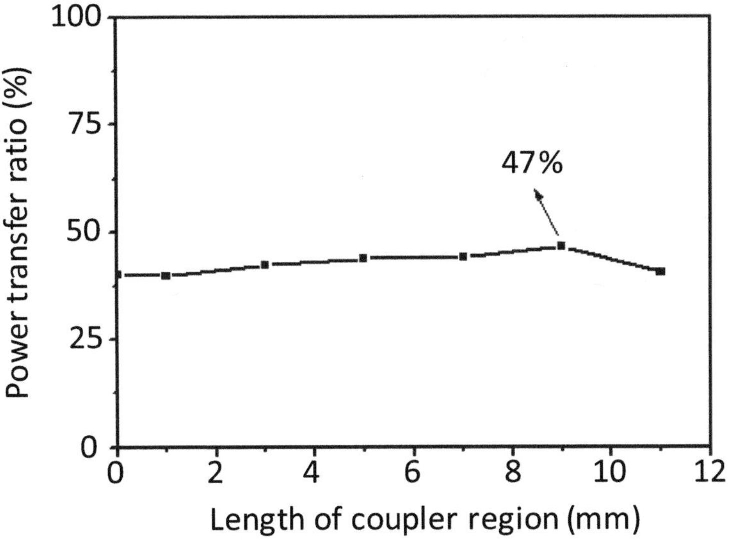

Figure 11.19 Power transfer ratio for the case of touching PE fibers.

11.4.1.2 THz couplers with pipe waveguides

After describing the subwavelength-fiber-based THz directional couplers, here we introduce couplers formed with THz pipe waveguides. For the convenience of discussion, we start with studying the directional coupler, which is composed of two slab-type hollow waveguides [41]. The structure is shown in Fig. 11.20, where two identical slab-type hollow waveguides are placed touching each other. The slab-type hollow waveguides can be treated as the 1D version of the pipe waveguide with the air core sandwiched by two thin dielectric cladding slabs. For each hollow waveguide, we assume the core width $w = 5$ mm, the cladding thickness $t = 0.5$ mm, and the refractive indices $n_2 = 1.4$ and $n_1 = 1$ (air). Antiresonant frequencies, which are midway between two adjacent resonant frequencies for a single hollow waveguide, can be obtained from Eq. 11.6. For the parameters assumed here, antiresonant frequencies of the first three passbands are 153.1, 459.3, and 765.5 GHz, respectively. In the following section, only the transverse electric (TE) polarization is considered.

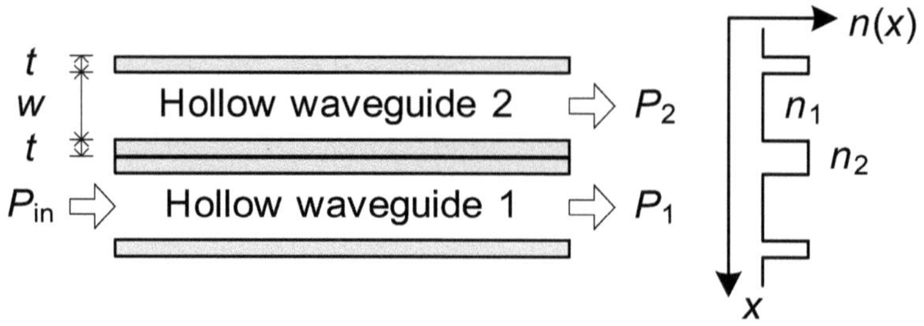

Figure 11.20 Structure of a THz directional coupler composed of touching slab-type hollow waveguides.

A Gaussian beam of unity power with beam waist $2w_0$= 3.6 mm is launched into the input end of hollow waveguide 1. Power evolution along the propagation distance z is simulated by using the finite-difference beam propagation method (BPM) [45]. When calculating the powers of each waveguide, P_1 and P_2, only the amount in the air core region is counted. We first observe the result calculated at 540 GHz, which is a common frequency between resonant and antiresonant frequencies. The result is shown in Fig. 11.21a. Clearly, owing to the leaky nature of the hollow waveguides, the total power $P_1 + P_2$ attenuates with the distance. The attenuation constant of the coupler is very low, and it can be estimated from the log plot of the total power (not shown here), which is 1.5×10^{-3} cm^{-1} in this case, about the same magnitude as that of a single hollow waveguide. If the coupling length L_c is defined as the distance at which P_2 obtains the maximum power, from Fig. 11.21a, L_c = 244.6 cm at 540 GHz, which is equivalent to 4400 wavelengths. Next, we examine the result calculated at 460 GHz, which is very close to the antiresonant frequency 459.3 GHz, and the result is shown in Fig. 11.21b. It can be seen that the power transfer between the hollow waveguides occurs much more rapidly and L_c reduces dramatically to 16.8 cm, equivalent to only 258 wavelengths, while the attenuation constant is almost the same; it is 1.1×10^{-3} cm^{-1}.

To further investigate the coupling characteristics of the directional coupler, modal analysis was conducted by using the FDFD mode solver. We still checked the results calculated at 540 GHz first. It is found that the field profiles in the air-core region of the even and odd system modes are both Gaussian-like. Hence, it is expected that both system modes would be equally excited by the Gaussian input, with each one having half the power. This is confirmed by calculating the overlap integrals, and the excited modal powers are 0.495 and 0.492 for the even and odd system modes, respectively.

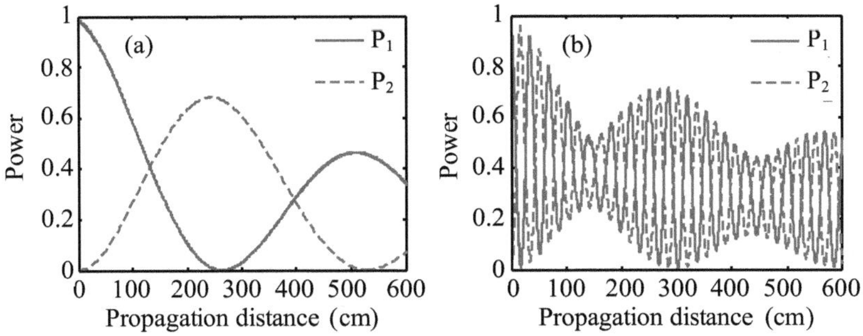

Figure 11.21 Power evolution along the propagation distance. (a) 540 GHz and (b) 460 GHz. Reproduced from Ref. [41] by permission of OSA, © 2011.

Then we turn to check the situation at 460 GHz. Figure 11.22a shows the dispersion spectra, that is, modal indices (β/k_0) as a function of frequency of the system modes. In Fig. 11.22a, three lowest-order system modes are shown; one is even (denoted as E1) and the other two are odd (denoted as O1 and O2, respectively). At most frequencies, only two system modes, one even and one odd, are close in the modal index. Hence, when the Gaussian input is launched, one even and one odd system mode would be largely and equally excited, like what was previously observed at 540 GHz. This is also similar to what happens in conventional directional couplers. However, as can be seen, near the antiresonant frequency (459.3 GHz here), both modal indices of the two odd modes are close to that of the even mode. As a result, there would be three system modes excited simultaneously in the coupler. For example, excited modal powers for the E1, O1, and O2 system modes at 460 GHz are 0.496, 0.298, and 0.189, respectively. Hence, the power evolution simulated at 460 GHz, as shown in Fig. 11.23b, is a consequence of the three-mode-beating behavior. We further examine modal profiles of the system modes near the antiresonant frequency. For the even system mode, it is found that the field distribution in the air core is still Gaussian-like, while for the odd system modes, either O1 or O2, the field distribution in the air core is much different from the Gaussian-like profile. Figure 11.22b shows field profiles of the two odd system modes, which clearly illustrates how the field distribution varies when the frequency approaches the antiresonant one.

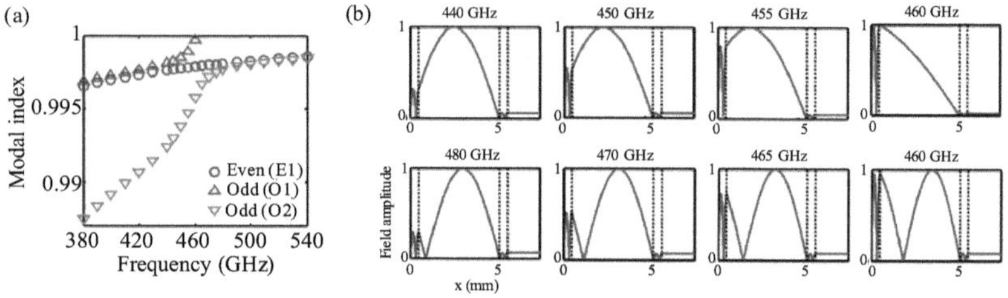

Figure 11.22 (a) Dispersion spectra of system modes. (b) Normalized field distributions (absolute values) of the odd system modes, where the top row is with O1 and the bottom row is with O2. Reproduced from Ref. [41] by permission of OSA, © 2011.

Figure 11.23a shows the coupling length L_c obtained by BPM as a function of frequency. Obviously, L_c at the antiresonant frequency (460 GHz) is much shorter than at other frequencies, owing to the three-mode-beating behavior. We also examine the situation where an air gap exists between the two hollow waveguides. Results shown in Fig. 11.23b indicate that L_c increases significantly when there is an air gap.

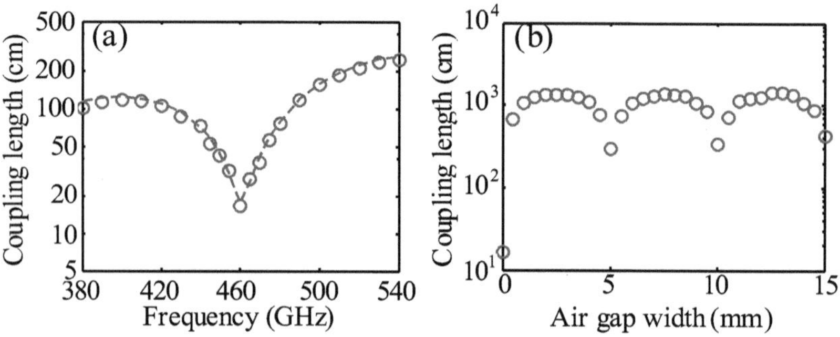

Figure 11.23 (a) Coupling length as a function of frequency. (b) Coupling length at 460 GHz with an air gap existing between the two hollow waveguides. Reproduced from Ref. [41] by permission of OSA, © 2011.

Having investigating the coupling characteristics theoretically, the following step is to implement a pipe-waveguide-based THz directional coupler [42]. Since the contact area between two touching circular pipe waveguides is limited, to reduce the coupling length, we constructed a directional coupler with two square pipe waveguides [34] instead of circular ones. We made two square PE pipe waveguides touch each other to establish the directional coupler, with an air-core size of 6 mm and a cladding thickness of 1 mm. In the central region of the coupler, the total thickness of the cladding is 2 mm, and this region is referred to as the "coupling cladding." Like what is shown in Fig. 11.20, after butt-coupling the directional coupler by a Gunn oscillator, the output power of each square pipe waveguide was measured by a Golay cell. The power transfer ratio is defined as $P_2/(P_1 + P_2)$.

The attenuation spectrum of single square PE pipe waveguide was measured, and it was found that 376 GHz is the antiresonant frequency. Then, the performance of the pipe-waveguide-based directional coupler was investigated, and the coupling characteristics are demonstrated in Fig. 11.24. Clearly, the directional coupler works most efficiently at the antiresonant frequency (376 GHz) of the pipe waveguides, while as the operating frequency is away from 376 GHz, it takes a longer overlap length to reach a high power transfer ratio. This phenomenon is well matched with the simulation results shown previously and can be attributed to the three-mode-beating behavior occurring at the antiresonant frequencies.

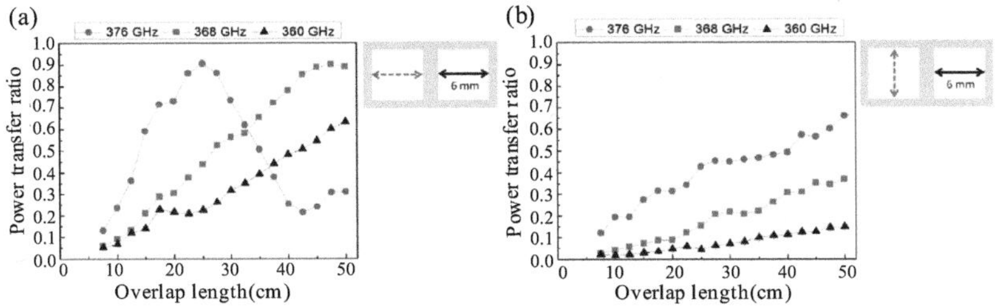

Figure 11.24 Measurement of a THz directional coupler for (a) TM polarization and (b) TE polarization to the coupling cladding. The red double-dashed arrow indicates the direction of polarization. Reproduced from Ref. [42] by permission of OSA, © 2011.

Figure 11.24a,b also demonstrates the coupling performance of the directional coupler for different polarizations. It is found that the directional coupler is polarization sensitive. When THz waves are transverse-magnetic (TM)-polarized with respect to the coupling cladding, as the direction of polarization shows in Fig. 11.24a, a higher power transfer ratio can be obtained. As we can see from Fig. 11.24, for 376 GHz, if the overlap length is 25 cm, the power transfer ratio reaches 90.5% for TM polarization but only 42.8% for TE polarization. Since the transmission of TM waves at the core–cladding interface is higher than that of TE waves, THz waves with TM polarization characteristics would more likely penetrate through the coupling cladding into another waveguide, thus leading to a higher power transfer ratio.

11.4.2 THz Fiber-Scanning Imaging Systems

THz radiation has a number of properties that make it an attractive clinical imaging technique. For example, THz wavelengths are much longer than infrared and optical radiation, so scattering in biologic tissues is much reduced and could be negligible [46]. Contrary to X-rays, THz radiation has a relatively low photon energy and so does not ionize tissues [47, 48]. Many materials of interest have unique spectral "fingerprints" in the THz range [49, 50]. This offers the possibility to combine spectral identification with THz imaging. THz radiation can also detect differences in the water content and density of a tissue, which could allow effective detection of cancer with a safer but less invasive means [47, 48]. However, due to high water absorption, previous THz imaging was limited in the reflection mode [51, 52], which could achieve a high degree of water content sensitivity, while transmission mode imaging was considered difficult due to the strong attenuation in most biological tissues. In this section, the transmission-type THz fiber-scanning imaging systems will be introduced [53, 54], which take advantage of subwavelength PE fibers for THz wave transmission. By employing compact, low-cost, and room-temperature-operated fiber-scanning imaging systems, we also demonstrate the high diagnostic performance of THz imaging on breast tumor diagnosis [55], as well as in vivo early breast cancer detection in a subcutaneous xenograft mouse model [56].

11.4.2.1 THz scanning imaging with a subwavelength plastic fiber

At the optical frequency, fiber-based imaging is attractive, which enables in vivo imaging or sensing in a less invasive manner and has been extensively applied in endoscopy [57] and microscopy [58]. For biomedical applications, many biomolecules have their characteristic absorption spectra in the THz frequency range, enabling marker-free and direct recognition of molecular distribution based on their distinct THz fingerprints. Therefore, if THz fiber-scanning imaging could be realized, it will not only allow remote coupling from a bulky THz-generation setup to the detected sample and allow flexible THz wave guidance but will also greatly expand THz technology to intravital molecular imaging just like its optical counterpart. The THz subwavelength plastic fiber is shown to be with a low attenuation constant and a high butt-coupling efficiency. Nonetheless, for fiber-scanning imaging applications, the THz fiber is required not only to be with low attenuation and high coupling efficiency for a high signal-to-noise ratio (SNR) but also to be with a low bending loss for large-area scanning. In the following section, the bending loss characteristic of the subwavelength plastic fiber is investigated and the plastic fiber is further employed to implement a transmission THz fiber-scanning imaging system [53].

We first conducted fiber-bending loss measurement. The adopted subwavelength PE fiber is 90 cm long with a diameter of 180 μm, and the fiber has a minimum attenuation constant around 380 GHz. A continuous-wave (CW) Gunn oscillator module was used as the source, where the free-running THz waves, collected and focused by a pair of parabolic mirrors, were coupled into the subwavelength plastic fiber. By holding the fiber straight and by laterally moving the output end of the fiber, Fig. 11.25 shows the measured THz transmitted power of the PE fiber as a function of the bending angle θ, which is normalized with respect to that of $\theta = 0°$. At a bending angle of 2°, the experimentally measured transmitted THz power after normalization is 0.86. By considering the total fiber transmission, including bending loss at 2° and fiber attenuation loss for a 90 cm long fiber, the total transmission is 0.65. Our study indicates that the total transmission loss in the 180-μm-diameter fiber is acceptable for scanning a 6×6 cm^2 area with a 90 cm fiber length.

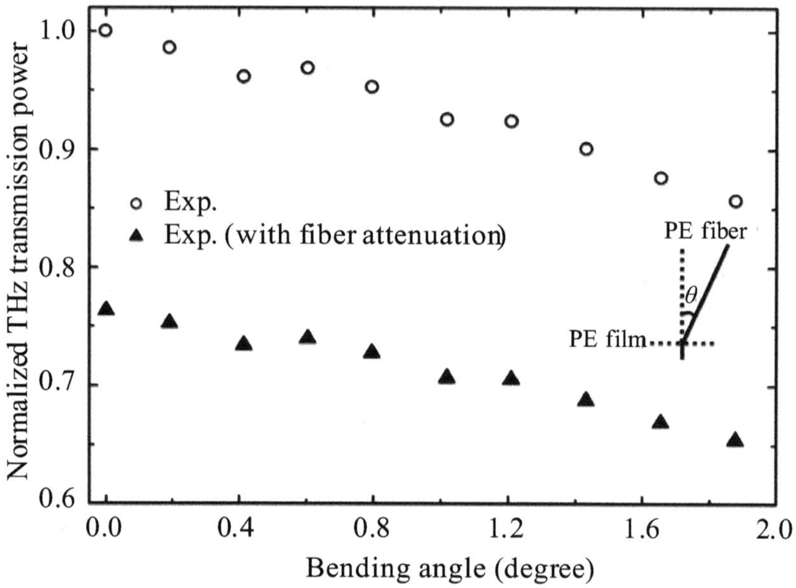

Figure 11.25 Normalized THz power transmission at different fiber-bending angles (open circles). The solid triangles are the THz power transmissions, including fiber attenuation. Inset shows the configuration for the bending loss measurement.

To further investigate the feasibility to perform fiber-scanning THz imaging by utilizing a low-loss subwavelength plastic fiber, we integrated the plastic fiber with a CW Gunn oscillator and a Golay cell to construct a THz fiber-scanning imaging system for imaging spatially fixed samples. Figure 11.26a shows the schematic diagram of the fiber-scanning imaging system. The CW THz output after the fiber was focused by a THz lens (picarin lens, numeric aperture ~0.42) to a stationary sample. The THz transmission image was accomplished by 2D scanning of the output end of the PE fiber, together with the picarin lens and Golay cell, which were all mounted on a motorized stage. The focused THz spot size after the picarin lens, measured by the knife-edge method [59], was of the order of 3 mm.

Figure 11.26b–e shows some examples of the acquired 380 GHz THz images, with a metal pattern (8×6.5 cm^2), a matchbox (5×3.5 cm^2), a dry seahorse (8.5×4 cm^2), and two dry fishes (5×6 cm^2). They were original THz images, with a scanned step size of 0.5×0.5

mm^2, that were not normalized to the background, that is, we did not correct the angle-dependent bending loss. These results reveal that the THz fiber-scanning imaging system is able to scan over a large area of metal materials, biological specimens, and concealed objects.

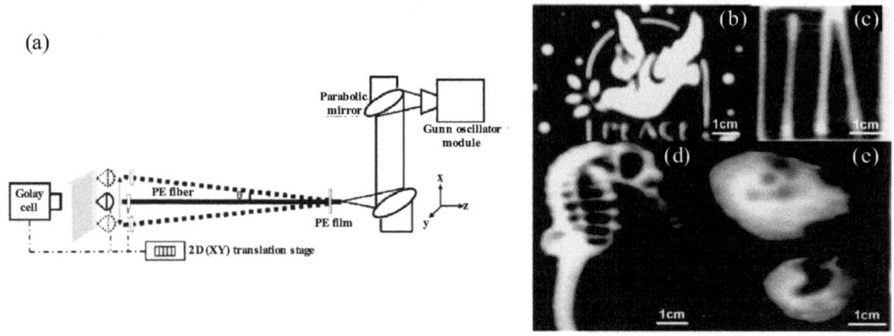

Figure 11.26 (a) Schematic diagram of the THz fiber-scanning imaging system. Fiber-scanning THz images of (b) a metal pattern with a word "peace" and a dove figure, (c) a paper-made matchbox with three matches inside, (d) a dry seahorse, and (e) two dry fishes. In (b), the yellow region is with high THz transmission, while the dark region is with low THz transmission. In (c–e), the yellow region is with low THz transmission, while the dark region is with high THz transmission. Reproduced from Ref. [53] by permission of AIP, © 2008.

11.4.2.2 THz fiber-scanning near-field microscope

Over the past few years, THz microscopy has attracted much attention owing to its intrinsic molecular imaging capability [60, 61]. However, the poor spatial resolution, limited by diffraction and at best close to the wavelength (300 μm at 1 THz), is the main drawback. One way to obtain a subwavelength resolution is to develop a near-field microscope. One technique for near-field imaging is to use a single aperture surrounded by concentric periodic grooves (referred to as a bull's-eye structure) [62]. It was demonstrated that a subwavelength aperture surrounded by a bull's-eye structure could result in not only a strongly enhanced transmission peak but also a strong angular confinement of the radiated THz waves [62, 63]. Nonetheless, samples are not always movable in most biomedical imaging applications. Moving the object will sometimes also disturb the sample. A THz scanning near-field imaging system is thus extremely needed.

On the basis of the aforementioned fiber-scanning THz imaging system and an optimal designed bull's-eye structure, we developed the transmission mode of an upright-type THz fiber-scanning near-field microscope with a compact size operating at room temperature and with a high spatial resolution [54]. Recently, several studies have reported on the measurements of the THz dielectric properties of human breast cancerous tissues [64, 65]. It indicates that there is a significant difference between absorption constants of breast cancers and those of normal tissues, especially around 0.3 THz. We thus used a 312 GHz wave as the source of the developed THz near-field microscope for breast cancer imaging and designed a bull's-eye structure with a resonant transmission peak at the same frequency. The optimized bull's-eye structure consisted of three periodic concentric grooves 475 μm wide and 100 μm deep with a 950 μm periodicity on a 500 μm thick PMMA substrate and was covered by a 200 nm thick gold film. A 200 μm diameter subwavelength circular aperture was arranged at the center of the concentric grooves. Both the grooves and aperture were then created by laser sculptured processing.

The schematic of the THz fiber-scanning near-field microscope is shown in Fig. 11.27a. The adopted subwavelength PE fiber had a diameter of 240 μm and a length of 25 cm. The source for the 312 GHz waves is a CW Gunn oscillator module. Behind the fiber output, we integrated the bull's-eye spatial filter to achieve both high transmission power and near-field spatial resolution. THz waves radiated from the fiber output end were normally incident onto the substrate side of the bull's-eye spatial filter, and the transmission THz power from the exit side of the aperture was collected by a Winston cone and detected by a room-temperature-operated Golay cell.

We applied this imaging system to examine breast cancer sections. The specimens under investigation were sectioned from excised human breast tissues into 20 μm thick slices and fixed on 150 μm thick coverglasses (shown in the inset of Fig. 11.27b,c). A coverglass with a breast section was placed on a microscope slide holder of the microscope stage. The distance in the z-direction between the bottom of the breast section and the surface of the metallic bull's-eye structure was about 250 μm. The THz near-field images were acquired by direct 2D (xy) scanning of the scanning head (including the fiber output end and bull's-eye structure). The obtained THz image has good contrast and a high signal-to-noise ratio (>800) with a 210 μm effective spatial resolution (<$\lambda/4$), as measured by the knife-edge method. THz near-field images shown were normalized to the background for correcting the angle-dependent bending loss. The background was measured by scanning a blank coverglass with no breast tissue on it. After THz near-field transmission illumination imaging, all samples were sent for routine pathological hematoxylin and eosin (H&E) staining and examination to identify the cancerous area in the THz-imaged breast sections.

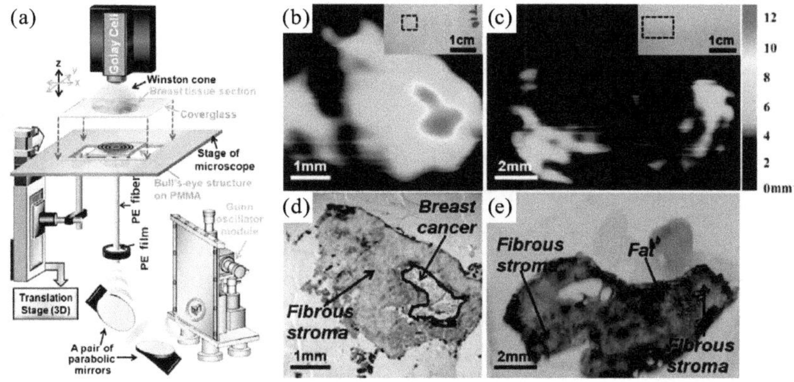

Figure 11.27 (a) Schematic of the THz fiber-scanning near-field microscope. Panels (b) and (c) are THz near-field absorption coefficient images of breast tissue sections (sample 1 and 2). Insets show actual photographs of samples 1 and 2. Panels (c) and (d) are photomicrographs of the breast tissue sections after H&E staining (samples 1 and 2). Reproduced from Ref. [54] by permission of OSA, © 2009.

The insets of Fig. 11.27b,c show the actual photograph of a 20 μm thick breast tissue section, and the dashed black box indicates regions scanned by the THz near-field microscope. As shown in Fig. 11.27b,c, we defined the color bar as follows: If the absorption coefficient was greater than 9 mm^{-1}, we showed the region in yellow and red (with a high cancer probability according to Ref. [65]). As a result, the color near-field image obtained by this transmission illumination system reflects a high capability of identifying tissues between breast cancer and normal tissues (such as fatty and fibrous tissue). Figure 11.27d displays the microscopic image of sample 1 after H&E staining. In this sample, the type of breast cancer was diagnosed as a papillary carcinoma and marked by a solid black boundary. The result clearly indicates that the distribution of breast cancer identified in the THz near-field image matches well with that seen afterward by pathologic examination. The absorption coefficient of fibrous tissue (shown in green) is obviously much less than that of the breast cancer tissues. In addition, the THz near-field image and the corresponding pathologic photomicrograph of sample 2 are shown in Fig. 11.27c and Fig. 11.27e, respectively. Obviously, the scanned THz image indicates that no region in sample 2 should be diagnosed as malignancy, and the fibrous stroma shown in green could be distinguished from the fatty tissue shown in blue. These results are also in excellent agreement with pathologic diagnosis followed by H&E staining.

11.4.2.3 Diagnosing breast tumors with THz fiber-scanning near-field microscope

The aforementioned THz fiber-scanning near-field microscope is further improved to study the performance of diagnosing unstained breast tumor slices without dehydration [55]. The system setup and corresponding experimental data analysis method are described in Section 11.4.2.2, except that the THz power was detected by a Schoktty diode detector. Compared to the Golay cell detector adopted previously, the imaging time (scanning area: 10 × 10 mm) was highly improved from 50 minutes/100 × 100 pixels to 3 minutes/100 × 100 pixels. The THz near-field images of the breast tissue sections are shown according to the measured absorption coefficient. The reference values we used for analyzing experimental results were taken from Ref. [65], that is, the mean absorption coefficients at 320 GHz of fatty tissues, fibrous tissue, and tumor are 2–4.5 mm^{-1}, 7.5–9 mm^{-1}, and 9–12 mm^{-1}, respectively. Therefore, as the THz near-field images show, we defined the color bar as follows: If the absorption coefficient inside a certain region of the THz image was greater than 9 mm^{-1}, we marked the regions as tumor and showed the region in red color (9–12 mm^{-1}), while fat was shown in blue (2–4.5 mm^{-1}) and fibrous was shown in green (7.5–9 mm^{-1}).

The acquired THz near-field microscopic images and the corresponding pathologic photomicrograph of H&E-stained sections for 16 patients are shown in Fig. 11.28. On the basis of the pathologic photomicrographs of H&E-stained sections, the cases were diagnosed by the pathologist as normal (cases 1–5), papillary tumors (cases 6–10), and invasive ductal carcinoma (cases 11–16). From Fig. 11.28, the diagnosis results of the THz near-field microscopy examinations of the same samples from 16 patients on the basis of the THz absorption coefficients are normal for cases 1–5 and with tumor for cases 6–16, all in excellent agreement with the corresponding pathologic diagnosis by H&E staining, to distinguish tumors from normal tissues. In clinical diagnosis, sensitivity

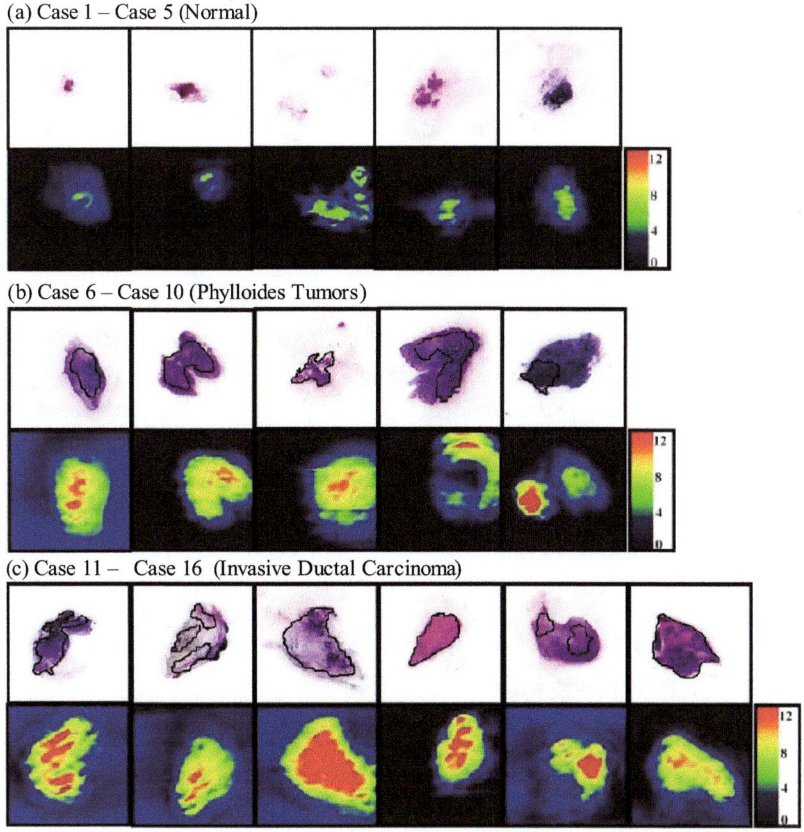

Figure 11.28 THz near-field microscopic images and corresponding pathologic photomicrograph of H&E-stained sections. The areas of breast tumor in the pathologic photomicrographs were marked by a solid black boundary. Republished from Ref. [55] by permission of OSA, © 2011.

and specificity are used to examine the accuracy of a diagnosis method. Sensitivity is defined as the probability of a positive diagnosis among patients with disease, and it can be expressed as Sensitivity = $\frac{\mathrm{TP}}{(\mathrm{TP}+\mathrm{FN})}$, where TP is true positive and FN is false negative. In contrast, specificity is the probability of a negative diagnosis among patients without disease, and it can be described as Specificity = $\frac{\mathrm{TN}}{(\mathrm{TN}+\mathrm{FP})}$, where TN is true negative and FP is false positive. Our clinical study indicates 100% sensitivity and 100% specificity for THz near-field microscopy to differential diagnose breast tumors from normal tissues.

Due to the extensive processing procedure and time-consuming analytical examination, pathologists can only analyze a small fraction of excised tissues. This current procedure risks the sampling error as only a few representative areas in a given lesion were taken for examination and only a few sections were observed. By using the easily quantifiable THz absorption images to differentiate breast tumors from normal tissues, automation on a faster and more complete prescreening program becomes possible, which makes THz near-field microscopy valuable to field surgeons so that pathologists can take advantage of the THz technology to choose slices or reduce the number of slices to be processed for pathological analysis. With the capability to reduce the load of the extensive and time-consuming pathological examination procedure, this THz near-field microscopy technique can be expected to economize the use of hospital and human resources.

11.4.2.4 Detection of early human breast cancer in a subcutaneous xenograft mouse model by using THz mammogram

Breast cancer is one of the most common cancers among women. Early detection of breast cancer allows treatment at an earlier stage and would significantly reduce associated mortality. In clinical practice, X-ray mammography is one of the methods of detecting breast cancer and is the only proven method for screening. However, mammography utilizes ionizing X-rays, which may induce malignancy. Therefore, a new noninvasive screening examination is needed for better early detection. As addressed previously, THz radiation has a number of properties that make it an attractive clinical in vivo imaging technique. Moreover, previous THz studies on ex vivo human breast cancer slices indicated that a THz wave is able to provide high absorption contrast between cancers, fatty tissues, and fibrous tissues with high specificity. Therefore, THz transmission imaging could be a valuable addition to X-ray mammography. However, due to high water absorption, in vivo transmission mode imaging was considered difficult in most biological tissues. To take advantage of the high absorption contrast between breast cancers, fatty tissues, and fibrous tissues, we demonstrated the possibility of performing in vivo THz transmission imaging of early human breast cancer in a subcutaneous xenograft mouse model [56] by lowering the THz frequency to the low-THz regime, for the THz absorption in living tissues tends to reduce with lower THz frequency.

X-ray mammography is performed in transmission mode with the examined breast tissue flattened with a uniform thickness. We developed a fiber-scanning transmission THz imaging system similar to the one described in Section 11.4.2.1 by replacing the source with a YIG oscillator module and the detector with a Schottky diode detector. The system was operated at 108 GHz. Figure 11.29a shows the schematic diagram of the fiber-scanning THz imaging system. The THz subwavelength PE fiber adopted had a diameter of 600 μm and a length of 40 cm.

To in vivo detect early cancer before it is visible, we developed a subcutaneous xenograft mouse model to test sensitivity. After the cancer cell injection, on the 6th day, we anesthetized the mouse and implanted mouse fatty tissue to embed the cancer cells (the place was marked after cancer cell implantation). Starting from the 13th day, we imaged the cancer implanted area every three to four days until the cancer was visible. During THz imaging, we anesthetized the mouse and sandwiched the dorsal cancer area between two coverglasses. In Fig. 11.29b, we show experimental images to demonstrate the possibility and sensitivity of our THz imaging system to detect breast cancer early. Figure 11.29b shows thus acquired 10 × 10 mm THz screening images on one mouse according to the imaging time from the 13th day to the 33rd day after cancer cell implantation. For the images acquired on the 23rd day to the 33rd day, the color bar is defined according to the maximum measured THz absorption coefficient of the sandwiched tissues, while from the 13th to the

17th day, the color bar is defined as high as 1.7 mm^{-1} according to the 23rd day for comparison. It is clear that the detected absorption change in early cancer development starts on the 23rd day and the absorption change $\Delta\alpha_a$ is 0.25 mm^{-1}. The early cancer development induced absorption change starts to be detected on the 18th day for another mouse, and the cancer cells embedded in fat induce an absorption change $\Delta\alpha_a$, which is as small as 0.15 mm^{-1}.

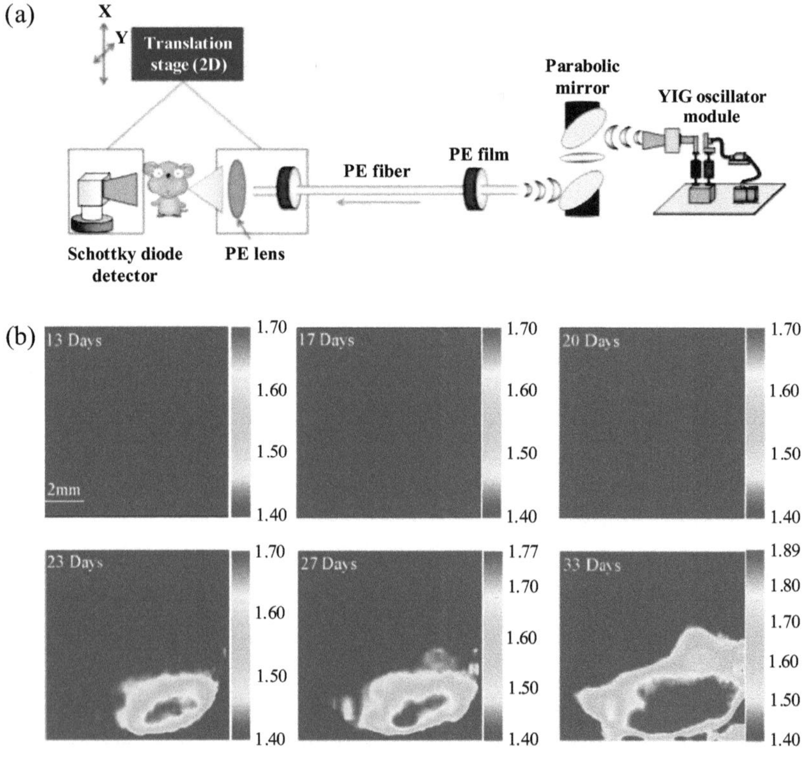

Figure 11.29 (a) Schematic diagram of the THz fiber-scanning transmission imaging system. (b) A series of THz images according to the screening time from the 13th day to the 33th day after cancer cell implantation. Early cancer development was detected on the 23th day, and the induced absorption coefficient change was 0.25 mm^{-1}, corresponding to a cancer volume of 1.3 mm^3. Reproduced from Ref. [56] by permission of OSA, © 2011.

The tumor volume can be estimated by introducing the concept of cell absorption cross section [56]. The volume of the cancerous tissue, as shown in Fig. 11.29b, in early cancer development on the 23rd day is 1.3 mm^3, while the early cancer development in another mouse is with 0.8 mm^3 on the 18th day. The minimum detectable volume on the breast cancer of our imaging system is estimated to be 0.05 mm^3. This high sensitivity of THz imaging on cancer cells is primarily contributed from the high absorption constant on the adopted wavelength range, which sacrifices the penetration capability. Current X-ray mammography can detect breast cancers as small as 2 mm in diameter [66]. Compared with the state-of-the-art clinical X-ray mammography, this noninvasive and nonionizing THz mammography system thus has the potential to serve as an alternative method to detect breast cancer early, with a much smaller size, especially suitable for preclinical smaller animal studies, surgical specimens, and drug screening.

References

1. G. Gallot, S. P. Jamison, R. W. McGowan, and D. Grischkowsky, Terahertz waveguides, *J. Opt. Soc. Am. B* **17**, 851–863 (2000).
2. R. Mendis and D. Grischkowsky, Undistorted guided-wave propagation of subpicosecond terahertz pulses, *Opt. Lett.* **26**, 846–848 (2001).
3. R. Mendis and D. Grischkowsky, Plastic ribbon THz waveguides, *J. Appl. Phys.* **88**, 4449–4451 (2000).
4. H. Han, H. Park, M. Cho, and J. Kim, Terahertz pulse propagation in a plastic photonic crystal fiber, *Appl. Phys. Lett.* **80**, 2634–2636 (2002).
5. K. Wang and D. M. Mittleman, Metal wires for terahertz wave guiding, *Nature* **432**, 376–379 (2004).
6. J. A. Harrington, R. George, P. Pedersen, and E. Mueller, Hollow polycarbonate waveguides with inner Cu coatings for delivery of terahertz radiation, *Opt. Express* **12**, 5263–5268 (2004).
7. B. Bowden, J. A. Harrington, and O. Mitrofanov, Silver/polystyrene-coated hollow glass waveguides for the transmission of terahertz radiation, *Opt. Lett.* **32**, 2945–2947 (2007).
8. T. Ito, Y. Matsuura, M. Miyagi, H. Miyagi, and H. Ito, Flexible terahertz fiber optics with low bend-induced losses, *J. Opt. Soc. Am. B* **24**, 1230–1235 (2007).
9. T. Hidaka, H. Minamide, H. Ito, J.-I. Nishizawa, K. Tamura, and S. Ichikawa, Ferroelectric PVDF cladding terahertz waveguide, *J. Lightwave Technol.* **23**, 2469–2473 (2005).

10. M. Skorobogatiy and A. Dupuis, Ferroelectric all-polymer hollow Bragg fibers for terahertz guidance, *Appl. Phys. Lett.* **90**, 113514 (2007).
11. J.-Y. Lu, C.-P. Yu, H.-C. Chang, H.-W. Chang, Y.-T. Li, C.-L. Pan, and C.-K. Sun, Terahertz air-core microstructure fiber, *Appl. Phys. Lett.* **92**, 064105 (2008).
12. A. Hassani, A. Dupuis, and M. Skorobogatiy, Porous polymer fibers for low-loss Terahertz guiding, *Opt. Express* **16**, 6340–6351 (2008).
13. S. Atakaramians, V. S. Afshar, B. M. Fischer, D. Abbott, and T. M. Monro, Porous fibers: a novel approach to low loss THz waveguides, *Opt. Express* **16**, 8845–8854 (2008).
14. L.-J. Chen, H.-W. Chen, T.-F. Kao, J.-Y. Kao, and C.-K. Sun, Low-loss subwavelength plastic fiber for terahertz waveguiding, *Opt. Lett.* **31**, 308–310 (2006).
15. C.-H. Lai, Y.-C. Hsueh, H.-W. Hsueh, Y.-J. Huang, H.-C. Chang, and C.-K. Sun, Low-index terahertz pipe waveguides, *Opt. Lett.* **34**, 3457–3459 (2009).
16. C. H. Chandler, An investigation of dielectric rod as wave guide, *J. Appl. Phys.* **20**, 1188–1192 (1949).
17. W. M. Elsasser, Attenuation in a dielectric circular rod, *J. Appl. Phys.* **20**, 1193–1196 (1949).
18. L. Tong, R. R. Gattass, J. B. Gattass, S. He, J. He, M. Shen, I. Maxwell, and E. Mazur, Suwavelength-diameter silica wires for low-loss optical wave guiding, *Nature* **426**, 816–819 (2003).
19. A. Yariv, *Optical Electronics in Modern Communications*, 5th ed. (Oxford University Press, UK, 1997).
20. J. R. Birch, G. J. Simonis, M. N. Afsar, R. N. Clarke, J. M. Dutta, H. M. Frost, X. Gerbaus, A. Hadni, W. F. Hall, R. Heidinger, W. W. Ho, C. R. Jones, F. Koniger, R. L. Moore, H. Matsuo, T. Nakano, W. Richter, K. Sakai, M. R. Stead, U. Stumper, R. S. Vigil, and T. B. Wells, An intercomparison of measurement techniques for the determination of the dielectric properties of solids at near millimetre wavelengths, *IEEE Trans. Microwave Theory Tech.* **42**, 956–965 (1994).
21. H.-W. Chen, Y.-T. Li, C.-L. Pan, J.-L. Kuo, J.-Y. Lu, L.-J. Chen, and C.-K. Sun, Investigation on spectral loss characteristics of subwavelength terahertz fibers, *Opt. Lett.* **32**, 1017–1019 (2007).
22. G. Brambilla, V. Finazzi, and D. Richardson, Ultra-low-loss optical fiber nanotapers, *Opt. Express* **12**, 2258–2263 (2004).
23. M. Sumetsky, How thin can a microfiber be and still guide light? *Opt. Lett.* **31**, 870–872 (2006).

24. C.-K. Sun, L.-J. Chen, H.-W. Chen, and J.-Y. Lu, Subwavelength plastic fiber for terahertz wave guiding, *Proc. SPIE* **6373**, 63730F-1–63730F-11 (2006).

25. C.-K. Sun, L.-J. Chen, and H.-W. Chen, Plastic waveguide for terahertz wave, US Patent No. US7,409,132 B2, 08/05/2008-07/26/2026 (2008).

26. D. Chen and H. Chen, A novel low-loss terahertz waveguide: polymer tube, *Opt. Express* **18**, 3762–3767 (2010).

27. M. Nagel, A. Marchewka, and H. Kurz, Low-index discontinuity terahertz waveguides, *Opt. Express* **14**, 9944–9954 (2006).

28. C.-H. Lai, B. You, J.-Y. Lu, T.-A. Liu, J.-L. Peng, C.-K. Sun, and H.-C. Chang, Modal characteristics of antiresonant reflecting pipe waveguides for terahertz waveguiding, *Opt. Express* **18**, 309–322 (2010).

29. M. A. Duguay, Y. Kokubun, T. L. Koch, and L. Pfeiffer, Antiresonant reflecting optical waveguides in SiO_2–Si multilayer structures, *Appl. Phys. Lett.* **49**, 13–15 (1986).

30. H. A. Haus, *Waves and Fields in Optoelectronics* (Prentice-Hall, USA, 1984).

31. C.-P. Yu and H.-C. Chang, Yee-mesh-based finite difference eigenmode solver with PML absorbing boundary conditions for optical waveguides and photonic crystal fibers, *Opt. Express* **12**, 6165–6167 (2004).

32. E. A. J. Marcatilli and R. A. Schmeitzer, Hollow metallic and dielectric waveguides for long distance optical transmission and lasers, *Bell Syst. Tech. J.* **43**, 1783–1809 (1964).

33. E. Nguema, D. Ferachou, G. Humbert, J.-L. Auguste, and J.-E. Blondy, Broadband terahertz transmission within the air channel of thin-wall pipe, *Opt. Lett.* **36**, 1782–1784 (2011).

34. J.-T. Lu, C.-H. Lai, T.-F. Tseng, H. Chen, Y.-F. Tsai, I.-J. Chen, Y.-J. Hwang, H.-C. Chang, and C.-K. Sun, Terahertz polarization-sensitive rectangular pipe waveguides, *Opt. Express* **19**, 21532–21539 (2011).

35. D. Grischkowsky, S. R. Keiding, M. van Exter, and Ch. Fattinger, Far-infrared time-domain spectroscopy with terahertz beams of dielectrics and semiconductors, *J. Opt. Soc. Am. B* **7**, 2006–2015 (1990).

36. R. A. Spanke, Architecture for guided-wave optical space switching systems, *IEEE Commun. Mag.* **25**, 42–48 (1987).

37. A. F. Fercher, W. Drexler, C. K. Hitzenberger, and T. Lasser, Optical coherence tomography: principle and applications, *Rep. Prog. Phys.* **66**, 239–303 (2003).

38. R. G. Lamont, D. C. Johnson, and K. O. Hill, Power transfer in fused biconical-taper single-mode fiber couplers, *Appl. Opt.* **24**, 327–332 (1985).

39. B. P. Pal, R. R. Chaudhuri, and M. R. Shenoy, Fabrication and modeling of fused biconical tapered fiber couplers, *Fiber Integr. Opt.* **22**, 97–117 (2003).

40. T.-F. Tseng, C.-H. Lai, J.-T. Lu, Y.-F. Tsai, Y.-J. Hwang, and C.-K. Sun, Investigation on strong coupling behaviors of THz subwavlength directional couplers, *IEEE Photon. J.* **4**, 2307–2314 (2012).

41. C.-H. Lai, C.-K. Sun, and H.-C. Chang, Terahertz antiresonant-reflecting-hollow-waveguide-based directional coupler operating at antiresonant frequencies, *Opt. Lett.* **36**, 3590–3592 (2011).

42. J.-T. Lu, C.-H. Lai, T.-F. Tseng, H. Chen, Y.-F. Tsai, Y.-J. Hwang, H.-C. Chang, and C.-K. Sun, Terahertz pipe-waveguide-based directional couplers, *Opt. Express* **19**, 26883–26890 (2011).

43. Z.-M. Mao, X.-S. Fang, and B.-H. Li, Mode excitation theory and experiment of single-mode fiber directional coupler with strong coupling, *J. Lightwave. Technol.* **4**, 466–472 (1986).

44. H.-W. Chen, C.-M. Chiu, C.-H. Lai, J.-L. Kuo, P.-J. Chiang, Y.-J. Hwang, H.-C. Chang, and C.-K. Sun, Subwavelength dielectric-fiber-based THz coupler, *J. Lightwave Technol.* **27**, 1489–1495 (2009).

45. J. Yamauchi, *Propagating Beam Analysis of Optical Waveguides* (Research Studies Press, UK, 2003).

46. A. J. Fitzgerald, B. E. Cole, and P. F. Taday, Nondestructive analysis of tablet coating thicknesses using terahertz pulsed imaging, *J. Pharm. Sci.* **94**, 177–183 (2005).

47. T. Löffler, T. Bauer, K. Siebert, H. Roskos, A. Roskos, and S. Czasch, Terahertz dark-field imaging of biomedical tissue," *Opt. Express* **9**, 616–621 (2001).

48. R. M. Woodward, B. E. Cole, V. P. Wallace, R. J. Pye, D. D. Arnone, E. H. Linfield, and M. Pepper, Terahertz pulse imaging in reflection geometry of human skin cancer and skin tissue, *Phys. Med. Biol.* **47**, 3853–3863 (2002).

49. K. Kawase, Y. Ogawa, Y. Watanabe, and H. Inoue, Non-destructive terahertz imaging of illicit drugs using spectral fingerprints, *Opt. Express* **11**, 2549–2554 (2003).

50. M. Yamaguchi, F. Miyamaru, K. Yamamoto, M. Tani, and M. Hangyo, Terahertz absorption spectra of L-, D-, and DL-alanine and their

application to determination of enantiometric composition, *Appl. Phys. Lett.* **86**, 053903 (2005).

51. Y. Sun, M. Y. Sy, Y.-X. J. Wang, A. T. Ahuja, Y.-T. Zhang, and E. Pickwell-Macpherson, A promising diagnostic method: terahertz pulsed imaging and spectroscopy, *World J. Radiol.* **3**, 55–65 (2011).
52. D. B. Bennett, Z. D. Taylor, P. Tewari, R. S. Singh, M. O. Culjat, W. S. Grundfest, D. J. Sassoon, R. D. Johnson, J.-P. Hubschman, and E. R. Brown, Terahertz sensing in corneal tissues, *J. Biomed. Opt.* **16**, 057003 (2011).
53. J.-Y. Lu, C.-M. Chiu, C.-C. Kuo, C.-H. Lai, H.-C. Chang, Y.-J. Hwang, C.-L. Hwang, and C.-K. Sun, Terahertz scanning imaging with a subwavelength plastic fiber, *Appl. Phys. Lett.* **92**, 084102 (2008).
54. C.-M. Chiu, H.-W. Chen, Y.-R. Huang, Y.-J. Hwang, W.-J. Lee, H. Y. Huang, and C.-K. Sun, All-terahertz fiber-scanning near-field microscopy, *Opt. Lett.* **34**, 1084–1086 (2009).
55. H. Chen, W.-J. Lee, H.-Y. Huang, C.-M. Chiu, Y.-F. Tsai, T.-F. Tseng, J.-T. Lu, W.-L. Lai, and C.-K. Sun, Performance of THz fiber-scanning near-field microscopy to diagnose breast tumors, *Opt. Express* **19**, 19523–19531 (2011).
56. H. Chen, T.-H. Chen, T.-F. Tseng, J.-T. Lu, C.-C. Kuo, S.-C. Fu, W.-J. Lee, Y.-F. Tsai, Y.-Y. Huang, E. Y. Chuang, Y.-J. Hwang, and C.-K. Sun, High-sensitivity *in vivo* THz transmission imaging of early human breast cancer in a subcutaneous xenograft mouse model, *Opt. Express* **19**, 21552–21562 (2011).
57. A. L. Polglase, W. J. McLaren, S. A. Skinner, R. Kiesslich, M. F. Neurath, and P. M. Delaney, A fluorescence confocal endomicroscope for in vivo microscopy of the upper- and the lower-GI tract, *Gastrointest. Endosc.* **62**, 686–695 (2005).
58. F. Helmchen, M. S. Fee, D. W. Tank, and W. Denk, A miniature head-mounted two-photon microscope: high-resolution brain imaging in freely moving animals, *Neuron* **31**, 903–912 (2001).
59. S. Hunsche, M. Koch, I. Brener, and M. C. Nuss, Terahertz near-field imaging, *Opt. Commun.* **150**, 22–26 (1998).
60. X.-C. Zhang, Terahertz wave imaging: horizons and hurdles, *Phys. Med. Biol.* **47**, 3667–3677 (2002).
61. W. L. Chan, J. Deibel, and D. M. Mittleman, Imaging with terahertz radiation, *Rep. Prog. Phys.* **70**, 1325–1379 (2007).
62. K. Ishihara, K. Ohashi, T. Ikari, H. Minamide, H. Yokoyama, J. Shikata, and H. Ito, Terahertz-wave near-field imaging with subwavelength

resolution using surface-wave-assisted bow-tie aperture, *Appl. Phys. Lett.* **89**, 201120 (2006).

63. A. Agrawal and A. Nahata, Time-domain radiative properties of a single subwavelength aperture surrounded by an exit side surface corrugation, *Opt. Express* **14**, 1973–1981 (2006).
64. A. J. Fitzgerald, V. P. Wallace, M. Jimenez-Linan, L. Bobrow, R. J. Pye, A. D. Purushotham, and D. D. Arnone, Terahertz pulsed imaging of human breast tumors, *Radiology*, **239**, 533–540 (2006).
65. P. C. Ashworth, E. Pickwell-MacPherson, E. Provenzano, S. E. Pinder, A. D. Purushotham, M. Pepper, and V. P. Wallace, Terahertz pulsed spectroscopy of freshly excised human breast cancer, *Opt. Express* **17**, 12444–12454 (2009).
66. A. C. M. Onuigbo, M. E. Cuffy-Hallam, N. A. Dunsmore, and E. S. Zinreich, Mammography reveals a 2-mm intraductal breast carcinoma, *Hosp. Physician* **37**, 61–64 (2001).

Chapter 12

Terahertz Information and Signal Processing

S. J. Ben Yoo, Ryan P. Scott, David J. Geisler, Nicolas K. Fontaine, and Francisco M. Soares
Department of Electrical and Computer Engineering,
University of California, Davis, CA 95618, USA
sbyoo@ucdavis.edu

12.1 Introduction

The desire to accurately generate, receive, and process information at high information rates, terahertz (THz) and beyond, has spurred recent efforts in radio frequency (RF) photonics to synergistically solve this recent challenge [1–3]. Exploiting benefits from both RF and photonics can yield high-fidelity and high-precision information generation, detection, and processing at possibly THz bandwidths and beyond [4]. A powerful combination of the optical technology offering wavelength parallelism, high bandwidth, and low energy consumption with electronic technology offers agile processing and storage capabilities as demonstrated by high-fidelity optical arbitrary waveform generation/measurement (OAWG/OAWM) at THz bandwidths [1, 5, 6]. Another benefit of RF photonics is the ability to operate in the other extreme of MHz ~ 10 GHz

Handbook of Terahertz Technologies: Devices and Applications
Edited by Ho-Jin Song and Tadao Nagatsuma

ISBN 978-981-4613-08-8 (Hardcover), 978-981-4613-09-5 (eBook)
www.panstanford.com

frequencies in which signal processing typically handled by RF electronics alone can be helped by optical lattice filters capable of creating reconfigurable filter shapes in the RF domain without involving power-hungry RF electronics [2, 7–10]. The combination of two regimes of RF and photonics offers agility, precision, and high resolution to THz-bandwidth signal generation, detection, and processing. Applications of such THz capabilities span a very wide range of new opportunities, including light detection and ranging (LIDAR)/laser detection and ranging (LADAR) [11, 12], THz analog-to-digital conversion (ADC) [13–15], THz digital-to-analog conversion (DAC) [16], and coherent communications for free-space, satellite, and optical fiber networks [1, 17, 18].

This chapter discusses THz bandwidth information and signal processing based on RF-photonic technologies. In particular, we will emphasize integrated circuit approaches to RF photonics where highly stable THz bandwidth information and signal processing can take place. Following this Introduction Section 12.2 provides an overview of OAWG, and Section 12.3 covers an overview of its inverse process, OAWM. Section 12.4 describes an RF photonic lattice filter useful for optical equalization, and Section 12.5 compares optical versus electrical equalization methods. Section 12.6 summarizes THz information and signal processing by integrated RF photonic methods and projects future directions.

12.2 OAWG Overview

12.2.1 Concept

Figure 12.1 illustrates how dynamic OAWG [1, 4, 19–23] can generate N spectral slices, each with bandwidth Δf_G, to form an aggregate output waveform with a total bandwidth of $N \times \Delta f_G$. Dynamic OAWG begins with a coherent optical frequency comb (OFC), which is spectrally demultiplexed with narrow passbands, placing each comb line at a separate spatial location. A set of in-phase and quadrature-phase modulators (I/Q modulators) each with a bandwidth of Δf_G apply temporal I/Q modulations to broaden the comb lines to create the spectral slices. Coherently combining the spectral slices using a gapless spectral multiplexer with broad overlapping passbands ensures a continuous-bandwidth output waveform. Further, incorporating compensation for the multiplexer transmission as a pre-emphasis of the modulation signals ensures high-fidelity waveform generation after the multiplexer [1, 24].

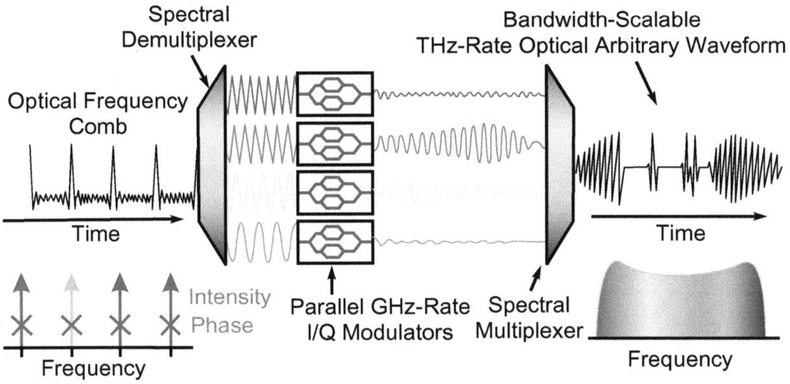

Figure 12.1 Dynamic optical arbitrary waveform generation principle.

12.2.2 Device Implementation

OAWG device implementation involved two configurations. The first relies on electrical excitation of optical modulators, and the second uses optical excitation of optical modulators. Figure 12.2 shows the mask layout and photograph of a 100 GHz (10 channel × 10 GHz) InP-based OAWG [25]. This device uses a matched pair of arrayed-waveguide gratings (AWGs) to implement the demultiplexing and multiplexing operations. Fabrication was performed on a semi-insulating InP substrate to enable high-speed operation. The chip employed two electrical layers, a DC layer to handle phase-error correction and an RF layer to drive amplitude and phase modulators via the electro-optic effect. The devices underwent a 14 mask–layer fabrication process to accommodate optical waveguides, high-speed RF transmission lines, and DC electrical lines.

Figure 12.2 illustrates the first type of device where a pair of AWGs of 10 channels and 10 GHz spacing demultiplexes and multiplexes the signals with 10 sets of amplitude and phase modulators placed in between them to provide dynamic coherent modulation.

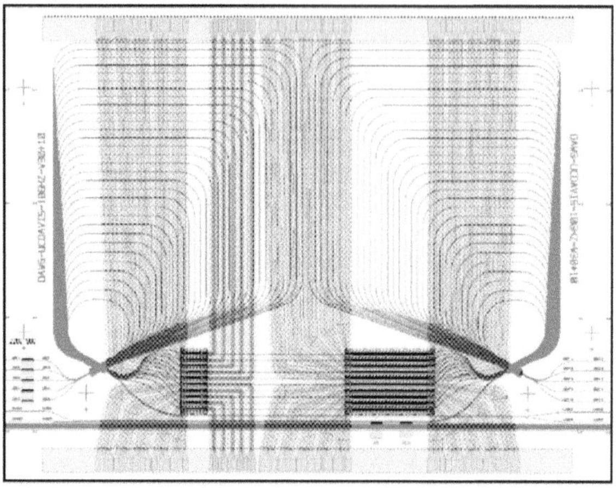

Figure 12.2 Mask layout of a 100 GHz (10 channel × 10 GHz) OAWG device showing both waveguide and metallization layers.

Figures 12.3a,b show the details of the traveling-wave optical phase modulator array where the phase velocity of the RF excitation and the group velocity of the optical wave are matched to achieve optimal interactions along the length of the modulator. Segmented H^+ ion implantation suppresses power losses by blocking the current flow along the 2 mm length of the waveguides. Figure 12.3c shows S_{21} measurement of electrical excitation and optical response (in dB) with >30 GHz bandwidth, and Fig. 12.3d shows electrical reflection (in dB) obtained from a 4 mm long test phase modulator [26].

RF excitation of all the amplitude and phase modulators casts challenges toward high-speed electrical interconnection and packaging while achieving low crosstalk. To provide scalability, we considered RF-over-fiber excitation into optical-to-optical Michelson modulators interfacing with a fiber ribbon cable on a silicon V-groove. Figure 12.4 shows a schematic drawing of a 1 THz (100 channel × 10 GHz) InP-based OAWG device with optically induced modulation. The device structure incorporates a reflection-based geometry in which a single AWG performs both spectral demultiplexing and multiplexing operations. Here, the AWG passband transmission (Fig. 12.4 inset on the right bottom) is optimized for isolation of individual comb lines for demultiplexing and for overlapping of spectral slices for multiplexing. After demultiplexing the individual comb lines of an OFC onto separate waveguide outputs, reflective-mode Michelson modulators apply amplitude and phase modulation using a push-pull configuration to each comb line. External optical control signals (e.g., 1310 nm) drive the reflective-mode modulators through the high-reflection/antireflection (HR/AR)-coated right facet configured for HR at 1550 nm and AR at 1310 nm. Next, the spectral multiplexer combines the spectral slices (i.e., modulated comb lines) to form the output waveform that has a continuous gapless spectrum.

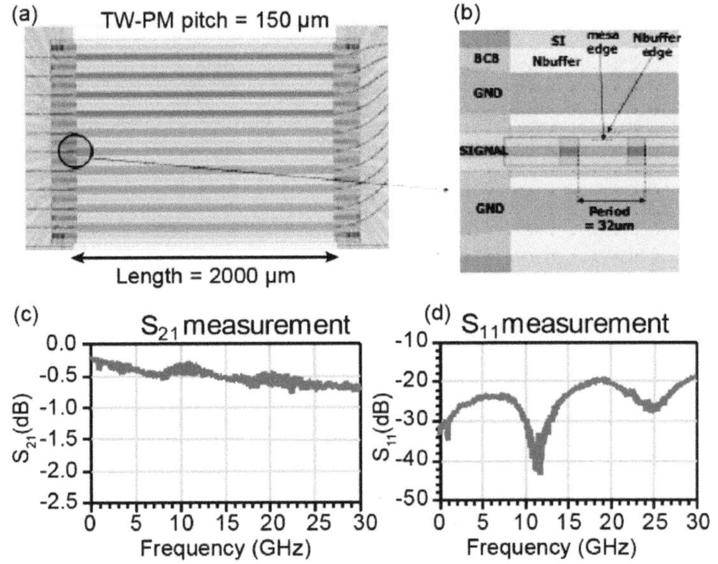

Figure 12.3 (a) Mask layout of a 10-channel traveling wave phase modulator (TW-PM) array and (b) the zoom-in view of one of the TW-PMs as part of the 10-channel × 10 GHz OAWG transmitter chip, including spectral demux, spectral mux, and arrays of 10 amplitude modulators and 10 PMs. Also shown are (c) S_{21} measurements of electrical excitation and optical response (in dB) with >30 GHz bandwidth and (d) the electrical reflection (S_{11} in dB) obtained from a 4 mm long test phase modulator fabricated in the same way as the TW-PM in the OAWG transmitter chip.

Figure 12.5 shows the actual device layout for a single 100 channel × 10 GHz InP-Based OAWG device that has dimensions of 30 × 35 mm [6]. The inset of Fig. 12.5 shows a photo of the fabricated device that consists of two devices on a single 50 mm wafer in which the upper and lower devices are mirror images. The cleaved left facet yields a 2 × 1 multimode interference (MMI) coupler to serve as the 1550 nm light input and output. This MMI coupler introduces a 6 dB penalty but avoids the need to use a circulator or other method in order to separate the input and output signals.

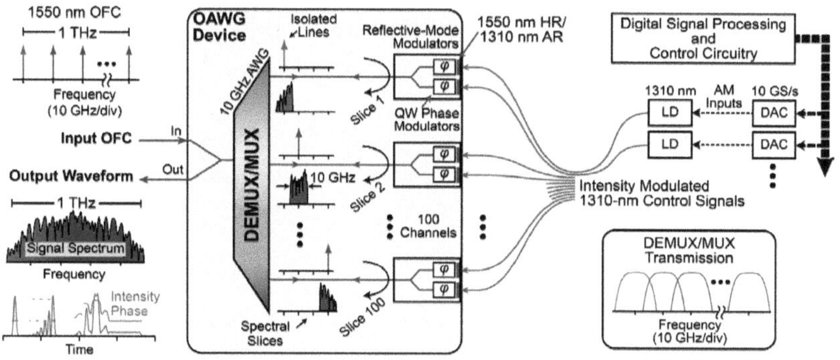

Figure 12.4 Schematic of dynamic OAWG using a reflective-mode device and off-chip optical control signals. DEMUX: demultiplexer; MUX: multiplexer; LD: laser diode.

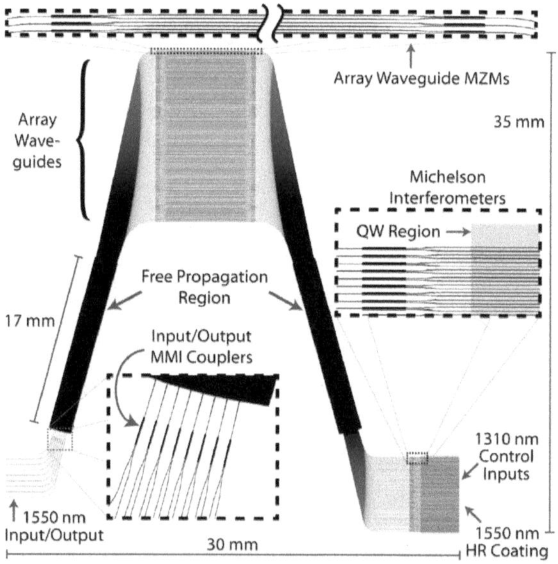

Figure 12.5 Device layout of a single 100-channel × 10-GHz OAWG device. Inset shows two fabricated InP devices from a single 50 mm diameter wafer with a flex circuit attached to one device [1, 29]

As Fig. 12.5 indicates the integrated OAWG contains 1200 independently addressable components and is currently the largest integrated photonic circuits on an InP platform. The AWG has a total of 400 array arms, and each pair of arms has an electro-optic Mach–Zehnder modulator (MZM). High-resolution AWGs are especially susceptible to phase errors that can result from imperfections in the fabrication process, causing variations in the array arm path length [27]. The MZMs enable implementing both phase error correction [28] and the passband shaping [22] necessary to ensure an AWG passband profile suitable for dynamic OAWG such as that shown in the inset of Fig. 12.4.

12.2.3 Device Characterization

Measurement of the complex transmission (i.e., amplitude and phase) of any component requires using a phase-sensitive measurement technique. Here, an optical vector network analyzer (OVNA) enabled complex transmission measurement by employing polarization-sensitive, swept-wavelength interferometry with only a single scan of a fast-sweeping tunable laser [29, 30]. The use of an OVNA with high optical spectral resolution (100 MHz here) provides the capability to measure phase errors with an OAWG device, in addition to providing a high dynamic range due to its balanced detection [31].

Additionally, the OVNA can be configured to provide real-time updates at several hertz for ease of measuring dynamic transmission variations or applying phase error correction.

The OAWG device supports propagation in both the quasi-transverse-electric (TE) and quasi-transverse-magnetic (TM) polarizations, but full functionality is only available in TE polarization. This is due to the fact that the [001] orientation of the InP crystal symmetry combined with the linear electro-optic effect for a vertical electric bias field influences only TE polarization. Many components within the OAWG device, such as MMI couplers and quantum well (QW) modulators, were also optimized for TE polarization. The spectral responses of the two polarizations differed slightly since TM polarization is faster than TE polarization by about 0.8 ps/cm due to birefringence in InP. This provided a means to optimize polarization into the OAWG device using a polarization controller at the input and measuring the delay of the impulse response on the OVNA [6].

Successful operation of high-resolution OAWG devices requires that the fabricated AWG have low channel crosstalk. To achieve a minimum of channel crosstalk, the phases between the array arms of the AWG need to be matched, which occurs when the path lengths all have a constant difference in length. Any deviations from the desired path length lead to phase errors between the array arms. The application of a reverse bias voltage to the phase error correction MZMs produces a large electric field across the intrinsic region of the p-i-n diodes. The electric field causes a change to the effective index along the affected array arms, causing the total accumulated phase on those arms to change. The electro-optic effect for phase error correction is based on a reverse-bias diode that draws only a small amount of leakage current of the order of several microamperes. Typically, a 2π radian phase change can be achieved using only an 8 V reverse bias over 4 mm long InGaAsP/InP waveguides with 1.5 μm thick depletion regions [6].

Figure 12.6 shows the relative phase for the central 200 array waveguides where the phase errors can be most relevant to the AWG crosstalk. The phase values between adjacent array arms are shown modulo 2π to emphasize the slow phase fluctuations that are important for AWG performance. The integrated OAWG device is capable of electro-optical phase tuning by more than 2π for phase error correction. Without phase error correction, the relative phase deviates significantly from the desired zero relative phase across all array arms. However, variations in core-layer composition and thickness and fluctuations in the waveguide width as a result of practical fabrication lead to the accumulation of phase errors. After phase error correction, the relative phase across the array arms is reduced, most notably in the range of array waveguides 150–200 in Fig. 12.6. The standard deviation of the phase errors also reduced from 1.03 radians to 0.33 radians after phase error correction, as a result of removing most of the linear phase and mitigating the slowly varying phase errors. Reducing the AWG phase errors has the effect of improving the crosstalk performance of all 100 OAWG channels.

After applying phase error correction, the passband profile is reduced to a single peak with ~17 dB of crosstalk suppression. This is especially good for high-resolution AWGs that are more susceptible to phase errors due to their larger physical size [6, 32]. Measurements of all 100 channels were performed by measuring each channel separately using an OVNA. The average channel bandwidth was 6 GHz, and the crosstalk between adjacent channels measured from the desired transmission peak to the next highest peak varied from a best case of −17 dB to a worst case of −11 dB, with an average value of −15.1 dB that has a standard deviation of 1.2 dB.

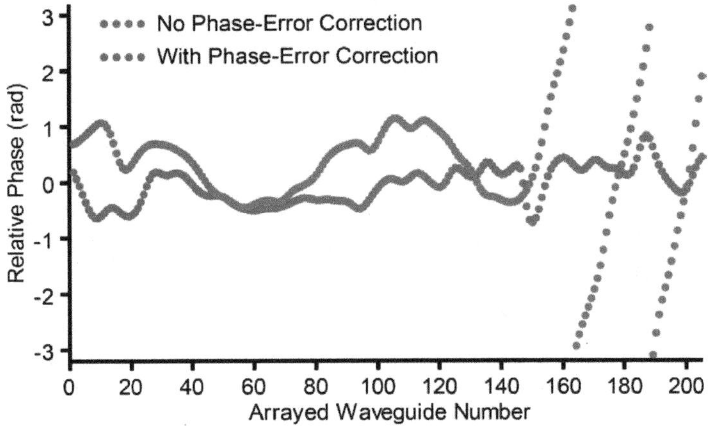

Figure 12.6 Measured relative phase of array waveguides (arms) in a 100-channel × 10 GHz AWG before and after phase error correction.

Figure 12.7 shows the single-pass TE polarization transmission of the full 100-channel × 10 GHz OAWG device. All 100 channels are distinguishable but have slightly different amplitudes since no electrical or optical control signals were present. The lack of control signals results in the Michelson interferometers having random attenuations. Using the Fabry–Pérot resonance method [33] to measure waveguide loss in test samples from the same wafer enabled estimation of the total loss through the OAWG device. The doping profile is graded from 1E18/cm^3 to 5E16/cm^3 to minimize optical losses, while providing sufficient reverse bias. The loss was estimated to be 1.7–2 dB/cm in which ~1 dB/cm is from doping p- and n-types for the p-n junction. This result agrees with the measured round-trip insertion loss at a channel peak of ~43 dB for the TE polarization after accounting for losses from all device components: ~11 cm round-trip path through the device (~22 dB), 2 × 3 dB loss for the MMI input/output coupler (~6 dB), 2 × 6 dB of coupling loss from the lensed fibers (~12 dB), and a minimal amount of excess loss (~3 dB).

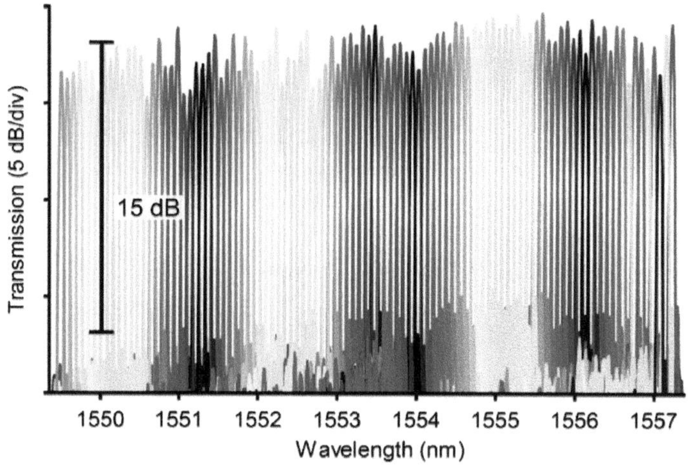

Figure 12.7 Single-pass transmission of a 100-channel × 10 GHz OAWG device after phase error correction.

The dynamics of the plasma dispersion effect primarily determine the performance of high-speed optically driven modulators. In multiple quantum wells (MQWs) the modulation depth is mostly independent of the modulator length, and instead the total phase change is proportional to the number of carriers. The phase modulation can also be increased as a result of the step-like density of states [34]. As an example, the penetration depth of 1310 nm light into the QW region is only a few micrometers and is completely absorbed within ~50 μm. As a result, the reflective-mode geometry enables modulators of ~10 GHz before the modulation efficiency is significantly degraded from a velocity mismatch (i.e., bidirectionality of optical signal propagation). Currently, optical-to-optical modulation response for a Michelson interferometer is limited by the photogenerated carrier lifetime and its sweepout from the waveguide region. The modulation speed can be improved through H^+ ion implantation, which affects the InP crystal lattice by adding defects and vacancies. These additional imperfections serve to reduce the carrier lifetime by providing additional recombination centers for free carriers. Researchers have already demonstrated the potential for carrier lifetimes of less than 1 ps at very high doses of H^+ ion implantation [35]. In our InP OAWG, the H^+ ion implantation at 8 × 10^{12} ions/cm^2 into the Michelson device resulted in −3 dB bandwidth exceeding 1 GHz. Higher doses of H^+ ion implantation can further improve the −3 dB bandwidth; however, the modulation efficiency degrades since the number of carriers is still constant. As a result, a compromise between modulation efficiency and bandwidth determines the optimal dose. The Michelson modulators are pumped by the 1310 nm pump lasers directly modulated by the RF electrical signal. Figure 12.8 shows the modulation response extending to 7 GHz. For IQ modulation, a Nyquist bandwidth of 5 GHz on each channel is sufficient to create a THz dynamic OAWG signal using the 10 GHz spacing optical frequency comb.

Various integrated OAWG devices based on InP electro-optic modulators, InP optical–optical modulators, and silica thermo-optical modulators have been successfully fabricated. High-quality device implementations allow generation of various waveforms carrying information in various modulation formats, data rates, and spectral widths. Figure 12.9 shows on-off-keying (OOK) and differential-phase-shifted-keying (DPSK) waveforms generated and detected at 360 Gb/s data rates. The agreement between the targeted (dots) versus measured (solid) data is reasonably good. Figure 12.10 shows a 1.2 Tb/s 16 quadrature amplitude modulation (QAM) signal with a 3 b/s/Hz spectral efficiency using a 120-bit sequence.

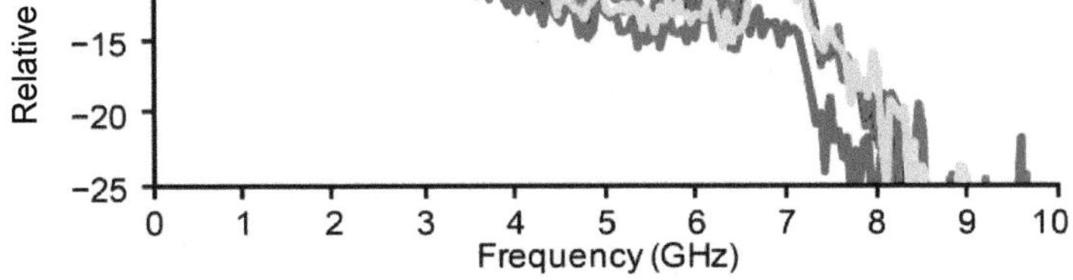

Figure 12.8 Response of the 1310 nm pump laser directly modulated by electrical excitation from 0 to 10 GHz.

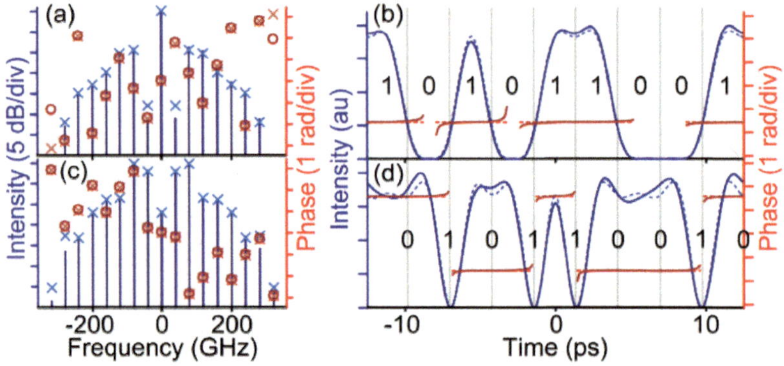

Figure 12.9 OOK (b) and DPSK (d) waveforms generated and detected at 360 Gb/s using line-by-line amplitude (*x*) and phase (*o*) manipulation of optical frequency comb lines for OOK (a) and DPSK (c). The agreement between the targeted (dots) vs. measured (solid) data is reasonably good [35].

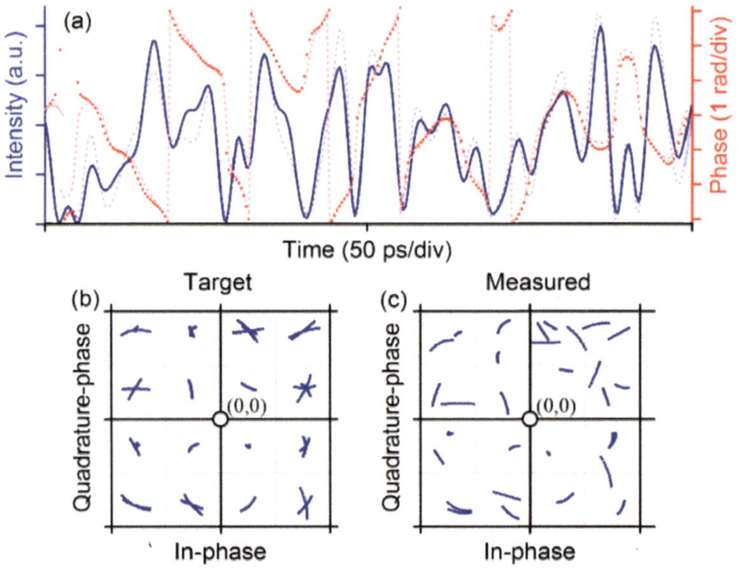

Figure 12.10 (a) 1.2 Tb/s 16 QAM signal with 3 b/s/Hz spectral efficiency. The constellation generated using a limited number of bits (120) showed a reasonably good agreement between (b) the targeted constellation and (c) the measured constellation [35]. The limitation in the constellation chart is due to the limited number of bits (120) of the word length used in waveform generation.

12.3 Optical Arbitrary Waveform Measurement

The inverse process of OAWG allows detection of the THz arbitrary waveform of full amplitude and phase information (or in-phase and quadrature information) by OAWM, which is a waveform characterization technique analogous to OAWG utilizing inverse Fourier transform and coherent detection (instead of Fourier transform and coherent modulation in OAWG). Figure 12.11a shows a schematic of OAWM, which is essentially the inverse of dynamic OAWG, where instead of producing a waveform by creating and combining spectral slices, OAWM splits the signal into spectral slices that are coherently measured. A key element in OAWM is the gapless demultiplexer for spectral slicing of the THz waveform signal so that no information is lost. The reference OFC is split into isolated lines by the spectral demultiplexer with high isolation. Then, these two signals provide inputs to an array of optical coherent receivers. A digital signal processing (DSP) algorithm reconstructs the signal waveform from the individual slice I/Q signals after removing the effects of optical and electrical filters in the measurement system (i.e., spectral and frequency responses of demultiplexers and photoreceivers). If the waveform data is recorded continuously, the spectral resolution is limited only by the reference OFC stability and the optical coherent receiver performance. Integrated circuit implementation provides the stability required for the coherent process.

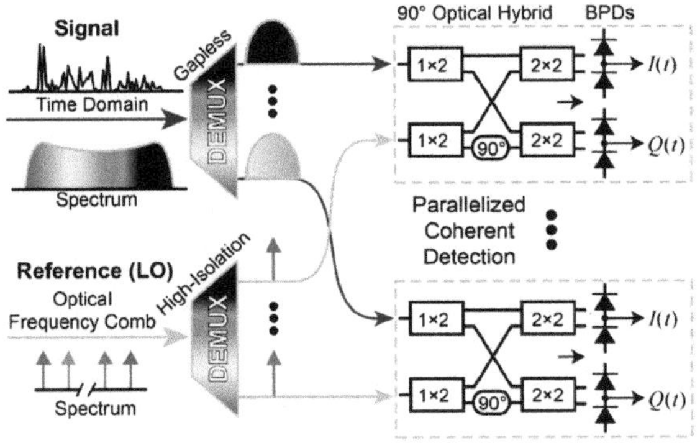

Figure 12.11 Optical arbitrary waveform measurement (OAWM). BPD: balanced photodiode.

Just as in OAWG, in OAWM it is critical that the spectral slices remain coherent with one another. Therefore the first generation of OAWM employed a specially designed silica planar lightwave circuit (PLC), which integrates all passive optical components necessary to implement the spectral slicing and optical coherent detection. Balanced detection (i.e., homodyne down-conversion) and digitizing occurs off-chip using DC-coupled high-speed balanced photodetectors (BPDs) connected to the inputs of high-speed digitizers. As Fig. 12.12 shows, the OAWM PLC incorporates two AWG spectral demultiplexers, both aligned so that their transmission peaks are matched to the reference OFC. The OAWM PLC can accommodate a total of 640 GHz of optical bandwidth. The AWGs have 16 outputs with a 40 GHz spacing, and there are sixteen 90° optical hybrids. Single-mode fiber arrays couple signals in and out of the PLC. The gapless AWG slices the signal into slightly overlapping spectral slices and directs each slice to a different optical hybrid. The high-isolation AWG demultiplexes and directs each comb line to the reference (LO) input of the corresponding optical hybrid. Then, each hybrid combines a spectral slice in four quadrature phases with a reference line. The hybrid outputs are coupled off chip to balanced photodiode pairs, which generate the I and Q signals. Then the signals are digitized at 20 GHz electrical bandwidth.

OAWM enables measurements with high temporal resolution over long record lengths. Using the fabricated 0.64 THz bandwidth silica OAWM device shown in Fig. 12.12 coupled to eight pairs of BPDs and eight 50 GS/s digitizers we demonstrated 160 GHz (i.e., four spectral slices) of instantaneous bandwidth for 2 μs record lengths. Figure 12.13a shows the temporal intensity of a long waveform composed of a sine^2-like waveform (160 GHz of optical bandwidth) repeating every 100 ps, modulated with a relatively slowly varying sawtooth intensity envelope. The envelope is created by driving the 2 × 2 MZM with a voltage ramp from an electronic arbitrary waveform generator, which is switched on in less than 100 ps and then slowly switched off over a 500 ns duration. Figure 12.13b shows the intensity and phase of the waveform for a 400 ps window of the time record at a fast-rising transition of the ramp. Within the window, the waveform abruptly switches on. Each pulse has subpulses, with the phase alternating between 0 and π radians (i.e., alternating between negative and positive fields). The 6.25 ps sampling interval (320 GS/s aggregate) determines the minimum resolvable temporal feature, as shown by the ~6 ps wide pulse peak. This measurement demonstrates a record-length-to-resolution ratio of over 320,000 (6.25 ps resolution and a 2 μs record length).

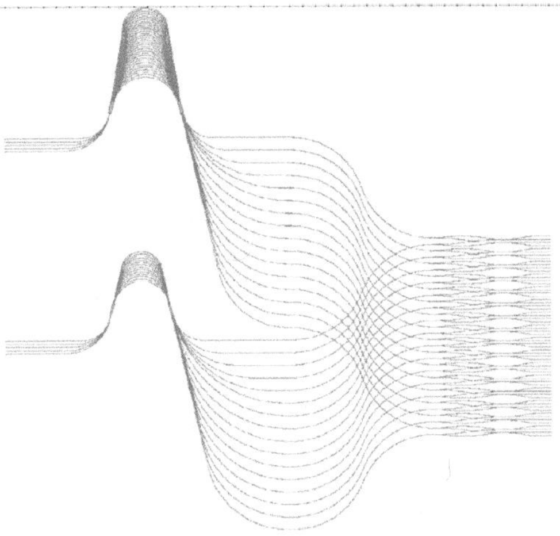

Figure 12.12 The planar lightwave integrated circuit design with 640 GHz spectral bandwidth (16 × 40 GHz) containing 2 spectral demultiplexers and 16 sets of optical hybrids for coherent detection. Balanced detectors are placed external to the silica PLC.

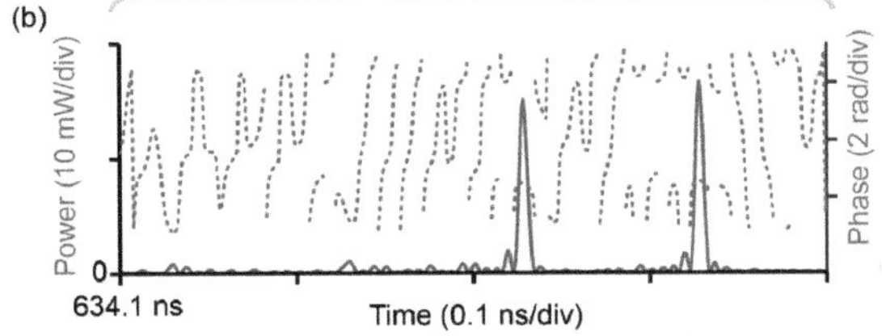

Figure 12.13 OAWM measurement of a train of transform-limited pulses with an overall sawtooth amplitude envelope (500 ns period). (a) The reconstructed signal waveform showing the full 2 μs record length. (b) Detail of a 400 ps portion from (a), showing the intensity (solid) and phase (dashed).

12.4 Reconfigurable RF Photonic Filters

One of the key challenges in THz signal synthesis and analysis is overcoming the nonuniform frequency response of the modulators and multiplexers in OAWG and detectors and demultiplexers in OAWM. Figure 12.14 illustrates the effect. Traditional methods rely on frequency-independent power splitters/combiners in place of optical frequency demux/muxes and sophisticated DSP, but these methods are power hungry and limited to low-bandwidth (<THz) applications. Figure 12.15 illustrates photonic lattice filters [2, 8, 36] to achieve RF domain signal processing using optical filtering of an arbitrary shape via reconfiguration of various components in the photonic lattice filter. The unit cell shown in Fig. 12.15a with a microresonator with a 10 GHz free spectral range (FSR) is repeated in series to synthesize the programmable cascaded lattice filter shown in Fig. 12.15b.

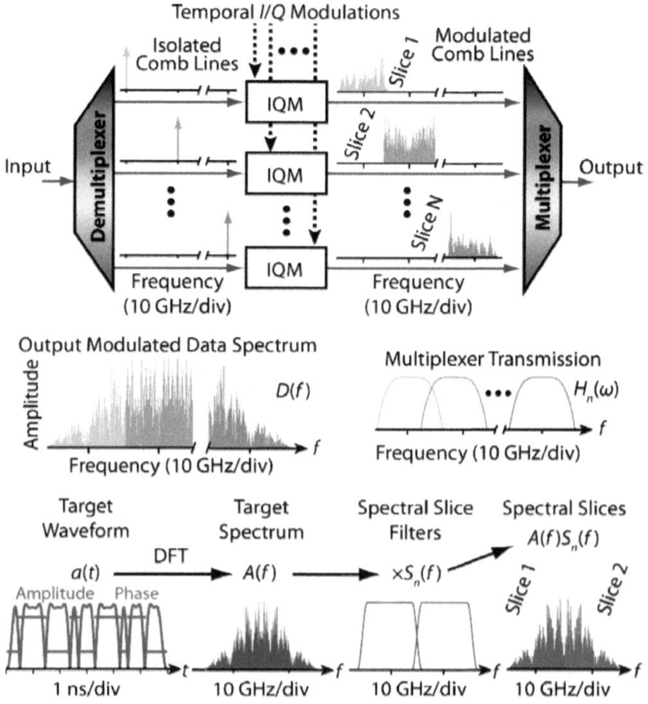

Figure 12.14 Synthesis of THz waveforms overcoming nonuniform responses of the frequency multiplexer.

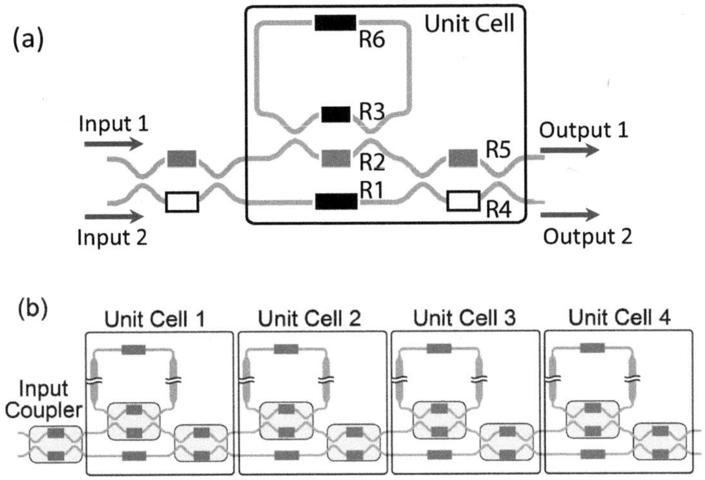

Figure 12.15 (a) Schematic and photo of a single unit cell, where boxes indicate phase shifter electrodes. (b) Schematic of a four-unit-cell filter.

Similarly to a digital filter, the optical lattice filter can be programmed and reconfigured by tuning each element on each unit cell, which will in turn control the positions of zeros and poles. Figure 12.16a shows measured transmission of a single unit cell configured as an optical equalizer for the H_{11} output (H_{12} is complementary). Inset shows a fitted pole and zero, and 12.16b shows a corresponding impulse response h_{11} output (h_{12} is complementary).

Instead of using electronic DSP, it is possible to utilize the programmable lattice filter, which has a matching FSR (10 GHz) with the OFC spacing of OAWG, to achieve channel equalization without consuming power in electronics.

12.5 Optical vs. Electrical Equalization

The variation of the Q-factor as a function of DAC effective number of bits (ENOB) for both optical and electrical equalization was investigated via simulations. These simulations started with a 5 ns, 100 GHz BPSK waveform (1500 bits) generated using a raised cosine modulation filter with the beta-rolloff factor = 0. Next, Gaussian white noise was added to both the in-phase and quadrature-phase components, which resulted in a signal-to-noise ratio (SNR) of 15 dB. At this point, the total 100 GHz spectrum was divided into ten 10 GHz spectral slices. For the electrical equalization case, multiplexer and I/Q modulator pre-emphasis was applied, the in-phase and quadrature-phase components of the waveform were quantized, and then the inverse multiplexer and I/Q modulator preemphasis was applied. The optical equalization case involved only quantization of the spectral slices. The Q-factor was then estimated from the constellation diagrams after reassembling the quantized spectral slices into the aggregate waveform.

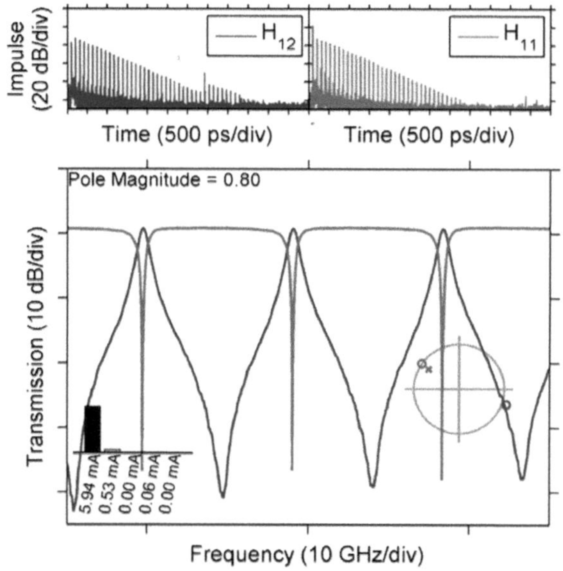

Figure 12.16 (Bottom) Measured transmission of a single unit cell configured as an optical equalizer for the H_{11} output (H_{12} is complementary). Inset shows a fitted pole and zero. (Top) The corresponding impulse response from time domain measurements [2, 8, 36].

Figure 12.17 shows simulation results for both optical and electrical equalization cases. For all DAC ENOB values, the optical equalization shows a higher Q-factor. This is since the electrical equalization case required accounting for the multiplexer and I/Q modulator pre-emphasis before quantization, which lowered the dynamic range achievable with the DAC outputs. The advantage of using optical equalization is more pronounced at lower ENOB values but is still present at high ENOB values. The presence of a "knee" around a DAC ENOB of 5 indicates a reduction in the improvement of the Q-factor for increases in DAC ENOB. Insets show progression of the constellation diagrams for the optical and electrical equalization cases.

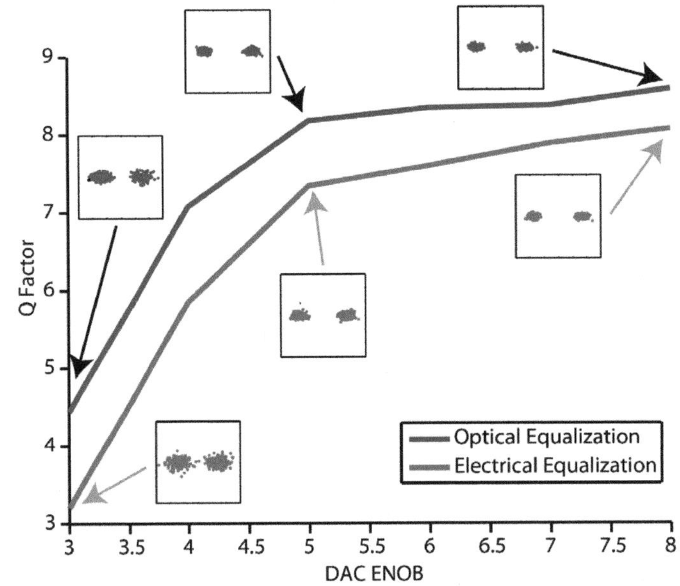

Figure 12.17 Simulation result showing the Q-factor as a function of DAC ENOB for optical and electrical equalization. Simulations assume a 10-channel × 10 GHz OAWG device and the BPSK modulation format.

12.6 Conclusion

Exploiting benefits from both RF and photonics can yield high-fidelity and high-precision information generation, detection, and processing at possibly THz bandwidths and beyond. A powerful combination of the optical technology offering wavelength parallelism, high bandwidth, and low energy consumption with electronic technology offers agile processing and storage capabilities, as demonstrated by high-fidelity OAWG and OAWM at THz bandwidths [1, 5, 6]. RF photonic signal processing utilizing the optical lattice filter allows precise and low-power RF processing at MHz–10 GHz frequencies covering the THz total bandwidth. The optical lattice filters are capable of creating reconfigurable filter shapes in the RF domain without involving power-hungry RF electronics. The combination of the two regimes of RF photonics offers agility, precision, and high resolution to THz bandwidth signal generation, detection, and processing. Applications of such THz capabilities span a very wide range of new opportunities, including LIDAR/LADAR [11, 12], THz ADC, THz DAC [16], and coherent communications for free-space, satellite, and optical fiber networks.

Acknowledgments

The author is grateful for the work from the Next Generation Networking Systems Group at the University of California, Davis, in particular Ryan Scott, Nicolas Fontaine, David Geisler, Francisco Soares, Tiehui Su, Stevan Djordjevic, Salah Ibrahim, Xinran Cai, Binbin Guan, Jonghwa Baek, Sang-Woo Seo, and collaborators at Multiplex, Inc., Royal Institute of Technology, Sweden, and Nistica. This work was supported in part by DARPA and SPAWAR under OAWG contract HR0011-05-C-0155, DARPA MTO Si-PhASER project Grant No. HR0011-09-1-0013, NSF ECCS grant 1028729, and the CISCO University Research Program.

References

1. D. J. Geisler, N. K. Fontaine, R. P. Scott, and S. J. B. Yoo, Demonstration of a flexible bandwidth optical transmitter / receiver system scalable to terahertz bandwidths, *IEEE Photon. J.* **3**, 1013–1022 (2011).
2. S. Ibrahim, N. K. Fontaine, S. S. Djordjevic, B. Guan, *et al.*, Demonstration of a fast-reconfigurable silicon CMOS optical lattice filter, *Opt. Express* **19**, 13245–13256 (2011).
3. R. C. Williamson and R. D. Esman, RF photonics, *J. Lightwave Technol.* **26**, 1145–1153 (2008).
4. S. J. B. Yoo, R. P. Scott, D. J. Geisler, N. K. Fontaine *et al.*, Terahertz information and signal processing by RF-photonics, *IEEE Trans. THz Sci. Technol.* **2**, 167–176 (2012).

5. R. P. Scott, N. K. Fontaine, Z. Linjie, F. M. Soares, *et al.*, Continuous, real-time, full-field waveform measurements via spectral slicing and parallel digital coherent detection, *2010 Conf. OFC 2010 Collocated OFC/NFOEC 2010*, 3 (2010).

6. F. M. Soares, N. K. Fontaine, R. P. Scott, J.-H. Baek, X. Zhou, T. Su, S. Cheung, Y. Wang, C. Junesand, S. Lourdudoss, K. Y. Liou, R. A. Hamm, W. Wang, B. Patel, L. A. Gruezke, W. T. Tsang, J. P. Heritage, and S. J. B. Yoo, Monolithic InP 100-channel × 10-GHz optical arbitrary waveform generation device, *IEEE Photon. Technol. Lett.* **3**, 975–985 (2011).

7. J. Capmany, B. Ortega, and D. Pastor, A tutorial on microwave photonic filters, *J. Lightwave Technol.* **24**, 201–229 (2006).

8. S. S. Djordjevic, L. W. Luo, S. Ibrahim, N. K. Fontaine, *et al.*, Fully reconfigurable silicon photonic lattice filters with four cascaded unit cells, *IEEE Photon. Technol. Lett.* **23**, 42–44 (2011).

9. E. M. Dowling and D. L. Macfarlane, Lightwave lattice filters for optically multiplied communication systems, *J. Lightwave Technol.* **12**, 471–486 (1994).

10. R. A. Minasian, Photonic signal processing of microwave signals, *IEEE Trans. Microwave Theory Tech.* **54**, 832–846 (2006).

11. P. Adany, C. Allen, and R. Q. Hui, Chirped Lidar using simplified homodyne detection, *J. Lightwave Technol.* **27**, 3351–3357 (2009).

12. S. M. Beck, J. R. Buck, W. F. Buell, R. P. Dickinson, *et al.*, Synthetic-aperture imaging laser radar: laboratory demonstration and signal processing, *Appl. Opt.* **44**, 7621–7629 (2005).

13. Y. Han and B. Jalali, Photonic time-stretched analog-to-digital converter: fundamental concepts and practical considerations, *J. Lightwave Technol.* **21**, 3085–3103 (2003).

14. J. Kim, M. J. Park, M. H. Perrott, and F. X. Kartner, Photonic subsampling analog-to-digital conversion of microwave signals at 40-GHz with higher than 7-ENOB resolution, *Opt. Express* **16**, pp. 16509–16515 (2008).

15. G. C. Valley, Photonic analog-to-digital converters, *Opt. Express* **15**, 1955–1982 (2007).

16. Y. Peng, H. Zhang, Y. Zhang, and M. Yao, Photonic digital-to-analog converter based on summering of serial weighted multiwavelength pulses, *IEEE Photon. Technol. Lett.* **20**, 2135–2137 (2008).

17. A. Belmonte and J. M. Kahn, Capacity of coherent free-space optical links using diversity-combining techniques, *Opt. Express* **17**, 12601–12611 (2009).

18. V. W. S. Chan, Free-space optical communications, *J. Lightwave Technol.* **24**, 4750–4762 (2006).
19. N. K. Fontaine, R. P. Scott, J. Cao, A. Karalar, *et al.*, 32 phase X 32 amplitude optical arbitrary waveform generation, *Opt. Lett.* **32**, 865–867 (2007).
20. R. P. Scott, N. K. Fontaine, J. Cao, K. Okamoto, *et al.*, High-fidelity line-by-line optical waveform generation and complete characterization using FROG, *Opt. Express* **15**, 9977–9988 (2007).
21. N. K. Fontaine, *Optical Arbitrary Waveform Generation and Measurement* (Univeristy of California, Davis, CA, 2010).
22. N. K. Fontaine, J. Yang, W. Jiang, D. J. Geisler, *et al.*, Active arrayed-waveguide grating with amplitude and phase control for arbitrary filter generation and high-order dispersion compensation, 34th Eur. Conf. Opt. Commun. (ECOC 2008), IEEE, Mo.4.C.31 (2008).
23. R. P. Scott, N. K. Fontaine, J. P. Heritage, and S. J. B. Yoo, Dynamic optical arbitrary waveform generation and measurement, *Opt. Express* **18**, 18655–18670 (2010).
24. S. J. B. Yoo, Flexible bandwidth terabit coherent optical communication networks by optical arbitrary waveform generation and measurement, *IEEE Photon. Soc. Newslett.* **25**, 5–12 (2011).
25. N. K. Fontaine, R. P. Scott, and S. J. B. Yoo, Dynamic optical arbitrary waveform generation and detection in InP photonic integrated circuits for Tb/s optical communications, *Opt. Commun.* **284**, 3693–3705 (2011).
26. S.-W. Seo, J. Yan, J.-H. Baek, F. M. Soares, *et al.*, Microwave velocity and impedance tuning of traveling-wave modulator using ion implantation for monolithic integrated photonic systems, *Microwave Opt. Technol. Lett.* **50**, 2151–2155 (2008).
27. H. Yamada, K. Yakada, and S. Mitachi, Crosstalk reduction in a 10-GHz spacing arrayed-waveguide grating by phase-error compensation, *J. Lightwave Technol.* **16**, 364–371 (1998).
28. S.-W. Seo, F. M. Soares, J.-H. Baek, W. Jiang, *et al.*, Monolithically integrated InP photonic micro systems on a chip for O-CDMA and OAWG applications, in *Photonics in Switching* (2007), 97–98, doi: 10.1109/PS.2007.4300754.
29. D. K. Gifford, B. J. Soller, M. S. Wolfe, and M. E. Froggatt, Optical vector network analyzer for single-scan measurements of loss, group delay, and polarization mode dispersion, *Appl. Opt.* **44**, 7282–7286 (2005).

30. G. D. Vanwiggeren and D. M. Baney, Swept-wavelength interferometric analysis of multiport components, *IEEE Photon. Technol. Lett.* **15**, 1267–1269 (2003).

31. K. Takada and S.-I. Satod, Method for measuring the phase error distribution of a wideband arrayed waveguide grating in the frequency domain, *Opt. Lett.* **31**, 323–325 (2006).

32. W. Jiang, K. Okamoto, F. M. Soares, S. Lourdudoss, *et al.,* 5 GHz channel spacing InP-based 32-channel arrayed-waveguide grating, in *Optical Fiber Communication* (2009), 1–3, IEEE, San Diego, CA.

33. T. Feuchter and C. Thirstrup, High precision planar waveguide propagation loss measurement technique using a Fabry–Pérot cavity, *IEEE Photon. Technol. Lett.* **6**, 1244–1246 (1994).

34. S. Jong-In, M. Yamaguchi, P. Delansay, and M. Kitamura, Refractive index and loss changes produced by current injection in InGaAs(P)-InGaAsP multiple quantum-well (MQW) waveguides, *IEEE J. Sel. Top. Quantum Electron.* **1**, 408–415 (1995).

35. K. F. Lamprecht, S. Juen, L. Palmetshofer, and R. A. Hopfel, Ultrashort carrier lifetimes in H^+ bombarded InP, *Appl. Phys. Lett.* **59**, 926–928 (1991).

36. S.-E. Ibrahim, L.-W. Luo, S. S. Djordjevic, C. B. Poitras, *et al.,* Fully reconfigurable silicon photonic lattice filters with four cascaded unit cells, *Opt. Fiber Commun. Conf.*, San Diego, California, Paper OWJ5 (2010).

Chapter 13

Low-Coherence Terahertz Signal Sources and Their Applications

Tadao Nagatsuma[a] **and Ho-Jin Song**[b]
[a]*Graduate School of Engineering Science, Osaka University, 1-3, Machikaneyama, Toyonaka, Osaka 560-8531, Japan*
[b]*NTT Device Technology Laboratories, NTT Corporation, 3-1, Morinosato Wakamiya, Atsugi, Kanagawa 243-0198, Japan*
nagatuma@ee.es.osaka-u.ac.jp, song.hojin@lab.ntt.co.jp

In this chapter, we present the photonic generation of noise signals or incoherence signals in the subterahertz (sub-THz) region. First, we describe how to control the bandwidth and output power of sub-THz noise signals using an optical amplified spontaneous emission noise, a wavelength-tunable laser, an optical filter, and a photodiode. Then, we demonstrate their applications to imaging not only to enhance the image quality by eliminating interference patterns but also to enable a tomographic imaging by adding depth resolution. Finally, we explain their other applications to spectroscopy and characterization of electronics devices.

Handbook of Terahertz Technologies: Devices and Applications
Edited by Ho-Jin Song and Tadao Nagatsuma

ISBN 978-981-4613-08-8 (Hardcover), 978-981-4613-09-5 (eBook)
www.panstanford.com

13.1 Introduction

Coherency in radio frequency (RF) signals is commonly related to the stability in amplitude, frequency, and phase of signals in the time domain. Deterioration of coherency can be identified as broadening of spectra in the frequency domain, which is quantitatively measured as the width of frequency spectra. The bandwidth can be transformed to the time duration, and it is called coherence time, Δt. Then, the coherence length is defined as $c\Delta t$, where c is the speed of light (3×10^{10} m/s).

In contrast, randomly fluctuated signals with respect to frequency are referred to as incoherent signals or noise signals. For example, thermal noise arises from vibrations of conduction electrons and holes due to their finite temperature. The noise spectrum produced by the thermal noise is nearly uniform over RF frequencies. The power delivered by a thermal source into an impedance-matched load is kTB watts, where k is Boltzmann's constant (1.38×10^{-23} joules/K), T is the temperature in Kelvin, and B is the system's noise bandwidth. Shot noise arises from the quantized nature of current flow and has a characteristic similar to thermal noise; the frequency spectrum is essentially uniform, producing equal power density across the entire RF frequency range.

Incoherent signal or broadband noise signal sources have been utilized in many RF applications such as noise characterization of electronic amplifiers and receivers, Fourier transform spectroscopy, and radars. Most popular noise sources are solid-state diode noise sources based on the mechanism of avalanche breakdown, which are commercially available at frequencies below around 100 GHz. At subterahertz (sub-THz) frequencies, a blackbody furnace or a mercury vapor lamp can radiate broadband noise signals, but it suffers from a decrease in output power.

Recently, photonic techniques for generating noise in the sub-THz region have been proposed [1–3]. The basic concept of noise generation is to convert a broadband optical noise to an electrical noise using photodiodes (PDs) with high frequency and high output-current operation [4]. In addition, use of optical filters placed before the PD can precisely control the bandwidth of the generated noise. In other words, the degree of coherence can be controlled from low to high.

In this chapter, we first describe schemes for photonically assisted noise signal generation and how to control coherency. Next, we show characteristics of generated signals with respect to the output power and coherence length. Then, we demonstrate applications of low-coherence sub-THz signals to imaging and spectroscopy.

13.2 Schemes for Noise Generation

Figure 13.1 shows the simplest schemes to generate RF noises or incoherent signals; they are converted with PDs from an amplified spontaneous emission (ASE) noise source, for example, by Er-doped fiber amplifiers (EDFAs) and by superluminescent diodes (SLDs). In Fig. 13.1a, the bandwidth of the generated noise is determined by the bandwidths of PDs and the packaged modules, which range from a lower-frequency cutoff, f_L, to an upper-frequency cutoff, f_U. Also, the noise bandwidth can be controlled with the use of optical filters.

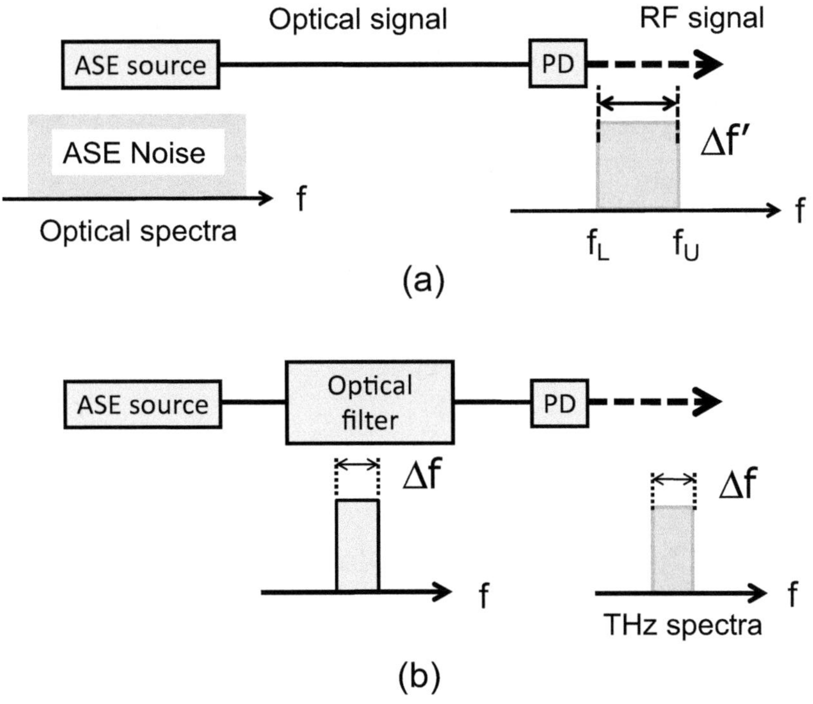

Figure 13.1 Schematic diagram of photonic generation of noise signals.

PDs convert every frequency difference of the ASE noise in the photomixing process, and noise signal components from DC to the lower-frequency cutoff and those above the upper-frequency cutoff are not emitted from PDs, being dissipated just as DC currents in PDs. By considering this issue, more efficient and bandwidth-controlled noise generation schemes are depicted in Fig. 13.2, where the center frequency of the noise signals emitted from PDs can be determined by the frequency difference in two optical lines, f_1 and f_2, with designated bandwidths, Δf, or with a single (narrow) frequency.

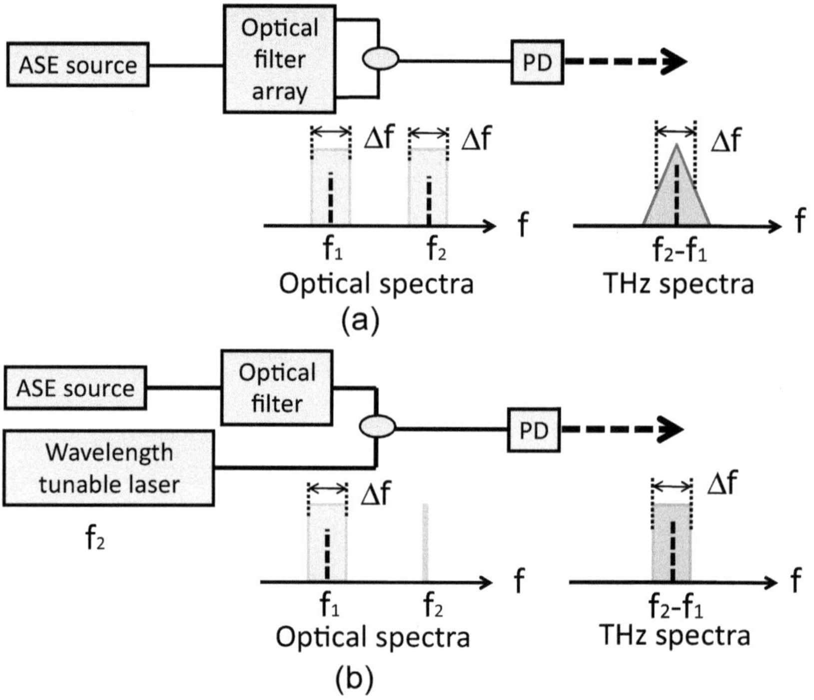

Figure 13.2 Schematic diagram of photonic generation of noise signals.

13.3 Characterization of Noise Signals

To characterize the noise power emitted from PDs, we measured the output power from a J-band uni-traveling-carrier photodiode (UTC-PD) module [5] for different noise generation schemes mentioned above. Figure 13.3 shows dependence of the output power from the

UTC-PD on the photocurrent for different generation schemes at a center frequency of 300 GHz: A: monochromatic single-frequency generation by mixing two wavelengths (frequencies) of lasers; B: noise generation by mixing a single-frequency laser and a noise signal filtered (Δf = 10 GHz), as depicted in Fig. 13.2b; C: noise generation by mixing two noise signals filtered (Δf = 10 GHz), as depicted in Fig. 13.2a; D: noise generation by directly converting ASE noise signals filtered (Δf = 10 GHz), as depicted in Fig. 13.1b; and E: noise generation by directly converting broadband ASE noise signals, as depicted in Fig. 13.1a. With respect to the efficiency of noise generation, the scheme of Fig. 13.2b is considered to be the best approach.

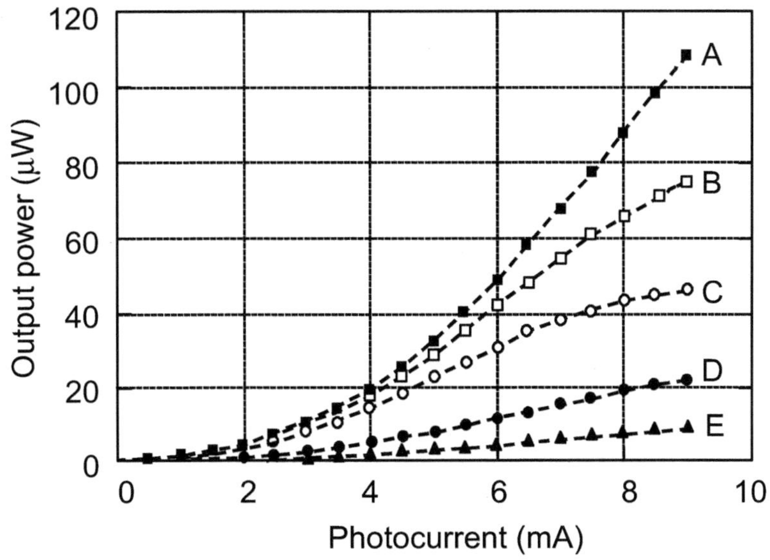

Figure 13.3 Relationship between the output power and the photocurrent for a monochromatic signal (A) and noise signals (B–E) generated with different schemes at a center frequency of 300 GHz.

Figure 13.4 shows typical optical spectra optimized to generate the highest output power from the PD for the scheme of Fig. 13.2b. It should be noted that peak powers for the monochromatic signal and the filtered noise signal are different; they should be controlled so that the integrated power for each signal is the same.

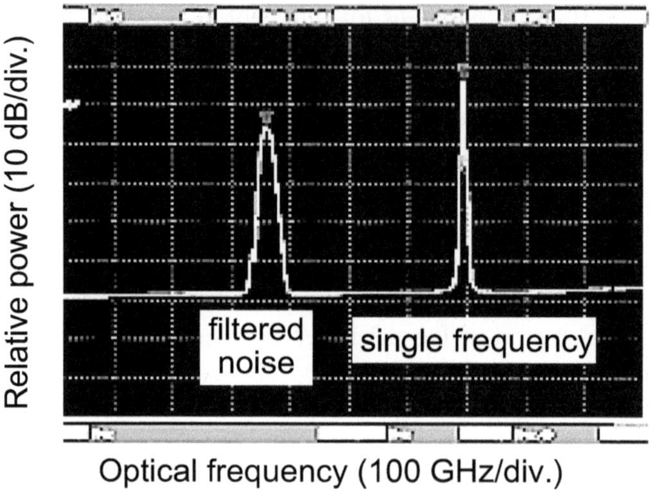

Figure 13.4 Typical optical spectrum optimized for the scheme of Fig. 13.2b. The frequency difference is 350 GHz.

13.4 Applications

13.4.1 Imaging

Millimeter-wave and THz-wave imaging have been extensively studied and deployed in numerous areas such as material inspection, nondestructive testing, and security applications [6–12]. Between continuous-wave (CW) and pulsed-wave imaging systems, CW imaging can be faster, more compact, cost effective, and simpler to operate.

One of the drawbacks the 2D CW imaging using monochromatic coherent sources is degradation in image quality due to interference patterns caused by reflection of waves at any boundary in objects under inspection and components used such as lenses and mirrors, which leads to a decrease in effective lateral spatial resolution [13].

CW THz waves have also been applied to 3D imaging by computed tomography (CT) techniques with efficient reconstruction methods [10, 11] and by frequency modulation (FM)-CW radars [12].

In this section, we describe a new approach to CW THz imaging systems using photonically generated broadband THz noise or low-coherence sources [1, 2], which not only eliminate the interference pattern in 2D imaging but also are applicable to 3D tomographic imaging.

13.4.1.1 2D imaging

Figure 13.5 shows a block diagram of a transmission-type THz-wave imaging system using a low-coherence illumination source or a THz noise source [3]. The noise is radiated into free space via a horn antenna from the UTC-PD module, and it is focused onto the object under test with dielectric lenses. The receiver consists of the horn antenna and a Schottky barrier diode (SBD) detector. An optical modulator is used to allow narrow-band lock-in detection. The typical modulation frequency was 10 kHz. The typical signal-to-noise ratio (SNR) of the system, which is measured when there is no object inserted between the transmitter (illuminator) and the receiver, exceeds 75 dB, with an output power of about 10 microwatts from the UTC-PD, as shown in Fig. 13.6.

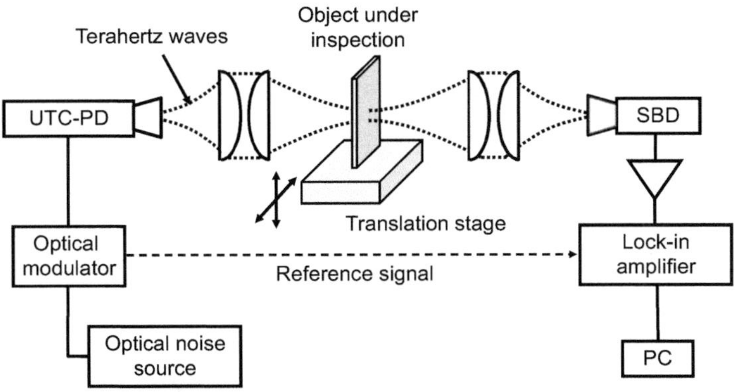

Figure 13.5 Block diagram of a THz imaging system using a low-coherence illumination source. PC: personal computer. The typical optical spectrum optimized for the scheme of Fig. 13.2b. The frequency difference is 350 GHz.

To verify the effectiveness of the proposed technique, the imaging of a clip hidden in a paper envelope was conducted, as shown in Fig. 13.7a. For comparison, the millimeter-wave images obtained by a monochromatic 320 GHz signal are also shown in Fig. 13.7b. The 320 GHz signal was generated by injecting two optical signals with a frequency difference of 320 GHz from two wavelength-tunable lasers into the UTC-PD. The linewidth of the generated THz-wave signal is about 10 MHz. Even in the case of the thin envelope, the interference effect caused an artifact near the clip. As clearly seen in Fig. 13.7, the interference effect has been dramatically reduced by low-coherence THz signals, which are generated with the scheme of Fig. 13.1a.

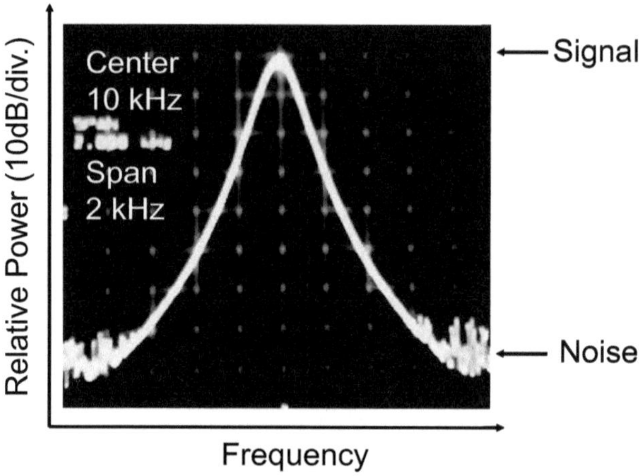

Figure 13.6 Typical signal-to-noise ratio (SNR) of the system.

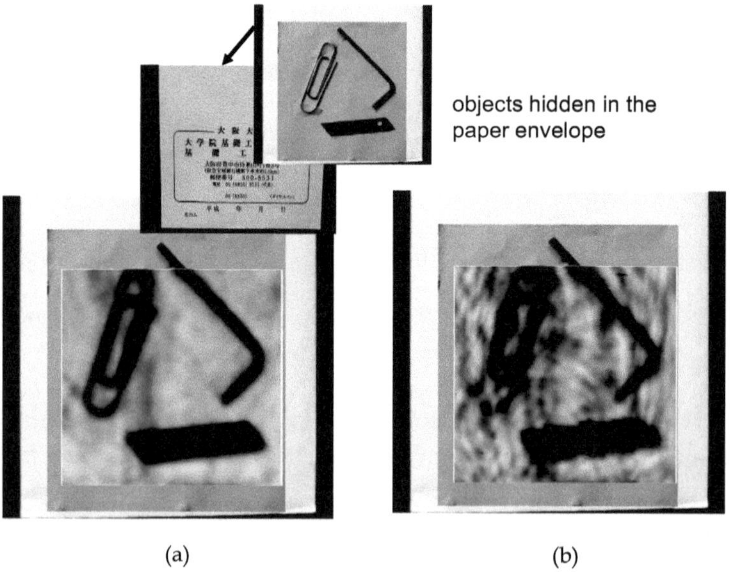

Figure 13.7 Comparison of images obtained by a noise source (a) and a coherent source at 320 GHz (b).

In general, there is a trade-off between the decrease in the interference effect and the spatial resolution, when there is an increase in the noise bandwidth. We performed 2D imaging to investigate the optimum noise bandwidth by changing the noise bandwidth, Δf, with the noise generation scheme of Fig. 13.2b, as shown in Fig. 13.8. Among measured images, the best one is with the noise bandwidth of 62.5 GHz.

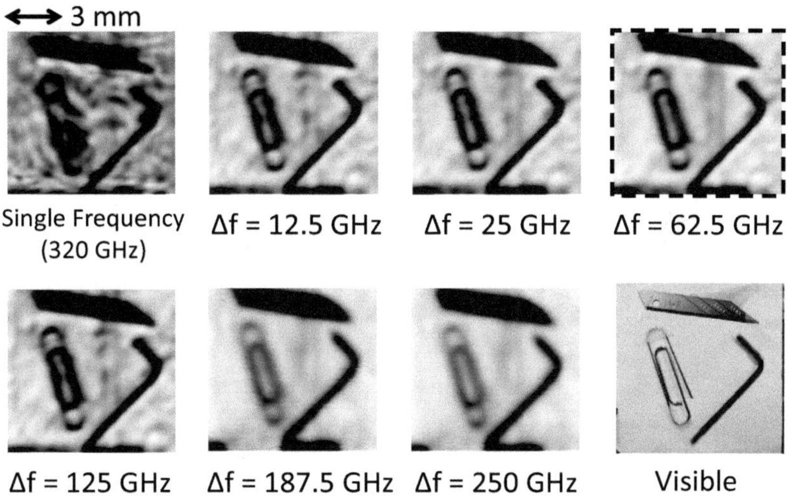

Figure 13.8 2D images measured by changing the noise bandwidth, Δf.

13.4.1.2 Tomographic imaging

Figure 13.9 shows a block diagram of the proposed tomographic imaging system using a Mickelson interferometer [14, 15]. This configuration is similar to that of optical coherence tomography (OCT) [16].

In the setup, first, a THz wave travels to the beam splitter after being collimated by a dielectric lens. The beam splitter divides the THz wave into two directions with a power ratio of 50/50. One goes to the reference mirror and the other to an object after being focused. Both reflected waves travel back to the beam splitter again and go to the SBD as a power detector. Finally, detected signals are amplified by a preamplifier and a lock-in amplifier. A personal computer (PC) controls positions of both the reference mirror and the objective lens.

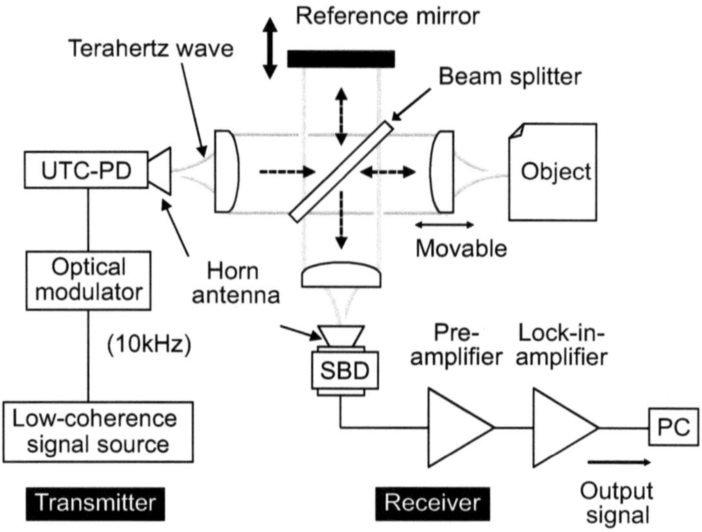

Figure 13.9 Block diagram of the tomography system using the Mickelson interferometer.

Figure 13.10 shows an interference signal of low-coherence (bandwidth: $\Delta f = 120$ GHz, center frequency: 350 GHz) THz sources, measured by placing a mirror as the object in Fig. 13.9. The low-coherence signal source generates an interference signal within the limited path length difference, which is referred to as "coherence length." In the application of low-coherence THz waves to tomography, the depth resolution is dominated by the coherence length. When we assume that the spectrum of the low-coherence THz wave has a Gaussian shape, the best-fitted full width at half maximum (FWHM), Δz, is 1.16 mm. The theoretical value of Δz is given as $2\ln2\ \lambda_c^2/(\pi\Delta\lambda)$, where λ_c is the center wavelength of the THz source and $\Delta\lambda$ is the spectral width and is calculated as 1.1 mm.

To investigate the effectiveness of the proposed technique, the imaging of letter-perforated plastic plates was conducted, as shown in Fig. 13.11. The size of each plate is 50 mm square, and it is 1 mm thick, and it is placed with a spacing of 11 mm. The letters of the three plates, "T," "H," and "Z," were clearly resolved.

As we can see the progress and evolution of OCT technologies, the above-mentioned "time domain THz OCT" can be extended to "frequency domain THz OCT" by using frequency-swept sources [17, 18].

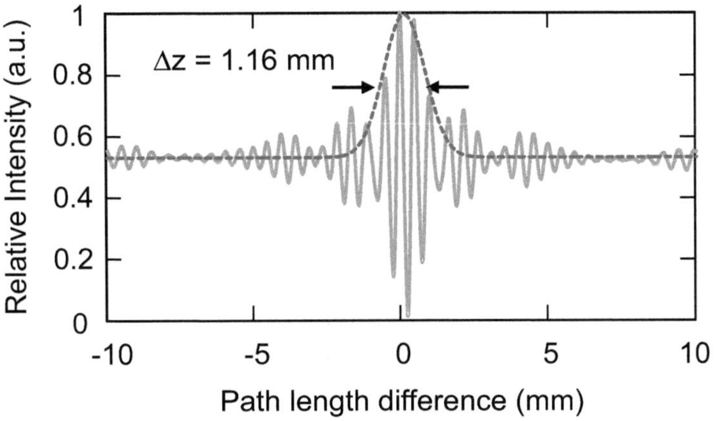

Figure 13.10 Interference signals (interferogram) by low-coherence THz sources.

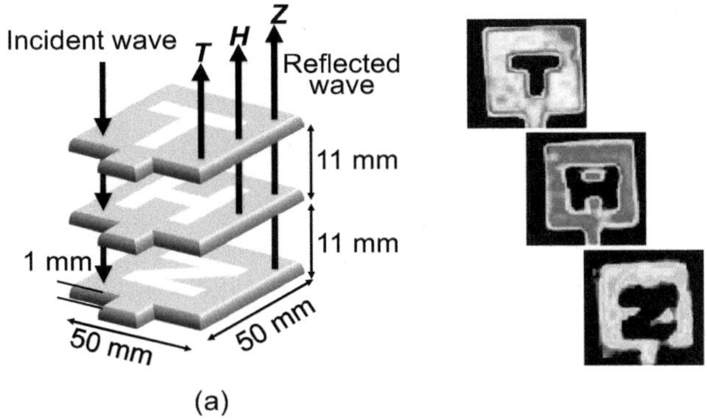

Figure 13.11 Tomography of three plastic plates with holes in shapes of different letters. (a) Object under test and (b) obtained images for each plate by moving the reference mirror.

13.4.2 Spectroscopy

Next, we describe spectroscopic measurements using photonically generated noise signals on the basis of the scheme of Fig. 13.2a, where intense narrow-band noise signals are generated by slicing the ASE noise with optical filters such as an arrayed waveguide grating (AWG) filter [2]. As is the case with imaging applications, we can avoid interference effects in the spectroscopy system by using noise signals.

In the present demonstration, as optical filters, we used two identical 25 GHz AWGs whose center wavelengths are shifted by half of the AWG channel spacing (Fig. 13.12). To tune the frequency band of the noise, we changed the AWG channels from one to another manually. Since the width of the noise spectral density of the optical noise sliced with the 25 GHz AWGs is 5.278 GHz, the 3 dB occupying bandwidth of the noise after a PD is expected to be around 17.6 GHz. Therefore, if only one 25 GHz AWG is used, the tuning step will be 25 GHz and there must be frequency gaps where the photonic noise source cannot produce sufficient noise power.

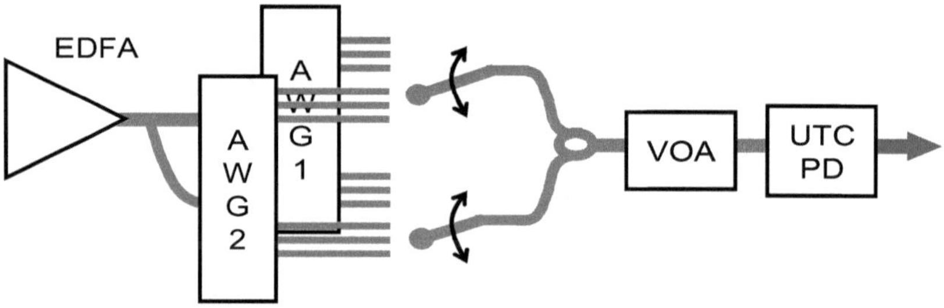

Figure 13.12 Configuration of a photonic noise source based on frequency-variable AWG filters and combiners. VOA: variable optical attenuator.

If we assume a Gaussian optical filter such as an AWG filter, noise spectral densities before and after the optical-to-electrical (OE) conversion at a desired THz center frequency, f_{TH}, can be expressed as

$$S_{\mathrm{ASE}}(v) = \frac{S_{\mathrm{ASE,0}}}{2}\left\{\exp\left(-\frac{(v-(v_0 - f_{\mathrm{TH}}/2))^2}{2\sigma_{\mathrm{OF}}^2}\right) + \exp\left(-\frac{(v-(v_0 + f_{\mathrm{TH}}/2))^2}{2\sigma_{\mathrm{OF}}^2}\right)\right\} \tag{13.1}$$

$$S_{\mathrm{a}}(f) = \frac{\Re_{\mathrm{n}}^2(f) I_{\mathrm{PD}}^2 R_0}{4\sigma_{\mathrm{ASE}}\sqrt{\pi}} \cdot \left\{\exp\left(-\frac{f^2}{4\sigma_{\mathrm{ASE}}^2}\right) + \frac{1}{2}\exp\left(-\frac{(f - f_{\mathrm{TH}})^2}{4\sigma_{\mathrm{ASE}}^2}\right)\right\} \tag{13.2}$$

where $S_{\mathrm{ASE}}(v)$ is the noise spectral density of the ASE noise and v is the optical frequency. $S_{\mathrm{ASE,0}}$ is the peak height of the noise density at the center of the noise spectra, σ_{OF} and σ_{ASE} represent the width of the noise spectral density, and v_0 is the center optical frequency of the optical noise. $S_a(f)$ is the noise spectral density of the THz noise, and f is the electrical frequency. $\Re_n(f)$ is the responsivity of the PD. I_{PD} is the DC photocurrent, and R_0 is the system impedance, 50 Ω.

Figure 13.13 shows a block diagram of the spectroscopy system using the above-mentioned frequency-tunable intense noise sources. The noise signal was radiated from the UTC-PD and passed through a 1 m long gas cell. The received signal was down-converted to the lower frequency centered at 13 GHz and then boosted with an intermediate-frequency amplifier and detected with an electrical spectrum analyzer. The noise temperature of the receiver, including the spectrum analyzer, is expected to be approximately 3500 K. Spectroscopic measurements were conducted in the frequency range between 300 GHz and 400 GHz. The sample under test was a mixture of N_2O and N_2 in the ratio of 3:1 (75%) atmospheric pressure.

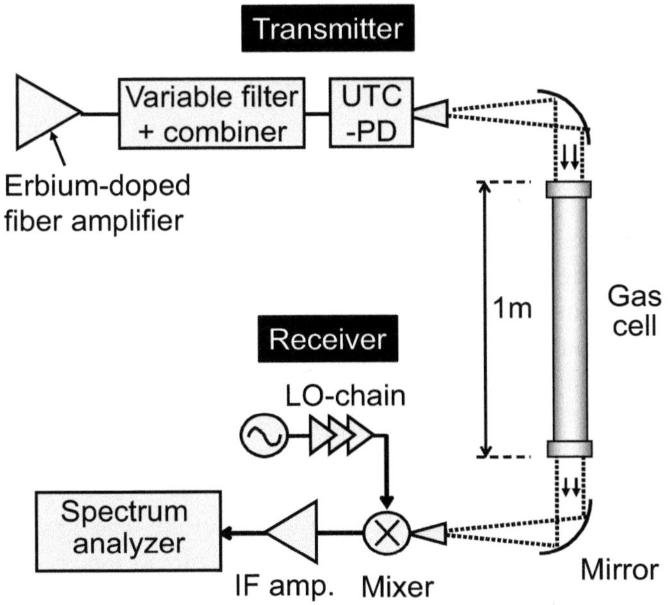

Figure 13.13 Experimental setup for spectroscopy measurement.

The measured transmittance for the gas is plotted in Fig. 13.14 with simulated results based on the high-resolution transmission molecular absorption (HITRAN) database [19]. As can be seen, the positions, tendency of the magnitude, and shape of absorption peaks from the measurement coincide well with those of HITRAN.

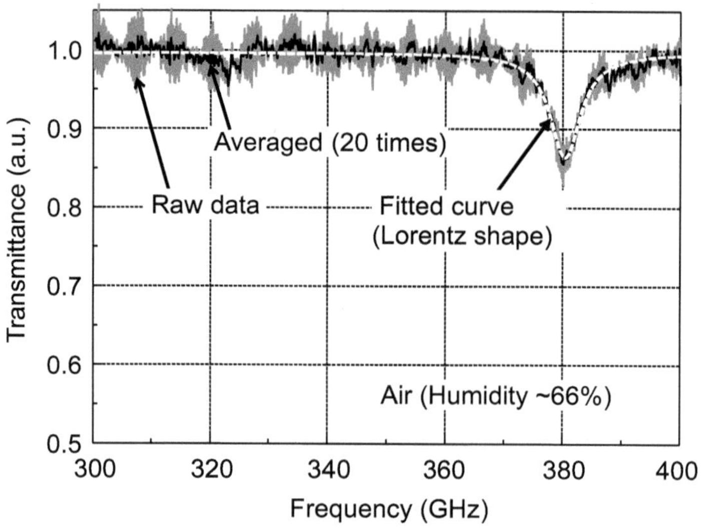

Figure 13.14 Measured spectroscopic trace of humid air with a humidity of 66%.

13.4.3 Amplifier Characterization

Noise measurement in the millimeter-wave and sub-THz-wave bands is not easy yet because of the lack of a noise source. The common solid-state diode noise source based on the avalanche breakdown mechanism is available only at frequencies lower than approximately 150 GHz. Though hot and cold loads based on Planck's law regarding blackbody radiation could be used as a noise source, even in the THz region, it is not easy to couple the noise power into waveguide structures.

Using the same photonic noise source that was used in the spectroscopy application, we measured the noise figure of a THz device in the frequency range of 295–355 GHz [1]. The experiment setup is shown in Fig. 13.15. In this measurement, a commercial second-harmonic mixer (SHM) (VDI-WR2.8SHM, Virginia Diodes, Inc.) was used as a device under test. The UTC-PD was directly connected to the SHM, and the THz noise was down-converted followed by a microwave amplifier.

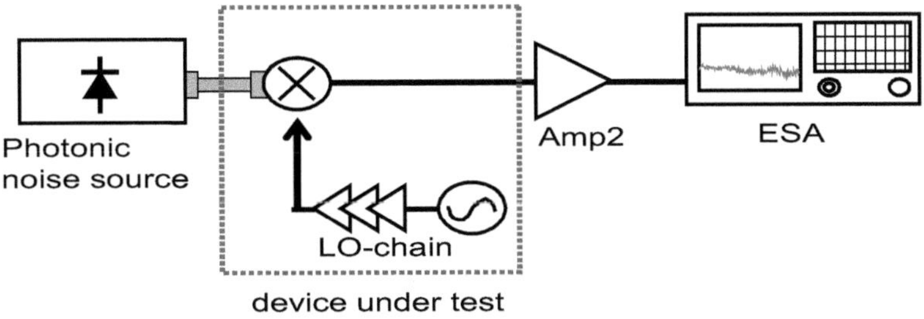

Figure 13.15 Block diagram of noise figure measurement of a second-harmonic mixer operating at 280–400 GHz.

The measurement was performed in the frequency range 295–355 GHz. For Y-factor measurement, the hot and cold states were also implemented with two different optical injection levels to the UTC-PD. For the hot state, the total optical power of the ASE noise was 3 dBm and the photocurrent was approximately 300 μA. Since the UTC-PD was turned off without illumination for the cold state, the output noise density in the cold state was assumed to be −174 dBm/Hz. The nominal excess noise ratio (ENR) is expected to be around 20 dB. The raw data of the measured noise densities for the hot and cold noise source states is plotted in Fig. 13.16. The measured noise densities include all contributions from the microwave amplifier and electrical spectrum analyzer.

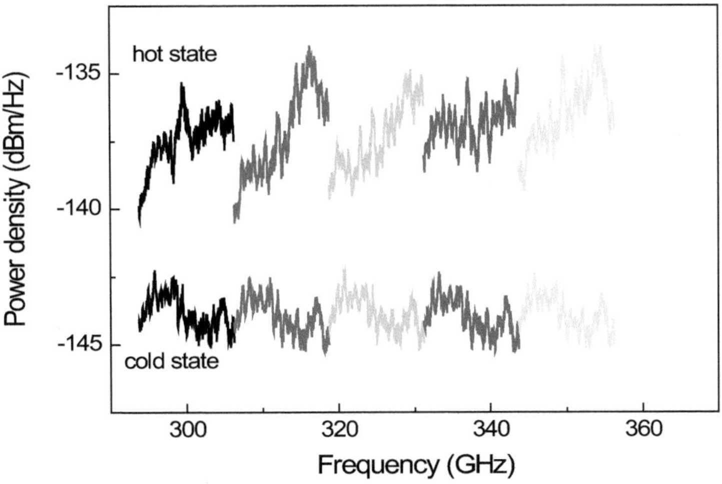

Figure 13.16 Measured output noise density after the microwave amplifier for hot and cold states.

To verify the result with the photonic noise source, we performed another measurement using the common hot/cold loads technique. Liquid nitrogen (N_2) was used for the cold load, and noise temperatures for hot and cold loads were assumed to be 290 K and 77 K, respectively. In this measurement, we used another 50 dB gain microwave amplifier at the output of the SHM because input noise powers from the holt/cold loads are much lower than those from the photonic noise source so that the output noise power is very small.

The measured results with two different techniques are plotted in Fig. 13.17. First, note that, although the measurement with the photonic noise source was separately performed in five different frequency bands, the results don't show any discontinuity around the frequency band edges. For the results with the hot/cold loads, the measured Y-factor in the experiment was a small value, 0.2–0.5 dB, and thus the extracted noise figure and gain show relatively large fluctuation. In both measurements, the noise figures and conversion gains vary in the ranges of 4 to 8 dB and –4 to –10 dB, respectively. For the noise figure, the results for the two different techniques fit each other well overall except at frequencies around 335 GHz. For gain at frequencies lower than 330 GHz, the two results show good agreement. However, at higher frequencies, results with the hot/cold loads are 2–4 dB less than those with the photonic noise source. It could be due to different noise-coupling methods between the two measurements.

13.5 Conclusion

We presented photonically assisted low-coherence noise sources, which consist of an optical ASE noise, an optical filter, and a PD. Proposed generation schemes provide not only very fine frequency resolution together with ultrabroad bandwidth but also high output power, resulting in a large SNR required for various applications.

Using photonic noise sources, we demonstrated practical applications such as imaging, spectroscopy, and noise figure characterization of electronic devices. In the imaging applications, low-coherence signals have proven to be useful to eliminate interference effects, resulting in significant improvement of image quality. In addition, they can be extended to the depth measurement of objects, or tomographic imaging, by introducing an interferometer.

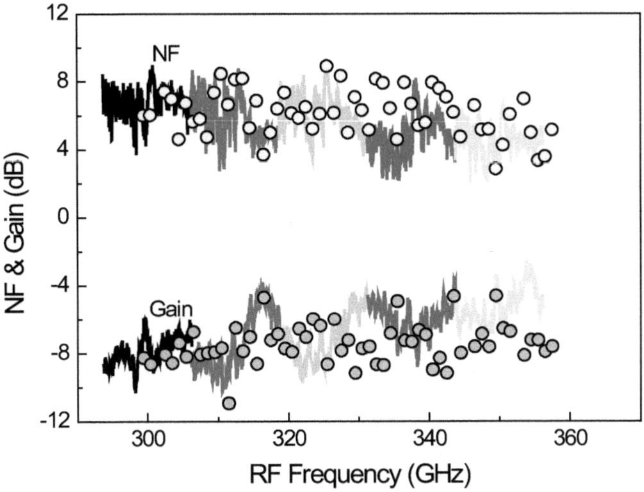

Figure 13.17 Measured noise figure and conversion gain of an SHM with the photonic noise source (solid line) and the conventional hot/cold load technique (scatter).

Bandwidth-controlled and intense noise signals were successfully applied to spectroscopic measurement of gases without the interference effect, which is always observed in CW THz spectroscopy systems. In this article, we used AWG filters, but frequency-variable optical filters with a bandwidth of less than 5 GHz, which are currently commercially available, will provide more practical measurement with respect to frequency continuity.

Finally, we demonstrated noise figure measurements for commercial electronic devices in the sub-THz-wave frequency bands. This scheme allows us to control the output noise level easily and can produce a noise signal at frequencies up to 1 THz. In addition, the optical filter with a well-known spectral shape allows us to concentrate the noise energy in a specific frequency band and to easily estimate the noise spectral density with reasonable accuracy.

References

1. H.-J. Song, N. N. Shimizu,Kukutsu, T. Nagatsuma, and Y. Kado, Microwave photonic noise source from microwave to sub-terahertz wave bands and its applications to noise characterizations, *IEEE Trans. Microwave Theory Tech.* **56**, 2989–2997 (2008).

2. H.-J. Song, N. Shimizu, T. Furuta, A. Wakatsuki, and T. Nagatsuma, Sub-terahertz noise generation with a photodetector and wavelength sliced optical noise signals and its application to spectroscopy measurement, *Appl. Phys. Lett.* **93**, 241113 (2008).
3. T. Nagatsuma, T. Kumashiro, Y. Fujimoto, K. Taniguchi, K. Ajito, N. Kukutsu, T. Furuta, A. Wakatsuki, and Y. Kado, Millimater-wave imaging using photonics-based noise source, *Proc. Int. Conf. IR mm THz Waves* **M3C05.0088** (2009).
4. T. Nagatsuma, H. Ito, and T. Ishibashi, High-power RF photodiodes and their applications, *Laser Photon. Rev.* **3**, 123–137 (2009).
5. A. Wakatsuki, T. Furuta, Y. Muramoto, T. Yoshimatsu, and H. Ito, High-power and broadband sub-terahertz wave generation using a J-band photomixier module with rectangular-waveguide output port, *Proc. Int. Conf. IR mm THz Waves,* 1999-1–1999-2 (2008).
6. B. B. Hu and M. C. Nuss, Imaging with terahertz waves, *Opt. Lett.* **20**, 1716–1718 (1995).
7. W. L. Chan, J. Deibel, and D. M. Mittleman, Imaging with terahertz radiation, *Rep. Prog. Phys.* **70**, 1325–1379 (2007).
8. A. J. Fitzgerald, B. E. Cole, and P. F. Taday, Non-destructive analysis of tablet coating thickness using terahertz pulsed imaging, *J. Pharm. Sci.* **94**, 177–183 (2005).
9. J. Takayanagi, H. Jinno, S. Ichino, K. Suizu, M. Yamashita, T. Ouchi, S. Kasai, H. Ohtake, H. Uchida, N. Nishizawa, and K. Kawase, High resolution time-of-flight terahertz tomography using a femtosecond fiber laser, *Opt. Express* **17**, 7549–7555 (2009).
10. X.-X. Yin, B. W.-H. Ng, B. S. Ferguson, P. Mickan, and D. Abbott, 2-D wavelet segmentation in 3-D T-ray tomography, *IEEE Sens. J.* **7**, 342–343 (2007).
11. B. Recur, A. Younus, S. Salort, P. Mounaix, B. Chassagne, P. Desbarats, J.-P. Caumes, and E. Abraham, Investigation on reconstruction methods applied to 3D terahertz computed tomography, *Opt. Express* **19**, 5106–5117 (2011).
12. http://www.synview.de/b2-synviewscan.php.
13. A. Redo-Sanchez, N. Karpowicz, J. Xu, and X.-C. Zhang, Damage and defect inspection with terahertz waves, *Proc. 4th Int. Workshop Ultrason. Adv. Methods Nondestruct. Test. Mater. Charact.* 67–77 (2006).
14. T. Isogawa, T. Kumashiro, H.-J. Song, K. Ajito, N. Kukutsu, K. Iwatsuki, and T. Nagatsuma, Tomographic imaging using photonically generated low-coherent terahertz sources, *Tech. Dig. IEEE Topical Meeting Microwave Photon.* Session S3, #2219 (2011).

15. T. Isogawa, T. Kumashiro, H.-J. Song, K. Ajito, N. Kukutsu, K. Iwatsuki, and T. Nagatsuma, Tomographic imaging using photonically generated low-coherence terahertz noise sources, *IEEE Trans. THz Sci. Tech.* **2**, 485–492 (2012).

16. D. Huang, E. A. Swanson, C. P. Lin, J. S. Schuman, W. G. Stinson, W. Chang, M. R. Hee, T. Flotte, K. Gregory, C. A. Puliafito, and J. G. Fujimoto, Optical coherence tomography, *Science* **254**, 1178–1181 (1991).

17. M. A. Choma, M. V. Sarunic, C. Yang, Joseph A. Izatt, Sensitivity advantage of swept source and Fourier domain optical coherence tomography, *Opt. Express* **11**, 2183–2189 (2003).

18. T. Ikeo, T. Isogawa, and T. Nagatsuma, Three dimensional millimeter- and terahertz-wave imaging based on optical coherence tomography, *IEICE Trans. Electron.* **E96-C**, 1210–1217 (2013).

19. http://cfa-www.harvard.edu/hitran/welcometop.html.

Chapter 14

Terahertz Cameras

François Simoens
CEA-Leti, 17 rue des Martyrs, Grenoble, 38054, France
francois.simoens@cea.fr

One dictionary definition says a camera is equipment used to convert the optical image of a scene into corresponding electrical signals. Referring to this statement, this chapter will consider terahertz (THz) cameras as devices that directly record and store images of THz scenes.

Emphasis will then be given on technologies that pixel detectors are possibly assembled in 1D or 2D focal plane arrays (FPAs) associated with a readout integrated circuit (ROIC) delivering real-time 2D imaging output. This chapter will attempt to give an overview of such video THz cameras being commercialized or under development. Special attention will be drawn to uncooled microbolometer array cameras, which is the field of R&D activity of the author.

14.1 Introduction

Terahertz (THz) radiation, loosely defined to be between 0.3 THz and 10 THz [1], combines many unique properties. Nonmetallic and

Handbook of Terahertz Technologies: Devices and Applications
Edited by Ho-Jin Song and Tadao Nagatsuma

ISBN 978-981-4613-08-8 (Hardcover), 978-981-4613-09-5 (eBook)
www.panstanford.com

nonpolar materials are transparent to THz radiation, whereas the latter are extremely sensitive to water content. The spatial resolution is approximately 10 times shorter than for millimeter-wave radiation (roughly defined between 30 GHz and 300 GHz) where commercial cameras and body scanners operate. Many substances exhibit characteristic spectral features in the THz frequency range (particularly above 1 THz). And finally this low-photon energy radiation (4 meV at 1 THz) will not cause harmful photoionization in biological tissues and so poses either no or minimal health risk. These features open the way to promising applications in numerous interdisciplinary fields [2–4], many of them requiring 2D imaging.

The first THz 2D imaging demonstration [5] was based on optoelectronic THz time domain spectroscopy (THz-TDS): transmission images were obtained with *xy* translation of leaves and packaged semiconductors. Since then, various different imaging modalities have been developed and demonstrated. However, most of these systems use a single detector and therefore require scanning in two dimensions of the samples to be imaged relatively to the fixed THz beam or vice versa. Mechanical scans limit inexorably the data acquisition rate, and point-to-point measurements are preferably applied to stationary objects.

The spread of THz imaging applications would be greatly increased by the availability of affordable, compact, easy-to-use, and highly sensitive cameras. These cameras would integrate large-format (i.e., many pixels) focal plane arrays (FPAs), and like any digital camera, they would operate in video-mode recording streams in real time without the need for raster scanning.

14.1.1 Proposed Classification of Cameras

Cameras can be classified on the basis of various features (refer to the following table) mainly derived from technology, operation conditions, and imaging methodology criteria.

First they can be divided into passive and active imaging systems. Passive systems image the radiometric contrasts arising from the differences in emissivity and physical temperature between objects through their blackbody emission. The radiation emitted by surrounding natural sources like the sun or the sky may contribute to the contrast. Conversely, in active systems an artificial light source—

or several sources—is used to illuminate the target and the image is formed by detecting the back-scattered and reflected radiation.

Alternatively, another distinction can be made between cooled and uncooled cameras. Cooled cameras integrate a cryocooler that reduces thermally induced noise. Closed-cycle refrigerators can achieve cryogenic temperatures, while thermoelectric coolers—like the Peltier module—are generally the preferred approach to control temperature. The most common cryocoolers used in cameras are the Joule Thompson and Brayton cycle refrigerators and the Stirling, Gifford–McMahon, and Pulse Tube regenerative cryocoolers.

Architectures used in FPAs may be classified as monolithic, hybrid, or mechanically assembled. A monolithic imager is a chip in which the base material substrate contains both detection of light and signal readout (to ensure, in particular, multiplexing), often manufactured in a continuous microelectronic flow. For example, most of the visible cameras integrate a monolithic single-chip sensor fabricated in a standard complementary metal oxide semiconductor (CMOS) process. Conversely, a hybrid device requires the bonding of the detector array to a separate readout circuit. The third architecture type refers to FPAs where the optical front-end receiver and readout circuits are all separated devices mechanically assembled (e.g., horn antennas with multistack detectors). As a general rule, increasing the number of processing and assembly steps leads to production yield degradation and higher costs; this trend explains that monolithic architecture is preferable for a THz FPA to be integrated in a commercial camera.

Another division of detection systems is coherent and incoherent criteria. Incoherent—or direct—detection acquires only the signal amplitude and as a rule is spectrally broadband. Coherent detection systems sense both amplitude and phase. In heterodyne mode, the detected signals are down-converted to much lower frequencies—named intermediate frequencies (IFs)—where they can then be amplified by low-noise amplifiers (LNAs). Basically these systems are selective (narrow band), and they preserve the amplitude and phase information of the incoming radiation. In the last decades these detectors are of choice for high-resolution spectroscopic studies and cosmic remote sensing. Relatively recently, they are applied to millimeter and submillimeter imaging, featuring potentially 3D reconstruction and time-of-flight features.

Finally photon and thermal detectors can be distinguished with respect to their principle of operation. Photon detectors create electrons through the absorption of photons, whereas thermal sensors create electronic signals by a change in heat created by the absorbed photons. These latter sensors are most of the time used in incoherent systems.

Table 14.1 below summarizes the classification criteria corresponding to the different camera technologies described in the following sections of this chapter.

Table 14.1 Classification of the camera technologies in reference to the sections of this chapter

	Passive imaging	**Cooled or uncooled**	**Coherent or incoherent**	**Assembled/ hybrid or monolithic FPA**
Cooled supraconductor cameras	Yes	Cooled	Incoherent (but coherent detection possible with cooled detectors)	Assembled/ hybrid
Cameras with a THz intermediate converter	No	Uncooled	Incoherent (coherent for electro-optics sampling bases)	Assembled
Radar cameras	Yes for single pixel	Uncooled	Both	Assembled
III–V semiconductor FPA cameras	No	Uncooled	Both	Hybrid
Monolithic silicon FET or SBD detectors	No	Uncooled	Both	Monolithic
Pyroelectric cameras	No	Uncooled	Incoherent	Hybrid
Thermopile cameras	No	Uncooled	Incoherent	Monolithic
Above-CMOS bolometers	No	Uncooled	Incoherent	Monolithic

14.1.2 Performance Parameter for Camera Comparison

The most relevant parameter for comparison of cameras in video mode is the noise-equivalent power (NEP). It is defined as the power that generates a signal output that is equal to the root-mean-square (rms) noise output in a 1 Hz bandwidth. Mathematically, it is the ratio between the output noise voltage spectral density v_n and the detector voltage responsivity R_v: $NEP = v_n/R_v$.

This quantity can be reported at various circuit levels, from the direct signal generated at the pixel output to the output of the overall FPA imaging chain operated in video mode. This latter NEP_{camera} requires only an a priori knowledge of the total impinging source power.

This chapter devoted to cameras will only consider NEP_{camera} expressed in [W] that can be computed as follows:

$$NEP_{camera}^{[W]} = \frac{v_n P_{In}}{\sum_{array} V_i} \tag{14.1}$$

where P_{in} is the total incident THz power and V_i is the signal of the ith pixel within the array.

NEP is often expressed in $W/\sqrt{Hz}$. It can be calculated from the analog video bandwidth B of the camera as $NEP_{camera}^{W/\sqrt{Hz}} = NEP_{camera}/\sqrt{B}$. This expression supposes a flat noise power spectral density; this assumption is seldom correct since $1/f$ noise is often generated by materials and ROICs. One has to note that NEP_{camera} refers to the optical impinging power and not to the electrical power absorbed in the detector. $NEP_{electrical} = \alpha_{opt} NEP_{camera}$, where $\alpha_{opt} \le 1$ is the optical coupling efficiency of the imaging system.

Specifically for passive imaging, one useful figure of merit is the noise-equivalent temperature difference (NETD). NETD is defined by a change in the equivalent blackbody temperature that corresponds to a change in radiance that will produce a signal-to-noise ratio of 1 in the imaging device. NETD can be related to the electrical NEP of the detector through

$$NETD = \frac{NEP}{\alpha_{opt} k_B \Delta v \sqrt{2\tau_{int}}} \tag{14.2}$$

where the Rayleigh–Jeans approximation of Planck's law is assumed to be valid ($hv << k_B T$), Δv is the radio frequency (RF) bandwidth, and τ_{int} is the readout integration time.

14.1.3 Outline of the Chapter

The first section reports on cryogenic supraconducting 1D or 2D FPAs that uniquely demonstrate incoherent THz passive real-time imaging. Commercial direct, passive millimeter-wave-range cameras (e.g., Ref. [6] or [7]) are not reviewed in order to limit the length of this chapter. These room-temperature cameras operate in a spectral domain lower than 0.3 THz, where LNAs are readily available to provide an NEP below 1 pW/sqrt(Hz). In comparison to the THz range, around 100 GHz, the spatial diffraction resolution is limited and requires large apertures. These features limit the exploitation of such cameras when small objects are to be identified.

The following sections will pertain only to active imaging where the commercial—or close to be industrially transferred—camera offer is continuously growing.

First we will present 2D real-time imaging systems using a transducer of THz radiation into an intermediate spatially distributed signal to be captured by a visible or infrared (IR) camera.

Then FPAs based on nonsilicon materials are addressed. Most of them need to be mechanically assembled or hybridized to the ROIC. This review of THz cameras will be completed with the monolithically integrated sensors relying on full-silicon CMOS technologies, including photon detectors and uncooled bolometers.

This chapter will describe only 1D or 2D FPA technologies that can possibly be implemented with an ROIC delivering real-time 2D imaging output. This overview cannot claim exhaustiveness, because the technological approaches and levels of maturity are very diverse, and in addition new players emerge every year in this active THz camera field.

14.2 Cooled Cameras

Passive cameras image the radiation emitted by objects deduced from their spectral emissivity and Planck's law. The advantages of a passive imaging system are its simplicity—no illumination source is to be included—and the easier societal acceptability since detection of "natural" radiation is employed.

However, in the THz range the total radiated optical power is low and moreover is substantially smaller than the thermal energy at room temperature, T ($h\nu \approx k_BT$). Consequently detectors have to be highly sensitive and with low thermal noise: the steep responsivity and cryogenic operation of superconducting sensors detection systems fulfil these requirements.

14.2.1 Superconducting Detector Technologies

For several decades, extremely cooled technology (i.e., as a rule at T < 1 K) has been supported by ambitious spatial programs because of their stringent requirements. For submillimeter-wave ground-based observation, atmospheric background is around 100 pW, and the background-limited NEP is about 10^{-16} W/sqrt(Hz). Observations from space requires an NEP of 10^{-18} to 10^{-19} W/sqrt(Hz) to achieve background-limited performance. Nowadays cryogenic FPAs [8] are playing a significant role in astrophysics, showing extreme sensitivities close to quantum limits. A large variety of cooled detectors can be found, like transition edge sensors (TESs), kinetic inductive detectors (KIDs), hot-electron bolometers (HEBs), or superconducting tunnel junctions (SJTs) like superconductor-insulator-semiconductors (SISs).

The TES is a bolometer that measures energy deposition as an increase of resistance of a superconducting film biased within the superconducting-to-normal transition [9]. A TES exploit the steepness of this transition to reach unprecedented high detection efficiency. Thermometers are made of low-T_C (T_C stands for temperature transition) superconductors such as Al, Ta, Sn, Nb, or NbN.

In spite of early demonstrations in the 1940s, impedance reading difficulties and operation instability slowed the use of the TES. Nowadays, these problems have been overcome by electrothermal feedback operation [10] and superconducting quantum interference devices (SQUIDs) or a metal–oxide–semiconductor field-effect transistor (MOSFET) current readout that can be impedance-matched to low-resistance TES detectors. Bolometer arrays now typically gather up to a few thousand pixels, such as APEX-SZ (55,150 GHz pixels [11]), SPT (160 Al/Ti bilayer TES at 95/150/220 GHz), SCUBA-2 (10,000 TES [12]), PACS (2560 ion-implanted silicon-on-insulator [SOI] thermometer array in the 60–130 μm range [13]; see Fig. 14.1).

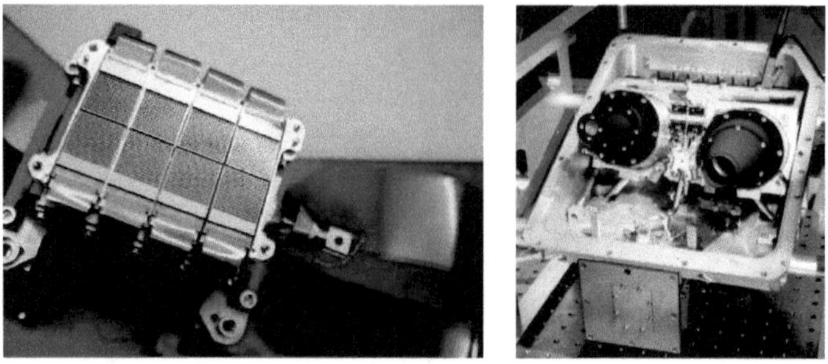

Figure 14.1 (Left) The PACS blue channel 32 × 64 pixels array developed at CEA for the ESA Herschel project. (Right) A photometer focal plane unit with two-bolometer array assemblies (60–130 μm short wave/130–201 μm long wave) and the 0.3 K sorption cooler. Reproduced from Ref. [14] by permission of SPIE, © 2004.

It is worth noting that this latter FPA, the PACS photometer, was including for the first time in cryogenic bolometers' history a CMOS ROIC cooled at 0.3 K and bonded to the array to ensure 16 × 1 multiplexing [14]. The achieved NEP of 2.2×10^{-16} W/sqrt(Hz) does not compete with the TES thermometer multiplexed by a SQUID. However, this combination of Si thermometers and MOS multiplexers offers a remedy to the bottleneck of data throughput, this imager being composed of 2560 pixels [15].

KIDs were first developed by scientists at the California Institute of Technology and the Jet Propulsion Laboratory in 2003 [16]. They are now used for different astronomy programs [17–20]. In these photon detectors, the absorbed radiation breaks Cooper pairs, changing the complex AC surface impedance of the superconductor. This change in impedance induces a shift of the resonant frequency of a superconducting resonant circuit that is coupled capacitively to a feedline. A probe signal on-resonance is sent through the feedline to deliver externally the impedance change information. Many different resonators at slightly different frequencies can be coupled to the same feedline, allowing for natural frequency multiplexing with expected negligible crosstalk. This multiplexibility of the readout is a major asset for FPA integration. Achieved NEPs are in the range of 1×10^{-15} W/Hz$^{1/2}$.

SIS junctions are frequently used as mixers in heterodyne mode [21], but some cameras use these photon detectors coupled to on-chip antennas for direct detection. They consist of two superconductors separated by a very thin layer of insulating material. The absorbed photons break the Cooper pairs and create quasi-particles that then tunnel through the junction via the process of photon-assisted tunneling. Heterodyne mode exploits the nonlinear current–voltage curve, while direct detection reads the current.

An NEP of the order of 10^{-16} W/sqrt(Hz) at 0.66 THz has been measured for a single direct detector based on a Nb SJT [22]. SIS FPAs are being developed for astronomical observations, like the SISCAM camera composed of a 32-channel SIS photon detector readout by GaAs-JFET cryogenic application-specific integrated circuits (ASICs) for imaging at 650 GHz [23].

Similarly to a TES, the HEB principle is based on the reading of a temperature-dependent electrical resistance biased near its superconducting transition. But unlike a TES, the radiation is absorbed by "hot" electrons driven out of thermal equilibrium with the phonon system, while phonons act as a thermal reservoir. HEBs use electrons as a subsystem in semiconductors or superconductors that interacts with the lattice phonons. Near the critical temperature, T_C, the superconductor's resistance is sensitive to the electron temperature. The processes of hot-electron effect below the transition temperature T_C that takes place in superconducting thin films result from electron–phonon interactions where energy relaxation times and times of escape into the substrate compete. Since the electron thermal capacity is much lower than the lattice, the heat induced by radiation at the sensitive element level can quickly escape through the thermal link, and thus ensures a faster response. Thermal relaxation can be made particularly fast using thin films and materials that favor large electron–phonon interaction or electron out-diffusion.

One interesting approach consists in using material like yttrium-barium-copper-oxygen (YBCO) as an HEB thermometer because it exhibits high-T_C detection, typically in the vicinity of 90 K. It can be operated at liquid nitrogen temperature achievable with more compact and simpler cryogenic systems—like Sterling modules—in comparison to He-cooled devices but at the price of sensitivity reduction.

HEBs are also preferentially used as mixers for heterodyne detection and may be implemented in arrays. However, the fabrication process and aging effects are still being investigated [24].

14.2.2 Discussion

All the cited research programs are dedicated to astronomy and astrophysics. Nowadays, many superconducting detector arrays are being developed for scientific research with explicit perspectives of commercial cameras.

The SPACEKIDS FP7 project aims at developing a KID camera for space missions but simultaneously considering commercial exploitation.

Pioneering works at the VTT Technical Research Center of Finland in collaboration with the National Institute of Standards and Technology (NIST) have allowed remarkable demonstrations of passive THz imaging of large scenes in the 0.3–0.65 THz range [25]. The camera is composed of 128 fully planar broadband superconducting TES microbolometers assembled in 8-pixel linear array modules [26]. Each pixel consists of one Nb bridge coupled to broadband lithographic circular spiral antennas: they are fabricated using a standard photolithographic process on a 100 mm diameter Si wafer. The readout electronics allows for detector-noise-limited performance without the use of cryogenic preamplifier stages like SQUIDs. The latest publications [27] report 10 fps video-rate THz passive imaging of a full-person field of view (FoV) with ~15 mm spatial resolution: the FPA is housed within a custom-built 7 K cryostat that integrates a turnkey commercial closed-cycle cryocooler.

Another project, Safe VISITOR [28], integrates a superconducting camera combined to charge-coupled device (CCD) visible and thermal IR bolometer imagers. Voltage-biased superconducting TES bolometers are made of a bilayer of molybdenum and gold–palladium. Pixels are arranged in 7-pixel arrays manufactured on silicon nitride–coated 4-inch Si substrates [29]. The first video imaging has been demonstrated but at quite a slow rate till now.

The two previous developments present common features. Both target security-screening application cameras, and the cryogenic temperature is provided by pulse-tube cryocoolers. Moreover, since the total number of pixels is small and a large FoV is to be attained with reasonable spatial resolution, the scene of interest is sampled by an optomechanical scanner. Deployed optics configurations include a folded Schmidt telescope with a conical scanner or a Cassegrain telescope with a rotating secondary mirror tilted periodically around the vertex. Such devices make the system quite cumbersome and with delicate fast-moving optical elements.

Cooled sensors offer significantly enhanced resolution and sensitivity as a result of the lower noise operation. They are particularly promising for direct-detection passive imaging in the atmospheric window close to 1 THz, found at 0.8 THz, 0.6 THz, 0.35 THz, and below. The transmission increases with lower frequencies, whereas the spatial resolution becomes insufficient for many applications unless oversized optical components are used.

However the extended use of these high-performance incoherent superconducting detectors[1] is hindered by the cost of fabrication and integration and by cryogenic operation. Reading and multiplexing cryogenic detectors are straightway challenging, especially because of impedance transformation and power consumption budget constraints. In addition cryogenic cooling significantly impacts the cost, the operation yield, and the size, especially when liquid helium is used.

14.3 Uncooled Cameras

There is a clear trend toward monolithic integration of sensors for uncooled THz real-time cameras. Such features aim at satisfying the common criteria that any large volume devices should meet, that is, low cost and low size, weight, and power (SWAP).

Uncooled operation implies a trade-off with performance, and it follows that all the cameras reviewed hereafter are based on active THz imaging to compensate for modest sensitivity. Active imaging systems are preferred in unfavorable environmental conditions, for example, to see reflective objects hidden behind a thick fabric. One issue of active imaging is the speckle effect, which is the result of diffraction between reflections of different details of an observed object within a single pixel. The most effective way of defeating this speckle effect is to use several spatially separated and mutually incoherent sources for the illumination.

[1]In addition to cryogenic operation, cooled coherent receivers present intrinsic higher complexity than incoherent presented detectors. It explains that in spite of their highest sensitivity they are deliberately not addressed by this section, being too far from meeting the criteria for a THz commercial camera.

14.3.1 Cameras Combining an Intermediate THz Transducer and a Video IR or Visible Camera

2D real-time THz imaging was historically first demonstrated by the transformation of THz radiation through the use of the Pockels effect and of a commercial CCD camera [30]. This section will report this technique and others based on intermediate transducing of the THz radiation into a spatial distribution of signals "readable" by a conventional camera.

14.3.1.1 Electro-optic transducer cameras

Electro-optic sampling relies on the linear Pockels effect where a polarization rotation in electro-optic crystals induced by a femtosecond probe pulse depends on the amplitude and direction of the THz field. The detected signal is the polarization rotation of the probe beam in the nonlinear crystal; therefore the imaged surface of the sample is defined by the size of the optical beam and of the crystal. The spatial distribution of the modulated beam intensity is imaged by a digital CCD camera, allowing one to collect information for each pixel in an image simultaneously.

The first demonstrated setup [30] uses ultrafast laser-driven optoelectronics devices. Since the modulation depth in the probe beam is somewhat small, usually frame averaging and amplified femtosecond lasers are needed. Alternative schemes can obviate these hurdles, especially some setups aimed at freeing from the use of relatively expensive and delicate laser amplifiers. For example, a pair of slightly detuned laser diodes can be used for both the generation and the detection process [31]. Coherent electro-optical detection can be realized with an optical parametric oscillator (OPO) pumped by a nanosecond laser [32].

This pioneering technique is still used in the laboratory with enriched innovative modalities but, to the best of the author's knowledge, with no commercial camera in perspective.

14.3.1.2 Photomechanical transducer-based camera

Another approach of intermediate transducing consists in converting THz radiation into a mechanical displacement. Such a method is being developed by Agiltron [33]. The photomechanical imager contains a microelectromechanical systems (MEMS)-based FPA that transduces THz radiation into a visible signal for capture by a high-performance CCD imager.

The prototyped array contains 130 × 90 pixels with 100 μm pitch that THz radiation absorption causes to heat up. The pixels mechanically deflect proportionally to the absorbed heat. The readout of the mechanical displacements is achieved by the projection of a light-emitting diode (LED) illumination beam onto a CCD imager through a 4*f* optical system. The involved thermal conversion requires that the sensor chip reside in a vacuum packaging. A time constant of the order of 120 ms is reported.

The authors have published an NEP of 76 pW/sqrt(Hz) at 30 fps and a sensitivity of 25 μW/cm^2: this latter parameter corresponds to a detection NEP of 2.5 nW per 100 μm pitch pixel. Several design changes are listed to improve this sensitivity, both on the THz photomechanical transducer and on the CCD imager side.

14.3.1.3 THz to IR transducer-based camera

A similar approach can be applied with the use of an IR camera. THz radiation is converted to heat spatially distributed on a specific film integrated in a vacuum-pumped housing. The resulting thermal distribution is then imaged by a standard thermal IR bolometer camera. The commercialized TERACAM implements this conversion principle [34]: it is specified at 0.1 THz with $NEP < 1$ μW/sqrt(Hz) and a minimum measurable signal of 50 μW. The image resolution is defined by a total of 40 × 40 pixels with a 1 mm pitch.

14.3.1.4 Discussion

For the time being these two later-commercialized cameras exhibit quite limited sensitivity (i.e., NEP in the range of microwatts to nanowatts per pixel). Moreover the acquisition speed is intrinsically limited by the thermal conversions (that occur twice when an IR camera is used). Indeed, in a common way, such camera architecture means that the performance is the result of the convolution of the time constants and of the conversion efficiencies of the cascaded elements.

The use of an intermediate planar transducer involves tackling the optomechanical implementation challenges, especially optical alignments and compactness. This is particularly critical in the Pockels effect or photomechanical-based cameras since a visible source and specific lenses have to be associated to the transducer and the CCD camera.

So most cameras under development now target FPAs providing direct conversion of THz radiation into a video output. All the cameras reviewed in the sections below integrate such a detection scheme.

The following section is related to hybrid technologies, that is, where FPAs are resulting from the assembly of different elements processed separately. An imaging radar can be composed of monolithic solid-state microwave integrated circuits (MMICs) that are issued from the microwave technology. Another type of hybrid sensors consists of on-chip front-end arrays that are hybridized on an ROIC, often a CMOS ASIC.

14.3.2 Uncooled Assembled and Hybrid FPA Cameras

14.3.2.1 Cameras based on radar techniques

When uncooled solid-state semiconductors are employed for very sensitive applications at high frequencies (typically >300 GHz), fast heterodyne schemes are necessary since so far proper THz amplifiers do not exist and they prevent direct detection. THz radar technologies using solid-state MMICs are a good candidate for THz cameras.

The radar method of frequency-modulated continuous-wave (FMCW) radar is generally applied. In an FMCW radar the frequency of the emitted signal is changed continuously. After reflection on a target at a range R, the frequency of the echoed signal is shifted by a quantity Δf, which is proportional to R. Such a technique allows range and velocity to be found with one radar set. Saw-tooth modulation of frequency is usually employed. The choice of frequency deviation per time defines the radar resolution, whereas the duration of the frequency shift determines the maximum range.

Thanks to the penetrating property of THz radiation, such a technique can be applied in a personnel screening checkpoint to measure how far hidden objects protrude from the body and possibly to distinguish through the echoed signals layers of clothing, threats, and skin. Combining fast-scanning optics and multibeam target illumination and transceiver arrays, such a radar system is potentially providing 3D real-time imaging.

Over the last four years the NASA Jet Propulsion Laboratory (JPL) has developed an ultrawideband 675 GHz radar for long-distance standoff security [35]. An FMCW prototype system operates over a 28.8 GHz bandwidth centered at 676.7 GHz. The radar can generate one circa 40 × 40 cm image per second, in real time, of targets at a standoff range of 25 m. The THz radar has been built largely from off-the-shelf components, but integrating them into a functioning system poses a substantial challenge. Ongoing development aims at multibeam video-rate imager with reduced size and complexity. The front-end components of a transceiver array would be integrated in a monolithic stack of gold-plated silicon wafers with micromachined waveguides and device cavities. For a radar transceiver array, an 8- or 16-element array would possibly be made of typically 26-wafer stacks that integrate GHz horns, waveguide hybrid couplers, Schottky diodes, doubler/tripler/subharmonic mixers, and MMIC power amplifiers. The authors have recently demonstrated radar imaging frame rates of 1 Hz at a 25 m standoff range, but still substantial development is needed.

Other researchers are also making rapid progress in THz imaging radars. Again near the 670 GHz window, an FMCW radar has been demonstrated primarily for industrial applications such as package inspection [36] but with the potential for person-borne concealed weapon detection as well. Similar imaging radars are also being developed near the next-lower-frequency transmission window at 340 GHz, where components are easier to obtain commercially and where clothing penetration is higher [37, 38].

An alternative interesting approach inspired by radar techniques relies on interferometric imaging. Interferometric imaging is motivated by the idea of emulating a large number of pixels with the radar synthetic aperture method. Over conventional imaging it needs fewer or no moving mechanical parts and does not include a focal plane. Each pair of detectors generates a spatial Fourier component in the THz image. By adding multiple Fourier components from all possible combinations of detector pairs, one can reconstruct the THz image by an inverse Fourier transform. In general, for *n* detectors, the number of spatial frequencies is $n (n - 1)/2$. The higher n is, the better the reconstructed image is.

Continuous-wave (CW) interferometric imaging [39] has been demonstrated at 0.1 THz, and high-power THz sources are expected before applying this system in the THz range. The limitation comes from the computing time of the reverse-transformed image but also from the overall optical system size and complexity.

A German group also develops such interferometric imaging [40]. A THz camera has been designed for operation at standoff distances of 8 m, with a scene area of 0.7×2 m^2 and to produce images close to real time. The system combines a linear array of 8 emitters (Tx) and 16 receivers (Rx) for synthetic imaging in the horizontal direction. Classical scanning optics is applied in the vertical direction. Proof-of-principle demonstration for synthetic-aperture imaging in the 300 GHz frequency regime has been provided, but further work needs to be carried out to operate in real time with fully operational elements.

The synthetic-aperture technique is relevant for 2D imaging with a limited number of Tx–Rx elements, but its bottleneck is the data acquisition hardware, which has to sustain the very high data flow of gigabytes per second.

14.3.2.2 Hybrid III–V semiconductor FPA cameras

Hybrid FPA cameras relate to imagers where the pixel array is processed separately from the ROIC and then hybridized to the ROIC with microelectronic techniques like wire bonding or flip-chip assembly. The latter method consists of using solder bumps, gold balls, or molded studs that have been deposited onto the chip pads to interconnect semiconductor devices, here the integrated circuit (IC) and the sensing chip. It allows compact direct interconnections of the pixel outputs with the ROIC and therefore ensures multiplexing of large arrays.

The separate process is often motivated by the noncompatibility between array manufacturing and ROIC constraints, such as maximum acceptable temperature or microelectronics aggressive process that would degrade the IC. This is, for example, the situation of detectors based on III–V semiconductors. Two examples of developments are described hereafter.

The first example of technology that would require hybrid FPA relates to InAsGa diodes. Such detectors are low noise because they can operate with zero bias voltage. They can be quite easily processed with one of the semiconductors leaves metallized in order to act as a bowtie antenna that collects the THz radiation in the vicinity of the apex [41]. In this way the resulting symmetry architecture combines the optical coupler and the semiconducting part. The operation principle relies on this asymmetric geometry that favors nonuniform electron heating, resulting in photovoltaic mode. Different geometries and materials are being investigated, and latest publications [42] describe tests of spectroscopic THz imaging. Heterodyne imaging has been demonstrated at frequencies in the range of 0.585–2.52 THz with an NEP of about 5 nW/sqrt(Hz) and a response time below the nanosecond time scale. To date no array has been, and the cited sensitivity does not correspond to the video-like camera operation that this chapter emphasizes.

Heterojunction backward diodes (HBD) can also exhibit high cut-off frequencies and reduced $1/f$ noise when operated at zero bias; intrinsic cutoff frequencies as high as 805 GHz have been measured for antimonide (Sb)-based HBD [43]. A 31 × 31 pixel camera designed to cover the 0.6–1.2 THz range has been developed at Ohio University [44]. The FPA consists of 100 μm pitch 80 × 64 zero-biased Sb HBDs monolithically integrated with planar antennas. Antennas couple the incoming radiation onto the HBDs, where the THz signal is rectified to a DC or baseband signal. Then the diode output signal is read out by a ROIC that is flip-chip-bonded to the planar FPA. An HBD THz camera is commercialized by TRaycer (http://www.traycer.com). It is specified to provide an unmodulated NEP of 200 nW/sqrt(Hz) with 25–100 fps for the 80 × 64 format. This camera operates in video-like mode with direct detection.

14.3.2.3 Discussion

The use of solid-state detectors has long been hampered by the transit time of charge carriers being larger than the time of one oscillation period of THz radiation. Impressive ongoing works push this limit with, for example, the recent usable frequency range of LNAs in the 600 GHz range as published by Fraunhofer Institute [45]. Ambitious III/V technology developments are running, like the terahertz imaging focal plane technology (TIFT) and submillimeter-wave imaging focal-plane technology (SWIFT) Defense Advanced Research Projects Agency (DARPA)-funded programs.

However, the high cost of MMICs limits the pixel count in imaging systems, and the use of LNAs is typically limited to below 200 GHz in practical applications. Heterodyne detection implies distributing the local oscillation (LO) signal to the multielement array; moreover LNAs are unavailable in silicon process technologies in the THz frequency band. Currently no THz camera based on an FMCW or a synthetic aperture is commercialized.

Detectors based on III–V semiconductor and waveguide technologies are currently dominating THz sources and receivers. But they are penalized in relation to full-silicon CMOS-compatible devices because of higher cost and lower level of integration. Moreover Si-based THz cameras would benefit from continued progresses of the CMOS technology that dominates the semiconductor industry.

In the two following sections, two categories of monolithic FPAs are described, photon and thermal detectors. Both detectors rely on Si technology and can be used for imaging in a video-mode camera with direct detection.

14.3.3 Uncooled Monolithic Silicon Detectors

14.3.3.1 Field-effect transistors

Two different principles of THz detection in a field-effect transistor (FET) are put forward.

According to Ref. [46], detection in FET devices makes use of plasma-wave propagation instead of electron drift in the channel. An upper frequency of operation of an FET is conventionally defined by the cutoff frequency[2] $f_T \approx 1/(2\pi T)$, where T is the electron transit time. Since the velocity of the plasma waves is much larger than the drift velocity of the 2D electrons in the FET channel, the THz radiation excites plasma waves in a two-dimensional electron gas (2DEG) that can operate at much higher frequencies than f_T.

Another approach is to consider the direct detection in FETs as a distributed resistive self-mixing process in a nonbiased (cold)

[2]The cut-off frequency of a transistor, f_T, is the unity current gain frequency. $f_{T_{max}}$ is the unity power gain frequency, also sometimes called maximum oscillation frequency.

transistor. The FET transistor channel can be seen as a distribution of elementary sections of a nonlinear transmission line. This non-quasi-static (NQS) model of the channel [47] leads to a partial differential equation (PDE) that can be used to investigate the intrinsic carrier transport dynamics inside the transistor channel and therefore to understand the mixing procedure.

Below f_{max}, where transistors exhibit power gain, silicon-based circuits may be operated fundamentally up to about 160–220 GHz with reasonable RF circuit performance. Above f_{max}, where transistors exhibit power loss, circuits may be operated subharmonically to further extent the RF operation region, reaching the THz range although with increased receiver noise.

The first detector realizations have been demonstrated in GaAs high-electron-mobility transistor (HEMT) technologies [Lisauskas 1992] [48]. Subsequently, this detection principle was implemented using commercial silicon CMOS technology [49].

Thanks to CMOS and SiGe bipolar CMOS (BiCMOS) process technologies, monolithic imaging FPAs can easily integrate many pixels that consist of MOS transistors coupled with an on-chip antenna. The readout of these currents is provided by the fully integrated analog or digital circuitry within the FPA. Multipixel integration in cameras is better suited to incoherent (direct) detection, where the power consumption per pixel needs to be small to enable large-pixel arrays.

A 32 × 32 FET array has been developed and demonstrated for real-time imaging [47]. The pixels consist of 1024 differential on-chip ring antennas coupled to n-type MOS (NMOS) that detect directly the incoming radiation at frequencies well beyond the FET cutoff frequency. At 0.856 THz and 25 fps, the voltage responsivity is 115 kV/W and the camera video NEP is 12 nW integrated over a 500 kHz bandwidth. This work corresponds to the first demonstration of a CMOS FET-based 2D camera operated to capture THz video streams.

Real-time imaging has also been demonstrated at 590 GHz with a 12 × 12 pixel CMOS array operated with lock-in detection technique [50]. Imaging streams have been acquired at a frame rate of 3 Hz while the source was modulated.

It is interesting to highlight the use of FETs in heterodyne mode. Ongoing works aim at exploiting the fast response of FETs—only limited by the readout electronics—with the implementation of coherent detection and time-of-flight detection. The latter is valuable for the determination of the surface topography of objects as well as for 3D volume reconstruction. Well-resolved coherent THz imaging at 591.4 GHz has been demonstrated with single or a few raster-scanned detectors [51]. Some developments are still to be carried out to implement a full monolithically integrated focal plane. For the time being a virtual 10-kilopixel camera is emulated by scanning a few detectors across the image plane.

The key issues for THz FETs are substrate-wave losses, the low-resistivity silicon substrate, and the low equivalent quantum efficiency of this square-law detector [47] that requires challenging low-noise THz camera readout. Parasitic capacitances tend to shunt down part of the coupled THz signals to the substrate. The antenna behavior is altered by substrate effects, resulting in considerable power carried by surface waves and multiple reflection effects at the Si–air boundary.

Radiation coupling can be improved by the use of silicon-on-insulator (SOI) substrates instead of bulk CMOS due to their lower device parasitics and their higher resistivity [52]. Different packaging configurations are being investigated (front-side or back-side illumination) to improve the on-chip dipole antenna efficiency, and best results are obtained with an antireflective-coated lens glued to the chip during the chip assembly process.

14.3.3.2 Si Schottky diodes

Similar to FETs, Schottky barrier diodes (SBDs) can be integrated in monolithic FPAs with full-silicon technologies. When a small AC signal is applied, a Schottky diode generates DC voltage proportional to the signal power due to its second- and other even-order terms associated with nonlinearity.

A 2 × 2 array of 280 GHz SBD detectors with an on-chip patch antenna (255 × 250 μm^2) has been fabricated in a 130 nm logic CMOS process [53]. The series resistance of the diode is minimized using polygate separation (PGS). These SBDs exhibit a cutoff frequency f_T = 2 THz. With a 16-pixel 280 GHz SBD imager prototyped array, imaging was demonstrated at 280 GHz in a setup including no mirror or lens. A single SBD detector operating at 860 GHz has also been tested, with the perspectives of being integrated in a multipixel array. This array has not been operated in video-mode camera, and so no corresponding NEP is available.

14.3.3.3 Discussion

These cameras based on monolithic CMOS technology are very promising both because of their foreseen moderate price and because of the achieved performances. They benefit from the industrially qualified CMOS process technology and can be mass-produced with significantly reduced cost. Moreover as demonstrated for the first time by the University of Wuppertal group CMOS FET–based cameras can acquire 2D images in video mode with a total NEP in the range of tens of nanowatts at 0.85 THz.

However, their optimal detection range is below 1 THz; at higher frequencies, their performance decreases dramatically because of the influence of parasitic frequency-dependent elements, such as shunt capacitance or series resistance. Conversely uncooled thermal cameras don't suffer loss of sensitivity with frequency because of their conversion principle that is independent of the spectral range. This family of uncooled thermal cameras is presented in the following section that concludes that review.

14.3.4 Uncooled Thermal Cameras

As a general principle, the sensitive element of a thermal sensor combines an absorber—to collect the incoming radiation—and a temperature transducer or thermometer. The absorption of the electromagnetic radiation by the active area of the thermal sensor results in an increase of its temperature. The thermometer converts the temperature variation into parameters that are measured electrically by a readout circuit.

14.3.4.1 Pyroelectric cameras

Pyroelectric cameras make use of electrical polarization dependence in temperature. Very broadband pyroelectric cameras are commercially available, like the PyrocamTM III sold by Ophir Optics (http://www.ophiropt.com). This camera integrates a 124 × 124 element array. Pixels are made of LiTa03 pyroelectric crystal sensors hybridized[3] with indium bumps to a solid-state readout multiplexer. Pixels are spaced by 100 μm with a sensitive area of 85 × 85 μm^2. The spectral response is very broadband since the coating of the crystal absorbs all wavelengths, including 1 μm to over 3000 μm (0.1 THz to 300 THz). However, the packaging window has to be interchanged to cover the whole wavelength range. The camera can operate with pulsed or CW sources. In the latter case, a chopper integrated in the camera has to be operated at 24 Hz or 48 Hz. For THz sources the sensitivity is said to be relatively low, with a published sensitivity around 300 mW/cm^2 at full output. With a signal-to-noise ratio (SNR) of 1000, beams of 30 mW/cm^2 are detectable. In addition, multiple frames can be summed to improve the SNR. Summing 256 frames enables viewing of beams as low as 1–2 mW/cm^2. So under the best condition, the video-mode NEP is in the range of 0.12–3 μW.

14.3.4.2 Thermopile cameras

Thermopiles are thermal sensors that exploit the Seebek effect generated when two different electrically conducting materials are joined. Several groups are investigating the use of thermopile cameras for THz 2D imaging.

The paper referenced as Ref. [54] reports the development of 8 × 8 pixel arrays sensitive on a large wavelength range of 50–500 μm. Different absorber (TiN, highly doped polySi) and thermopile materials (p-polySi/n-polySi, p-polySi/Al) have been implemented in pixels of 1 × 1 cm^2 size. The best combination has led to values of $NEP \approx 5 \times 10^{-9}$ W/sqrt(Hz) with a bandwidth of ~50 Hz. $NEP \approx 35$ nW at 2.52 THz.

Honeywell, which has a long history in thermal detectors, also publishes development of thermopile-based cameras optimized for THz sensing [55]. A 1 × 8 linear array of pixels has been prototyped, where the incoming radiation is collected by a large planar spiral antenna (500 × 500 μm^2) and converted into an electrical signal in a central, much smaller microbridge supporting the thermopile metal film sandwich. The chip has been mounted in a custom vacuum

[3]One has to note that unlike the monolithic detectors described in Sections 14.3.3 and 14.3.4, this pyroelectric technology corresponds to a hybrid FPA. This breach on the categorization was motivated by a desire to gather all the thermal sensors in a unique chapter.

packaging with a silicon window. A single pixel output *NEP* ≈ 16pW/sqrt(Hz) has been measured with a 900 K blackbody source. Integration with a backplane is predicted to improve significantly the NEP. To the best of the writer's knowledge, no array has been demonstrated in video camera operation.

14.3.4.3 Standard IR bolometer cameras

In bolometer cameras, the temperature variation is measured by a resistance change.

Figure 14.2 shows a schematic drawing of a microbolometer pixel commonly implemented in uncooled thermal IR commercial cameras. Each pixel consists of a microbridge structure where a membrane is suspended above the substrate—the heat sink—by means of thermal insulation arms and metal studs. The membrane supports at the same time the absorber and the thermometer made of a thin film—amorphous Si or vanadium oxide (VO_x) in general. The resistance of the thermometer layer is read out by a full custom CMOS ROIC located below the microbridge. A reflective metallic film is deposited on the upper surface of the ROIC to build a quarter-wavelength cavity between this reflector and the membrane in order to enhance the optical collection efficiency. Such an architecture corresponds to a monolithic FPA.

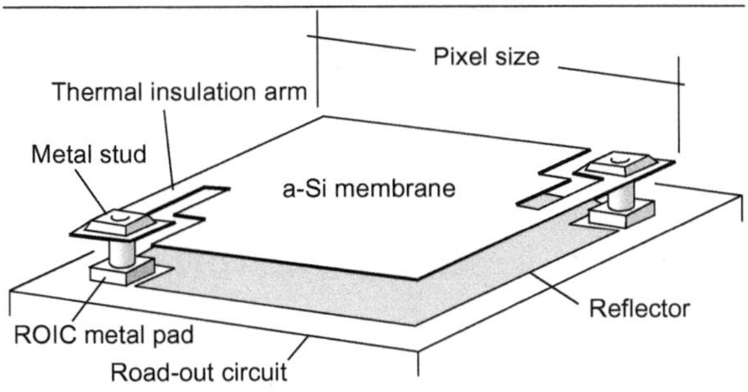

Figure 14.2 Scheme of a standard IR microbolometer pixel.

Actually first real-time THz 2D imaging has been demonstrated in 2005 by the MIT group [56] using a commercial uncooled bolometer IR camera in association with quantum cascade lasers (QCLs). Since then, several other institutes and companies have tested this imaging setup configuration, including the author's group [57]. In a common way, these tests showed that those cameras can be promisingly tweaked for optimized sensing of THz radiations.

14.3.4.4 Modified resistive layer bolometer

The company NEC (Japan), one of the industrial players in the thermal IR market, has developed a bolometer camera customized in the THz range with moderate modification of pixels [58]. Actually the equivalent impedance has been optimized with the deposition of a resistive metallic layer at the top of the microbridge structure described in the previous section. As a result, the sheet resistance of the absorber layer can be adjusted, whereas the pixel dimensions are unchanged.

Such a modified 23.5 μm pitch 320 × 240 bolometer has been tested to acquire real-time images of THz beam profiles delivered by QCLs at 3.1 THz with NEP = 41 ± 4 pW at 3.1 THz. Since then NEC has commercialized the first real-time handy THz bolometer camera. The NEP is specified to be smaller than 100 pW at 4 THz, the complete imaging chain being considered [59].

This impedance-tuning approach has been followed by the R&D institute INO (Canada). An NEP of 70 pW has been characterized at 3 THz with a 160 × 120 bolometer array with a nominal pitch of 52 μm [60].

However, the lack of a resonant cavity appropriate to THz wavelengths necessarily hampers the optical coupling efficiency. In addition, when THz frequencies smaller than 3 THz are being targeted, the mismatch between the absorber size and the wavelength will quickly degrade the sensitivity.

14.3.4.5 Antenna and resonant cavity–coupled camera

CEA-LETI initiated in 2005 [61], the development of a dedicated real-time camera: its proprietary thermal IR bolometer pixel architecture has been substantially modified for optimized THz sensing (Fig. 14.3). This architecture relies on the separation between the two main functions of the bolometer pixel (i.e., the absorber and the thermometer), while achieving a fully silicon microelectronic monolithic FPA.

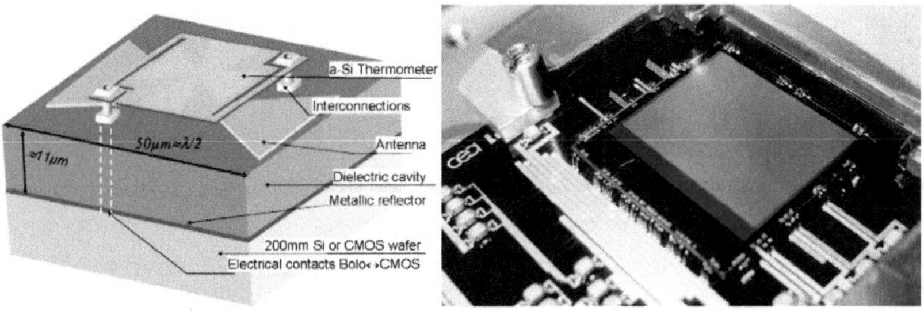

Figure 14.3 (Right) Scheme of an antenna and resonant-cavity THz microbolometer. (Left) A 320 × 240 chip (Leti documents).

Optical absorption is ensured by crossed bowtie antennas combined to a quarter-wavelength cavity matched to THz wavelengths. The thermometer function is provided by a central microbridge derived from standard IR technology.

The crossed polarized antennas have been implemented in a two-storied structure to make the bolometer sensitive to both transverse electric (TE) and transverse magnetic (TM) polarizations. One antenna is directly located on the microbridge, while the other impinging polarization is coupled by an antenna below. The surface currents induced in this antenna are coupled via a capacitive mechanism to metallic planes deposited on the suspended membrane. This unique two-storied antenna architecture permits the tuning of the optical coupling of both polarizations independently and with no mutual interferences. The antenna loads are integrated on the bolometer microbridge.

The absorber function can be designed independently from the thermometer to match the illumination characteristics, and consequently the bolometer microbridge can be kept small, so the response time is preserved. This is particularly essential when detection is desired in the low-frequency THz range.

The antenna-coupled microbridge is built above a metallic reflector that is buried in a thick oxide layer in order to define a quarter-wavelength backplane below the antenna floor that is optimized to THz detection. The pixels that consist of this cavity and the complete stack above are collectively processed above a CMOS ROIC, ending up with a monolithic 2D sensor fully compatible with standard microelectronics silicon technology. The Joule power dissipated in antenna loads induces bolometer resistance changes that are read out at the pixel level by a front-end CMOS circuit. The row-wise reading scheme of the ROIC allows the integration of the bolometer current during 125 μs per pixel in the 5 pF capacitance of its associated capacitive transimpedance amplifier (CTIA).

One major technological key process is the etching and metallization of vias through this thick dielectric cavity. These through-oxide vias (TOVs) are essential to interconnect electrically the bolometer resistance with the CMOS upper metallic contacts. Thanks to Si technology, 11 μm high and 3 μm diameter TOVs are manufacturable with yields as demonstrated by the prototyped FPAs.

Two formats of THz FPAs have been designed, processed, tested, and used for imaging demonstrations. First a large 320 × 240 pixel chip has been developed and integrated in a compact video camera housing [62]. A second chip integrates THz bolometers with visible and IR pixels to act as a trispectral monolithic detector [63].

As common features, the THz bolometric pixels of these arrays present thermal time constants between 20 ms and 40 ms that is compliant with a 25 Hz video frame rate. Real-time active imaging capabilities of the prototyped 2D FPAs have been tested in several setups, both in reflection and in transmission configurations [64].

The 320 × 240 array is composed of 50 μm pitch pixels designed for an optimized sensitivity in the 2–4 THz range [65]. Extensive 3D finite-element method (FEM) simulations [66] were carried out with the commercial HFSS® and Comsol® codes to design two kinds of pixels. They are optimized at 1.7 THz and 2.4 THz for one polarization, whilst the maximum absorption is set at 1.35 THz for the crossed direction.

These arrays have been integrated in a camera, including a first front-end electronics card delivering low-noise analog voltages and digitizing the chip output voltage. Then a field-programmable gate array (FPGA) board drives the sensor and outputs synchronization signals as well as the video stream through an Ethernet port. The camera NEP operating in video mode at 25 Hz has been characterized with a THz QCL as the source. *NEP* = 32 pW has been measured at 2.5 THz, placing this camera at a state of the art above 1 THz uncooled cameras.

Imaging at a frequency below 1 THz can be addressed, while keeping the thickness of the resonant cavity and microbridge dimensions by integration of metallic structures in the vertical technological stack. Preliminary simulations of a 70 μm pitch antenna-coupled bolometer show better than 80% optical absorption, between 0.8 THz and 0.9 THz. Such modeling allows us to expect an NEP in the range of tens of picowatts with such an uncooled bolometer array camera [67].

14.4 Conclusion

Very highly sensitive cameras are being developed, based either on cooled bolometers or solid-state semiconductor devices assembled in radar architecture systems. The high cost of production and maintenance of these cameras will restrain their deployment to high-end applications like astronomy and astrophysics and probably to some security and defense systems.

Actually, like for any camera market, there is a clear trend toward monolithic integration of uncooled sensors for THz real-time imaging:

- Commercial and consumer applications are stymied by the cost, size, and weight of the imagers.
- T-ray imagers are close to meeting all those requirements for short-range scanners, as long as the additional feature of spectroscopy is excluded. That feature may take a few more years of work, requiring smaller, brighter T-ray sources and more sensitive detectors.
- A codesign procedure among the detector, antenna, and optics is essential for wideband detector operation.

Like illustrated by this chapter, a large panel of technologies is being investigated, with newcomers and new commercial cameras emerging every year.

The main uncooled 2D FPAs can be distinguished with respect to the level of maturity and the demonstrated NEP. The CMOS FET 32 × 32 pixel camera exhibits the 12 nW at 0.856 THz and 25 fps. In the 2–4 THz range, an NEP of the order of 30 pW is detected by a

320 × 240 pixel array bolometer camera specifically tailored for THz detection.

Both technologies are fully silicon CMOS compatible and benefit from the semiconductor industry's low cost and yield. Supposing a customized design below 1 THz of the optical absorber of this bolometer FPA, NEPs in the same range could be achieved. On the other hand FET FPAs can provide incoherent detection but do not require vacuum packaging.

As a general rule, one has to consider that the camera specifications are not self-consistent for real-life applications. The most important parameter relies on the compatibility between the market needs and the system specifications combining the source and the camera for a real application. The choice of the camera by the user will be mainly driven by this analysis that includes the price of production and maintenance as well as its ease of use.

References

1. P. H. Siegel, Terahertz technology, *IEEE Trans. Microwave Theory Tech.* **50**(3), 910–928 (2002).
2. W. Withayachumnankul, G. M. Png, Y. Xiaoxia, S. Atakaramians, I. Jones, L. Hungyen, S. Y. U. Benjamin, J. Balakrishnan, B. W.-H. Ng, B. Ferguson, S. P. Mickan, B. M. Fischer, and D. Abbott, T-ray sensing and imaging, *Proc. IEEE* **95**(8), 1528–1558 (2007).
3. M. Tonouchi, Cutting-edge THz technology, *Nat. Photon.* **1**, 97–105 (2007).
4. C. Jansen, S. Wietzke, O. Peters, M. Scheller, N. Vieweg, M. Salhi, N. Krumbholz, C. Jördens, T. Hochrein, and M. Koch, Terahertz imaging: applications and perspectives, *Appl. Opt.* **49**(19), E48–E57 (2010).
5. B. B. Hu and M. C. Nuss, Imaging with terahertz waves, *Opt. Lett.* **20**, 1716–1718 (1995).
6. *L-3 Communications, ProVision® Advanced Imaging Technology*, http://www.sds.l-3com.com/advancedimaging/provision.htm (Accessed August 5, 2013).
7. Smiths Detection, *Smiths Detection Products and Solutions*, http://www.smithsdetection.com/eqo.php (Accessed August 5, 2013).
8. J. Zmuidzinas and P. L. Richards, Superconducting detectors and mixers for millimeter and submillimeter astrophysics, *Proc. IEEE* **92**, 1597–1616 (2004), doi:10.1109/JPROC.2004.833670.

9. K. D. Irwin and G. C. Hilton, Transition-edge sensors, *Appl. Phys.* **99**, 63–150 (2005).

10. A. T. Lee, P. L. Richards, S. W. Nam, B. Cabrera, and K. D. Irwin, A superconducting bolometer with strong electrothermal feedback, *Appl. Phys. Lett.* **69**, 1801–1803 (1996).

11. J. Mehl, P. A. R. Ade, K. Basu, D. Becker, A. Bender, F. Bertoldi, H. M. Cho, M. Dobbs, N. W. Halverson, W. L. Holzapfel, R. Gusten, J. Kennedy, R. Kneissl, E. Kreysa, T. M. Lanting, A. T. Lee, M. Lueker, K. M. Menten, D. Muders, M. Nord, F. Pacaud, T. Plagge, P. L. Richards, P. Schilke, D. Schwan, H. Spieler, A. Weiss, and M. White, TES bolometer array for the APEX-SZ camera, *J. Low Temp. Phys.* **151**, 697–702 (2007).

12. D. Bintley, P. Friberg, M. J. MacIntosh, W. S. Holland, D. Berry, J. T. Dempsey, H. S. Thomas, T. Jenness, D. Kelly, G. Xiaofeng, P. A. R. Ade, W. Grainger, L. Moncelsi, K. D. Irwin, G. C. Hilton, M. Niemack, C. D. Reintsema, M. I. Hollister, A. Woodcraft, M. Amiri, B. Burger, E. Chapin, A. Gibb, M. Halpern, M. Hasselfield, and D. Scott, Characterising and calibrating the SCUBA-2 superconducting bolometer arrays for science observing, *35th Int. Conf. IRMMW-THz*, 1–3 (2010).

13. N. Billot, P. Agnèse, O. Boulade, C. Cigna, E. Doumayrou, B. Horeau, J. Lepennec, J. Martignac, J.-L. Pornin, V. Reveret, L. Rodriguez, M. Sauvage, F. Simoens, and L. Vigroux, Recent achievements on the development of the HERSCHEL/PACS bolometer arrays, *Nucl. Instrum. Methods Phys. Res. A* **567**(1), 137–139 (2006).

14. F. Simoens, P. Agnese, A. Béguin, J. Carcey, J.-C. Cigna, J.-L. Pornin, P. Rey, A. Vandeneynde, L. Rodriguez, O. Boulade, J. Lepennec, J. Martignac, E. Doumayrou, V. Reveret, and L. Vigroux, Submillimeter Bolometers arrays for the PACS/Herschel spectro-photometer, *Proc. SPIE* **5498**, 177–186 (2004).

15. V. Revéret, P. André, M. Talvard, L. R. Rodriguez, O. Boulade, E. Doumayrou, P. Gallais, B. Horeau, J. Le Pennec, M. Lortholary, J. Martignac, P. Agnèse, HERSCHEL—PACS bolometer arrays for submillimeter ground-based telescopes, *J. Low Temp. Phys.* **151**(1–2), 32–39 (2008).

16. P. K. Day, H. G. LeDuc, B. A. Mazin, A. Vayonakis, and J. Zmuidzinas, A broadband superconducting detector suitable for use in large arrays, *Nature* **425**, 817–821 (2003), doi:10.1038/nature02037.

17. J. Schlaerth, A. Vayonakis, P. Day, J. Glenn, J. Gao, S. Golwala, S. Kumar, H. LeDuc, B. Mazin, J. Vaillancourt, and J. Zmuidzinas, A millimeter and submillimeter kinetic inductance detector camera, *J. Low Temp. Phys.* **151**, 684–689 (2008), doi 10.1007/s10909-008-9728-3.

18. M. I. Hollister, N. G. Czakon, P. K. Day, R. Duan, J. Glenn, S. Golwala, H. Leduc, P. R. Maloney, B. Mazin, H. T. Nguyen, O. Noroozian, J. Sayers, J. Schlaerth, S. Siegel, J. E. Vaillancourt, A. Vayonakis, P. Wilson, and J. Zmuidzinas, An update on MUSIC: a kinetic inductance detector camera for sub/millimeter astrophysics at the Caltech submillimeter observatory, *21st Int. Symp. Space THz Technol.*, 263–269 (2010).

19. A. Monfardini, L. J. Swenson, A. Bideaud, F. X. Désert, S. J. C. Yates, A. Benoit, A. M. Baryshev, J. J. A. Baselmans, S. M. Doyle, B. Klein, M. Roesch, C. E. Tucker, P. A. R. Ade, M. Calvo, P. Camus, C. Giordano, R. Guesten, C. Hoffmann, S. Leclercq, P. D. Mauskopf, and K. F. Schuster, NIKA: a millimeter-wave kinetic inductance camera, *Astron. Astrophys.* **521**, A29 (2010), doi:10.1051/0004-6361/201014727.

20. A. Monfardini, A. Benoit, A. Bideaud, L. Swenson, A. Cruciani, P. Camus, C. Hoffmann, F. X. Désert, S. Doyle, P. Ade, P. Mauskopf, C. Tucker, M. Roesch, S. Leclercq, K. F. Schuster, A. Endo, A. Baryshev, J. J. A. Baselmans, L. Ferrari, S. J. C Yates, O. Bourrion, J. Macias-Perez, C. Vescovi, M. Calvo, and C. Giordano, A dual-band millimeter-wave kinetic inductance camera for the IRAM 30 m telescope, *Astrophys. J., Suppl. Ser. 194*, **24** (2011), doi:10.1088/0067-0049/194/2/24.

21. A. Karpov, D. Miller, F. Rice, J. A. Stern, B. Bumble, and H. G. LeDuc, Low noise 1 THz–1.4 THz mixers using Nb/Al-AlN/NbTiN SIS junctions, *IEEE Trans. Appl. Supercond.* **17**(2), 343–346 (2007).

22. S. Ariyoshi, C. Otani, A. Dobroiu, H. Sato, K. Kawase, H. M. Shimizu, T. Taino, and H. Matsuo, Terahertz imaging with a direct detector based on superconducting tunnel junctions, *Appl. Phys. Lett.* **88**, 203503 (2006).

23. H. Matsuo, H. Nagata, J. Kobayashi, S. Ariyoshi, M. Fujiwara, M. Yuko, M. Yu, N. Misato, and C. Otani, Focal plane array technologies for SISCAM, *Conf. IRMMW-THz*, 113–114 (2007).

24. M. Aurino, A. J. Kreisler, I. Türer, A. Martinez, A. Gensbittel, and A. F. Dégardin, YBCO hot-electron bolometers dedicated to THz detection and imaging: embedding issues, *J. Phys.: Conf. Ser.* **234**(4), 042002 (2010), doi:10.1088/1742-6596/234/4/042002.

25. A. Luukanen, M. Gronholm, P. Lappalainen, M. Leivo, A. Rautiainen, A. Tamminen, and J. Ala-Laurinaho, Passive real-time submillimetre-wave imaging system utilizing antenna-coupled microbolometers for stand-off security screening applications, *Int. Workshop Antenna Technol. (iWAT)* (2010).

26. E. Grossman, Passive terahertz camera for standoff security screening, *Appl. Opt.* **49**, E106–E120 (2010).

27. A. Luukanen, J. Ala-Laurinaho, J. Hakli, D. Gomes-Martins, T. Kiuru, P. Koivisto, M. Leivo, A. Rautiainen, J. Saily, A. Tamminen, H. Toivanen, R. Tuovinen, and Raisanen, Towards video rate imaging at submillimetre-waves: Finnish developments of passive multi-band imaging and holographic submm-wave beam steering at VTT, *Proc. APMC*, 782–784 (2013).

28. T. May, G. Zieger, S. Anders, V. Zakosarenko, H.-G. Meyer, M. Schubert, M. Starkloff, M. Rößler, G. Thorwirth, and U. Krause, Safe VISITOR: visible, infrared, and terahertz object recognition for security screening application, *Proc. SPIE* **7309**, 73090E (2009), doi:10.1117/12.818524.

29. T. May, V. Zakosarenko, R. Boucher, E. Kreysa, and H.-G. Meyer, Superconducting bolometer array with SQUID readout for submillimetre wavelength detection, *Supercond. Sci. Technol.* **16**, 1430–1433 (2003). doi:10.1088/0953-2048/16/12/024.

30. Q. Wu, T. D. Hewitt, and X.-C. Zhang, Two-dimensional electro-optic imaging of THz beams, *Appl. Phys. Lett.* **69**, 1026–1028 (1996).

31. A. Nahata, J. T. Yardley, and T. F. Heinz, Two-dimensional imaging of continuous-wave terahertz radiation using electro-optic detection, *Appl. Phys. Lett.* **81**, 963 (2010), doi:10.1063/1.1497190.

32. F. Z. Meng, M. D. Thomson, D. Molter, T. Löffler, J. Jonuscheit, R. Beigang, J. Bartschke, T. Bauer, M. Nittmann, and H. G. Roskos, Coherent electro-optical detection of terahertz radiation from an optical parametric oscillator, *Opt. Express* **18**(11), 11316–11326 (2010).

33. M. Erdtmann, L. Zhang, S. Radhakrishnan, S. Wu, T. M. Goyette, and A. J. Gatesman, Uncooled photomechanical terahertz imagers, *Proc. SPIE* **8363**, 83630C (2012), doi:10.1117/12.919058.

34. C. Pradere, J. P. Caumes, D. Balageas, S. Salort, E. Abrahamd, B. Chassagne, and J. C. Batsale, Photothermal converters for quantitative 2D and 3D real-time terahertz imaging, *Quant. IR Thermogr. J.* **7**(2), 217–235 (2010), doi : 10.3166/qirt.7.

35. K. B. Cooper, R. J. Dengler, N. Llombart, B. Thomas, G. Chattopadhyayand, and P. H. Siegel, Imaging radar for standoff personnel screening, *IEEE Trans. THz Sci. Technol.* **1**(1), 169–182 (2011).

36. C. A. Weg, W. von Spiegel, R. Henneberger, R. Zimmermann, T. Loeffler, and H. G. Roskos, Fast active THz cameras with ranging capabilities, *J. IR mm THz Waves* **30**(12), 1281–1296 (2009).

37. D. M. Sheen, D. L. McMakin, T. E. Hall, and R. H. Severtsen, Active millimeter-wave standoff and portal imaging techniques for personnel

screening, *Proc. IEEE Conf. Technol. Homeland Security*, 440–447 (2009).

38. R. Arusi, Y. Pinhasi, B. Kapilevitch, D. Hardon, B. Litvak, and M. Anisimov, Linear FM radar operating in the Terahertz regime for concealed objects detection, *IEEE Int. Conf. COMCAS* (2009).
39. K. Su, Z. Liu, R. B. Barat, D. E. Gary, Z.-H. Michalopoulou, and J. F. Federici, Two-dimensional interferometric and synthetic aperture imaging with a hybrid terahertz/ millimeter wave system, *Appl. Opt.* **49**(19), E13–E19 (2010).
40. F. Friederich, W. von Spiegel, M. Bauer, M. Fanzhen, M. D. Thomson, S. Boppel, A. Lisauskas, B. Hils, V. Krozer, A. Keil, T. Loffler, R. Henneberger, A. K. Huhn, G. Spickermann, P. H. Bolivar, and H. G. Roskos, THz Active imaging systems with real-time capabilities, *IEEE Trans. THz Sci. Technol.* 1(1), 183–200 (2010).
41. L. Minkevičius, V. Tamošiūnas, I. Kašalynas, D. Seliuta, G. Valušis, A. Lisauskas, S. Boppel, H. G. Roskos, and K. Köhler, Terahertz heterodyne imaging with InGaAs-based bow-tie diodes, *Appl. Phys. Lett.* **99**, 131101 (2011), doi : http://dx.doi.org/10.1063/1.3641907.
42. I. Kašalynas, R. Venckevicius, and G. Valusis, Continuous wave spectroscopic terahertz imaging with InGaAs bow-tie diodes at room temperature, *IEEE Sens. J.* **13**(1), 50–54 (2013).
43. Z. Zhang, R. Rajavel, P. Deelman, and P. Fay, Sub-micron area heterojunction backward diode millimeter-wave detectors with 0.18 pW/Hz^1/2 noise equivalent power, *IEEE Microwave Wireless Compon. Lett.* **21**(5), 267–269 (2011).
44. G. C. Trichopoulos, H. L. Mosbaker, D. Burdette, and K. Sertel, A broadband focal plane array camera for real-time THz imaging applications, *IEEE Trans. Antennas Propag.* **61** (4), 1733–1740 (2013).
45. A. Tessmann, A. Leuther, H. Massler, V. Hurm, M. Kuri, M. Zink, and M. R. Losch, High-gain submillimeter-wave mHEMT amplifier MMICs, *IEEE Intl. Microwave Symph.*, 53–56 (2010).
46. M. Dyakonov and M. Shur, Plasma wave electronics: novel terahertz devices using two dimensional electron fluid, *IEEE Trans. Electron. Devices* **43**(10), 1640–1645 (1996).
47. R. Al Hadi, H. Sherry, J. Grzyb, Z. Yan, W. Forster, H. M. Keller, A. Cathelin, A. Kaiser, and U. R. Pfeiffer, A 1 k-pixel video camera for 0.7–1.1 terahertz imaging applications in 65-nm CMOS, *IEEE J. Solid-State Circuits* **47**(12), 2999–3012 (2012).

48. W. Knap, F. Teppe, Y. Meziani, N. Dyakonova, J. Lusakowski, F. Boeuf, T. Skotnicki, D. Maude, S. Rumyantsev, and M. S. Shur, Plasma wave detection of sub-terahertz and terahertz radiation by silicon field-effect transistors, *Appl. Phys. Lett.* **85** (4), 675–677 (2004).

49. A. Lisauskas, U. Pfeiffer, E. Öjefors, P. Bolìvar, D. Glaab, and H. G. Roskos, Rational design of high-responsivity detectors of terahertz radiation based on distributed self-mixing in silicon field-effect transistors, *J. Appl. Phys.* **105**(11), 114511 (2009).

50. A. Lisauskas, S. Boppel, M. Saphar, V. Krozer, L. Minkevičius, R. Venckevičius, D. Seliuta, I. Kašalynas, V. Tamošiūnas, G. Valušis, and H. G. Roskos, *Proc. SPIE* **8496**, 84960J (2012), doi:10.1117/12.930086.

51. S. Boppel, A. Lisauskas, A. Max, V. Krozer, and H. G. Roskos, CMOS detector arrays in a virtual 10-kilopixel camera for coherent terahertz real-time imaging, *Opt. Lett.* **37**(4), 536–538 (2012), doi: 10.1364/OL.37.000536.

52. H. Sherry, R. Hadi, J. Grzyb, E. Öjefors, A. Cathelin, A. Kaiser, and U. Pfeiffer, Lens-Integrated THz Imaging Arrays in 65nm CMOS Technologies, *IEEE Radio Frequency Integr. Circuits Symp.* **6**, 365–368 (2011).

53. R. Han, Y. Zhang, Y. Kim, D. Y. Kim, H. Shichijo, E. Afshari, and O. Kenneth, 280 GHz and 860 GHz image sensors using Schottky barrier diodes in 0.13 m digital CMOS, *IEEE CICC Conf.*, 254–256 (2012).

54. I. Kašalynas, A. J. L. Adam, T. O. Klaassen, N. J. Hovenier, G. Pandraud, V. P. Iordanov, and P. M. Sarro, Some properties of a room temperature THz detection array, *Proc. SPIE* **6596**, 65960J (2007).

55. J. A. Cox, R. Higashi, F. Nusseibeh, K. Newstrom-Peitso, and C. Zins, Uncooled MEMS-based detector arrays for THz imaging applications, *Proc. SPIE* **7311**, 73110R-1 (2009).

56. W. Lee and Q. Hu, Real-time, continuous-wave terahertz imaging by use of a microbolometer focal-plane array, *Opt. Lett.* **30**(19), 2563–2565 (2005).

57. F. Simoens, T. Durand, J. Meilhan, P. Gellie, W. Maineult, C. Sirtori, S. Barbieri, H. Beere, and D. Ritchie, Terahertz imaging with a quantum cascade laser and amorphous-silicon microbolometer array, *Proc. SPIE* **7485**, 74850M (2009).

58. N. Oda, Uncooled bolometer-type terahertz focal plane array and camera for real-time imaging, *C. R. Phys.* **11**, 496–509 (2010), doi:10.1016/j.crhy.2010.05.001.

59. N. Oda, T. Ishi, T. Morimoto, T. Sudou, H. Tabata, S. Kawabe, K. Fukuda, A. W. Lee, and Q. Hu, Development of terahertz focal plane arrays and handy camera, *Proc. SPIE* **8012**, 80121B (2011).

60. M. Bolduc, M. Terroux, B. Tremblay, L. Marchese, E. Savard, M. Doucet, H. Oulachgar, C. Alain, H. Jerominek, and A. Bergeron, Noise-equivalent power characterization of an uncooled microbolometer-based THz imaging camera, *Proc. SPIE 2011* **8023**, 80230C (2011).

61. E. Peytavit, P. Agnèse, J.-L. Ouvrier-Buffet, A. Beguin, and F. Simoens, Room temperature terahertz microbolometers, *Proc. IRMMW-THz* **1**, 257–258 (2005).

62. F. Simoens, J. Meilhan, S. Pocas, J.-L. Ouvrier-Buffet, T. Maillou, P. Gellie, and S. Barbieri, THz uncooled microbolometer array development for active imaging and spectroscopy applications, *Proc. IR mm THz Waves*, 1–2 (2010).

63. M. Perenzoni, N. Massari, D. Stoppa, S. Pocas, B. Delplanque, J. Meilhan, F. Simoens, and W. Rabaud, A 160 × 160-pixel image sensor for multispectral visible, infrared and terahertz detection, *ESSCIRC*, 93–96 (2012).

64. J. Meilhan, B. Dupont, V. Goudon, G. Lasfargues, J. Lalanne Dera, D. T. Nguyen, J. L. Ouvrier-Buffet, S. Pocas, T. Maillou, O. Cathabard, S. Barbieri, and F. Simoens, Active THz imaging and explosive detection with uncooled antenna-coupled microbolometer arrays, *Proc. SPIE* **8023** (1), 80230E-1–80230E-13 (2011).

65. F. Simoens and J. Meilhan, Terahertz real-time imaging uncooled array based on antenna- and cavity-coupled bolometers, *Phil. Trans. R. Soc. A* **372**, (2014).

66. D. T. Nguyen, F. Simoens, J.-L. Ouvrier-Buffet, J. Meilhan, and J.-L. Coutaz, Simulations and measurements of the electromagnetic response of broadband THz uncooled antenna-coupled microbolometer array, *IEEE Trans. THz Sci. Technol.* **2**(3), 299–305 (2012).

67. J. Meilhan, F. Simoens, J. Lalanne-Dera, S. Gidon, G. Lasfargues, S. Pocas, D. T. Nguyen, and J.-L. Ouvrier-Buffet, Terahertz frequency agility of uncooled antenna-coupled micro-bolometer arrays, *Proc. IRMMW-THz*, 1–2 (2012).

Chapter 15

Terahertz Chemical Spectroscopy

Yuko Ueno[a] **and Katsuhiro Ajito**[b]
[a]*NTT Basic Research Laboratories, NTT Corporation, 3-1, Morinosato-Wakamiya, Atsugi, Kanagawa 243-0198, Japan*
[b]*NTT Device Technology Laboratories, NTT Corporation, 3-1, Morinosato-Wakamiya, Atsugi, Kanagawa 243-0198, Japan*
ueno.yuko@lab.ntt.co.jp, ajito.katsuhiro@lab.ntt.co.jp

The unique and characteristic properties of Terahertz (THz) waves offered the opportunity to develop THz chemical spectroscopy, which has been attracting great interest not only in basic research but also for practical applications in a wide variety of fields. The success of THz chemical spectroscopy is characterized by THz time domain spectroscopy (THz-TDS) and THz spectroscopic imaging (TSI) based on TDS. It has proved to be an effective method for the safe and nondestructive analysis of chemical compounds that contain weak intermolecular interactions such as hydrogen bonds. This chapter introduces recent studies in THz chemical spectroscopy to show its potential for detecting and identifying chemical compounds in unknown mixture samples such as a real drug tablet. Some brief remarks are provided on future developments, the main issues, and the prospects for THz chemical spectroscopy.

Handbook of Terahertz Technologies: Devices and Applications
Edited by Ho-Jin Song and Tadao Nagatsuma

ISBN 978-981-4613-08-8 (Hardcover), 978-981-4613-09-5 (eBook)
www.panstanford.com

15.1 Introduction

Terahertz (THz) technology (1 THz = 10^{12} Hz) has been making rapid progress following recent advances in ultrashort pulse lasers and semiconductor coherent sources emitting in the THz frequency range [1]. The development of brighter sources and detectors that are more sensitive to THz waves is now helping us to employ THz waves in new analytical methods such as THz spectroscopy [2]. One of the attractive properties of THz waves is their ability to pass through a wide variety of materials [3]. This property allows nondestructive inspection by enabling us to see through many types of packaging material, including paper, plastics, ceramics, leather, and wood.

THz waves cover a frequency range that corresponds to the lower end of the far-infrared region, and thus THz spectroscopy is useful for characterizing far-infrared vibrational modes such as rotational, torsional, phonon, intermolecular, and intramolecular modes (Fig. 15.1). Therefore, THz spectroscopy enables us to achieve selective detection of weak intermolecular and intramolecular vibrational modes, leading to the identification or analysis of molecular structures by using unique information obtained from the THz spectra. It is noteworthy that THz spectroscopy is generally regarded as being different from conventional far-infrared spectroscopy, which commonly employs blackbody radiation sources and photon detectors. The fact that THz spectroscopy uses bright sources and sensitive detectors means that it can provide a higher signal-to-noise (S/N) ratio and a wider dynamic range than far-infrared spectroscopy [4].

Of the many techniques related to THz spectroscopy, THz time domain spectroscopy (THz-TDS) [5, 6] is one of the most widely studied. It is noteworthy that recent advances in THz technology have led to successful development of commercially available spectrometers for THz-TDS, which in turn has popularized this method as a new tool for chemical analysis. A compact [7] and portable spectrometer [8] has also been realized, and this offers great potential for real-time and on-site THz spectroscopic analysis. Another widely used technique is THz spectroscopic imaging (TSI) based on TDS techniques, which measures the 2D properties of the THz spectra of a sample caused by intermolecular and intramolecular interactions. TSI is fundamentally different from conventional X-ray imaging. Specifically, TSI has a great advantage in relation to detecting fragile biological samples, because THz waves cause no serious damage to materials [9–12]. This approach is useful for analyzing the distribution and concentration of chemicals inside the targeted materials [13, 14]. Many applications using the unique merits of THz spectroscopy are expected to be developed, and these include security inspection, industrial process control, medical diagnosis, food inspection, and biological and biomedical imaging.

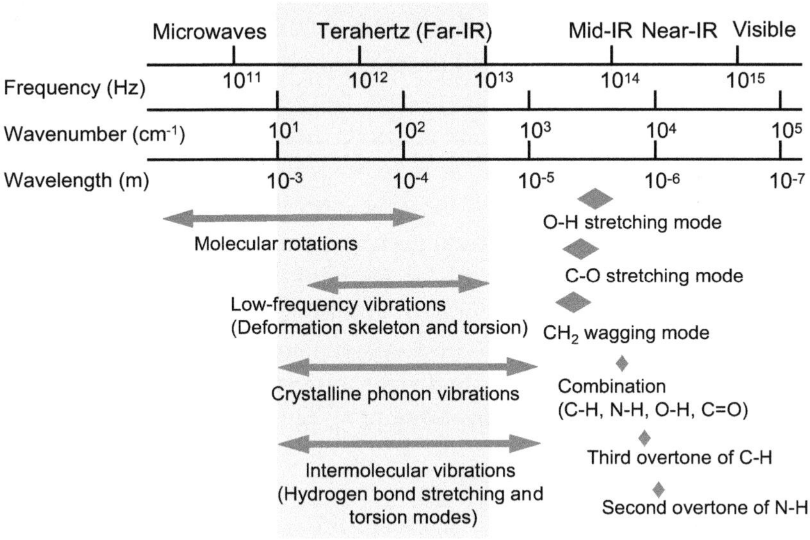

Figure 15.1 Characteristic infrared adsorption frequencies of chemical compounds.

15.2 THz-TDS

15.2.1 Overview

THz-TDS is a useful analytical technique since it enables the detection and identification of molecules or materials that have strong absorption in the THz frequency region. THz-TDS was proposed as a coherent emission and detection system that emits single-cycle THz pulses and detects them at a repetition rate of around 100 MHz. The spectroscopic method that employs the above emission and detection system with a femtosecond laser is called THz-TDS. Figure 15.2 shows a typical optical setup for transmission THz-TDS. The femtosecond laser pulses are divided into a pump light to actuate the THz emitter and a probe light to drive the detector. The generated subpicosecond THz pulses are transmitted through the sample, and the change in the electric field intensity of the THz pulses over time is directly measured instead of detecting the power of the light. Then the directly measured electric field intensity of a repetitive THz pulse in the time domain is converted to both amplitude and phase spectra in the frequency domain by using Fourier transformation. Since the electric field is monitored without being disturbed by the atmospheric temperature [15, 16], the coherent detection obtained when a photoconductive (PC) antenna is switched by femtosecond pulses gives THz-TDS a higher S/N ratio than conventional far-infrared spectroscopy, which employs blackbody radiation as a light source and photon detectors to measure the light power. Since the THz frequency matches the low vibration frequency and the vibration modes of weak interactions such as hydrogen bonds, THz-TDS constitutes a unique tool for studying various molecules and materials whose molecular and/or crystal structures are stabilized by these weak interactions. The steric structures of many biological molecules are stabilized by hydrogen bonds, and THz-TDS studies have been undertaken on proteins [17], polypeptides [10, 18, 19], and DNA [20].

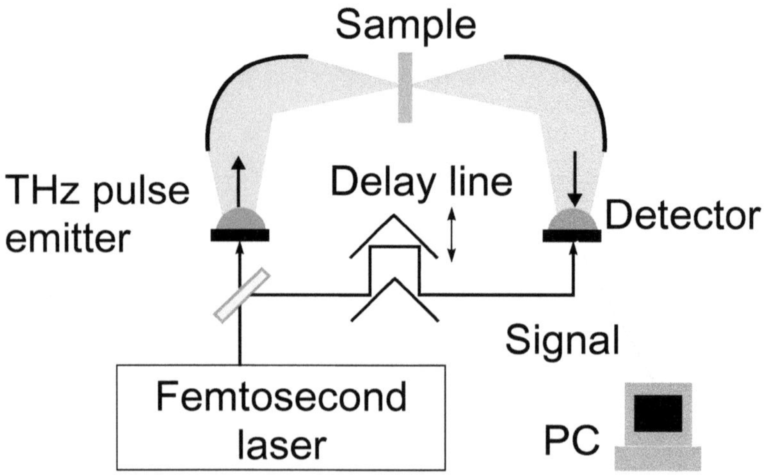

Figure 15.2 Schematic example of the optical setup for transmission THz-TDS.

It is worth mentioning that aqueous samples of these molecules are difficult to observe with transmission THz-TDS, owing to the high absorption of THz waves by polar liquids such as water. Thus, molecular vibrational modes of molecules have been mainly investigated in their solid form by, for example, pressing powder into pellets. Important small biological molecules such as pharmaceutical molecules [21, 22], amino acids [10, 12, 23–26], and sugars [27, 28] are also studied by THz-TDS because these molecules have sharper spectral features in the THz frequency region because they have fewer normal modes and are thus much easier to characterize.

15.2.2 Application: Quantitative Analysis of Amino Acids in Supplements

In this section, we introduce our study related to the quantitative analysis of amino acids in supplements, which reveals the potential for quantitative detection and identification of THz-sensitive chemicals in an unknown sample. We have shown the usefulness of employing THz-TDS for the quantitative analysis of chemicals that have characteristic absorption in the THz region, such as amino acids [25]. By using a similar technique, we successfully demonstrated the quantitative analysis of five amino acids in an actual dietary amino acid supplement [29]. The supplement contained other ingredients such as sugars and vitamins and thus was a suitable sample for testing the selective detection of an amino acid using THz-TDS. The merit of using THz-TDS for the direct quantitative analysis of packaged samples was also shown by an analysis of tablets of a commercially available dietary amino acid supplement wrapped in paper, and the accuracy of the direct quantitative analysis of the packaged samples was compared to that of unwrapped samples. Since one characteristic of THz waves is their ability to pass through commonly used packaging materials such as paper and plastics, the accuracy with which they can provide a nondestructive quantitative analysis of packaged samples is of interest.

Commercial dietary amino acid supplements were purchased from two different makers. The samples of the two different kinds of supplements are denoted by supplement A and supplement B. The concentrations of amino acids and other ingredients were calculated on the basis of nutrition fact sheets and are summarized in Table 15.1. The supplement granules were ground into powder and compressed into a tablet (10 mm in diameter and <1 mm thick). Wrapped samples for use in the nondestructive inspection of a packaged tablet were prepared by placing the above tablets between sheets of weighing paper (~30 μm thick). Prior to the quantitative analysis of the supplements, we obtained the standard THz spectra of the major constituents—L-arginine (L-Arg), L-glutamine (L-Gln), L-isoleucine (L-Ile), L-leucine (L-Leu), L-valine (L-Val), maltitol, and citric acid monohydrate—and calculated the molar absorption coefficient in a certain THz region. We used the molar absorption

Table 15.1 Concentration of amino acids and other ingredients in the supplements

	Concentration (%)	
	Supplement A	**Supplement B**
Amino acids		
L-Arg	17.3	7.5
L-Glu	18.0	7.5
L-Ile	12.3	10.0
L-Leu	15.3	15.0
L-Val	10.3	10.0
Other ingredients		
Maltitol	Unknown	Unknown
Citric acid	Unknown	Unknown
Sugars	Unknown	Unknown
Natural flavors	Unknown	Unknown
Vitamin C	0.63	1.25
Niacin	0.13	0.21
Pantothenic acid	0.08	0.06
Vitamin B6	0.01	0.02
Vitamin B2	0.01	0.02
Vitamin B1	0.01	0.01
Vitamin A	<0.01	<0.01
Vitamin D	<0.01	<0.01
Vitamin E	0.54	0.13

coefficient to analyze the most likely mixing ratio of the amino acids in supplement A in the observed THz spectrum. The solid line in Fig. 15.3a shows the THz absorption spectrum of supplement A in the 0.5–3.0 THz region. Several characteristic peaks were clearly observed. The dotted line shows the calculated spectrum obtained by adding the standard spectra, which were scaled in proportion to the calculated concentrations of five amino acids and maltitol. The spectra of the five amino acid constituents and maltitol are overlaid to show their contributions to the calculated spectra. The experimental spectrum was well reproduced by the calculated spectrum.

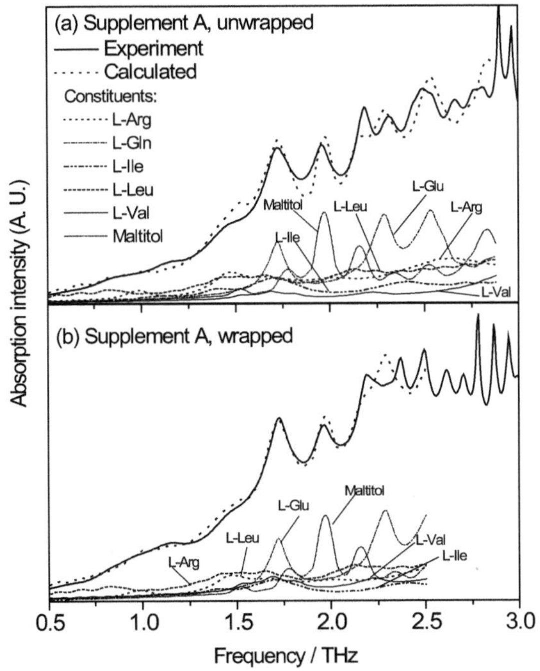

Figure 15.3 Spectra of an unwrapped supplement A tablet (a) and a tablet wrapped in weighing paper (b). Experimental (bold solid line) and calculated (bold dotted line) spectra are compared. The spectra of the constituents are overlaid to show their contributions to the calculated spectra. Reproduced from Ref. [29] by permission of JSAC, © 2011.

Table 15.2 summarizes the concentrations of the five amino acids shown in the nutrition fact sheet and the calculated results for the supplements. The recovery ratios (the calculated concentration over the concentration given in the nutrition fact sheet) revealed that the estimation errors, defined by (recovery ratio) – 1) × 100, were within ±12% for the three constituents (L-Arg, L-Gln, and L-Ile). For L-Leu and L-Val, the estimation errors were 59% and 29%, respectively. The concentration order was correctly predicted except as regards L-Leu. Thus, despite employing a rough analysis that did not take account of the contribution of minor components such as vitamins, we successfully analyzed an actual supplement by using THz-TDS.

Table 15.2 Calculated concentrations and recovery ratios of amino acids in the supplements [29]

	Concentration (%)				
	Nutrition sheet	Calculated			
		Unwrapped	Recovery ratio	Wrapped	Recovery ratio
Supplement A					
L-Arg	17.3	15.3	0.88	16.7	0.97
L-Glu	18.0	19.6	1.09	17.8	0.99
L-Ile	12.3	12.2	0.99	16.4	1.33
L-leu	15.3	24.4	1.59	7.7	0.50
L-Val	10.3	13.3	1.29	1.7	0.17
Supplement B					
L-Arg	7.5	19.5	2.60	19.9	2.65
L-Glu	7.5	10.3	1.37	10.5	1.40
L- Ile	10.0	9.0	0.90	12.8	1.28
L-leu	15.0	6.6	0.44	38.4	2.56
L-Val	10.0	10.3	1.03	11.7	1.17

We found that wrapping the sample in paper had very little effect on the spectral shape because the differential spectra calculated by subtracting the absorption intensity of a wrapped sample from that of an unwrapped sample was satisfactorily flat. Thanks to the ability of THz waves to pass through paper, we can perform a quantitative analysis by using the same standard spectra as those used for the unwrapped sample (Fig. 15.3b). The experimental spectrum was well reproduced by the calculated spectrum. The recovery ratios in Table 15.2 show that the estimation errors were within ±33% for the three constituents (L-Arg, L-Gln, and L-Ile). For L-Leu and L-Val, the underestimation errors were 50% and 83%, respectively. Since L-Ile and L-Val both have peaks at around 1.7 THz, the estimation of L-Ile and the underestimation of L-Val can occur simultaneously by confusing the two components, which have similar THz absorption profiles. Using a wider spectral region for an unwrapped sample or changing the measurement temperature to see the difference in the spectral shapes will help to avoid the confusion and allow us to improve the accuracy. It is noteworthy that THz-TDS can realize almost the same level of accuracy as quantitative analysis for both packaged and unwrapped samples.

15.3 Terahertz Spectroscopic Imaging

15.3.1 Overview

THz imaging is a technique that provides 2D properties of the THz absorption of a sample [30]. TSI is a THz imaging technique that obtains spectroscopic information at each measured position. Thus, TSI is very informative and allows us to detect and identify chemicals and also determine the distribution and concentration of the chemicals in targeted materials [13, 14]. As with THz spectroscopy, THz imaging can be used to analyze samples covered with invisible materials that are transparent to THz waves. This is because THz waves can pass through a wide variety of nonpolar materials such as paper and plastics [3], which are commonly used for packaging. Therefore, THz imaging is often compared with X-ray imaging, which has been a widely used technique for nondestructive inspection. The two techniques are fundamentally different. The main advantage of using THz waves must be that they cause very little damage to materials. This makes THz imaging a particularly promising tool for analyzing biological and medical samples such as cancer tissue [31].

A THz image can be obtained by using a system designed for THz-TDS, depending on picosecond THz pulses [32], and also a system based on continuous-wave (CW) THz sources [33]. While THz pulse imaging has the advantage of providing an image with broad frequency information from 0.1 THz to 5 THz, the advantage of the CW imaging technique lies in its real-time operation, frequency-sensitive measurement, and higher dynamic range, which result from the fact that the power of its spectral density is of the order of higher magnitude [34]. A TSI image can be obtained by mechanically scanning the sample in two dimensions using a THz spectrometer [30]. Currently, a shortcoming of TSI is that it takes long to measure a large image. Thus, a high-speed imaging system for TSI has been investigated. Another problem with overall THz imaging is the poor spatial resolution, which is estimated in terms of the diffraction limit, namely it is submillimeter at best. To overcome this resolution limitation, near-field microscopy has been applied to the THz region to achieve subwavelength resolution [35].

15.3.2 Measurement System

Figure 15.4 shows the TSI system we used for the demonstrations that we describe later [37]. It is composed of a spectrometer of THz-TDS with a vacuum chamber mounted on a 3D translational stage. For multicomponent chemical analysis, a cryostat was added to the vacuum chamber in the TSI system. The bottom of the sample holder inserted into the chamber is a quartz plate, and the chamber can be cooled to a temperature of 77 K with the cryostat using liquid nitrogen and an electric heater. The cryostat is stable within 2 K. The THz time domain spectroscope consists of a 9 fs near-infrared pulse laser (Integral Pro, Femtolasers), two gallium arsenide PC antennas (AISPEC), a mechanical stage delay line, and mirrors. One PC antenna is an emitter, and the other is a detector. A 13 fs near-infrared pulse laser (Fusion, Femtolasers) is also used for TDS measurement. The delay line is used to obtain a time domain waveform, which is converted to a frequency domain spectrum by Fourier transformation. The 3D translational stage with a 0.1 mm step resolution and THz-TDS are controlled by a personal computer to obtain a THz time domain spectrum at each point in a sample tablet. A 12×12 mm^2 image takes about seven hours to acquire.

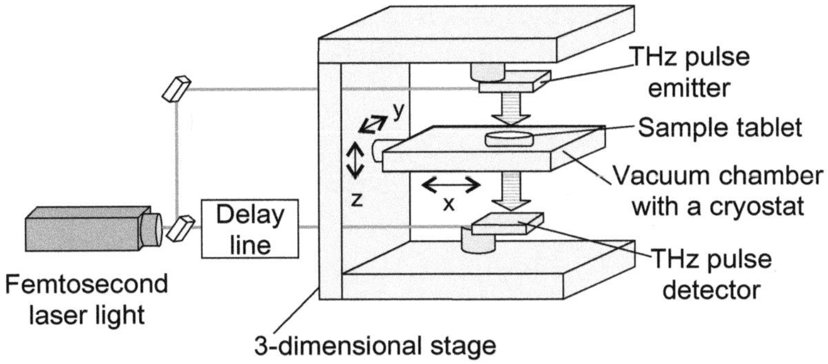

Figure 15.4 Diagram of the THz imaging system.

15.3.3 Application (1): Chemical Mapping of Pharmaceuticals in Medicine

In this section, we introduce our recent study on chemical mapping of a pharmaceutical, famotidine, an over-the-counter drug tablet. Famotidine is a histamine H2-receptor antagonist for the prevention and treatment of stomach and intestinal ulcers. It is a polymorph, that is, its molecular crystals have different crystal structures. A study related to the polymorphic forms of molecular crystals has attracted considerable interest because it is significant as regard the chemical properties of the crystals, such as solubility, melting point, and bioavailability as drugs. For example, the most stable form of famotidine is form A, which is an inactive drug, whereas the metastable form B is an active drug. Since many polymorphic forms of molecular crystals have different types of hydrogen bonds between their molecules, THz spectroscopy is a promising tool for the detection and identification of polymorphic forms of molecular crystals. We use TSI to detect the famotidine polymorphic forms A and B and D-mannitol crystals, which is sometimes used in famotidine pills to control solubility in the body.

Figure 15.5 shows a set of temperature-dependent THz spectra of famotidine forms A and B and D-mannitol, which were measured at 298 K and 220 K, respectively. The concentrations of all the pharmaceutical tablets were 20 wt%. The peaks in the spectra of both famotidine forms were sharpened at low temperature, and their frequencies shifted very little. However, the peaks in the D-mannitol spectrum shifted greatly, indicating that the neighboring peaks of famotidine form B and D-mannitol can be separated at 220 K.

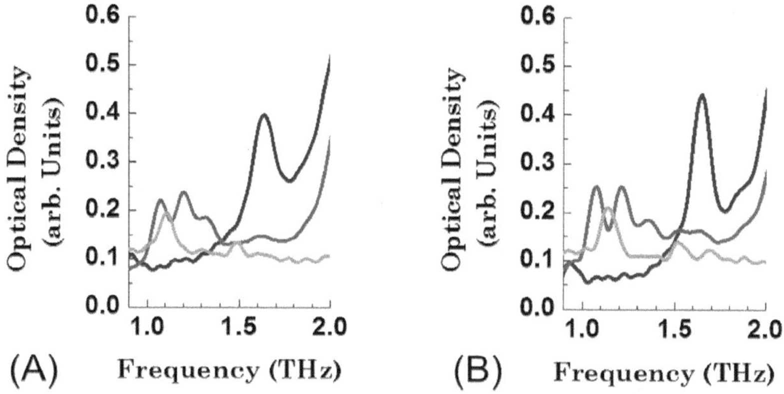

Figure 15.5 A set of temperature-dependent THz spectra of famotidine forms A and B and D-mannitol measured at 298 K (A) and 220 K (B). Reproduced from Ref. [36] by permission of Electrochemical Society, © 2013.

Figure 15.6 shows a set of temperature-dependent THz spectroscopic images of a compressed pellet containing famotidine polymorphic forms A and B and D-mannitol crystals measured at 220 K and 298 K. The peak frequencies of famotidine B and D-mannitol spectral peaks are close to each other at 298 K (room temperature). Therefore, the 298 K image at 1.07 THz, which is one of the peak positions of famotidine form B, shows both famotidine form B and D-mannitol. However, only a famotidine form B distribution is observed at 1.08 THz in the image measured at 220 K. As a result, 2D distribution of the three pharmaceuticals is almost completely determined at 220 K. This result indicates that the multicomponent chemical analysis of pharmaceuticals is possible when a cryostat is added to the vacuum chamber in a TSI system.

Figure 15.7 shows a set of temperature-dependent THz spectroscopic images of an over-the-counter tablet containing 10 mg of famotidine. It also shows a set of THz absorbance images at 1.08 THz and 1.65 THz obtained at 220 K after calibration with an image at 0.8 THz. The image acquisition time for the tablet was about three hours with a 0.2 mm spatial resolution. Since the over-the-counter tablet mainly contained form B, the absorbance of the image at 1.08

is much larger than that of the 1.65 THz image. Furthermore, the inhomogeneity in the distribution of form B was detected, which was less than 20%. The frequency-dependent images clearly show the molecular distribution of famotidine form B and other chemicals in the tablet for the first time.

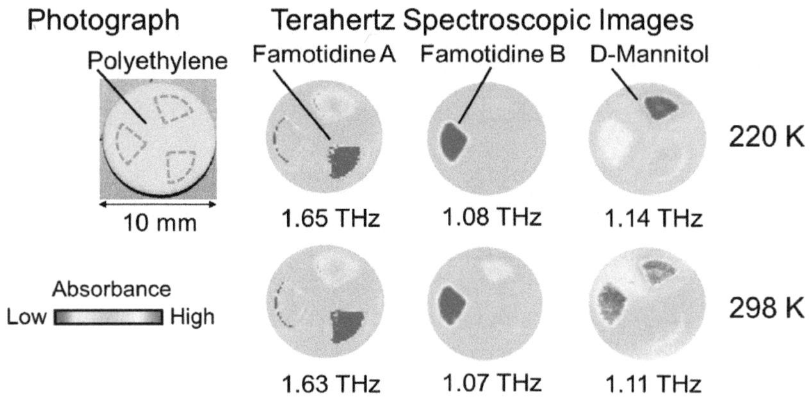

Figure 15.6 A set of temperature-dependent THz spectroscopic images of a compressed pellet containing pieces of famotidine polymorphic forms A and B and D-mannitol crystals measured at 220 K and 298 K. Reproduced from Ref. [36] by permission of Electrochemical Society, © 2013.

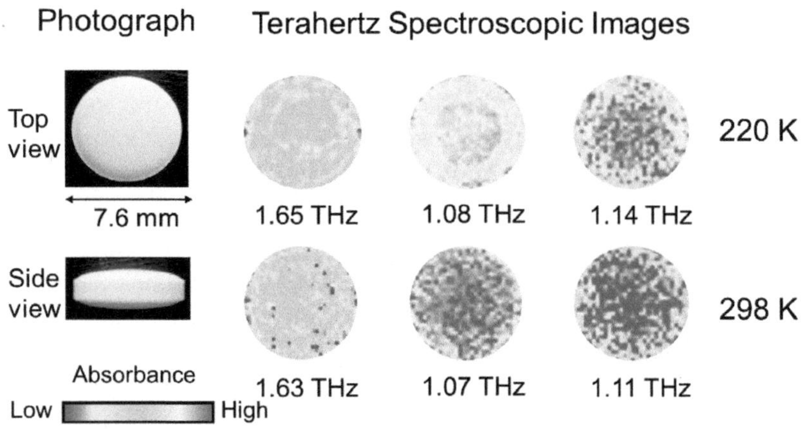

Figure 15.7 A set of temperature-dependent THz spectroscopic images of an over-the-counter tablet containing 10 mg of famotidine. Reproduced from Ref. [36] by permission of Electrochemical Society, © 2013.

The results indicate that TSI is a unique and powerful tool for obtaining the distribution of different polymorphic forms of molecular crystals in opaque samples, which is useful as a pharmaceutical evaluation technique. Its potential to utilize multicomponent chemical analysis could lead to a new biomedical imaging analysis technique.

15.3.4 Application (2): Chemical Mapping of Pharmaceutical Cocrystals

In this section, we introduce another example of TSI, namely, the first demonstration of the chemical mapping of a pharmaceutical cocrystal in a tablet. Cocrystals are solid molecular complexes of two or more neutral components bound together in stoichiometric amounts by noncovalent intermolecular interactions such as hydrogen bonds. Cocrystallization plays a similar role to polymorphism in pharmaceutical development. Pharmaceutical cocrystals contain an active pharmaceutical ingredient (API). In its pure form, the API is itself crystalline, but after cocrystallization the API molecules are interspersed within the new crystal structure. This new structure consists of the API molecules and one or more additional component molecules. These additional components are called coformers, and their selection plays a key role in dictating the physicochemical properties of the cocrystal, such as solubility, solution stability, and processability [38]. The physicochemical properties of crystalline materials are dependent not only on their molecular species but also on their crystal structures. By adjusting the synthesis conditions, cocrystals may take multiple polymorphic forms owing to their conformational flexibility [39]. In the pharmaceutical industry, the purpose of crystallizing cocrystals and polymorphs is to improve drug properties. The aim is to enhance bioavailability without altering the chemical nature, and thus bioactivity, of an API. Therefore, differences between the crystal structures of cocrystallized and pure pharmaceuticals affect the dosage response of a drug. Furthermore, chemical imaging methods are required for observing and evaluating the distribution of cocrystallized pharmaceuticals in tablets. Chemical mapping has been used to distinguish genuine and counterfeit drugs, as the API distribution reflects the tablet processing conditions [40]. In addition, spatial nonuniformity affects drug release kinetics and therapeutic efficiency [41]; therefore, chemical mapping is important in process monitoring and control.

We undertook the TSI measurement of a test tablet that included nonuniform caffeine (API)–oxalic acid (conformer) cocrystal powders. The TSI system was described in the previous section (Fig. 15.4).

Cocrystallization was readily confirmed by THz-TDS. Caffeine–oxalic acid exhibits three strong absorption peaks (ν_i) at 1.35 THz, 1.43 THz, and 1.56 THz at around 120 K, whereas caffeine and oxalic acid exhibit no absorption peaks within the 1–2 THz range that was studied (Fig. 15.8). Figure 15.9a shows visible light photographs of two nonuniform sample tablets. Regions of concentrated chemicals are indicated in Fig. 15.4b, though their distribution cannot be seen in the photographs or in the THz pulse intensity images (Fig. 15.9b). THz spectral data were used to map these unknown regions. Cocrystal pixels within the two nonuniform tablets were identified by the presence of ν_3. A frequency range is used for identification instead of the absolute peak frequency to take account of high noise in the image spectra and local temperature fluctuations. The presence of this peak was determined within $\nu_3 \pm 0.5\Delta\nu_3$ THz. Within

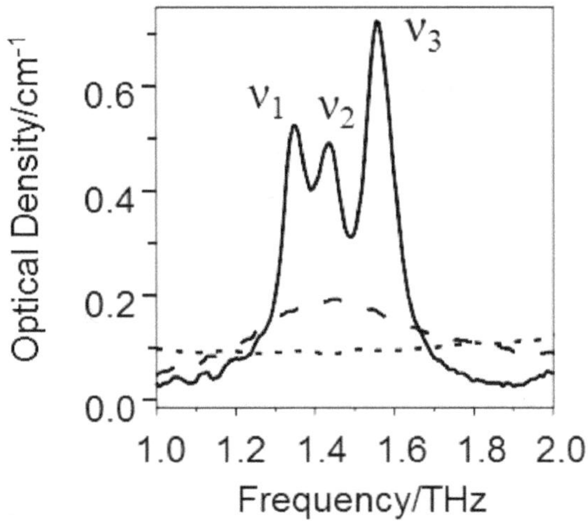

Figure 15.8 THz spectra of caffeine (dashed line), oxalic acid (dotted line), and their cocrystal (solid line) measured at 120 K. Reproduced from Ref. [37] by permission of ACS, © 2013.

this range, the minimum transmittance is assumed to be the peak intensity only if the intensities at the two neighboring frequencies are comparable. If this is not the case, the minimum transmittance within the studied range is assumed to be due to noise, and instead, the absolute transmittance at ν_3 is taken as the peak intensity. The cocrystal concentration (Fig. 15.9d) for each pixel previously identified as containing cocrystal in the two nonuniform tablets was determined using the previously calculated THz imaging concentration calibration. The absolute error in the calculated concentration due to variations in the homogeneous calibration standards is given in Fig. 15.9e.

The propagated error is proportional to the calculated concentration, not to the concentration-dependent S/N ratio of ν_3. For the tablet containing a nonuniform distribution of caffeine–oxalic acid in polyethylene, the total cocrystal concentration was found to be 5.25% by weight, which is only 0.352% more than the known concentration. Similarly, for the tablet containing cocrystal in caffeine, oxalic acid, and polyethylene, the total cocrystal concentration was calculated to be 8.61%, an overestimate of 1.24%. This overestimation may result from an error in the concentration calibration or from fluctuations in temperature and THz pulse intensity during imaging. In addition, at the interface between the tablet and the metal sample holder, THz radiation is scattered independent of frequency due to the change in the densities of the two materials. Therefore, transmittance near the edge of the imaged tablet may be lower than without this halo effect. For both nonuniform tablets, the cocrystal was successfully mapped on the basis of its spectral information separately from the diluent polyethylene and its pure precursor components, caffeine, and oxalic acid. For a more practical investigation of commercially available pharmaceutical tablets, various kinds of tablet matrix materials should be tested. In addition, more complex distributions of cocrystals should be analyzed, including an evaluation of smaller-scaled domains and differences between the surfaces and inner regions of tablets. Assembling a THz spectrum database of cocrystals and other excipients will allow us to analyze each component quantitatively, even if several kinds of chemicals are mixed in the focal spot of the THz beam. This result broadens the prospective applications of THz spectroscopy in pharmaceutical chemical imaging.

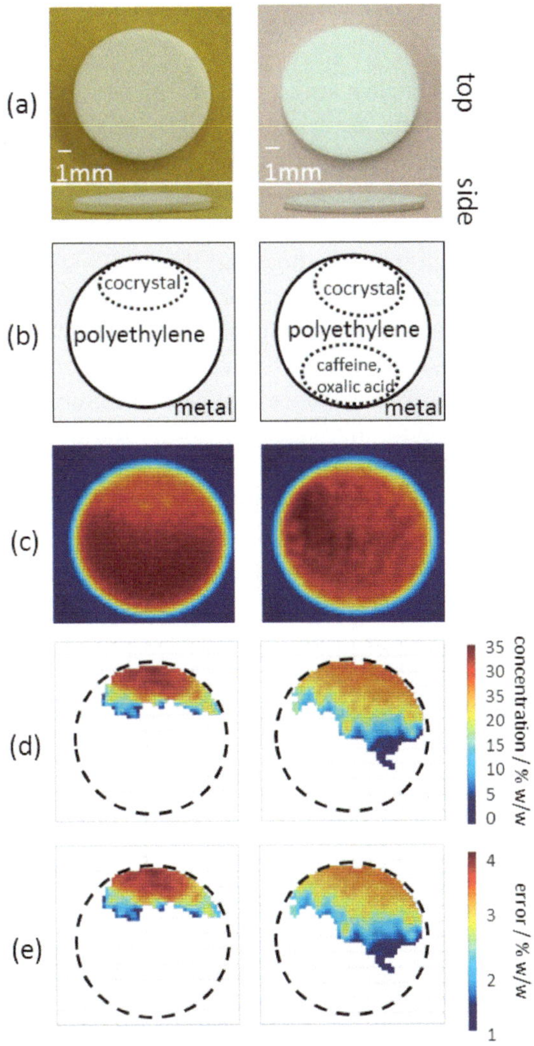

Figure 15.9 Chemical mapping of a caffeine–oxalic acid cocrystal in nonuniform tablets with the THz imaging system set at 120 K. Left: Cocrystal in polyethylene; right: cocrystal in polyethylene with caffeine and oxalic acid. (a) Photograph of an imaging tablet, (b) diagram of high sample concentration regions within the tablet, (c) THz pulse amplitude image, (d) cocrystal fraction by weight at a ν_3 peak for each identified pixel, and (e) absolute error in calculated cocrystal concentration. The approximate tablet position is outlined by a dashed black line in (d) and (e). Reproduced from Ref. [37] by permission of ACS, © 2013.

15.4 Summary

We introduced pioneering studies on THz-TDS and TSI, which are the two most successful approaches in THz chemical spectroscopy, to show the noteworthy aspects of THz spectroscopy and to reveal its potential. Advances in optical design, brighter light sources, and more sensitive detectors may expand the effective spectral region toward a higher frequency, and this will make THz chemical spectroscopy a more powerful tool for detection, identification, and analysis of intermolecular and intramolecular modes in many different substances. Although we have described only experimental studies in this chapter, the theoretical calculation of the vibrational modes in the THz region increases our understanding of molecular/crystal structures. In the near future, we expect THz spectroscopy to be admitted to this group of well-established analytical spectroscopic methods such as Fourier-transformed infrared spectroscopy, Raman spectroscopy, microwave spectroscopy, and dielectric spectroscopy.

References

1. Y. Ueno and K. Ajito, Analytical terahertz spectroscopy, *Anal. Sci.* **24**, 185–192 (2008).
2. D. H. Auston, K. P. Cheung, and P. R. Smith, Picosecond photoconducting hertzian dipoles, *Appl. Phys. Lett.* 45, 284–286 (1984).
3. Y. Watanabe, K. Kawase, and T. Ikari, Component spatial pattern analysis of chemicals using terahertz spectroscopic imaging, *Appl. Phys. Lett.* **83**, 800–802 (2003).
4. D. Grischkowsky, S. Keiding, M. P. van Exter, and Ch. Fattingert, Far-infrared time-domain spectroscopy with terahertz beams of dielectrics and semiconductors, *J. Opt. Soc. Am. B* **7**, 2006–2015 (1990).
5. M. P. van Exter and D. Grischkowsky, Carrier dynamics of electrons and holes in moderately doped silicon, *Phys. Rev. B* **41**, 12140–12149 (1990).
6. M. Hangyo, T. Nagashima, and S. Nashima, Spectroscopy by pulsed terahertz radiation, *Meas. Sci. Technol.* **13**, 1727–1738 (2002).
7. Product brochure of terahertz transmission spectrometer by TeraView Ltd., http://www.teraview.com/products/tpsspectra1000.html.
8. Product brochure of time domain spectrometers by Zomega Terahertz Corp., http://dl.z-thz.com/brochures/miniz_1301.pdf18; product

brochure of PB7200 portable frequency domain terahertz (THz) spectrometer by EMCORE corp., http://www.emcore.com/terahertz-thz-frequency-domain-spectrometer/.

9. E. Pickwell and V. P. Wallace, Biomedical applications of terahertz technology, *J. Phys. D: Appl Phys.* **39**, R301–R310 (2006).
10. K. Yamamoto, K. Tominaga, H. Sasakawa, A. Tamura, H. Murakami, H. Ohtake, and N. Sarukura, Terahertz time-domain spectroscopy of amino acids and polypeptides, *Biophys. J: Biophys. Lett.* L22–L24 (2005).
11. B. M. Fischer, M. Hoffmann, and H. Helm, Terahertz time-domain spectroscopy and imaging of artificial RNA, *Opt. Express* **13**, 5205–5215 (2005).
12. K. Yamamoto, M. H. Kabir, and K. Tominaga, Terahertz time-domain spectroscopy of sulfur-containing biomolecules, *J. Opt. Soc. Am. B* **22**, 2417–2426 (2005).
13. M. Yamashita, M. Usami, K. Fukushima, R. Fukasawa, C. Otani, and K. Kawase, Component spatial pattern analysis of chemicals by use of two-dimensional electro-optic terahertz imaging, *A Opt.* **44**, 5198–5201 (2005).
14. H. Zhong, A. Redo-Sanchez, and X.-C. Zhang, Identification and classification of chemicals using terahertz reflective spectroscopic focalplane imaging system, *Opt. Express* **14**, 9130–9141 (2006).
15. M. V. Exter and D. R. Grischkowsky, Characterization of an optoelectronic terahertz beam system, *IEEE* **38**, 1684–1691 (1990).
16. B. Yo, F. Zeng, Y. Yang, Q. Xing, A. Chechin, X. Xin, I. Xeylikovich, and R. R. Alfano, Torsional vibrational modes of tryptophan studied by terahertz time-domain spectroscopy, *Biophys. J.* **86**, 1649–1654 (2004).
17. C. Zhang, E. Tarhan, A. K. Ramdas, A. M. Weiner, and A. M. Durbin, Broadened far-infrared absorption spectra for hydrated and dehydrated myoglobin, *J. Phys. Chem. B* **108**, 10077–10082 (2004).
18. M. R. Kutteruf, C. M. Brown, L. K. Iwaki, M. B. Campbell, T. M. Korter, and E. J. Heiweil, Terahertz spectroscopy of short-chain polypeptides, *Chem. Phys. Lett.* **375**, 337–343 (2003).
19. K. Siegrist, C. R. Bucher, I. Mandelbaum, A. R. H. Walker, R. Balu, S. K. Gregurick, and D. F. Plusquellic, High-resolution terahertz spectroscopy of crystalline trialanine: extreme sensitivity to β-sheet structure and cocrystallized water, *J. Am. Chem. Soc.* **128**, 5764–5775 (2006).

20. A. G. Markelz, A. Roitberg, and E. J. Heilweil, Pulsed terahertz spectroscopy of DNA, bovine serum albumin and collagen between 0.1 and 2.0 THz, *Chem. Phys. Lett.* **320**, 42–48 (2000).
21. J. A. Zeitler, D. A. Newnham, P. F. Taday, C. J. Strachan, M. Pepper, K. C. Gordon, and T. Rades, Temperature dependent terahertz pulsed spectroscopy of carbamazepine, *Thermochim. Acta* **436**, 71–77 (2005).
22. J. A. Zeitler, K. Kogermann, J. Rantanen, T. Rades, P. F. Taday, M. Pepper, J. Aaltonen, and C. J. Strachan, Drug hydrate systems and dehydration processes studied by terahertz pulsed spectroscopy, *Int. J. Pharm.* **334**, 78–84 (2007).
23. M. Yamaguchi, F. Miyamura, K. Yamamoto, M. Tani, and M. Hangyo, Terahertz absorption spectra of L-, D-, and DL-alanine and their application to determination of enantiometric composition, *Appl. Phys. Lett.* **86**, 053903 (2005).
24. R. Rungsawang, Y. Ueno, I. Tomita, and K. Ajito, Angle-dependent terahertz time-domain spectroscopy of amino acid single crystals, *J. Phys. Chem. B* **110**, 21259–21263 (2006).
25. Y. Ueno, R. Rungsawang, I. Tomita, and K. Ajito, Quantitative measurements of amino acids by terahertz time-domain transmission spectroscopy, *Anal. Chem.* **78**, 5424–5428 (2006).
26. T. M. Korter, R. Balu, M. B. Campbell, M. C. Beard, S. K. Gregurick, and E. J. Heilweil, Terahertz spectroscopy of solid serine and cysteine, *Chem. Phys. Lett.* **418**, 65–70 (2006).
27. M. Walther, B. M. Fischer, and P. U. Jepsen, Noncovalent intermolecular forces in polycrystalline and amorphous saccharides in the far infrared, *Chem. Phys.* **288**, 261–268 (2003).
28. R. Rungsawang, Y. Ueno, I. Tomita, and K. Ajito, Terahertz notch filter using intermolecular hydrogen bonds in a sucrose crystal, *Opt. Express* **14**, 5765–5772 (2006).
29. Y. Ueno, K. Ajito, N. Kukutsu, and E. Tamechika, Quantitative analysis of amino acids in dietary supplements using terahertz time-domain spectroscopy, *Anal. Sci.* **27**, 351–356 (2011).
30. B. B. Hu and M. C. Nuss, Imaging with terahertz waves, *Opt. Lett.* **20**, 1716–1718 (1995).
31. S. Nakajima, H. Hoshina, M. Yamashita, and C. Otani, Terahertz imaging diagnostics of cancer tissues with a chemometrics technique, *Appl. Phys. Lett.* **90**, 41102 (2007).
32. P. R. Smith, D. H. Austone, and M. C. Nuss, Subpicosecond photoconducting dipole antennas, *IEEE J. Quantum Electron.* **24**, 255–260 (1988).

33. A. Nahata, J. T. Yardley, and T. F. Heinz, Two-dimensional imaging of continuous-wave terahertz radiation using electro-optic detection, *Appl. Phys. Lett.* **81**, 963–965 (2002).

34. J. Darmo, V. Tamosiunas, G. Fasching, J. Kröll, K. Unterrainer, M. Beck, M. Giovannini, J. Faist, C. Kremser, and P. Debbage, Imaging with a terahertz quantum cascade laser, *Opt. Express* **12**, 1879–1884 (2004).

35. S. Hunsche, M. Koch, I. Brener, and M. C. Nuss, THz near-field imaging, *Opt. Commun.* **150**, 22–26 (1998).

36. K. Ajito, Y. Ueno, H.-J. Song, J.-Y. Kim, E. Tamechika, N. Tamechika, W. Limwikrant, K. Yamamoto, and K. Moribe, Nondestructive multicomponent Terahertz chemical imaging of medicine in tablets, *J. Electrochem. Soc.* **161**, B171–B175 (2014).

37. D. M. Charron, K. Ajito, J.-Y. Kim, and Y. Ueno, Chemical mapping of pharmaceutical cocrystals using terahertz spectroscopic imaging, *Anal. Chem.* **85**, 1980–1984 (2013).

38. N. Qiao, M. Li, W. Schlindwein, N. Malek, A. Davies, and G. Trappitt, Pharmaceutical cocrystals: an overview, *Int. J. Pharm.* **419**, 1–11 (2011).

39. K. Schultheiss and L. S. Taylor, Pharmaceutical cocrystals and their physicochemical properties, *Cryst. Growth Des.* **9**, 2950–2967 (2009).

40. K. Kwok and L. S. Taylor, Analysis of counterfeit cialis® tablets using raman microscopy and multivariate curve resolution, *J. Pharm. Biomed. Anal.* **66**, 126–135 (2012).

41. F. Franch-Lagea, J. M. Amigod, E. Skibstedc, S. Maspocha, and J. Coelloa, Fast assessment of the surface distribution of API and excipients in tablets using NIR-hyperspectral imaging, *Int. J. Pharm.* **411**, 27–35 (2011).

Chapter 16

Industrial Applications of Terahertz Waves

Enrique Castro-Camus,[a] Karin Wiesauer,[b] Ralf Gente,[c] and Martin Koch[c]

[a]*Centro de Investigaciones en Óptica A.C., Loma del Bosque 115, Col. Lomas del Campestre, 37150, León, GTO, México*

[b]*RECENDT - Research Center for Non-Destructive Testing GmbH, Altenberger Str. 69, Science Park 2, 4040 Linz, Austria*

[c]*Faculty of Physics and Material Sciences Center, Philipps-Universität Marburg, Renthof 5, 35032 Marburg, Germany*

ralf.gente@physik.uni-marburg.de, enrique@cio.mx, wiesauer.k@gmail.com, martin.koch@physik.uni-marburg.de

16.1 Introduction

Terahertz (THz) technology made tremendous progress during the last 20 years. Meanwhile a rich variety of sources, detectors, and passive components to guide and manipulate THz waves have emerged. In particular basic research did benefit tremendously from these developments. A clear highlight in this respect is the manipulation of matter through intense THz fields [1–3]. Yet, a lot of the effort to further develop THz devices was driven by the search for real-world applications. The identification of application fields

Handbook of Terahertz Technologies: Devices and Applications
Edited by Ho-Jin Song and Tadao Nagatsuma

ISBN 978-981-4613-08-8 (Hardcover), 978-981-4613-09-5 (eBook)
www.panstanford.com

that require a large number of THz units is crucial for the long-term survival of suppliers of THz systems and components, as the scientific market alone might be too small to rely on.

One market that did already take off is security applications. Body scanners, which work with sub-THz waves, are already routinely used at airports to screen passengers for weapons or explosives. Another market that is still several years down the road is wireless short-range communications at THz frequencies. As THz communication systems could become an integral part of consumer products such as laptop computers or personal digital assistants, this will turn out to be a real mass market. Yet, a lot of development still needs to be done as this will require new concepts on several levels [4–6].

This chapter addresses the third market that has been identified: nondestructive testing (NDT) of industrial goods. This market is expected to grow soon. Of course, industrial NDT is already a wide field, and a plethora of different sensors come to use. THz systems have to offer an advantage over existing sensors, that is, they have to measure a crucial parameter that is not accessible up to now or measure a parameter with greater precision or more cost efficiency. In recent years there have been several successful demonstrations of industrial applications for THz systems. These demonstrations are summarized in this chapter. The five industrial branches that are most promising are paper, polymer, food, pharmaceuticals, and crops. Yet, demonstrating that it is in principle possible to measure a particular property of an industrial good does not guarantee market success. Hurdles are often the speed with which industrial items can be controlled and the price of THz systems. For this reason the last part of this chapter deals with recent attempts to increase the speed of THz imaging systems for NDT and with strategies to reduce system cost. Altogether it is expected that this market for THz systems will soon come to life.

16.2 The Different Kinds of THz Systems

There is a variety of different schemes to generate and detect far-infrared radiation. THz waves can be generated purely electronically, purely optically, or optoelectronically. In the following section we will very briefly review these schemes and rate their relevance for industrial applications. Of particular importance are optoelectronic schemes that use photoconductive antennas. We will dedicate one subsection to such devices.

A part of the electronic sources is based on free-space electron beams. This class includes synchrotrons [7], free-electron lasers [8], Smith–Purcell emitters [9], and backward wave oscillators [10]. All of these sources are too spacious and expensive for industrial applications. The other part of electronic THz emitters relies on classical microwave sources, for example, Gunn, Impatt, or resonant tunneling diodes. As the fundamental frequencies of these devices are typically not high enough for many THz applications they need to be multiplied in special mixers. These microwave-based THz sources fit well into a shoe box. They will soon approach the cost level that is acceptable for industrial applications.

Purely optical sources are based on the laser principle. One type is the molecular gas laser, which relies on transitions between different rotational levels of molecular species [11]. These lasers can emit several megawatts of power at discrete THz frequencies ranging from less than 300 GHz to more than 10 THz. Yet, this kind of laser has to be optically pumped by a tunable carbon dioxide laser, making the entire system rather bulky, expensive, and, hence, unattractive for industrial applications.

Another class of optical THz sources, which has been introduced by Watanabe et al. in the mid-nineties, uses parametric down-conversion to generate THz waves. The emission of two laser lines is mixed in a nonlinear crystal outside the laser cavity. Most commonly nanosecond laser pulses from Q-switched, diode-pumped Nd:YAG lasers are used. The emitted THz radiation is narrow band, widely tuneable, and so powerful that it can be detected with a room-temperature-operated pyroelectric detector [12]. Since the first demonstration this Japanese group has constantly improved the working parameters of their systems [13]. In 2003 it extended its system for imaging and made an impressive demonstration on the identification of different chemicals arranged in a pattern using spectroscopic information [12].

In 2010 Scheller et al. presented a continuous-wave THz source based on intracavity difference frequency generation [14]. They used a dual-color vertical external cavity surface emitting laser and a nonlinear crystal with a surface emitting phase-matching scheme.

A similar scheme involving an optical parametric oscillator was recently presented by Kiessling et al. [15].

Let's now turn to optoelectronic schemes. Quantum cascade lasers (QCLs) represent a very promising technology toward the realization of room-temperature, monolithic, and compact current-injected sources in the 1–5 THz range. The first QCL was demonstrated in 1994 by Faist et al. [16]. These lasers rely on the emission from transitions between subbands in quantum wells. When a voltage is applied to a complicated sequence of repeating quantum well structures, electrically injected electrons stream down a potential staircase and emit a THz photon at each step. Early QCLs needed cryogenic cooling, worked only in pulsed mode, and emitted in the mid-infrared [16]. Since then tremendous progress has been made. In 2002 Tredicucci et al. presented the first real THz QCL working at 4.4 THz [17]. This device still required cryogenic cooling. However, in 2013 Lu et al. reported the first high-power, room-temperature THz QCL based on difference frequency generation [18].

16.2.1 Optoelectronic Generation and Detection of THz Waves Using Photoconductive Antennas

There are a number of schemes to produce and detect THz radiation based on photoconductive antennas. In this section we will briefly discuss three of these methods. Firstly we will describe terahertz time domain spectroscopy (THz-TDS), which is probably the most widely used spectroscopic technique for the far-infrared band [19]. We will also present the generation and detection of continuous-wave THz radiation by photomixing in photoconductive antennas [20]. Finally we will describe a novel and cost-effective technique known as terahertz quasi–time domain spectroscopy (THz QTDS) [21].

TDS is a powerful tool that provides access to the entire THz band by using broadband single- or few-cycle pulses of radiation [19, 22, 23]. A schematic of a THz-TDS system is shown in Fig. 16.1. An ultrashort pulsed laser is used to produce ~100 fs pulses of light either in the vicinity of 800 nm (Ti:sapphire technology) or around 1550 nm (Er:fiber technology). The femtosecond laser pulses are used to excite two photoconductive switches, which serve as the THz emitter and detector, respectively (see inset to Fig. 16.1). These devices are fabricated by depositing metallic contacts on top of a semiconducting substrate with a separation between them that varies typically from a few microns to about a millimeter. When the gap between the contacts is illuminated by a femtosecond pulse, a transient increase of the conductivity between the two contacts is produced. In the case of an emitter device, a bias voltage varying from a few volts to a few hundreds of volts is applied, producing an electric field of the order of several kilovolts per centimeter; therefore the carriers optically generated by the pulse are accelerated, creating a transient photocurrent that persists for a few hundred femtoseconds. This short current surge is the source for a THz pulse that is radiated into free space [22, 24, 25].

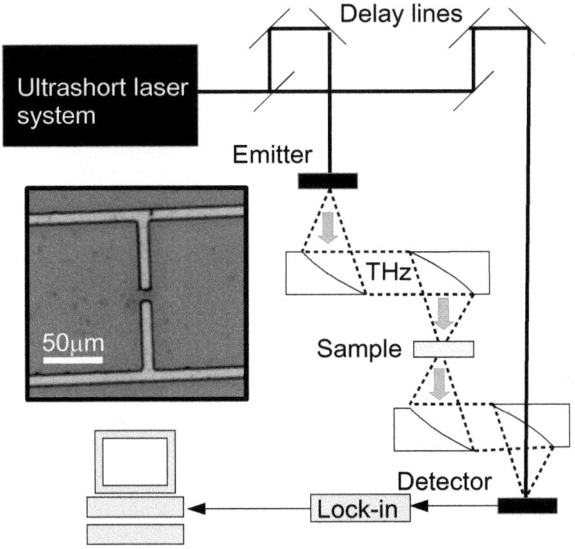

Figure 16.1 Schematic representation of a terahertz time domain system. The inset shows a micrograph of a photoconductive antenna structure.

As mentioned before, photoconductive switches can also be used as detectors. The metallic structure deposited on detector antennas looks rather similar to that of the emitter antenna. Yet, photoconductive detectors require being made on very short-carrier-lifetime substrates, such as low-temperature-grown GaAs or ion-implanted InP. If the ultrafast optically pumped carriers are quickly trapped in the substrate, then the high transient photoconductivity decays quickly and can be assumed to be almost instantaneous. If the THz transient and the femtosecond pulse arrive to the detector simultaneously, then the carriers will be accelerated only by the electric field that corresponds to the very short duration of the high photoconductivity; this way the electric field can be instantaneously probed by measuring the current between the detector contacts. By changing the delay between the ultrashort laser pulse gating the receiver and the THz transient, the waveform of the transient can be obtained [19]. One must bear in mind that the instantaneous appearance and disappearance of photocarriers is just an approximation and that therefore the waveform recorded will differ slightly from the one really arriving at the receiver antenna [26–28].

Figure 16.2a shows typical THz pulses obtained with a time domain spectrometer. The solid line shows a reference pulse that has propagated through free space. It should be noted that the spectrometer had been purged by nitorgen in order to aviod resonances due to water vapor [29].

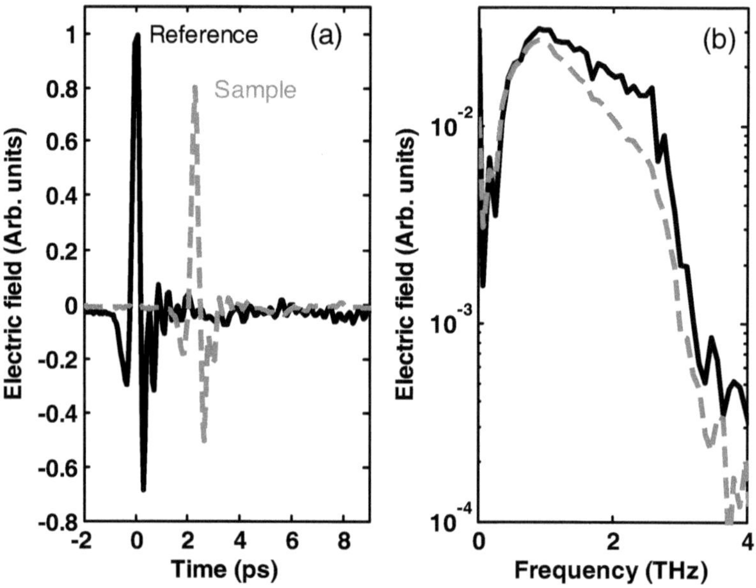

Figure 16.2 (a) Terahertz emission waveform of a photoconductive antenna as a function of time after ultrafast photoexcitation (at $t = 0$). (b) Spectra of the transient simulated covering the terahertz band.

As this pulse is temporally sharp with subpicosecond duration it contains a broad range of frequencies. Its spectrum, which is plotted in Fig. 16.2b, is obtained via Fourier transformation.

In this case the useful bandwidth covers from approximately 50 GHz to 3 THz, which is typical for TDS systems.

If a sample is inserted into the beam the THz pulse shifts in time (dashed line in Fig. 16.2a) as it has to propagate now through a medium that has a higher refractive index than air. Simultaneously its amplitude drops as a result of absorption in the sample and reflection losses from its interfaces. The ability to record the time-dependent electric field waveform makes this technique a phase-sensitive one; therefore complex dielectric properties, such as the complex refractive index of materials, can be directly extracted from the measurement without the need to apply Kramers–Kronig transformations.

In the past all states of matter have been investigated using this method. This includes solids [19, 30], liquids [31, 32], liquid crystals [33], and gases [29, 34]. Furthermore it was used to characterize devices and artificial structures, including lenses [35], reflectors [36], filters [37], and waveguides [38, 39].

To produce THz transmission images by TDS, the sample needs to be scanned through a focal point in the THz radiation path (between the emitter and the detector). This way the point-by-point information about the transmission amplitude and phase is collected and an image can be reconstructed. This process is typically performed using a pair of computer-controlled translational stages that move the sample across the focus and acquire the transmitted waveform automatically. Since the waveforms contain spectroscopic data, the image can be reconstructed using the amplitude, phase or a specific spectral pattern [40].

Another method to produce and detect THz radiation is to perform photomixing of two lasers of slightly detuned wavelengths. If the frequency difference is Δf and the two beams are superimposed this will produce a beating with a carrier frequency of the average of the two lasers and an envelope with a frequency Δf [20]. If a photoconductive antenna fabricated on a short-carrier-lifetime substrate is excited by the combined lasers' radiation, the resulting photocurrent varies at a frequency equal to Δf, and as dictated by Maxwell's equations electromagnetic radiation is generated at that frequency. This method allows changing the frequency of the radiation produced by controlling the detuning of the lasers. Detection of THz radiation can be achieved using photoconductive receivers just like in TDS by cross-correlating the superimposed beating radiation of the lasers with the THz continuous-wave radiation in a similar fashion to TDS [20, 41–44].

An alternative to these techniques is the technique called THz QTDS, which was recently demonstrated [21]. This technique uses a setup identical to the one for traditional TDS but replaces the ultrafast laser by a multimode laser diode. This laser is used to produce a large number of modes (~40) spectrally equispaced by a relatively small frequency (~25 GHz). This way a bandwidth of several hundred gigahertz is covered. All these modes are then mixed using a photoconductive emitter in order to produce broadband THz radiation. The photoconductive detector is gated in a similar way as in TDS. The result is a waveform that contains the amplitude and relative phase information of all the THz modes in the photomixed signal [21]. This technique is a promising alternative for industrial applications, given that it can be implemented at a much lower cost.

16.3 Polymer Industry

There are several applications that are foreseen and have been demonstrated for the polymer industry. Two of these applications will be highlighted in this chapter. The first is the use of THz systems to monitor the additive content of polymeric compounds. The second is the ability to analyze anisotropies in polymeric systems, which, for example, can result from injection molding. Other applications are the control of plastic weld joints [45] or glued interfaces [46] and determination of the moisture content [47] or glass transition temperature of polymers [48].

16.3.1 Polymeric Compounds

In our daily life plastic materials with different mechanical properties and colors surround us. One could get the impression that these represent dozens of different base materials with different molecular compositions. But in fact most plastic devices are made from about half a dozen different base polymers such as polyethylene, polypropylene, polyamide, and a few others. The physical properties of these base materials are then modified by mixing them with so-called additives. Additives can be pigments, glass fibers, nanotubes, flame retardants, or many other particles that can be quite diverse in size and shape. To add the particles, the base polymer is heated above its melting temperature. A dosing unit adds a defined quantity of additives. Subsequently, the mixture is homogenized in an extruder. This process is called polymeric compounding. Plastic devices are fabricated out of this polymeric compound in a subsequent process, mostly by injection molding.

The ability of THz spectroscopy to determine the additive content in polymeric compounds is demonstrated in Ref. [49]. These measurements were performed offline but yet on a variety of additive–polymer combinations with differing additive contents. Namely, the investigated combinations of materials were magnesium hydroxide in linear low-density polyethylene; glass fibers in polyamide; and calcium carbonate, silicon, or silicon dioxide in polypropylene. The underlying idea of applying THz spectroscopy to all these combinations of different materials is that the dielectric properties of the additives differ significantly from those of the polymers. This implies that the dielectric properties of a mixture of a polymer and an additive will depend on the fractions in which these two components are mixed [50]. The reverse of this relation also holds true. So the additive content can be determined from the dielectric properties of the mixture. Prior to THz-TDS measurements, the samples were prepared in a laboratory extruder setup. Concentration series of the material systems were produced in the form of granulates. From this material, flat samples with a thickness of up to 2 mm and smooth planar surfaces were produced by the means of injection molding. Measurements in these samples were performed in a THz-TDS transmission setup. From the recorded time domain data, the refractive index and absorption coefficient of the samples were derived. The refractive index and weight percentage of the additive content show a nonlinear relation. Given the specific gravity of the components the volumetric additive content can be derived from the weight percentage of the additive. After this conversion, a nearly linear correlation between the volumetric additive content and the refractive index emerges for these particular samples.

After offline measurements, in-line measurements on melted compounds inside an extruder were performed [51]. To ensure the success of such measurements several hurdles needed to be cleared. First of all, two windows are needed, through which the THz beam can enter the extruder and leave it again after interacting with the melted material. The window material has to withstand the conditions inside the extruder, where the melt is processed at temperatures of nearly 300°C and under a pressure of up to 26 bar. While being appropriately robust, the windows must of course still be transparent for THz radiation. These conditions were met by 15 mm thick fused silica. These two glass plates were arranged as a duct with an adjustable height of 0.5 mm to 10 mm. This construction allows for in-line measurements on the melt while it is piped through the duct. The window material was characterized at a wide range of temperatures of 20°C to 280°C, and no temperature-dependent effects were detected within a frequency range of 0.1–0.4 THz. Now that the technical preconditions for the in-line measurements are fulfilled, it is still necessary to take into account that the behavior of the melted material can differ from its solid form, which was characterized in the offline measurements before. Parameters, which can also have an effect on the results of the measurements, are temperature and pressure of the melt. Measurements on pure polypropylene were performed at different temperatures and different pressures. Within the range of pressures, which could be realized inside the extruder, no influence of the melt pressure on the refractive index and absorption coefficient of the melt was observed. As for the melt temperature, a decrease of the refractive index is observed with increasing temperature. This relation between temperature and refractive index is attributed to be actually a relation between density and refractive index, as the density changes with temperature. After these preparations measurements on a melted polypropylene–$CaCO_3$ compound were performed. Provided that the measurements are performed at a constant temperature, the measurements on the melted material show the same connection between refractive index and volumetric additive content as offline measurements on the solid material. Yet, as the refractive index of molten polypropylene at 250°C differs from its refractive index at room temperature by about 0.1, the results do not match exactly. Moreover, it is also important to notice that the refractive index of

$CaCO_3$ stays almost constant, as it is still in the solid state, while the polypropylene is already melted. This results in slightly different slopes of the two curves for the offline and in-line measurements. As with increasing additive content, the influence of the $CaCO_3$ on the properties of the compound becomes more pronounced, the curves lie closer together at higher additive contents.

The measurements, which have been described so far, have been performed in-line in the extruder setup but still with the slow data acquisition times of a laboratory setup. The actual aim to meet the requirements of modern production lines is fast real-time in-line measurements. Production planning tends to minimize storage times, which sets a strict limit to the time that is available for process control. Sometimes, even several production steps are integrated into one single unit, meaning that the only chance for controlling the product between these steps is to perform measurements in real time.

In order to reach this aim, one has to use high-speed mechanical delay lines or a fiber stretcher (see Section 16.8.1.). An easy option that was used here is not scanning the waveform of the pulse at all. Instead, the delay line is held at a fixed position on the edge of the pulse. Any change of the refractive index of the sample will lead to a change of the temporal position of the THz pulse. As the delay line still stays in its fixed position, this results in a change of the amplitude of the detected signal. Even small shifts of the THz pulse can be detected with this simple method. The range of values, which can be detected with this scheme, depends on the length of the pulse. A longer pulse will allow for a broader range of values, while a shorter pulse will increase sensitivity. So, the length of the pulse should be adjusted accordingly to the measurement problem. While the measurement was running, the additive content, which was fed into the extruder, was changed every few minutes. A comparison of the desired weight percentage of $CaCO_3$ and the recorded amplitude of the THz signal shows that the THz signal follows the additive content. A delay of about one to two minutes before a change in the additive content shows up in the THz signal is noticed. This delay accounts for the time that is needed to transport the material in the extruder from the dosing unit to the point where the measurement takes place. When the additive content is increased, a steep rise in the THz signal is noticed. A decrease of the additive content appears in the THz signal as an exponential decay, which accounts for the fact that the additive is gradually washed out (Fig. 16.3).

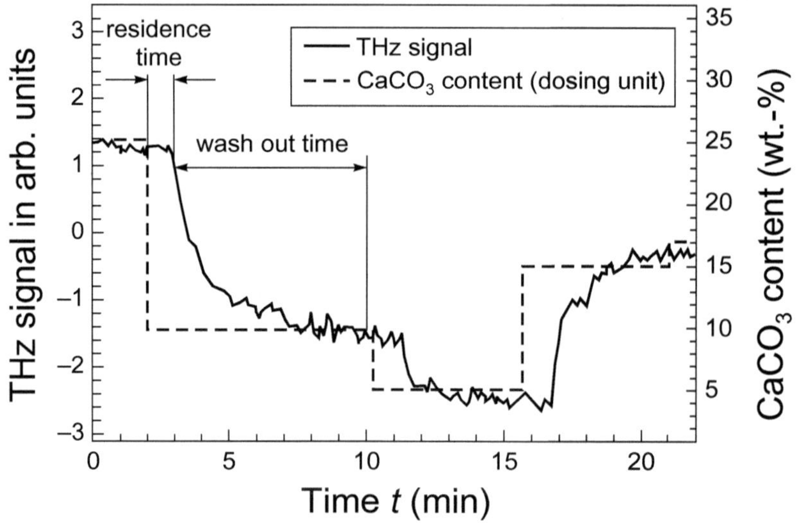

Figure 16.3 THz measurement compared to additive content. The dashed line represents the additive content set at the dosing unit. This correlates very well with the measurement. As the measurement was performed at the other end of the extruder, there is a delay of a few minutes between the two curves [51, 52].

Another class of materials that can benefit considerably from a convenient and reliable in-line process control is elastomers [53]. Compared to most additives, the refractive indices and absorption are relatively low. Thus, a compound with a higher additive content will have a higher refractive index and absorption than one with a lower additive content. In the elastomer industry, carbon black is the most important additive. At the same time, this additive has the highest influence on the THz signal, resulting in very high measurement sensitivity. Similar as with the measurements on polymeric compounds, the in-line measurements on elastomers were preceded by offline laboratory measurements to determine the refractive index and absorption of the materials. The next step was to develop a specialized system design for in-line measurements in the extrusion process. For this purpose an extruder head was built, which incorporates a bypass of the material flow, which can be monitored in transmission geometry through a glass window. The THz-TDS system, which was used for these measurements, was completely fiber-coupled and utilized a fiber stretcher as a delay unit. Using the fiber stretcher, full waveforms could be recorded at a high repetition rate. For the evaluation of the capabilities of the system to monitor the additive content in elastomers, measurements with a mixture based on natural rubber were performed. Mixtures with different concentrations of zincoxide and carbon black were injected into the extruder while THz waveforms were continually recorded. Changes in the concentration of the additives were detectable with a delay of some minutes, which corresponds to the time it takes for the material to be processed through the extruder. It was observed that a harder material pushed a softer material out of the extruder, leading to a quite sharp change in the THz signal. Contrary to that a softer material following a harder material mixed into the harder one, which resulted in a much slower change in the THz signal. As expected the measurements had the best sensitivity for carbon black due to its high dielectric contrast. Small variations as small as 1% could be detected. For zinc oxide, variations of about 2% could be monitored. To validate the results of the in-line measurements, some control samples were taken from the extruded material and conventional laboratory measurements were performed. The results of these measurements agree with the results of the in-line measurements.

16.3.2 Analysis of Anisotropies

Another interesting application of THz systems for NDT of polymeric products is the analysis of anisotropies. Anisotropies or more specific anisotropic dielectric material parameters can be investigated by polarization-sensitive (PS) THz techniques whenever they result in birefringence in the THz spectral range. Birefringence means that the refractive index loses its isotropy but depends on the polarization and propagation direction of the probing radiation. PS techniques in general are based on the use of polarized radiation and PS detection. Well-established PS techniques employing visible light, such as photoelasticity measurements (e.g., [54]), can be applied to optically transparent components and models only. In contrast, PS THz techniques also allow an investigation of optically opaque anisotropic polymers due to the large penetration depth of the THz waves and are therefore highly promising in the field of NDT.

One example for the analysis of anisotropies by PS THz techniques with great practical significance is the investigation of glass-fiber-reinforced polymers (GFRPs). This class of materials plays an important role in the automotive and aeronautic sector, as well as in the field of renewable energies, for example, for rotor blades of wind turbines. The orientation of the fibers is an intrinsic parameter as it influences the mechanical properties of the GFRP components, such as their tensile strength (e.g., [55, 56]). Hence, analysis of the fiber orientation is essential. For this purpose, commonly destructive techniques are applied, where parts are cut out of the sample, polished, and investigated by microscopy. It has been shown that GFRPs exhibit THz birefringence when the fibers are aligned along a preferential direction [57–59]. The underlying physical effect is form birefringence (e.g., [60]): as in most cases the fibers have diameters smaller than the THz wavelengths in the range of 100 μm, the sample can be described as an anisotropic effective medium birefringent at THz frequencies. Consequently, PS THz techniques can be applied for orientation analysis of the fibers in GFRP materials. Being nondestructive, their use clearly constitutes an advantage for orientation analysis. THz birefringence has been proven not only for GFRPs but also for other fibrous materials, such as wood [61, 62], paper [61, 63, 64], and textiles [64], further broadening the range of applications of PS THz techniques for NDT. An overall review on THz birefringence studies can be found in Ref. [65]. In many cases the preferential fiber orientation can vary over the sample, such as in injection-molded short-fiber-reinforced polymers, where the fiber alignment is influenced, among others, by the flow direction of the mold. In that case, THz imaging would be the method of choice, giving a spatially resolved map of the fiber orientation across the sample. It should be noticed that the large penetration depth of THz waves can only be exploited for dielectric materials, such as polymers reinforced by glass- or other nonconductive fibers. In the case of carbon-fiber-reinforced materials, the penetration depth of the THz waves is restricted to the first few hundred microns of the sample only [66].

An early approach for PS THz imaging based on THz-TDS was demonstrated in Ref. [67]. Thereafter, different more or less complex THz imaging modalities providing polarization contrast were developed and applied [57, 58, 68–75], with PS THz near-field imaging [72] or a PS THz system with potential for real-time imaging [73] as examples. The benefit of exploiting the polarization information was demonstrated, for example, by showing the rotation of the THz electric field vector at sample edges [68, 69], for visualizing the anisotropic conductivity of an organic conductor [72], or for detecting anisotropic texture in a sprayed polyurethane insulating foam [70, 71]. In these studies, however, only polarization contrast was provided, but no further analysis of anisotropy was done.

The analysis of anisotropies by PS THz measurements requires appropriate algorithms for data extraction. Two physical parameters can be of interest for a birefringent sample, namely, the orientation of the optical axis and the value of birefringence. The orientation of the optical axis represents the direction of anisotropy, given by the preferential fiber orientation in GFRPs, and is therefore of intrinsic interest for the characterization of these materials. In addition, the value of birefringence is a measure for the magnitude of anisotropy. For GFRPs, higher values of birefringence are a result of a higher density of aligned fibers, either due to a general higher fiber density in the sample or due to a higher fraction of fibers that are aligned [57]. Thus, the value of birefringence gives additional information on the anisotropy of a sample. Next, PS THz imaging for the investigation of GFRPs and corresponding data evaluation algorithms with a focus on orientation analysis will be discussed in more detail.

First PS THz imaging of GFRPs was done for a part of a helicopter stator blade in [74]. The sample was a glass-fiber-reinforced liquid-crystal polymer (LCP) containing 30% of short glass fibers. On the basis of a standard THz-TDS system in transmission geometry, birefringence of the material was revealed. Here, the first qualitative statement on the orientation of anisotropy was obtained, but no quantitative orientation analysis was performed. Later, the same LCP was investigated in Ref. [75], where a quantitative orientation analysis was performed on the basis of a similar THz-TDS setup. The aim was to extract the orientation angle φ between the optical axis of the sample and the probing THz polarization. Two different approaches were presented:

- The first, so-called basic algorithm can be applied when birefringence of a sample results in two separated pulses for the ordinary (o) and the extraordinary (e) wave. Then, φ can be extracted from the detected peak amplitudes of the two pulses, e_1 and e_2, in the time domain:

$$\varphi = \arctan\left(\sqrt{\frac{e_1}{e_2}}\right) \tag{16.1}$$

  This approach is fast and easy to use and can therefore be ideally applied to imaging data. However, it is restricted to samples without diattenuation.

- The second, so-called advanced algorithm presented in Ref. [75] is more complex but also more versatile. It can be applied to any type of birefringent sample, also with a very small birefringence where no separate pulse maxima occur, and it considers effects such as dispersion or different absorption and reflection for the e- and o-axes of the material. Nevertheless, a priori knowledge or separate measurement of the e and o refractive indices of the sample is required. The measured time-dependent THz signal $e(t)$ at a certain orientation angle φ can be described as

$$e(t) = \sin^2(\varphi)e_o(t) + \cos^2(\varphi)e_e(t) \tag{16.2}$$

The ordinary and extraordinary components $e_o(t)$ and $e_e(t)$ can be expressed as functions of the pulse responses of the sample, $h_o(t)$ and $h_e(t)$, respectively, and of the reference pulse, $e_{\text{ref}}(t)$:

$$e_{o,e}(t) = e_{\text{ref}}(t) \cdot h_{o,e}(t) \tag{16.3}$$

Details of how to calculate the pulse responses from the refractive indices are given in Ref. [75]. Subsequently, angle φ is obtained from a numerical optimization process, where the measured sample signal $e_{\text{meas}}(t)$ is compared with the signal $e_{\text{sim}}(t)$ simulated for a certain value of φ.

These algorithms were applied to different anisotropic materials, as further described in Ref. [75]. For the LCP sample, an orientation image was obtained using the basic approach, with pointwise verification by the advanced algorithm. The results of both algorithms agreed well, covering an angular range of the orientation angle of the fibrils between 20° and 70°.

Another method for orientation analysis was presented in Ref. [57]. This method does not require the knowledge or measurement of the *e* and *o* refractive indices of the sample and can therefore be universally applied. However, it requires the measurement at three different orientation angles of the sample, with the best results obtained for 0°, 45°, and 90°. From these measurements, the three corresponding refractive indices $n_{0°,45°,90°}$, are obtained, which are auxiliary parameters for extracting the refractive indices $n_{e,o}$ and the angle φ from the following equations:

$$n_{0^\circ} = n_e \cdot \cos^2(\varphi) + n_o \cdot \sin^2(\varphi) \tag{16.4}$$

$$n_{45^\circ} = n_e \cdot \cos^2(\varphi + 45^\circ) + n_o \cdot \sin^2(\varphi + 45^\circ) \tag{16.5}$$

$$n_{90^\circ} = n_e \cdot \cos^2(\varphi + 90^\circ) + n_o \cdot \sin^2(\varphi + 90^\circ) \tag{16.6}$$

In Ref. [57], additionally the additive content and the degree of fiber orientation were derived, depending on a model using effective medium theory. For details see the reference.

The developed approach was applied to different injection-molded GFRP samples for spatially resolved orientation analysis, namely, a shouldered test bar (high-density polyethylene [HDPE] matrix, 50 wt.% glass fibers with diameter 14 μm and length 4.5 mm) and a flat plate sample (polyamide 66 matrix, 30 wt.% glass fibers with diameter 10 μm and length 200–220 μm). Both samples showed anisotropic behavior in the THz frequency range. The subsequent THz orientation analysis was performed at single frequencies of 0.42 THz and 0.6 THz, respectively. Comparison of the results with a MoldFlow® simulation revealed good agreement, indicating that the fibers were mainly aligned along the mold flow.

The THz systems applied for orientation analysis in the above studies are based on standard THz-TDS systems, using linearly polarized THz radiation and detecting a single polarization component. Thus, they can easily be realized. However, they rely either on additional information on the sample or on multiple measurements. A method circumventing these problems is presented in Ref. [58], which in turn requires a more complex setup. The system (see Fig. 16.4) is based on a scanning THz-TDS imaging setup in reflection geometry developed in Ref. [76], where the THz beam is scanned through a scanning lens by an *xy* galvanometer scanner.

In its PS variant, the setup is extended by a Fresnel prism made from high-resistivity silicon. This Fresnel prism acts as an achromatic quarter-wave plate and converts the initially linearly polarized into a quasi-circularly-polarized THz pulse, which is used for probing the sample. After reflection by the sample, the probing pulse passes the Fresnel prism a second time. In the case of an isotropic sample, again linear polarization is obtained after the Fresnel prism, whereas for a birefringent sample, the polarization becomes elliptic, carrying information on the anisotropy of the sample. For full determination of the resulting signal, both polarization components of the THz wave are detected, using a three-contact photoconductive antenna with perpendicular gaps for PS detection, similar to the one in Ref. [77].

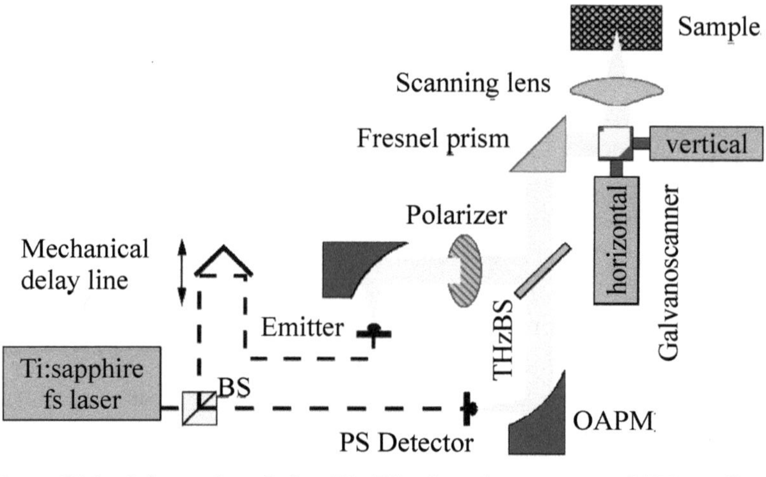

Figure 16.4 Schematic of the PS THz imaging system. OAPM: off-axis parabolic mirror; BS: beam splitter.

The extraction of the anisotropic material parameters of the sample requires a detailed mathematical description of the problem. On the basis of Jones matrix formalism in the frequency range, relatively simple equations are derived:

$$g_1(f) = \frac{|S_a + S_b|}{A} \tag{16.7}$$

$$g_2(f) = \frac{|S_a|^2 + |S_b|^2}{2A^2} \tag{16.8}$$

S_a and S_b are the complex spectral components of the two polarization components of the signal that is reflected from the rear surface of the sample, depending on frequency f. The parameter A^2 used for normalization corresponds to the spectral energy of the signal. Details of the derivation of the equations can be found in Ref. [58]. From these equations, the value of birefringence as well as the orientation of the optical axis can be obtained from the measurement data. It was shown that the relation given by Eq. 16.7 will exhibit a minimum at a frequency f_{min} if the phase shift between the two polarization components equals π. On the basis of these parameters, the average birefringence of a sample with known thickness can be calculated, as described in detail in Ref. [58]. Furthermore, it was shown that Eq. 16.8 exhibits an amplitude equal to $\sin(2\varphi)$, directly yielding the orientation of the optical axis with regard to the horizontal.

The algorithm was tested for PS THz imaging data of a similar HDPE shouldered test bar with 50 wt.% glass fiber as in Ref. [57] obtained by measurements with the described setup. A photo of the sample is shown in Fig. 16.5a, and the results of the data evaluation are displayed in Fig. 16.5b. The short lines indicate the direction of the optical axis, corresponding to the average fiber orientation of the sample. Now, in addition information on the birefringence is also obtained, displayed by the shading. In the left part of the sample, the fiber orientation appears to be slightly asymmetric with regard to the shape of the sample. There, also lower birefringence is observed, indicating a reduced degree of orientation of the fibers. This is consistent with the fact that the mold was injected from the left side, potentially leading to higher turbulences. On the right side in the narrow part of the sample, the fibers show nearly parallel alignment with regard to the sample edges and the highest values of birefringence, indicating a high degree of orientation of the fibers.

An advantage of the presented approach is that it requires only a single time domain depth scan using a PS detector. Thus, neither a reference measurement nor multiple scans with varying orientation of the sample or of the polarization of the probing THz beam are needed. Under the restriction that the algorithm is only valid for linear nondichroitic samples, it is a versatile method for orientation analysis and measurement of birefringence, which can easily be applied to PS THz imaging. The minimum detectable birefringence depends on the inverse of the product of THz bandwidth and sample thickness. More details can be found in Ref. [58].

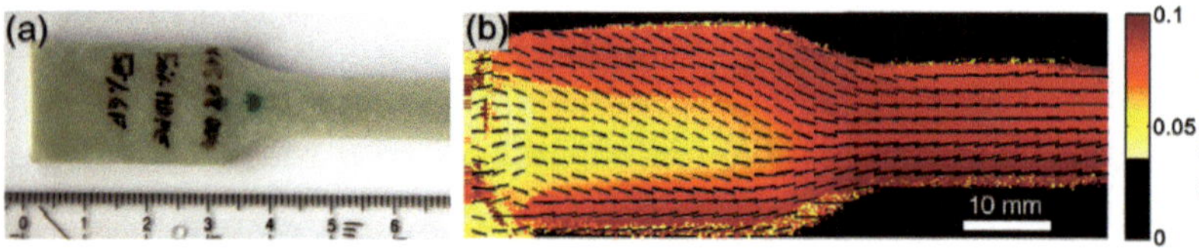

Figure 16.5 (a) Photo of HDPE shouldered test bar with 50 wt.% of short glass fibers. (b) Results of PS THz measurements. Short lines: average direction of optical axis; shading: birefringence.

Altogether, PS THz measurements are highly promising for the investigation of anisotropies in dielectric materials that are opaque for visible light. No other nondestructive method based on nonionizing radiation is capable of delivering comparable information. As was shown for GFRP materials, the average fiber orientation can be determined by different algorithms, and in addition, information on birefringence can be extracted. These proofs of principle show the high potential of PS THz methods for NDT of anisotropic polymeric materials that are produced in various industrial branches. For future application in an industrial environment, the development of robust and mobile PS THz systems is required, paving the way for their becoming a novel tool for NDT.

16.4 Paper Industry

Another commercial branch that can benefit from an in-line control with THz systems is the paper industry. Here, the task is to perform an in-line monitoring of the thickness and moisture content of paper during production. The first off-line tests were carried out by Banerjee et al. They demonstrated that the paper moisture content can be quantitatively determined on the basis of extinction and phase measurements in the lower THz range. The spatial resolution was in the millimeter range and scanning times below two minutes for an area of 40 × 40 mm [78]. Besides, Mousavi et al. used the Bruggemann effective medium theory to simultaneously determine thickness and moisture content of paper sheets [79]. Their results are a first indication that the uncertainties obtained with THz systems are comparable to what other existing sensor technologies offer.

Also as a laboratory experiment, measurements on a multilayer system consisting of paper and glue, that is, a sticker, were performed with a THz time domain spectrometer. First, the paper was characterized in a separate measurement without the glue. Thickness and material parameters of the paper were obtained from this measurement according to Scheller et al. [80]. After the glue was applied to the paper, another measurement on the complete multilayer system was performed. Now the glue is the only unknown component, while the properties of the paper are already known. In the situation of a multilayer system with only one unknown component, the algorithm proposed in Ref. [81] can be used. This way, also the thickness of the glue can be determined.

An in-line measurement in a production environment could be performed by two independent measurement systems, one directly before and one after the application of the glue or paint. Then, by matching the spatial position of these two measurements on the paper, the thickness of both paper and glue could be monitored.

In current paper production lines, the area density of the paper is determined in-line by measuring the transmission of ionizing radiation from a radioactive sample trough the paper. While the precision of this technique is acceptable, the handling of radioactive substances is always a security issue. As THz waves are not ionizing THz systems could circumvent this problem.

As a first test of in-line THz measurements on paper, a fiber-coupled THz time domain spectrometer was integrated into a real paper production line in transmission geometry. Unfortunately, the performance of the fiber-coupled antennas that were used in this measurement was inferior to what can be expected as the state of the art. Nevertheless, the results of these measurements clearly show the capability of this technique. Even without a very sophisticated data analysis the phase of the recorded THz signal is directly correlated to the results of the traditional measurements, for which beta radiation from krypton 85 (KR-85) was used in this case (Fig. 16.6).

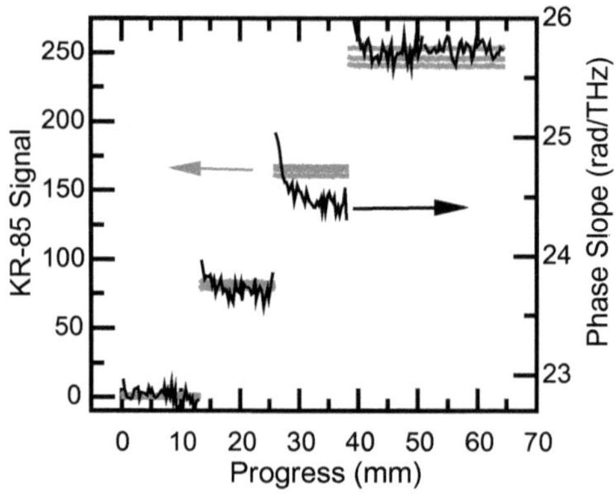

Figure 16.6 Results of the first test of a THz-TDS system in a paper production line. The results from the traditional KR-85 measurements are compared to the THz signal.

16.5 Food Industry

Contamination is a major problem in the food industry. The two most established methods to detect contamination are metal detectors and X-ray scanners. Unfortunately, the former can only detect metallic contamination, and the use of X-ray scanners implies trained personnel. Besides, it is not openly communicated as the industry fears that consumers may not feel comfortable with the idea of testing eatables with ionizing radiation. As mentioned above, THz waves are nonionizing. Hence, THz systems may offer an alternative. Yet, most eatables cannot be penetrated by THz radiation as they contain considerable amounts of water. Nevertheless, THz systems may have some applications in the food industry for eatables that are low in moisture or fatty. One example is chocolate.

Recently, it was demonstrated that different metallic and nonmetallic contaminants within chocolate bars can be detected with THz waves. Jördens and Koch showed that it is possible to distinguish between foreign bodies and ingredients that belong in the product, such as nuts, by analyzing the shape of the transmitted THz pulse [82]. If the transmitted pulse is still a single pulse it is an indication of pure chocolate or nuts in the chocolate. If the pulse breaks into two pulses this indicates nonmetallic contamination, like a stone, a piece of glass, or a plastic particle. The double-pulse structure results from differences in the refractive index between the chocolate and the contaminants. Yet, as local variations in the thickness of the chocolate bar may also lead to a double-pulse structure, the height profile of the chocolate has to be determined using optical sensors. This information must be measured and taken into account as well in the image-processing step. In the thickness-corrected delay image it is then possible to detect a glass splinter within a chocolate bar containing hazelnuts.

Herrmann et al. carried out another demonstration [83]. They showed the detection of foreign bodies in powdered sugar and wheat flour. They could detect an eggshell and different plastic pieces. The powdered samples were, however, only 1 cm thick. It is not clear if THz waves can penetrate thicker powder layers. This would be required for practical, real-life cases.

16.6 Pharmaceutical Industry

An industry that has very high quality standards is the pharmaceutical one. These standards are predominantly ensured by laboratory analysis. An entire batch is rejected if the final product does not meet the specification. This is the reason why in-line tests are highly desirable throughout the entire manufacturing process. Recently, Taday et al. studied the potential of THz systems for the pharmaceutical sector. They demonstrated the following applications: It is known that an agent can come in different polymorphic forms that often differ in their physicochemical properties and pharmaceutical performance. As different isomers of a molecule lead to different crystalline structures with different spectral fingerprints in the THz range, Taday et al. could distinguish between two polymorphic forms of pharmaceuticals [84].

Besides, Taday et al. could demonstrate that one can determine the concentration of substances in pharmaceutical mixtures [85] and that drug hydrate systems and dehydration processes can be analyzed using THz-TDS [86]. Furthermore, the quality control of tablet coatings is important. Coatings on tablets are used to release active pharmaceutical ingredients in a controlled way. Recently, it was shown [87] that it is possible to distinguish between two brands of over-the-counter ibuprofen tablets by their coating.

16.7 Crops Industry

According to the United Nations' Food and Agriculture Office water consumption for agriculture accounts for around 70% of the total freshwater withdrawal in the world [88]. This fact added to the continuous decline of availability and quality of this liquid makes optimization of crop irrigation methods a challenge of potential global impact. To minimize water consumption without compromising agricultural productivity it is very important to have a method to monitor the water status of plants in real time to provide feedback to irrigation systems. Currently the water status and water dynamics of plants are determined using techniques such as measurement of their osmotic potential [89] or the gravimetric method [90]. All these techniques are destructive as the separation of tissue from the organism is required. The gravimetric method refers to the technique of comparing fresh and dry weights of leaves or other excised tissues. This is the most commonly used method by plant scientists today. However, this method has two major disadvantages. Firstly, as mentioned before, it is destructive. Secondly, the separation of the organs produces additional stress on the plants, which means that only one unperturbed measurement is possible on each plant. Therefore continuous monitoring of the plants' water status under controlled conditions in a single living organism is impossible and turns complicated if multiple organisms are to be used.

The strong attenuation of THz signals by water is usually seen as a disadvantage for the study of samples in aqueous solution, given that the radiation is attenuated by water to the point of being undetectable. Yet, when it comes to detecting moisture content of a sample, THz radiation can be used as a highly sensitive noncontact, nondestructive water probe.

The first demonstration of the feasibility of this approach was reported by Mittleman et al. [91] and Hu et al. [92]. Leaves are among the first samples of which images were taken with THz-TDS. In such images, the leaf veins are clearly visible owing to the higher amount of water contained in them. The images recorded by Hu et al. show a significant THz transmission change between a freshly cut leaf and the same leaf after drying at ambient conditions for 48 hours demonstrating the possibility to detect moisture changes. In

Ref. [91] measurements on a leaf of a living plant are shown. After watering it, the plant rehydrates, and therefore, an evolving decrease of the THz transmission through the leaf is observed. In Refs. [93, 94] similar measurements are shown, which exhibit the same behavior.

Breitenstein et al. [95] present measurements of the transmission of continuous THz waves through leaves of coffee plants at a frequency of about 0.2 THz. For comparison destructive gravimetric measurements of the water content are performed. While the plants are exposed to drought stress, it is observed that loss of water and THz transmission show a consistent evolution over time. Also the dynamics of rehydration after watering a plant can be monitored with such a measurement system.

When taking into account the whole spectrum consisting of amplitude and phase information over a broad frequency range, a more detailed data analysis is possible.

Hadjiloucas et al. [96] give an overview of different possibilities to employ linear transforms for this purpose. Their method allows the classification of measured spectra into "wet," "dry," and "very dry" categories with the aid of principal component analysis. This method allows the estimation of leaves' water content, provided that their thickness and roughness are known.

More recently Jördens et al. [97] presented a physical model for the transmission of THz radiation through a leaf on the basis of effective medium theory. Using this model, which will be described in more detail later in this chapter, Gente et al. [98] developed a computational algorithm capable of calculating both the leaf thickness and its water content from a single THz-TDS transmission measurement.

To take advantage of the THz spectroscopy capabilities in plant breeding and plant physiology, appropriate spectrometers are required. In a closed-laboratory environment, it is possible to automate this process. When the leaves under investigation are held at well-defined positions, a fiber-coupled measurement head can be moved between these positions by a motorized mechanism allowing automated measurement of many plants in sequence (Fig. 16.7a). For measurements outside a closed laboratory, for example, in a greenhouse or even outside on a field, a portable measurement system is needed. Such systems have been built incorporating all components of the system except photoconductive THz antennas and

THz optics in a closed case. The measurement head, which contains photoconductive THz antennas and THz optics, can be moved to the plant by hand. The optical fibers and cables, which connect it to the rest of the system, are combined in one flexible hose (Fig. 16.7b). Further developments regarding size and power consumption of THz systems will simplify the use of this technology, opening the possibility of THz spectrometers being used on the field without the need for an external power supply [21].

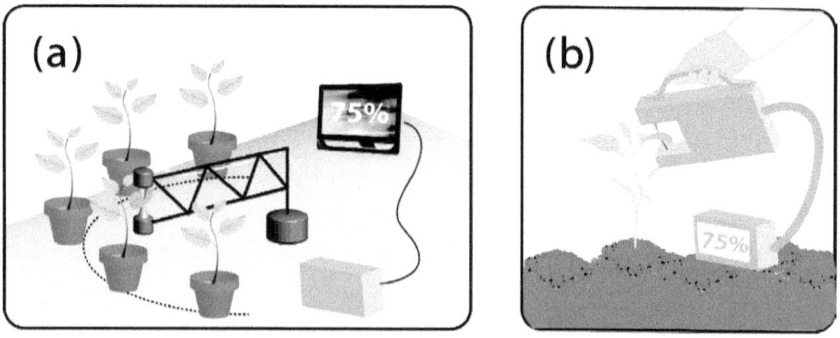

Figure 16.7 (a) A fiber laser-based system with a motorized arm that holds both the photoconductive emitter and the detector. This system is capable of monitoring many plants in sequence automatically. (b) A portable fiber-based system with a handheld emitter and detector head that can be used for measurements on the field. Illustrations courtesy of *Manzana Azul Publicidad+Diseno*.

16.7.1 Effective Medium Theory Applied to Plant Leaves

Vegetative tissue is a mixture of many substances such as proteins, sugars, water, and air arranged in a relatively complex structure. In Ref. [97] it was demonstrated that the dielectric function of a leaf, ε_L, at THz frequencies can be modeled as a combination of the dielectric functions ε_W, ε_A, and ε_S of water, air, and dry tissue, respectively, given by

$$\sqrt[3]{\varepsilon_L} = a_W \sqrt[3]{\varepsilon_W} + a_A \sqrt[3]{\varepsilon_A} + a_S \sqrt[3]{\varepsilon_S} \quad (16.9)$$

where a_W, a_A, and a_S are the volumetric fractions of water, air, and dry tissue, respectively. Generally speaking if these volumetric fractions and dielectric functions of the three components are known it is relatively simple to calculate the dielectric function of the leaf. However, if the dielectric functions of the leaf and the components are known, it is not trivial to find the relative volumetric fraction coefficients. In Ref. [98] an iterative algorithm that finds the coefficients is presented. This algorithm uses experimental time domain data to calculate the complex transmission of the leaf:

$$T^{\exp} = \frac{E_{\text{sam}}(\omega)}{E_{\text{ref}}(\omega)} \tag{16.10}$$

Subsequently the algorithm assumes values for a_W, a_A, and a_S and calculates a theoretical transmission

$$\left|T^{\text{theo}}\right| = t_{12}t_{21}\exp(-\omega d\kappa/c) \tag{16.11}$$

where t_{12} and t_{21} are the Fresnel transmission coefficients through the front and back surfaces of the leaf, d is the thickness of the leaf, and κ is the imaginary part of its refractive index corrected for surface scattering:

$$\kappa = \text{Im}(\sqrt{\varepsilon_L}) + \left[\left(\text{Real}(\sqrt{\varepsilon_L}) - 1\right)\frac{4\pi\tau\cos(\theta)}{\lambda}\right]^2 \frac{c}{2\omega d} \tag{16.12}$$

Here τ is the surface roughness measured as the standard deviation of the height profile, θ is the angle of incidence, and ε_L is the leaf's dielectric function calculated from Eq. 16.9. The algorithm varies the coefficients a_W, a_A, and a_S and the thickness d until the difference between $T^{\exp}$ and T^{theo} is smaller than a user-defined tolerance. This way the a_i and the thickness values are found.

16.7.2 Application to Barley

In Ref. [98] the algorithm described in the previous section was tested experimentally. Both gravimetric and THz transmission measurements were carried out on barley leaves that were cut (at time zero) and dried at 60°C for over ~2 hours. During this period measurements were taken every 10 minutes in order to follow the dehydration process of the tissue. The gravimetric and processed THz measurements are shown in Fig. 16.8. The two measurements correlate very well with a deviation between each other lower than 4%. This demonstrates that the algorithm is able to extract the water content of the leaves successfully. It is worth mentioning that the THz spectra for each measurement was recorded with a fast scan delay line in 2.5 seconds, so the accuracy of the measurement could be significantly improved if the measurement was taken over a longer averaging period, if required.

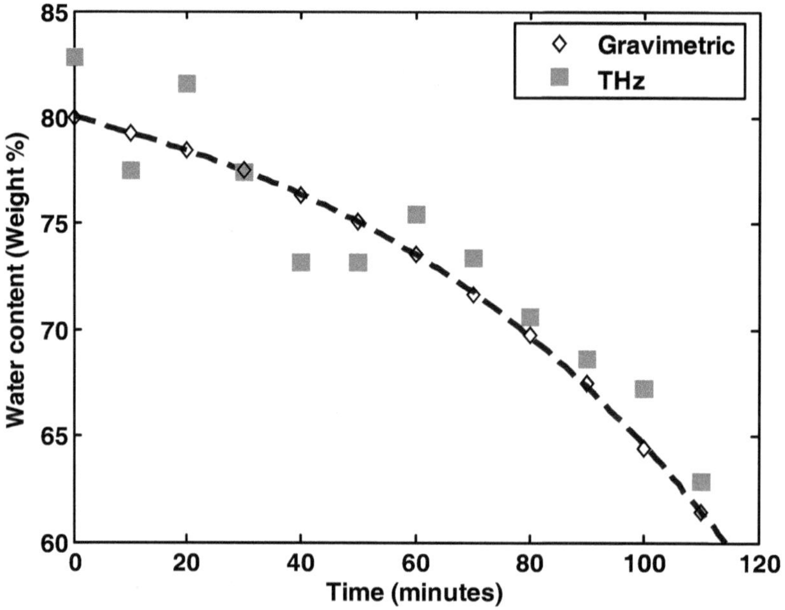

Figure 16.8 Water content of barley leaves determined both by gravimetric measurement (diamonds) and by THz transmission combined with the effective medium theory (EMT) inversion algorithm (squares). The two measurements correlate very well. The line is a "guide to the eye."

16.7.3 Perspectives

The ability to determine the amount of water in vegetative tissue in a nondestructive and noncontact way opens new possibilities for direct field application in crops as it could be the key to feedback intelligent irrigation systems or evaluation of certain aspects of the "health" state of vegetables in vivo. It also provides plant scientists with an unprecedentedly powerful tool for the investigation of both naturally occurring drought-resistant varieties and genetically modified crops under drought, saline, or other forms of stress, which influence their moisture content [99]. Given that most packaging of food products, including vegetables, is plastic based and that most plastics are THz transparent, this technology could also become a tool for the objective evaluation of "on-shelf" vegetables from the harvest moment to the final sell point, which could represent significant economic savings and a reduction of disposal of food in a good state—in a sense THz radiation could redefine what we understand by "best before" for vegetables.

16.8 Approaches to Increase the Speed of THz Systems

THz-TDS and imaging technology have presented enormous advances over the last three decades. Some applications for such technologies have been reviewed in this chapter, and there are many more that have been investigated and reported in the literature [52]. Currently the main technical aspect preventing the use of this technology for quality control in industrial environments is the slow data acquisition of THz systems that operate in research laboratories, which is incompatible with production time scales. Such systems normally use motorized delay lines to record the THz waveform as a function of time, and it, typically, takes between a few seconds to a few minutes to perform one scan. Additionally, if images are to be acquired the sample has to be scanned across a mesh of points normally using two additional motorized stages, which means that the data required for an image takes from several minutes to a few hours to be recorded. To overcome this problem there have been a number of publications, particularly over the last five years, where potential solutions were presented.

16.8.1 Faster Waveform Acquisition

Probably the first solution to the long waveform acquisition time problem was the proposal of rotatory delay lines that employ various spiral or helical geometries [100–104]. These devices are still mechanical and allow continuous scanning at a repetition rate between a few hertz to a few kilohertz. The components are still relatively hard to make with good enough quality, given that a relatively complicated surface has to be machined and polished to optical quality.

Given that fiber-coupled spectrometers, usually operating around 1550 nm, present significant advantages for industrial applications, another possible solution to the delay line problem for such systems is the implementation of a fiber stretcher [105]. These alternative delay lines use a piezoelectric actuator to stretch the optical fiber of one of the arms in the spectrometer within the elastic limit, producing a longer (or shorter) path in that arm and therefore scanning the relative delay of the pulses.

In recent years three innovative solutions based on tuning the optical cavity length of the ultrafast laser used were presented. The first one is a method called asynchronous optical sampling (ASOPS) [106, 107], in which two femtosecond lasers are used, one for the emitter and one for the detector. The pulse repetition rate of the two lasers is actively controlled so that these two frequencies, typically in the hundreds of megahertz, differ by Df, usually a few kilohertz. This will produce a relative delay of the two pulses to be scanned at a rate equal to the frequency difference, and extremely long scans (equal to the speed of light divided by the repetition rate) and therefore high spectral resolution can be achieved. The second method is called electronically controlled optical sampling (ECOPS) [108, 109], in which the repetition rate difference between the two lasers is actively changed from a certain value $+\Delta f$ to $-\Delta f$. With this a sweep "back and forward" of the mutual delay is achieved. As a consequence ECOPS will not produce such a long scan as ASOPS, but it will be able to acquire waveforms significantly faster. Finally, the third method is called optical sampling by laser cavity tuning (OSCAT) [110, 111]. This method is based on a single ultrafast laser with a tunable repetition rate. In this system a beam splitter is used to separate a THz generation and a detector gate beam. Subsequently the optical path of one of the arms is sent to travel an integer multiple of the distance required to produce a delay equal to the laser's repetition period, usually through an optical fiber. By doing this the pulse used for THz generation and the one used to trigger the detector are first or second or in general n^{th} neighbors on the pulse train. If the cavity repetition rate is then slightly modified, the relative delay of the two pulses is changed; therefore by continuously tuning the laser repetition rate it is possible to scan the delay between the two arms.

16.8.2 Faster Imaging

The generation of THz images is a complex process. The need to scan one temporal and two spatial variables makes data acquisition extremely slow. In the previous section we summarized some methods to accelerate temporal scanning; however, two degrees of spatial freedom have to be scanned quickly. Fortunately some methods that either avoid mechanical scanning or speed up this process have been proposed recently.

The idea of using THz f-θ lenses has been proposed. With these, large surface areas can be scanned by making relatively small angular deflections of the THz beam that can be achieved by using a pair of galvanometric scanners [112]. In the case of production line quality control, one possibility is to use the motion of samples along the production line conveyor in order to avoid one degree of scanning, reducing the problem to the production of 1D scans. This can be addressed in various ways. Firstly, an array of emitters and detectors can acquire the transmission or reflection of points across a line simultaneously [113]. Alternatively a THz f-θ lens with a single galvanometric scanner can be used [114].

If spectral resolution is not of interest, and only power reflection or transmission is important, it is also possible to produce a 1D image by dispersing a THz pulse along a line. This idea has been explored, and images have been recorded in this fashion [115].

16.8.3 Alternatives to Time Domain Imaging

Although the pulsed THz technique presents many advantages for imaging, other possibilities have also been explored. Compact monochromatic THz sources such as QCLs have improved significantly, and their operation at room temperature has been demonstrated [116]. On the detection side there has been enormous progress. Currently 2D THz sensor arrays have been reported, opening the possibility of fabricating THz cameras [117–120]. These technologies will be appropriate for the acquisition of images that do not require spectral information and that need being produced nearly instantaneously.

16.9 Approaches to Reduce the Cost of THz Systems

In addition to the technical limitations for THz systems to become useful in industrial environments, the cost of such systems is preventing THz technology from being accessible. Fortunately, recent developments are also addressing this issue. The most expensive component of a THz time domain system is the ultrafast laser whose cost is in the several tens of thousands of US dollars to a few hundred thousand US dollars. Additionally, most THz systems use off-axis parabolic mirrors, which are costly to fabricate, and high-resistivity silicon optics, which is an expensive material to produce and polish in comparison to glass or other standard optical materials. Most photoconductive emitters and detectors rely on the low-temperature growth of GaAs or other III–V semiconductors, which is still a relatively difficult process to implement on a large scale. This impedes the production of reliable photoconductive devices at low cost.

The introduction of THz QTDS [21], which was discussed earlier in this chapter, is probably one of the most important advances to reduce the cost of THz spectroscopy. This technique replaces the expensive ultrafast laser by a multimode diode laser, which is available commercially at a cost of a few hundred dollars. This technique has the advantage of maintaining magnitude and phase information; therefore, in that sense it is just as powerful as traditional TDS. However, the bandwidth achievable by QTDS is typically limited to several hundred gigahertz and the frequency resolution is limited to around 20 GHz [21] Nevertheless, this will be sufficient for several of the applications discussed in this chapter.

New methods to fabricate THz optical components at lower costs are being developed and reported frequently. Off-axis paraboloids are now being replaced by refractive plastic optics that can be molded by compression of micropowders [121, 122] or by mold injection [123] at extremely competitive costs. Plastic transmission [124] and reflection [125] gratings, as well as prisms [126], have been reported supplying dispersive components that can be mass-produced at a very low price. Variable focal length lenses for THz radiation have been reported as well [127]. Perhaps even cheaper than plastic is paper. Diffractive paper lenses for the THz band have been demonstrated [128]. Another interesting report demonstrates the possibility to produce a THz birefringent material by stacking sheets of paper. This way an economical waveplate can be produced [129] to manipulate the polarization of THz radiation.

The reliable industrial production of photoconductive devices is still probably one of the main problems to be resolved. Yet, the introduction of ion implantation to replace low-temperature growth processes is a promising approach, given that this technique produces substrates with repeatable optoelectronic properties. Emitters and detectors using these substrates have been reported and show comparably good performance [130–133].

QCLs are devices that are still relatively expensive to fabricate, but they can be mass-produced, and their cost is expected to drop dramatically in the coming years. The possibility of using them at room temperature also contributes to considerably reducing the operation cost of such devices. Planar detector arrays for THz frequencies, which were discussed in the previous section, are still research devices, but the integrated circuit industry could incorporate these detectors in production lines in the coming years, making them available for purchase in the near future at reasonable prices.

16.10 Conclusion

Since the first efforts to produce and detect radiation in the THz band about three decades ago, enormous progress has been achieved. Technologies from the low-frequency "electronic" world and from the high-frequency "optical" band have improved significantly. Additionally other forms of combined technology, such as photoconductive emitters, have progressively helped to close the so-called THz gap.

Over the last decade there has been particular interest in the development of industrial applications. In this chapter we reviewed the use of THz spectroscopy and imaging in the polymer, paper, food, pharmaceutical, and crop industries. Although these are fields in which THz radiation has demonstrated great potential, this is in no way an exhaustive list. In the last few years there has been enormous progress to adapt THz systems to industrial environments. Such systems have become more reliable and compatible with temperature and humidity fluctuations that are common in factories. The cost of THz equipment has dropped significantly and is expected to keep decreasing. Finally data acquisition has speeded up, making THz a realistic tool for quality control in production lines.

In a sense, we see THz industrial applications as an airplane at the end of its runway already with enough momentum to take off. Given that the technology is quickly maturing to a usable state, we foresee that THz will become a standard NDT technique applicable to many industrial processes in the years to come.

Acknowledgments

The majority of the data shown and discussed in this chapter has been obtained by the research group of Prof. Martin Koch. Many PhD students and collaborators have contributed to these results. Regarding the development of THz systems and devices, we acknowledge valuable contributions from N. Vieweg, R. Wilk, T. Hasek, B. Scherger, O. Peters, M. Schwerdtfeger, S. Busch, C. Jansen, S. Wietzke, M. Scheller, K. Altmann, T. Probst, N. Krumbholz, C. Jördens, K. Baaske, N. Born, and N. Voß. Applications for THz systems have been explored in collaboration with G. Gärtner, P. Plew, S. Sostmann, T. Hochrein, M. Bastian, S. Katlez, M. Pfleger, H. Pühringer, J. Léon, W. Sannemann, M. Hohmann, R. Snowden, and D. Selmar.

References

1. S. Leinß, T. Kampfrath, K. V. Volkmann, M. Wolf, J. T. Steiner, M. Kira, S. W. Koch, A. Leitenstorfer, and R. Huber, Terahertz coherent control of optically dark paraexcitons in Cu_2O, *Phys. Rev. Lett.* **101**(24), 246401 (2008).
2. H. Hirori, M. Nagai, and K. Tanaka, Excitonic interactions with intense terahertz pulses in ZnSe/ZnMgSSe multiple quantum wells, *Phys. Rev. B* **81**(8), 081305 (2010).
3. W. D. Rice, J. Kono, S. Zybell, S. Winnerl, J. Bhattacharyya, H. Schneider, M. Helm, B. Ewers, A. Chernikov, M. Koch, S. Chatterjee, G. Khitrova, H. M. Gibbs, L. Schneebeli, B. Breddermann, M. Kira, and S. W. Koch, Observation of forbidden exciton transitions mediated by coulomb interactions in photoexcited semiconductor quantum wells, *Phys. Rev. Lett.* **110**(13), 137404 (2013).

4. R. Piesiewicz, M. Jacob, M. Koch, J. Schoebel, and T. Kurner, Performance analysis of future multigigabit wireless communication systems at THz frequencies with highly directive antennas in realistic indoor environments, *IEEE J. Sel. Top. Quantum Electron.* **14**(2), 421–430 (2008).
5. T. Kleine-Ostmann and T. Nagatsuma, A review on terahertz communications research, *J. IR mm THz Waves* **32**(2), 143–171 (2011).
6. H. J. Song and T. Nagatsuma, Present and future of terahertz communications, *IEEE Trans. THz Sci. Technol.* **1**(1), 256–263 (2011).
7. R. Müller, A. Hoehl, A. Serdyukov, G. Ulm, J. Feikes, M. Ries, and G. Wüstefeld, The metrology light source of PTB: a source for THz radiation, *J. IR mm THz Waves* **32**(6), 742–753 (2011).
8. G. R. Neil, Accelerator sources for THz science: a review, *J. IR mm THz Waves* (online first) (2013).
9. D. D. Smith and A. Belyanin, Room-temperature semiconductor coherent Smith–Purcell terahertz sources, *Appl. Phys. Lett.* **98**(6), 063501–063501 (2011).
10. H. Yuan, X. Ge, and C. Zhang, Terahertz wave two-dimensional transmission imaging with a backward wave oscillator, in 2008 *International Conference on Optical Instruments and Technology: Microelectronic and Optoelectronic Devices and Integration*, edited by X. Zhang, W. J. Bock, X. Bao, and P. Shum, *Proc. SPIE* **7158**, 71580 (2009).
11. M. A. Salhi, I. Pupeza, and M. Koch, Confocal THz laser microscope, *J. IR mm THz Waves* **31**(3), 358–366 (2010).
12. Y. Watanabe, K. Kawase, T. Ikari, H. Ito, Y. Ishikawa, and H. Minamide, Component spatial pattern analysis of chemicals using terahertz spectroscopic imaging, *Appl. Phys. Lett.* **83**(4), 800–802 (2003).
13. K. Suizu, K. Koketsu, T. Shibuya, T. Tsutsui, T. Akiba, and K. Kawase, Extremely frequency-widened terahertz wave generation using Cherenkov-type radiation, *Opt. Express* **17**(8), 6676–6681 (2009).
14. M. Scheller, J. M. Yarborough, J. V. Moloney, M. Fallahi, M. Koch, and S. W. Koch, Room temperature continuous wave milliwatt terahertz source, *Opt. Express* **18**(26), 27112–27117 (2010).
15. J. Kiessling, F. Fuchs, K. Buse, and I. Breunig, Pump-enhanced optical parametric oscillator generating continuous wave tunable terahertz radiation, *Opt. Lett.* **36**(22), 4374–4376 (2011).
16. J. Faist, F. Capasso, D. L. Sivco, C. Sirtori, A. L. Hutchinson, and A. Y. Cho, Quantum cascade laser, *Science* **264**(5158), 553–556 (1994).

17. R. Köhler, A. Tredicucci, F. Beltram, H. E. Beere, E. H. Linfield, A. G. Davies, D. A. Ritchie, R. C. Iotti, and F. Rossi, Terahertz semiconductor-heterostructure laser, *Nature* **417**(6885), 156–159 (2002).
18. Q. Y. Lu, N. Bandyopadhyay, S. Slivken, Y. Bai, and M. Razeghi, Room temperature terahertz quantum cascade laser sources with 215 μW output power through epilayer-down mounting, *Appl. Phys. Lett.* **103**(1), 011101–011101 (2013).
19. D. Grischkowsky, S. Keiding, M. Van Exter, and C. Fattinger, Far-infrared time-domain spectroscopy with terahertz beams of dielectrics and semiconductors, *J. Opt. Soc. Am. B* **7**(10), 2006–2015 (1990).
20. S. Verghese, K. A. McIntosh, S. Calawa, W. F. Dinatale, E. K. Duerr, and K. A. Molvar, Generation and detection of coherent terahertz waves using two photomixers. *Appl. Phys. Lett.* **73**(26), 3824–3826 (1998).
21. M. Scheller and M. Koch, Terahertz quasi time domain spectroscopy, *Opt. Express* **17**(20), 17723–17733 (2009).
22. P. R. Smith, D. H. Auston, and M. C. Nuss, Subpicosecond photoconducting dipole antennas, *IEEE J. Quantum Electron.* **24**(2), 255–260 (1988).
23. S. Hunsche, D. M. Mittleman, and M. Koch, New dimensions in T-ray imaging, *IEICE Trans. Electron.* **81**(2), 269–276 (1998).
24. E. Castro-Camus, J. Lloyd-Hughes, and M. B. Johnston, Three-dimensional carrier-dynamics simulation of terahertz emission from photoconductive switches. *Phys. Rev. B* **71**(19), 195301 (2005).
25. K. Ezdi, B. Heinen, C. Jördens, N. Vieweg, N. Krumbholz, R. Wilk, M. Mikulics, and M. Koch, A hybrid time-domain model for pulsed terahertz dipole antennas, *J. Eur. Opt. Soc.: Rapid Pub.*, **4** (2009).
26. E. Castro-Camus, M. B. Johnston, and J. Lloyd-Hughes, Simulation of fluence-dependent photocurrent in terahertz photoconductive receivers, *Semicond. Sci. Technol.* **27**(11), 115011 (2012).
27. P. U. Jepsen, R. H. Jacobsen, and S. R. Keiding, Generation and detection of terahertz pulses from biased semiconductor antennas, *J. Opt. Soc. Am. B* **13**(11), 2424–2436 (1996).
28. Z. Piao, M. Tani, and K. Sakai, Carrier dynamics and terahertz radiation in photoconductive antennas, *Jpn. J. Appl. Phys.* **39**, 96 (2000).
29. M. van Exter, C. Fattinger, and D. Grischkowsky, Terahertz time-domain spectroscopy of water vapor, *Opt. Lett.* **14**(20), 1128–1130 (1989).
30. D. M. Mittleman, J. Cunningham, M. C. Nuss, and M. Geva, Noncontact semiconductor wafer characterization with the terahertz Hall effect, *Appl. Phys. Lett.* **71**(1), 16–18 (1997).

31. C. Rønne, P. O. Åstrand, and S. R. Keiding, THz Spectroscopy of liquid H_2O and D_2O, *Phys. Rev. Lett.* **82**(14), 2888 (1999).

32. J. P. Laib and D. M. Mittleman, Temperature-dependent terahertz spectroscopy of liquid n-alkanes, *J. IR mm THz Waves* **31**(9), 1015-1021 (2010).

33. N. Vieweg, M. K. Shakfa, and M. Koch, BL037: A nematic mixture with high terahertz birefringence, *Opt. Commun.*, **284**(7), 1887–1889 (2011).

34. D. M. Mittleman, R. H. Jacobsen, R. Neelamani, R. G. Baraniuk, and M. C. Nuss, Gas sensing using terahertz time-domain spectroscopy, *Appl. Phys. B: Lasers Opt.* **67**(3), 379–390 (1998).

35. B. Scherger, C. Jördens, and M. Koch, Variable-focus terahertz lens, *Opt. Express* **19**(5), 4528–4535 (2011).

36. R. Wilk, N. Vieweg, O. Kopschinski, and M. Koch, Liquid crystal based electrically switchable Bragg structure for THz waves, *Opt. Express* **17**(9), 7377–7382 (2009).

37. C. Jansen, I. A. Al-Naib, N. Born, and M. Koch, Terahertz metasurfaces with high Q-factors., *Appl. Phys. Lett.* **98**(5), 051109–051109 (2011).

38. V. Astley, J. Scheiman, R. Mendis, and D. M. Mittleman, Bending and coupling losses in terahertz wire waveguides, *Opt. Lett.* **35**(4), 553-555 (2010).

39. M. Ahmadi-Boroujeni, K. Altmann, B. Scherger, C. Jansen, M. Shahabadi, and M. Koch, Terahertz parallel-plate ladder waveguide with highly confined guided modes, *IEEE Trans. THz Sci. Technol.* **3**(1), 87–95 (2013).

40. M. Herrmann, M. Tani, and K. Sakai, Display modes in time-resolved terahertz imaging, *Jpn. J. Appl. Phys.* **39**, 6254 (2000).

41. R. Wilk, F. Breitfeld, M. Mikulics, and M. Koch, A continuous wave THz spectrometer as a non-contact thickness measuring device, *Appl. Opt.* **47**, 3023 (2008).

42. B. Sartorius, M. Schlak, D. Stanze, H. Roehle, H. Künzel, D. Schmidt, H. G. Bach, R. Kunkel, and M. Schell, Continuous wave terahertz systems exploiting 1.5 μm telecom technologies, *Opt. Express* **17**(17), 15001-15007 (2009).

43. A. Roggenbuck, H. Schmitz, A. Deninger, I. C. Mayorga, J. Hemberger, R. Güsten, and M. Grüninger, Coherent broadband continuous-wave terahertz spectroscopy on solid-state samples, *New J. Phys.* **12**(4), 043017 (2010).

44. H. C. Ryu, N. Kim, S. P. Han, H. Ko, J. W. Park, K. Moon, and K. H. Park, Simple and cost-effective thickness measurement terahertz system based on a compact 1.55 μm λ/4 phase-shifted dual-mode laser, *Opt. Express* **20**(23), 25990–25999 (2012).

45. S. Wietzke, C. Jördens, N. Krumbholz, B. Baudrit, M. Bastian, and M. Koch, Terahertz imaging: a new non-destructive technique for the quality control of plastic weld joints, *J. Eur. Opt. Soc.: Rapid Pub.*, **2**, 07013 (2007).

46. C. Jansen, S. Wietzke, H. Wang, M. Koch, and G. Zhao, Terahertz spectroscopy on adhesive bonds, *Polym. Test.* **30**(1), 150–154 (2011).

47. C. Jördens, S. Wietzke, M. Scheller, and M. Koch, Investigation of the water absorption in polyamide and wood plastic composite by terahertz time-domain spectroscopy, *Polym. Test.* **29**(2), 209–215 (2010).

48. S. Wietzke, C. Jansen, T. Jung, M. Reuter, B. Baudrit, M. Bastian, S. Chatterjee, and M. Koch, Terahertz time-domain spectroscopy as a tool to monitor the glass transition in polymers, *Opt. Express* **17**(21), 19006–19014 (2009).

49. S. Wietzke, C. Jansen, F. Rutz, D. M. Mittleman, and M. Koch, Determination of additive content in polymeric compounds with terahertz time-domain spectroscopy, *Polym. Test.* **26**(5), 614–618 (2007).

50. M. Scheller, S. Wietzke, C. Jansen, and M. Koch, Modelling heterogeneous dielectric mixtures in the terahertz regime: a quasi-static effective medium theory, *J. Phys. D: Appl. Phys.* **42**(6), 065415 (2009).

51. N. Krumbholz, T. Hochrein, N. Vieweg, T. Hasek, K. Kretschmer, M. Bastian, M. Mikulics, and M. Koch, Monitoring polymeric compounding processes inline with THz time-domain spectroscopy, *Polym. Test.* **28**(1), 30–35 (2009).

52. P. U. Jepsen, D. G. Cooke, and M. Koch, Terahertz spectroscopy and imaging: modern techniques and applications, *Laser Photon. Rev.* **5**(1), 124–166 (2011).

53. O. Peters, M. Schwerdtfeger, S. Wietzke, S. Sostmann, R. Scheunemann, R. Wilk, R. Holzwarth, M. Koch, and B. M. Fischer, Terahertz spectroscopy for rubber production testing, *Polym. Test.* **32**(5), 932–936 (2013).

54. K. Ramesh, *Digital Photoelasticity*, 1st ed. (Springer, Berlin, Heidelberg, Germany, 2000).

55. S. R. Doshi and J. M. Charrier, A simple illustration of structure-properties relationships for short fiber reinforced thermoplastics, *Polym. Compos.* **10**, 28–38 (1989).

56. S.-Y. Fu and B. Lauke, Effects of fiber length and fiber orientation distributions on the tensile strength of short-fiber-reinforced polymers, *Compos. Sci. Technol.* **56**, 1179–1190 (1996).

57. C. Jördens, M. Scheller, S. Wietzke, D. Romeike, C. Jansen, T. Zentgraf, K. Wiesauer, V. Reisecker, and M. Koch, Terahertz spectroscopy to study the orientation of glass fibres in reinforced plastics, *Compos. Sci. Technol.* **70**, 472–477 (2010).

58. S. Katletz, M. Pfleger, H. Pühringer, M. Mikulics, N. Vieweg, O. Peters, B. Scherger, M. Scheller, M. Koch, and K. Wiesauer, Polarization sensitive terahertz imaging: detection of birefringence and optical axis, *Opt. Express*, **20**, 23025–23035 (2012).

59. F. Rutz, S. Wietzke, M. Koch, H. Richter, S. Hickmann, V. Trappe, and U. Ewert, Non-destructive testing of glass-fibre reinforced polymers using terahertz spectroscopy, *Proc. Eur. Conf. Non-destruct. Test. (ECNDT)*, Paper We2.8.2 (2006).

60. M. Born and E. Wolf, *Principles of Optics*, 7th ed., 237 (Cambridge University Press, Cambridge, 1999).

61. M. Reid and R. Fedosejevs, Terahertz birefringence and attenuation properties of wood and paper, *Appl. Opt.* **45**, 2766–2772 (2006).

62. T. M. Todoruk, I. D. Hartley, and M. E. Reid, Origin of birefringence in wood at terahertz frequencies, *IEEE Trans. THz Sci. Technol.* **2**, 123–130 (2012).

63. T. Hattori, H. Kumon, and H. Tamazumi, Terahertz spectroscopic characterization of paper, *Proc. 35th Int. Conf. IRMMW-THz*, 1–2 (2010).

64. S.-I. Ebara, Y. Hirota, M. Tani, and M. Hangyo, Highly sensitive birefringence measurement in THz frequency region and its application to stress measurement, *Proc. Joint 32nd Int. Conf. IR mm Waves 15th Int. Conf. THz Electron. (IRMMW-THz)*, 666–667 (2007).

65. K. Wiesauer and C. Jördens, Recent advances in birefringence studies at THz frequencies, *J. IR mm THz Waves* **34**, 663–681 (2013).

66. J.-W. Park, K.-H. Im, D. K. Hus, J.-A. Jung, and I.-Y. Yang, Terahertz spectroscopy approach of the fiber orientation influence on CFRP composite solid laminates, *J. Meas. Sci. Technol.* **26**, 2051–2054 (2012).

67. D. M. Mittleman, J. Cunningham, and M. C. Nuss, Noncontact semiconductor wafer characterization with the terahertz Hall effect, *Appl. Phys. Lett.* **71**, 16–18 (1987).

68. N. C. H. van der Valk, W. A. M. van der Marel, and P. C. M. Planken, Terahertz polarization imaging, *Opt. Lett.* **30**, 2802–2804 (2005).

69. R. Zhang, Y. Cui, W. Sun, and Y. Zhang, Polarization information for terahertz imaging, *Appl. Opt.* **47**, 6422–6426 (2008).

70. L. Zhang, H. Zhong, C. Deng, C. Zhang, and Y. Zhao, Terahertz polarization imaging with birefringent materials, *Opt. Commun.*, **283**, 4993–4995 (2010).

71. L. Zhang, H. Zhong, C. Deng, C. Zhang, and Y. Zhao, THz wave polarization-controlled spectroscopic imaging for anisotropic materials, *Opt. Commun.*, **284**, 4356–4359 (2011).

72. B. Gompf, N. Gebert, H. Heer, and M. Dressel, Polarization contrast terahertz-near-field imaging of anisotropic conductors, *Appl. Phys. Lett.* **90**, 082104 (2007).

73. X. Wang, Y. Cui, W. Sun, J. Ye, and Y. Zhang, Terahertz polarization real-time imaging based on balanced electro-optic detection, *J. Opt. Soc. Am.* 27, 2387–2393 (2010).

74. F. Rutz, T. Hasek, and M. Koch, Terahertz birefringence of liquid crystal polymers, *Appl. Phys. Lett.* **89**, 221911 (2006).

75. Jördens, C., M. Scheller, M. Wichmann, M. Mikulics, K. Wiesauer, and M. Koch, Terahertz birefringence for orientation analysis, *Appl. Opt.* **48**, 2037–2044 (2009).

76. S. Katletz, M. Pfleger, H. Pühringer, N. Vieweg, B. Scherger, B. Heinen, M. Koch, and K. Wiesauer, Efficient terahertz en-face imaging, *Opt. Express* **19**, 23042–23053 (2011).

77. E. Castro-Camus, J. Lloyd-Hughes, and M. B. Johnston, Polarisation-sensitive terahertz detection by multicontact photoconductive receivers, *Appl. Phys. Lett.* **86**, 254102 (2005).

78. D. Banerjee, W. Von Spiegel, M. D. Thomson, S. Schabel, and H. G. Roskos, Diagnosing water content in paper by terahertz radiation, *Opt. Express* **16**(12), 9060–9066 (2008).

79. P. Mousavi, F. Haran, D. Jez, F. Santosa, and J. S. Dodge, Simultaneous composition and thickness measurement of paper using terahertz time-domain spectroscopy, *Appl. Opt.* **48**(33), 6541–6546 (2009).

80. M. Scheller, C. Jansen, and M. Koch, Analyzing sub-100-μm samples with transmission terahertz time domain spectroscopy, *Opt. Commun.*, **282**(7), 1304–1306 (2009).

81. C. Jansen, S. Wietzke, H. Wang, M. Koch, G. Zhao, Terahertz spectroscopy on adhesive bonds, *Polym. Test.* **30**(1), 150–154 (2011).

82. C. Jördens and M. Koch, Detection of foreign bodies in chocolate with pulsed terahertz spectroscopy, *Opt. Eng.* **47**(3), 037003–037003 (2008).

83. M. Herrmann, M. Tani, M. Watanabe, and K. Sakai, Terahertz imaging of objects in powders, *IEE Proc.: Optoelectron.* **149**(3), 116–120 (2002).

84. P. F. Taday, I. V. Bradley, D. D. Arnone, and M. Pepper, Using terahertz pulse spectroscopy to study the crystalline structure of a drug: A case study of the polymorphs of ranitidine hydrochloride, *J. Pharm. Sci.* **92**(4), 831–838 (2003).

85. P. F. Taday, Applications of terahertz spectroscopy to pharmaceutical sciences, *Philos. Trans. R. Soc. London A* **362**(1815), 351–364 (2004).

86. J. A. Zeitler, K. Kogermann, J. Rantanen, T. Rades, P. F. Taday, M. Pepper, J. Aaltonen, and C. J. Strachan, Drug hydrate systems and dehydration processes studied by terahertz pulsed spectroscopy, *Int. J. Pharm.* **334**(1), 78–84 (2007).

87. A. J. Fitzgerald, B. E. Cole, and P. F. Taday, Nondestructive analysis of tablet coating thicknesses using terahertz pulsed imaging, *J. Pharm. Sci.* **94**(1), 177–183 (2005).

88. Food and Aquaculture Organization of the United Nations (2013) (cited August 27, 2013), http://www.fao.org/nr/water/aquastat/water_use/index.stm.

89. N. P. Money, Osmotic pressure of aqueous polyethylene glycols relationship between molecular weight and vapor pressure deficit, *Plant Physiol.* **91**(2), 766–769.

90. Nautiyal, P. C., Rachaputi, N. R. and Joshi, Y. C. (2002). Moisture-deficit-induced changes in leaf-water content, leaf carbon exchange rate and biomass production in groundnut cultivars differing in specific leaf area, *Field Crops Res.* **74**(1), 67–79 (1989).

91. D. M. Mittleman, R. H. Jacobsen, and M. C. Nuss, T-ray imaging, *IEEE J. Sel. Top. Quantum Electron.* **2**(3), 679–692 (1996).

92. B. B. Hu and M. C. Nuss, Imaging with terahertz waves, *Opt. Lett.* **20**(16), 1716–1718 (1995).

93. M. Koch, THz imaging: fundamentals and biological applications, *Proc. SPIE* **3828**, 202–208 (1999).

94. M. Koch, Bio-medical applications of THz imaging, in *Sensing with Terahertz Radiation*, 295–316 (Springer, Berlin Heidelberg, 2003).

95. B. Breitenstein, M. Scheller, M. K. Shakfa, T. Kinder, T. Müller-Wirts, M. Koch, and D. Selmar, Introducing terahertz technology into plant biology: a novel method to monitor changes in leaf water status, *J. Appl. Botany Food Qual.* **84**(2), 158 (2012).

96. S. Hadjiloucas, R. K. Galvão, and J. W. Bowen, Analysis of spectroscopic measurements of leaf water content at terahertz frequencies using linear transforms, *J. Opt. Soc. Am. A* **19**(12), 2495–2509 (2002).

97. C. Jördens, M. Scheller, B. Breitenstein, D. Selmar, and M. Koch, Evaluation of leaf water status by means of permittivity at terahertz frequencies, *J. Biol. Phys.* **35**(3), 255–264 (2009).

98. R. Gente, N. Born, N. Voß, W. Sannemann, J. Léon, M. Koch, and E. Castro-Camus, Determination of leaf water content from terahertz time-domain spectroscopic data, *J. IR mm THz Waves* 34(3–4), 316–323 (2013).

99. E. Castro-Camus, M. Palomar, and A. A. Covarrubias, Leaf water dynamics of Arabidopsis thaliana monitored in-vivo using terahertz time-domain spectroscopy, *Sci. Rep.* **3**, 2910 (2013).

100. J. Xu, Z. Lu, and X. C. Zhang, Compact involute optical delay line, *Electron. Lett.* **40**(19), 1218–1219 (2004).

101. Y. Wang, C. Wang, Q. Xing, F. Liu, Y. Li, L. Chai, Q. Wang, F. Fang, and X. Zhang, Periodic optical delay line based on a tilted parabolic generatrix helicoid reflective mirror, *Appl. Opt.* **48**(11), 1998–2005 (2009).

102. G. J. Kim, S. G. Jeon, J. I. Kim, and Y. S. Jin, High speed scanning of terahertz pulse by a rotary optical delay line, *Rev. Sci. Instrum.* **79**(10), 106102–106102 (2008).

103. G. J. Kim, S. G. Jeon, J. I. Kim, and S. T. Han, A novel optical delay line using a rotating planar reflector for fast measurement of a terahertz pulse, *J. Korean Phys. Soc.* **56**, 1763 (2010).

104. M. Salhi, F. Rutz, T. Kleine-Ostmann, V. Petukhov, C. Metz, and M. Koch, Spiral optical delay line, in *Optical Terahertz Science and Technology, Technical Digest (CD)*, Paper TuD5 (Optical Society of America, 2005).

105. N. Krumbholz, M. Schwerdtfeger, T. Hasek, B. Scherger, and M. Koch, A fiberstretcher operating as an optical delay line in a fiber-coupled THz spectrometer, *Proc. 33th Int. Conf. IRMMW-THz*, 1–2 (2008).

106. G. Klatt, R. Gebs, C. Janke, T. Dekorsy, and A. Bartels, Rapid-scanning terahertz precision spectrometer with more than 6 THz spectral coverage, *Opt. Express* **17**, 22847 (2009).

107. T. Yasui, K. Kawamoto, Y. D. Hsieh, Y. Sakaguchi, M. Jewariya, H. Inaba, K. Minoshima, F. Hindle, and T. Araki, Enhancement of spectral resolution and accuracy in asynchronous-optical-sampling terahertz time-domain spectroscopy for low-pressure gas-phase analysis, *Opt. Express* **20**(14), 15071–15078 (2012).

108. Y. Kim and D. S. Yee, High-speed terahertz time-domain spectroscopy based on electronically controlled optical sampling, *Opt. Lett.* **35**(22), 3715–3717 (2010).

109. S. Kray, F. Spöler, T. Hellerer, and H. Kurz, Electronically controlled coherent linear optical sampling for optical coherence tomography, *Opt. Express* **18**(10), 9976–9990 (2010).

110. T. Hochrein, R. Wilk, M. Mei, R. Holzwarth, N. Krumbholz, and M. Koch, Optical sampling by laser cavity tuning, *Opt. Express* **18**(2), 1613–1617 (2010).

111. R. Wilk, T. Hochrein, M. Koch, M. Mei, and R. Holzwarth, OSCAT: novel technique for time-resolved experiments without moveable optical delay lines, *J. IR mm THz Waves* **32**(5), 596–602 (2011).

112. S. Katletz, M. Pfleger, H. Pühringer, N. Vieweg, B. Scherger, B. Heinen, M. Koch, and K. Wiesauer, Efficient terahertz en-face imaging, *Opt. Express* **19**(23), 23042–23053 (2011).

113. C. Jördens, M. Salhi, T. Hasek, G. Thorwirth, U. Böttger, and M. Koch, Micro-mirrors for a multifocus terahertz imaging system, *Proc. Eur. Microwave Asso.* **2**(4), 300–304 (2006).

114. B. Scherger, C. Jordens, D. Stanze, N. Krumbholz, and M. Koch, Increasing the speed of THz TDS imaging, *Proc. 34th Int. Conf. IRMMW-THz*, 1–2 (2009).

115. S. Schumann, C. Jansen, M. Schwerdtfeger, S. Busch, O. Peters, M. Scheller, and M. Koch, Spectrum to space transformed fast terahertz imaging, *Opt. Express* **20**(17), 19200–19205 (2012).

116. M. A. Belkin, F. Capasso, F. Xie, A. Belyanin, M. Fischer, A. Wittmann, and J. Faist, Room temperature terahertz quantum cascade laser source based on intracavity difference-frequency generation, *Appl. Phys. Lett.* **92**(20), 201101–201101 (2008).

117. F. Schuster, D. Coquillat, H. Videlier, M. Sakowicz, F. Teppe, L. Dussopt, B. Giffard, T. Skotnicki, and W. Knap, Broadband terahertz imaging with highly sensitive silicon CMOS detectors, *Opt. Express* **19**(8), 7827–7832 (2011).

118. S. Boppel, A. Lisauskas, V. Krozer, and H. G. Roskos, Performance and performance variations of sub-1 THz detectors fabricated with 0.15 μm CMOS foundry process, *Electron. Lett.* **47**(11), 661–662 (2011).

119. N. Oda, Uncooled bolometer-type Terahertz focal plane array and camera for real-time imaging, *Comptes Rendus Phys.* **11**(7), 496–509 (2010).

120. R. Al Hadi, H. Sherry, J. Grzyb, Y. Zhao, W. Forster, H. M. Keller, A. Cathelin, A. Kaiser, and U. R. Pfeiffer, A 1 k-pixel video camera for 0.7–1.1 terahertz imaging applications in 65-nm CMOS, *IEEE J. Solid-State Circuits* **47**(12) (2012).

121. B. Scherger, M. Scheller, C. Jansen, M. Koch, and K. Wiesauer, Terahertz lenses made by compression molding of micropowders, *Appl. Opt.* **50**(15), 2256–2262 (2011).

122. B. Scherger, S. Wietzke, M. Scheller, N. Vieweg, M. Wichmann, M. Koch, and K. Wiesauer, Characterization of micro-powders for the fabrication of compression molded THz lenses, *J. IR mm THz Waves* **32**(7), 943–951 (2011).

123. M. Wichmann, A. S. Mondol, N. Kocic, S. Lippert, T. Probst, M. Schwerdtfeger, S. Schumann, T. Hochrein, P. Heidemeyer, M. Bastian, G. Bastian, and M. Koch, Terahertz plastic compound lenses, *Appl. Opt.* **52**(18), 4186–4191 (2013).

124. B. Scherger, N. Born, C. Jansen, S. Schumann, M. Koch, and K. Wiesauer, Compression molded terahertz transmission blaze-grating, *IEEE Trans. THz Sci. Technol.* **2**(5), 556–561 (2012).

125. S. F. Busch, N. Born, M. Koch, and B. Fischer, Terahertz reflection gratings made by room-temperature high-pressure molding, *J. IR mm THz Waves* **34**, 413 (2013).

126. C. Goy, M. Scheller, B. Scherger, V. Wallace, and M. Koch, Terahertz waveguide prism, *Opt. Express* **21**(16), 19292–19301 (2013).

127. B. Scherger, C. Jördens, and M. Koch, Variable-focus terahertz lens, *Opt. Express* **19**(5), 4528–4535 (2011).

128. A. Siemion, M. Makowski, J. Suszek, J. Bomba, A. Czerwiński, F. Garet, J.-L. Coutaz, and M. Sypek, Diffractive paper lens for terahertz optics, *Opt. Lett.* **37**(20), 4320–4322 (2012).

129. B. Scherger, M. Scheller, N. Vieweg, S. T. Cundiff, and M. Koch, Paper terahertz wave plates, *Opt. Express* **19**(25), 24884–24889 (2011).

130. E. Castro-Camus, J. Lloyd-Hughes, L. Fu, H. Tan, C. Jagadish, and M. Johnston, An ion-implanted InP receiver for polarization resolved terahertz spectroscopy, *Opt. Express* **15**, 7047–7057 (2007).

131. M. Suzuki and M. Tonouchi, Fe-implanted InGaAs photoconductive terahertz detectors triggered by 1.56μm femtosecond optical pulses, *Appl. Phys. Lett.* **86**(16), 163504–163504 (2005).

132. T. Liu, M. Tani, M. Nakajima, M. Hangyo, K. Sakai, S. Nakashima, and C. Pan, Ultrabroadband terahertz field detection by proton-bombarded InP photoconductive antennas, *Opt. Express* **12**, 2954–2959 (2004).

133. T. A. Liu, M. Tani, M. Nakajima, M. Hangyo, and C. L. Pan, Ultrabroadband terahertz field detection by photoconductive antennas based on multi-energy arsenic-ion-implanted GaAs and semi-insulating GaAs, *Appl. Phys. Lett.* **83**(7), 1322–1324 (2003).

Chapter 17

Terahertz Wireless Communications

Ho-Jin Song,[a] **Sebastian Priebe,**[b] **and Thomas Kürner**[b]
[a]*NTT Device Technology Laboratories, NTT Corporation,*
3-1, Morinosato-Wakamiya, Atsugi, Kanagawa 243-0198, Japan
[b]*Instiut für Nachrichtentechnik, Technische Universität Braunschweig,*
Schleinitzstraße 22, 38106 Braunschweig, Germany
song.hojin@lab.ntt.co.jp, priebe@ifn.ing.tu-bs.de, kuerner@ifn.ing.tu-bs.de

17.1 Introduction

Ever since various multimedia services based on the Internet were introduced in the 1990s, traffic related to such services has been increasing steadily worldwide, usually via wired networks. However, with the introduction of new mobile devices and new multimedia services delivered wirelessly, we have been seeing changes in the media consumption behavior over the last few years. Users consume many more digital services and a lot of digital information with their own handheld mobile devices such as smartphones or tablet PCs. Moreover, the popularity of personal multimedia devices, such as digital cameras and camcorders, and inexpensive large-capacity personal storage media enable every individual to perform a more active role in the circulation of multimedia information. People are

Handbook of Terahertz Technologies: Devices and Applications
Edited by Ho-Jin Song and Tadao Nagatsuma

ISBN 978-981-4613-08-8 (Hardcover), 978-981-4613-09-5 (eBook)
www.panstanford.com

not just enjoying multimedia services provided by service providers but also generating digital contents and distributing them with their smartphones via the Internet. Users, having become sources of digital information, are demanding new ways to manage and share data on PCs and mobile devices and to store data more easily and quickly. This includes the replacement of wired connections by wireless links.

To satisfy the needs of users, the data capacity of wireless communications has improved with progress much faster than for wired systems over the past couple of decades, and this trend is extremely likely to continue [1]. Extrapolating the evolution of wireless data rates from Ref. [1] results in a prediction of around 100 Gbps required in 2020.

In the race toward such ultrafast wireless communications systems far beyond 1 Gbps, several candidates, including ultrawide band (UWB), 60 GHz radio, free-space optical (FSO) communications, and the Infrared Data Association (IrDA), have been investigated, and terahertz (THz) waves [2] have recently joined the competition. For many current wireless communications systems operating at microwave frequencies, data capacity has been improved by increasing spectral efficiency with advanced modulation schemes and signal-processing techniques [3, 4]. However, realization of such high data rates becomes quite challenging because of a fundamental limitation of current technologies—limited bandwidth. The largest chunk of the worldwide-available spectrum allocated to mobile services provides a bandwidth of 7 GHz at around 60 GHz. Transmitting 100 Gbps in a 7 GHz narrow radio channel would require unrealistically high spectral efficiencies of around 15 bit/s/Hz. Hence, for future wireless communications systems, it is obvious that more spectral resources are necessary, which are available beyond 300 GHz in the so-called THz frequency band. Wireless data transmission at THz frequencies is referred to as THz communications.

Over the past 10 years, significant progress has been made in basic propagation research and semiconductor technology, paving the way toward realizing THz communications systems. Moreover, activities toward the standardization of such systems as well as toward spectrum regulations above 300 GHz have been initiated in 2008.

In this chapter, an overview is given of potential applications for THz communications, the state of the art in semiconductor technology, and the radio wave propagation research with respect to THz communications. The chapter is rounded off with a summary of the current status in standardization and spectrum regulations, giving the inclined reader a fundamental understanding of the challenges, chances, and concepts of wireless communications at THz frequencies.

17.2 Why THz Waves for Communications

The reason many researchers around the world are interested in THz waves for wireless communications is that these unexploited frequency bands can offer very huge bandwidth, which is essential for increasing data capacity. Figure 17.1 shows the atmospheric attenuation up to 3 THz, including the effects of rain at rates of 25 mm/h and 5 mm/h, computed with the International Telecommunication Union Radio Communication Sector (ITU-R) [5, 53] models. Obviously, attenuation in the atmosphere at frequencies above 100 GHz is much larger than that in the microwave frequency band. In particular, the frequency region above 1 THz seems to be unsuitable for wireless communications because of the many absorption lines of water vapor and other atmospheric gases. Though there are also several water absorption lines below 1 THz, the attenuation level in the frequency windows may not be a big problem for short-range applications, where the range is likely to be a few tens of meters or less. In Table 17.1, possible windows at frequencies above 275 GHz, which are not allocated yet in many countries, are listed for attenuation less than 1 dB/m. As can be seen, for example, we can have approximately a 40 GHz bandwidth at around 300 GHz. For the sixth and seventh bands in Table 17.1, more than a 100 GHz spectral span, at the maximum, is expected to be available, which allows a 100 Gbps throughput even with a simple modulation scheme. Note that this is much larger than the entire spectral resources that we are currently using for all kinds of wireless systems on this planet.

To get an idea of how much data capacity is available with THz waves, a simple link budget analysis has been conducted using the Friis formula. The received power P_r and the signal-to-noise ratio (SNR_{dB}) at the receiver are given as

$$P_r = P_t + G_t + G_r + 20\log\left(\frac{\lambda_c}{4\pi d}\right) - \{\alpha_a(f_c)\cdot d\} - L_{ex} \quad (17.1)$$

$$SNR_{\text{dB}} = P_r - (N_0 + 10\log(B) + NF + M) \tag{17.2}$$

where P_t is the power input to the transmitting antenna. G_t and G_r are antenna gains of the transmitting and receiving antennas, respectively. d and λ are the distance and wavelength, respectively. (a_a) is the atmospheric attenuation at frequency f_c. L_{ex} is the excess loss not included in free-space loss (FSL). NF and M are the total noise figure of the receiver and the system margin in decibels. N_0 is the noise power spectral density. B is the system noise bandwidth.

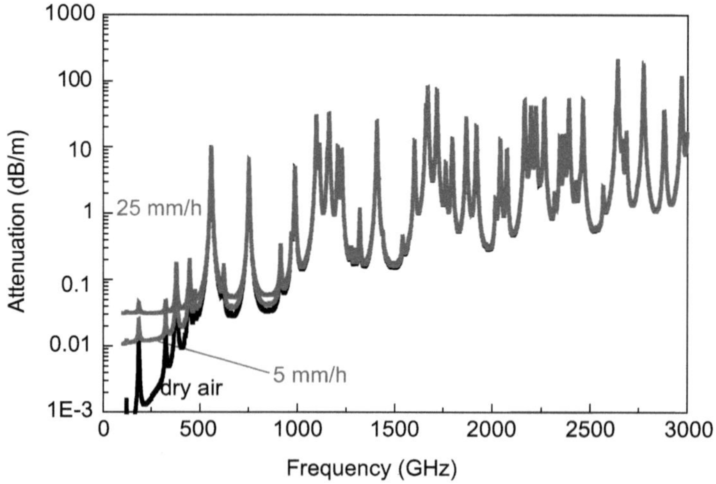

Figure 17.1 Atmospheric attenuation computed, including effects of rain at rates of 25 mm/h and 5 mm/h (sea-level altitude, water content of 7.5 g/m^3, and atmospheric temperature of 20°C).

Table 17.1 List of potential atmospheric windows at THz frequencies (atmospheric attenuation < 1 dB/m)

No.	Bands (GHz)	Bandwidth (GHz)
1	275–315	40
2	330–365	35
3	390–435	45
4	450–515	65
5	625–725	100
6	790–900	110
7	925–950	25

In this analysis, amplitude-shift keying (ASK) with incoherent detection and binary phase-shift keying (BPSK) with coherent detection are assumed with additive white Gaussian noise. Spectral efficiency was fixed at 1 bps/Hz, and though unrealistic, it was assumed that the system uses sufficient bandwidth for a specific data rate. To simplify the discussion, G_t and G_r are assumed to be equal. The parameters used in the calculation are summarized in Table 17.2. In Fig. 17.2, Shannon's maximum data capacity and available data rates with ASK and BPSK modulations for a bit error rate (BER) of 10^{-6} are plotted as a function of antenna gain in the transmitter and receiver. As can be seen, for 10 Gbps, an antenna gain of 28–31 dBi is necessary, depending on the modulation format, and these results show good agreement with those in Ref. [6]. Since the antenna gain is linked to the SNR of the system, traces in Fig. 17.2 will move horizontally if some parameters affecting the SNR or the modulation format change. An antenna with a gain of 30 dBi would be impractical for indoor applications at microwave frequencies because of its huge size. However, at THz frequencies, size won't be a problem at all. In the case of a 30 dBi antenna at 300 GHz, an effective antenna aperture would be around 80 mm^2, which is comparable to the cross-sectional area of a pencil.

Table 17.2 Parameters for link budget calculation

Quantity	Symbol	Value
Transmitting power	P_t	0 dBm
Carrier frequency	f_c	300 GHz
Wavelength	λ_c	1 mm
Distance	d	5 m
Atmospheric attenuation	α_a	0.01 dB/m @ f_c
Excess loss	L_{ex}	0 dB
Noise spectral density	N_0	–174 Bm/Hz
Spectral efficiency	h	1 bps/Hz
Noise bandwidth	B	Data rate × η
Total noise figure	NF	15 dB
System margin	M	10 dB

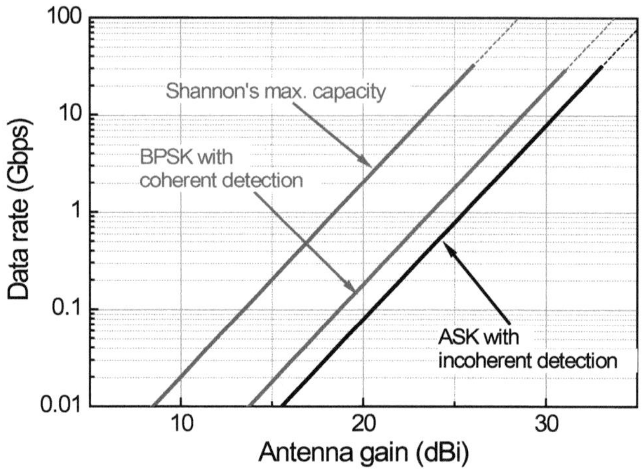

Figure 17.2 Calculated Shannon's maximum data capacity for spectral efficiency of 1 bps/Hz and available data rates at a BER of 10^{-6} for ASK with incoherent detection and BPSK with coherent detection.

However, since the half-power beam width for a 30 dBi antenna gain is quite narrow, around 6–7 degrees, the transmitter and receiver should be on the line of sight (LOS) for proper operation. To avoid a link failure due to an obstacle in the LOS path, walls, ceilings, or floors can be used as reflectors so that the link can be kept alive. For such a non-LOS operation, the additional loss due to the reflection should be considered in the link budget calculation as well.

Note that the values of the parameters used in this link budget calculation are not beyond those achievable with current device technologies. Advanced devices and a more sophisticated modulation scheme with a better understanding of THz wave propagation characteristics would enable us to achieve extremely high throughput even over 100 Gbps.

17.2.1 Application Scene of THz Communications

Wireless local area network (WLAN) and wireless personal area network (WPAN) systems are the most common indoor communication systems these days, and we can conceive that THz communications will be used in a similar way at first. The Wi-Fi Alliance, an organization promoting WLAN technologies worldwide, has categorized the usage models of future WLAN into (1) wireless displays, (2) in-home distribution of HDTV and other content, (3) rapid uploading and downloading of large files to and from a server, (4) backhaul traffic in a mesh or point-to-point configuration, (5) campus/auditorium deployments, and (6) manufacturing-floor automation [7]. As can be seen in the list above, home wireless networks for digital video would be one of the most common applications. Though the current HDTV (1080p60) video format requires "just" around a 3 Gbps data rate, the required data rate for the future video format will dramatically increase with increasing resolution. For example, one future video format called Super Hi-Vision, which has a resolution of 7680 × 4320, 16 times larger than the current 1080p format, requires more than a 24 Gbps data rate, depending on the frame rate and color depth. Therefore, future WLAN and WPAN systems will have to provide at least a couple of 10 Gbps, and for this purpose, THz wave communications will play an important role, though beam-forming techniques for avoiding obstacles, such as people moving around will be an issue.

Britz has proposed a little more advanced concept, the so-called "triple-stack nanocellular architecture" [8]. In this concept, three different radios for cellular, Wi-Fi, and THz communications are installed in a mobile handset in parallel. The cellular and Wi-Fi radios provide wide coverage with a slower rate, while the THz communications link offers extremely high throughputs where THz communications services are available. In addition, Koch [9] and Federici and Moeller [10] have shown another possible use of THz communications: a secure wireless system, which is likely to be used on the battlefield. Since the THz beams are highly directional and atmospheric attenuation is very high, unauthorized persons would have to be on the same narrow beam path to intercept messages. Another approach is to spread the data signal over a wide spectral span in THz bands using a long code sequence, for example, at a chip rate of 100 Gbps. The high-order encryption with the long code would make the system strong against a jamming attack.

17.3 Current Technologies

In this section, we review present technologies for THz communications, with the focus on device technologies for the front end and demonstrations for testing the feasibility of THz communications.

17.3.1 Device Technologies for the Front End

Recently, various kinds of device technologies that can be implemented at the front end of THz communications have been extensively reported and demonstrated. To avoid overlap with other related chapters in this book, we will briefly cover only a few selected results here.

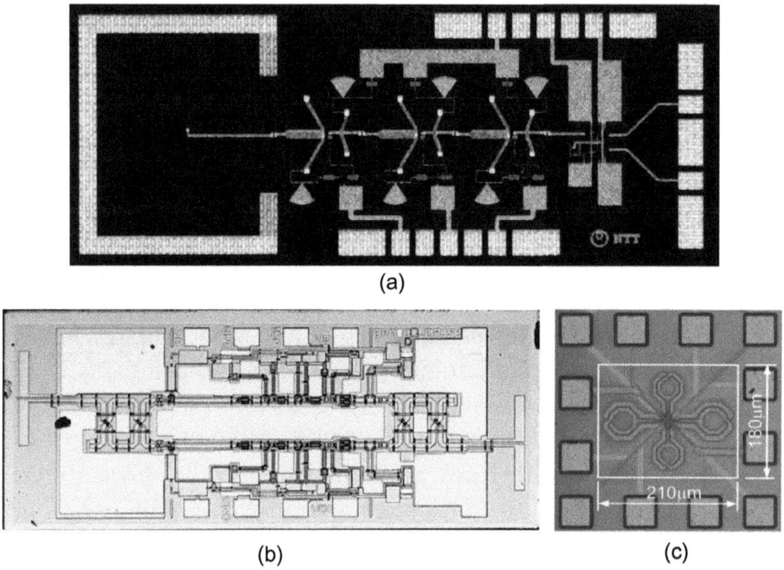

Figure 17.3 Photos of several functional ICs operating at THz frequencies. (a) Fully integrated ASK receiver, (b) 10 mW power amplifiers, and (c) THz CMOS voltage-controlled oscillator (VCO). Republished from Refs. [17, 19, and 22] by permission of IEEE, © 2013, 2012, and 2008, respectively.

Compound semiconductor transistors have always been considered to be the strongest candidates for high-frequency applications, especially for millimeter-wave and THz wave bands. Recently, InP high-electron-mobility transistor (HEMT) devices with a cutoff frequency of over 1 THz have been reported [11, 12] and demonstrated for amplifiers and other elementary components up to around 500 GHz [13–15], and 10 Gbps quadrature phase-shift keying (QPSK) modulator and demodulator chipsets for 120 GHz band have recently been demonstrated [16]. In addition, though not for communications applications, a single-chip receiver integrated with an on-chip antenna, radio frequency (RF) amplifier, detector, and postlimiting amplifier operating at around the 300 GHz band have already been demonstrated [17]. State-of-the-art solid-state power amplifiers can provide more than 50 mW and 10 mW at 220 GHz [18] and 338 GHz [19], respectively. Note that even with current technologies, 10 mW at 300 GHz looks feasible, which is 10 times higher than the power assumed in the link budget calculation in the previous section.

It has been assumed that Si-based technologies such as complementary metal oxide semiconductor (CMOS) and SiGe would not be suitable for THz wave applications because of the relatively high loss of the base material. However, recently, several components operating at over 100 GHz have been reported, and the results show great potential for the development of cost-effective THz communications systems [20]. Though the output is still too low for practical applications, oscillators producing THz waves at 324 GHz [21] and 410 GHz [22] have been developed with standard Si-CMOS technologies. Fully integrated transmitter and receiver chipsets with state-of-the-art SiGe bipolar complementary metal oxide semiconductor (BiCMOS) technology have also been presented for a 160 GHz band application [23]. The transmitter and receiver provide output power of more than 3 mW and a noise figure of less than 14 dB, respectively.

Asada's group at the Tokyo Institute of Technology has reported simple THz wave generators with resonant tunneling diodes (RTDs) [24]. They obtained fundamental oscillation up to 1.04 THz at room temperature. RTD oscillators have a very simple structure consisting of a single RTD and a planar antenna for radiating THz waves. The oscillation frequency is determined by the LC resonance, where the inductance L is from the antenna and the capacitance C is from both the RTD and the antenna. An output power of 7 μW has been demonstrated at 1.04 THz.

In addition to electronic devices, photodiodes (PDs) and other photonic devices, originally developed for fiber-optic communications, have been used for generating millimeter and THz waves. Especially for THz wave applications, uni-travelling carrier photodiodes (UTC-PDs) have shown promising performance in output power and operation frequency compared with conventional p-i-n-type PDs. Nippon Telegraph and Telephone Corporation (NTT) has demonstrated various types of UTC-PDs for different applications. For a J-band UTC-PD module packaged in a WR-3 waveguide, the maximum output power is more than 1 mW at 300 GHz and the 3-dB bandwidth is around 100 GHz [25].

17.3.2 Demonstrations at over 100 GHz

To date, a few experiments for demonstrating the feasibility of THz communications have been conducted. They have normally done in the frequency band below 400 GHz because of limited output power and sensitivities of transmitters and receivers at higher frequencies, respectively.

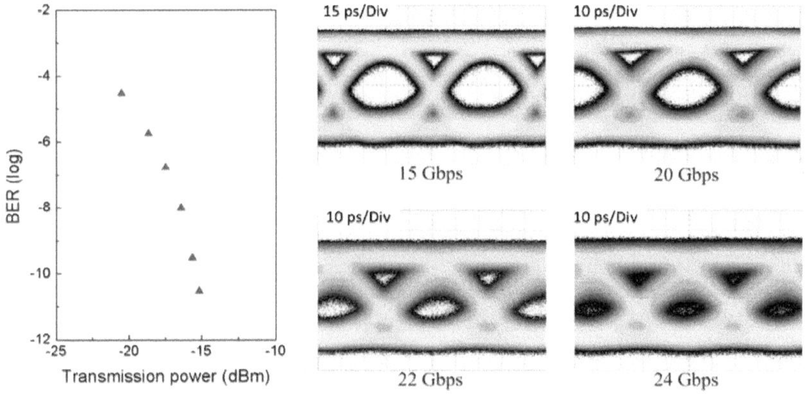

Figure 17.4 BER at 12.5 Gbps with respect to the transmitted power from a photonic source and eye diagrams for several data rates. Reported in Ref. [17].

One of the most successful demonstrations is a wireless data transmission system using the 120 GHz band developed by NTT, Japan. The aim of this system is to relay a large volume of digital data, where a high-data-rate link over a few kilometers is temporarily necessary. In the early version of the system, photonic technologies were used for generating and modulating a carrier signal [26]. Later, an InP HEMT power amplifier was added to compensate for the limited output power from the UTC-PD [10], which extended the link distance to a few hundred meters. To improve mobility outdoors, all the photonic components were replaced with InP HEMT monolithic microwave integrated circuits (MMICs) [28]. During the Beijing Olympic Games in 2008, this system was utilized by Fuji Television Network Inc., one of the Japanese broadcasting networks, to relay HD video signals of live TV shows from a studio to the International Broadcast Center over approximately a 1 km distance. Recently, the data rate has been increased to 20 Gbps by using polarization multiplexing techniques [29].

NTT and Osaka University have demonstrated a THz communications link in the 250 GHz and 300 GHz bands as well [30, 31] using photonic technologies. In the first preliminary experiment, 8 Gbps and 2 Gbps error-free transmission was achieved at 250 GHz and 300 GHz, respectively, over a 0.5 m distance with ASK modulation. Taking the carrier frequency into consideration, the link should provide higher data capacity, and it was found that the maximum data rates of 8 Gbps and 2 Gbps were limited by the bandwidth of the Schottky barrier diode (SBD) detectors used as the receiver, not by the limited power and sensitivity of the transmitter and receiver. Using the improved version of the detector with a wider bandwidth characteristic, error-free transmission at over 20 Gbps was demonstrated in the 300 GHz band [32]. Recently, the NTT group has demonstrated a fully integrated ASK receiver MMIC operating at 300 GHz, and utilizing the receiver, error-free transmission at over 20 Gbps was achieved over a 30 cm distance with a very low transmitting power of around 100 uW [17].

Using transmitter and receiver MMICs based on 50 nm metamorphic high-electron-mobility transistor (mHEMT) technology, Kallfass et al. demonstrated wireless data transmission in the 220 GHz band [33]. Up to 20 Gbps data has been delivered with fully packaged transmitter and receiver devices. In addition, Jastrow et al. have demonstrated transmission of analog and digital video signals in the 300 GHz band [34, 35] using harmonic mixers. For digital video transmission, they used two different modulation formats, DVB-S2 with a single carrier 8 PSK (96 Mbps) and DVB-T with multicarrier 64 quadrature amplitude modulation (QAM) (31.668 Mbps). Quasi-error-free transmission (BER $< 2 \times 10^{-4}$) over a 52 m distance was achieved for the DVB-S2 signal with two dielectric lenses for the transmitter and receiver. For multicarrier modulation, link failure was observed when the number of orthogonal frequency division multiplexing (OFDM) carriers was increased. The authors assumed that this was due to intercarrier interference caused by the phase noise of the 300 GHz carrier signal.

17.4 THz–Radio Wave Propagation

Prior to the actual design of any radio communications system, the specifics of radio channel conditions have to be taken into close consideration. The final system specifications must appropriately address the characteristics and specialties of the underlying radio channel. As an example, it appears intuitive that cellular mobile communication networks with cell ranges up to a few kilometers will be subject to entirely different channel conditions than short-range links, such as Bluetooth or WLAN, which are predominantly operated in indoor environments. System parameters merely dependent on the physics of electromagnetic radio wave propagation, such as link attenuation and richness of potential multipath propagation, clearly dictate properties of the physical layer system conception. No link budget, modulation scheme, or equalization method can be set up or chosen without any knowledge of the radio channel characteristics. In the end, the wave propagation even affects higher layers of the Open Systems Interconnect model beyond the physical layer, for instance, the access procedure. For this reason, a thorough understanding of the underlying wave propagation becomes essential for any new radio telecommunications technology. This holds even more for THz communications than for evolving systems in conventional frequency ranges at 5 GHz or below for two reasons. First, the carrier frequency is higher by a factor of 150 at 300 GHz compared to, for example, the widely used industrial, scientific, and medical (ISM) band at 2.4 GHz. Second, the envisaged bandwidths of 10 GHz and beyond exceed the 40 MHz wide bands of WLANs according to IEEE 802.11n by a factor of at least 250. Such huge differences induce entirely different wave propagation conditions and, ultimately, require entirely different concepts to tackle the arising challenges. Hence, THz wave propagation aspects are inseparably connected to the overall THz communications idea.

Aiming at a comprehensive understanding of THz wave propagation, this section provides a brief overview of the relevant individual propagation phenomena with the focus on indoor environments. Given the very high frequencies, wave propagation in the THz range can in general be considered quasi-optical. Exemplary propagation paths are illustrated in Fig. 17.6 using the example of a laptop connection to an access point (AP) in an office scenario. The relevant propagation phenomena comprise:

- free-space attenuation;
- atmospheric absorption;
- transmission;
- specular reflections;
- diffuse rough surface scattering;
- diffraction;
- frequency dispersion; and
- ray shadowing due to human movement.

These are detailed in the following section.

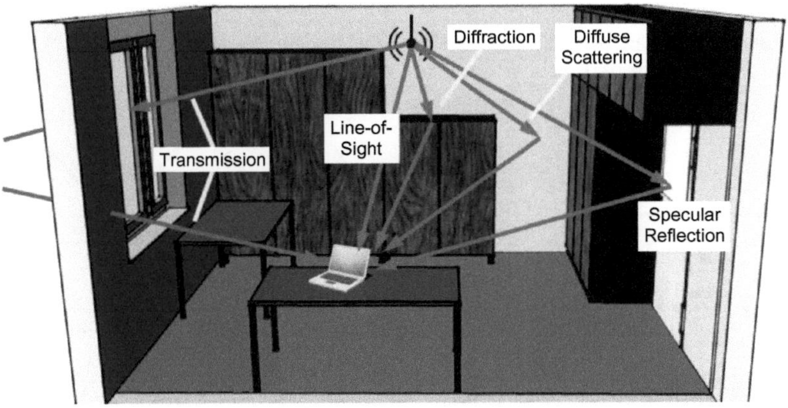

Figure 17.5 Examples of propagation paths in an indoor environment for a wireless connection of a laptop to an access point.

17.4.1 Free-Space Propagation

As for any radio wave propagating in free space, the Friis transmission equation [36] also holds for the THz range to determine the attenuation under LOS conditions. Of course, all other propagation phenomena described subsequently are also subject to the FSL.

Due to the quadratic increase of the FSL with frequency, THz waves are subject to significantly higher attenuations than conventional radio systems. Figure 17.7 (left) illustrates the FSL over distance d for 900 MHz (GSM), for 2.4 GHz (WLANs), and for 300 GHz. The general curve shapes are very similar. Given in absolute values, the FSL at d = 10 m amounts to 51.5 dB, 60.1 dB, and 102 dB in order of increasing frequency. FSLs higher by about 41.9 dB and 50.5 dB at 300 GHz are observed compared to 2.4 GHz and 900 MHz, respectively.

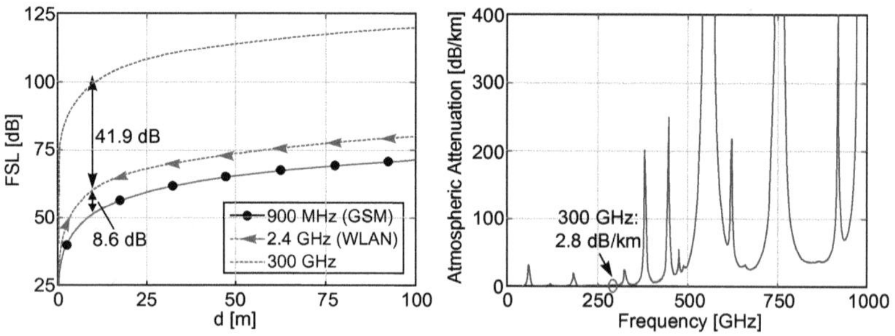

Figure 17.6 Free-space loss over distance at different frequencies (left) and atmospheric attenuation over frequency (right).

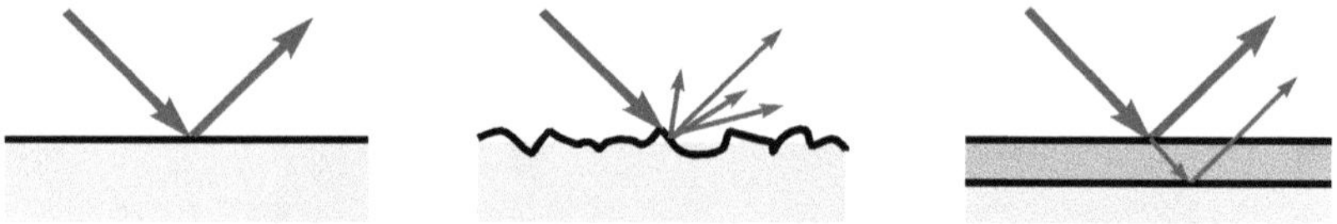

Figure 17.7 Reflection from surfaces at THz frequencies. (Left): Smooth material; (middle): rough surface; and (right): layered medium.

These enormous differences between FSLs bring about an utmost significant implication for THz communications. Whereas conventional WLANs and WPANs or mobile phones can operate with nearly omnidirectional antenna elements, highly directive antennas become mandatory for THz communications at any rate to counteract the free-space attenuation. Moreover, low output powers of THz components of the order of at maximum a few decibel-milliwatts are feasible at the moment (see the previous section). In contrast, the radiated power is restricted to comparatively high values of, for example, 30 dBm in the U.S. for the 2.4 GHz band [37] by regulation, not by technology. This double disadvantage requires antenna gains of the order of at least 20–25 dB [38], preferably more. Attainable transmission distances will reach a few meters up to at maximum a few tens of meters, provided that improved THz amplifiers and/or higher-gain antennas are used. Hence, typical use cases like WLANs or WPANs can be realized with acceptable ranges. Similar, but not that severe, problems have to be addressed at 60 GHz, where a paradigm change from omnidirectional architectures toward directive antenna arrays is underway also.

As an exception to the distance limitations above, highly focusing radiators like Cassegrain or dish antennas with gains of 50 dB and more will enable long-range fixed radio links over 1 km and beyond [39]. Then, the atmospheric impact may contribute to the overall attenuation despite LOS conditions, as detailed in the following section.

17.4.2 Atmospheric Attenuation

Molecular resonances cause very high atmospheric absorptions at certain frequencies throughout the entire THz range. Figure 17.7 (right) demonstrates atmospheric attenuation over frequency as simulated with the MODerate resolution atmospheric TRANsmission (MODTRAN) model. This attenuation has to be considered, in addition to the FSL. Losses of up to several 100 dB/km clearly prohibit data transmission, even over distances of a few tens of meters around the resonances, but multiple windows remain, which exhibit tolerable absorption. For instance, the atmospheric impact amounts to no more than 2.8 dB/km at 300 GHz and can be neglected for short-range communications, as discussed previously. Ultrashort-range transmissions over, at maximum, a few centimeters would even be thinkable in highly attenuated ranges.

17.4.3 Transmission

The absorption behavior of materials becomes relevant to determine transmission of THz waves through walls, windows, or objects and hence to assess the radio coverage or interference outside confined environments. Lots of material parameter measurements have been conducted with the help of THz time domain spectroscopy. A comprehensive characterization of typical building materials such as different kinds of wood, brick, and glass can be found in Ref. [40], for which absorption coefficients of, for example, $\alpha = 11.4\ cm^{-1}$, $9.1\ cm^{-1}$, and $10\ cm^{-1}$ have been determined at 350 GHz. Converted into transmission attenuations, these α values mean losses of several 10 dB for a few-centimeter-thick glass window up to several 100 dB for a brick wall. On the other hand, electromagnetic waves in the lower GHz range are subject to a far lower absorption of just a few decibels with typical $\alpha < 0.5\ cm^{-1}$ for the same materials as before.

Whereas conventional WLAN APs can easily cover a wider indoor as well as an adjacent outdoor area, THz cells will clearly remain restricted to single rooms. The resulting disadvantage is the need for an extensive deployment of THz APs, which can still be realized in high-demand areas such as offices or living rooms. As a significant advantage, this allows for efficient frequency reuse and hence for full system capacity in every environment, being especially beneficial to ad hoc WPANs. Moreover, THz indoor cells become very robust against intercell interference and can be considered inherently secure due to the impossibility of interception from the outside.

17.4.4 Specular Reflections

Apart from a direct ray, specular reflections from various objects, the walls, the ceiling, or the floor contribute to the received power in indoor environments so that the reflection modeling is of considerable relevance. With the material properties known from separate measurements [40], the Fresnel equations suffice for the determination of reflection factors in the THz range, provided that the surface is smooth and sufficiently thick compared to the wavelength. Otherwise, the reflection behaves more complex, as indicated schematically in Fig. 17.8. For a smooth homogeneous surface (Fig. 17.8, left), typical reflection losses amount to 5–15 dB, just similar to other frequency ranges. Should the surface be rough (Fig. 17.8, middle), a significant amount of the incident power is scattered diffusely and is not reflected specularly anymore, which effectively corresponds to an increase in the reflection loss. Then, the conventional Fresnel reflection coefficients must be corrected by the Rayleigh roughness factor [41]. This effect is likely to occur in indoor environments for the reason that the height deviations of typical surface materials range in the order of the wavelength at

THz frequencies. Examples are walls covered with plaster or ingrain wallpaper, where the roughness can easily cause an attenuation of 10 dB and more, in addition to the reflection loss of the material itself. Such materials appear as smooth at conventional WLAN frequencies so that this effect is particular to the THz range.

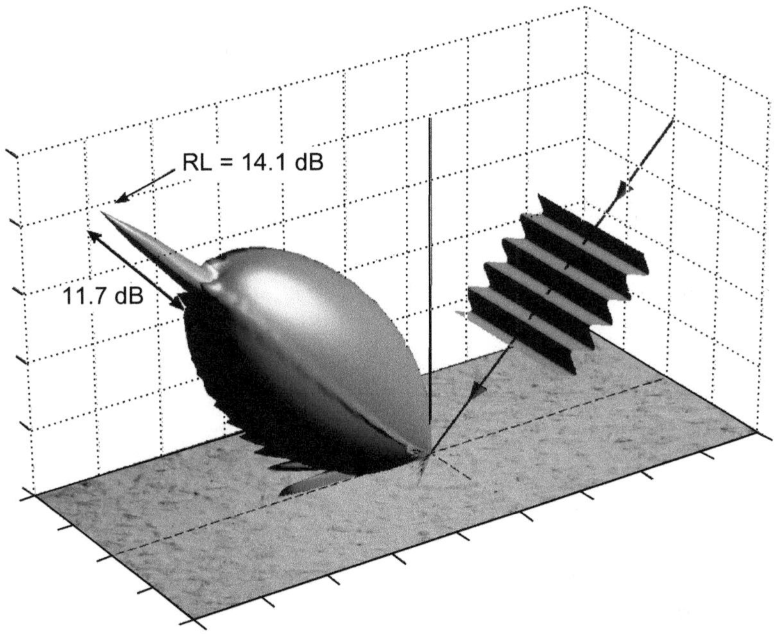

Figure 17.8 Simulated scattering behavior of a rough plaster surface for an incident plane wave.

In the case of stratified materials with layer thicknesses of the order of the wavelength (Fig. 17.8, right), for example, paint on plaster or double glass, the interference of multiple rays reflected at the different interfaces induces a highly frequency-dependent reflection behavior [42]. Variations of the reflection losses by a few decibels over the several 10 GHz wide bands can occur, causing non-negligible channel frequency selectivity apart from fading effects due to mere multipath propagation. Alike the roughness impact, this effect does not have to be taken into account in the case of conventional radio systems because of the much greater wavelength and the far narrower utilized bands.

17.4.5 Rough Surface Scattering

Apart from the impact on specular reflections, as discussed previously, rough surface scattering may additionally lead to multipath propagation due to diffuse components. The diffuse scattering from typical rough building materials, such as plaster, can be modeled appropriately by means of Kirchhoff scattering theory at THz frequencies, which has been validated with dedicated measurements in Ref. [43].

Figure 17.9 illustrates the simulated behavior of an incident planar wave being scattered at a plaster element (please refer to Ref. [44] for details). A sharp spike can be identified in the direction of the specular reflection with a loss of 14.1 dB. Around the spike, the diffuse components are apparent as a broad lobe with attenuations of 25.8 dB and more. The specular part is attenuated increasingly with increasing surface roughness and can even disappear completely, whereas the amplitudes of the diffuse components rise at the same time. In consequence, the diffuse scattering becomes more relevant in te case of higher roughnesses from surface-height standard deviations of about $\sigma_h > 0.2$ mm onward in the lower THz range [43] and must not be disregarded in propagation modeling. This holds true especially for non-line-of-sight (NLOS) conditions, where the received power solely depends on indirect rays. Should a THz communication system be operated under such circumstances, scattering may significantly impair the overall system performance [43]. Otherwise, the direct ray clearly dominates the propagation conditions so that the diffuse scattering can be disregarded.

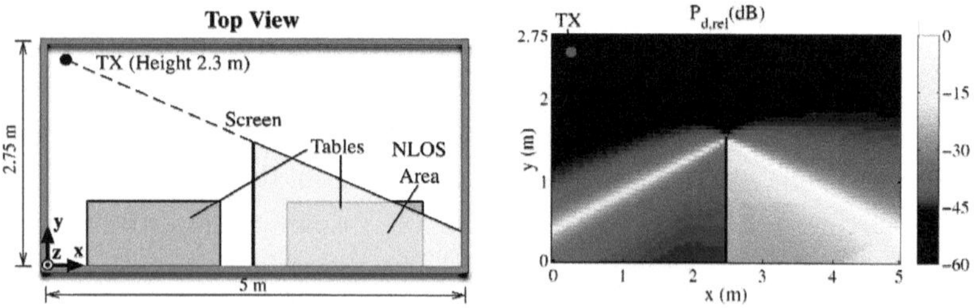

Figure 17.9 Simulated relative diffracted powers (right) with respect to the overall received power in a small-office scenario (left) at a height of 0.75 m for a frequency of 300 GHz. Republished from Ref. [45] by permission of IEEE, © 2012.

17.4.6 Diffraction

Diffraction effects can occur at a wide variety of discontinuities in indoor environments, for example, edges of tables, wedge-like corners of wardrobes, doors left open, or room dividers. The assessment of power diffracted at such objects becomes crucial to determine the overall received power, especially in the shadowing region behind the obstacles. An extensive measurement campaign has proven the uniform theory of diffraction applicable in the THz range to model the diffraction at edges or wedges [45]. Both shapes can be used to approximate furniture or other furnishings.

Due to the short wavelength, very little power is diffracted at 300 GHz and above in general compared to the lower GHz frequency ranges. In practical setups, typical diffraction attenuations of THz waves amount to at least 20 dB and easily reach up to 40 dB or more, whereas the corresponding attenuations can be expected to be lower by around 10 dB to 20 dB for conventional WLANs.

Reference [45] studies a small office room by means of ray tracing, as shown in Fig. 17.10 (please refer to Ref. [45] for details). There, the power diffracted into the shadowing region behind a screen is found around 15 dB below the overall receiver power, which is determined by reflections rather than diffraction. Under LOS conditions, the power is clearly dominated by the direct ray. A noticeable impact of diffraction occurs very close to the reflection shadow boundary only, that is, the border between LOS and NLOS conditions. However, this contribution of diffraction becomes relevant at less than 1% of the receiver positions in the considered scenario so that diffraction can be seen as negligible in terms of the overall radio channel conditions.

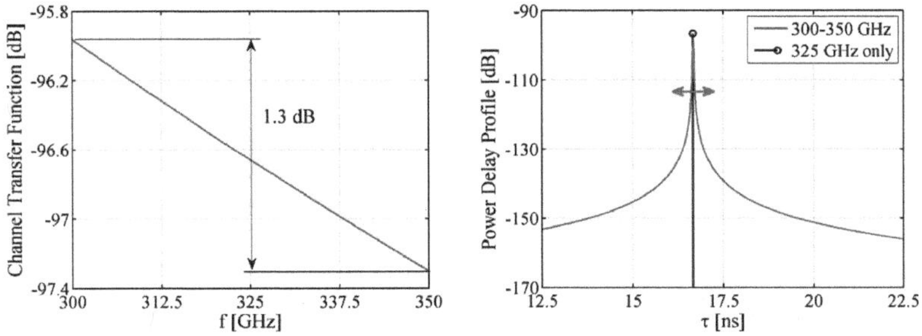

Figure 17.10 Channel transfer function (left) and power delay profile (right) for a free-space connection over 5 m.

17.4.7 Frequency Dispersion

Despite high carrier frequencies, huge absolute bandwidths greater than 10 GHz easily correspond to fractional bandwidths of the order of 0.1 or more. Due to the wave propagation being a function of frequency, the propagation conditions may vary non-negligibly over the band of interest.

To illustrate this aspect, Fig. 17.11 (left) shows the channel transfer function of a simple LOS connection covering a distance of 5 m without any multipath propagation from 300 GHz to 350 GHz. The channel transfer function drops by about 1.3 dB from the beginning to the end of the 50 GHz wide band due to the frequency dependency of the FSL, which can also be considered as the frequency-dependent variation of the effective antenna aperture. This results in a temporal pulse broadening in the power delay profile compared to a theoretically discrete pulse for a discrete frequency observed with an infinite bandwidth, as illustrated in Fig. 17.11 (right). Ultimately, the dispersive propagation causes a certain distortion of the transmitted pulse shape. Such an effect has already been observed in experiments for ultrawide band systems from 3.1 GHz to 10.6 GHz [46].

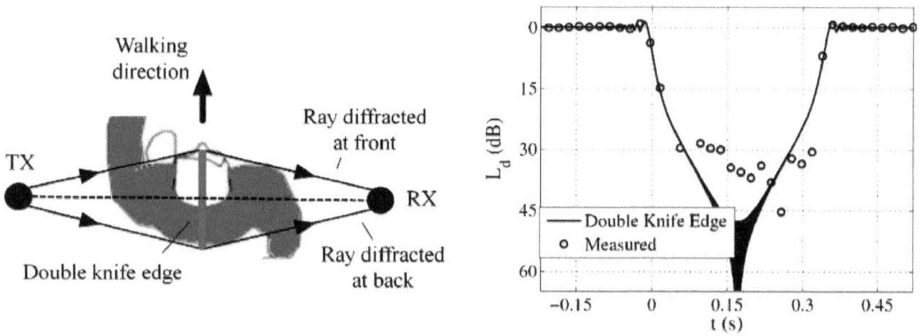

Figure 17.11 Modeling of a human body with knife edges (left) and measured as well as simulated shadowing attenuation for a person moving through a ray at 300 GHz. Republished from Ref. [45] by permission of IEEE, © 2012.

Further consequences of frequency dispersion on the THz channel properties are discussed separately in Ref. [47] without any detailing at this point. Please be aware that not only the FSL but also all other propagation phenomena have to be treated as dispersive.

Hence, simulation of wave propagation at single frequencies does not suffice in case of ultrabroadband THz radio channels, unlike for conventional radio systems with far narrower bands. Here, ray-based channel simulations in the frequency domain allow for consideration of propagation dispersion, as described with regard to the THz range in detail in Ref. [47], which is also referred to as frequency domain ray tracing.

17.4.8 Ray Shadowing by Moving Persons

So far, all propagation phenomena have always been considered under static conditions. In practice, dynamics may be introduced by two different aspects, by moved transmitters and/or receivers and by moving objects or persons. Among these, ray shadowing due to human movement becomes of utmost importance in indoor environments because people crossing the direct ray path may cause an unintended connection interruption. Other effects such as Doppler shifts of reflections at moving persons as well as Doppler shifts caused by moved devices have not been subjected to detailed studies yet and remain to be assessed with regard to their relevance to the impairment of THz radio systems.

Due to the large dimensions of a human body with respect to wavelengths of 1 mm and less, a considerable additional path attenuation can be expected in the case of ray shadowing by persons. The high water content makes the torso and limbs effectively impenetrable so that the shadowing attenuation is not determined by transmission through but by diffraction around the body. A person crossing a ray path can be effectively approximated as a diffracting object with one knife edge at the front and one at the back of the torso, which is illustrated in Fig. 17.12 (left). This modeling methodology has been validated against measurements of the induced human blockage attenuations at 300 GHz in Ref. [45]. An excerpt of the results can be seen in the right part of Fig. 17.12, where the temporal course of the shadowing attenuation is shown for a person who crosses the 1 m long direct path between the transmitter and the receiver perpendicularly, just as depicted in the left part of the figure. The person begins to obstruct the direct ray path at the point of time $t = 0$ s and leaves it approximately at $t = 0.35$ s. Before and after the shadowing event, the reflection at the body causes a slight interference pattern. Upon shadowing, the loss increases rapidly and reaches values of 35 dB and more, which even exceed the dynamic range of the measurement device. Predicted attenuations amount to more than 45 dB in the deep shadowing region, definitely causing connection interruptions of hypothetical data transmission.

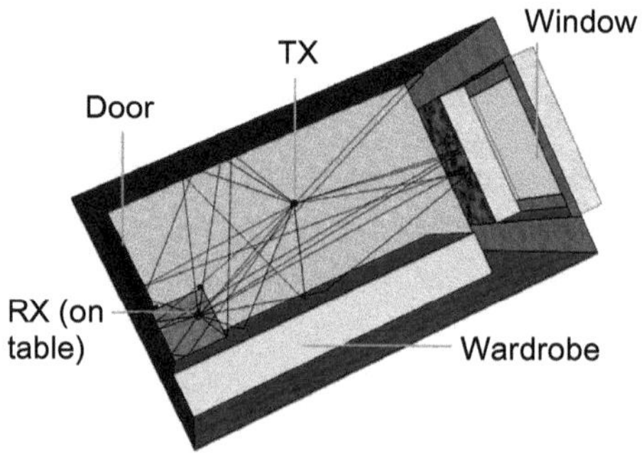

Figure 17.12 Geometric ray-tracing results in a small indoor scenario.

As a countermeasure, the directive antennas will have to dynamically react to such a ray-shadowing event with an aimed beam redirection toward a strong reflected ray path. Smart antennas will hence be one key component of THz radio systems. First, an imminent shadowing event has to be recognized sufficiently early so that control data can still be exchanged and the beam switch can be initiated. Next, the best possible path with the lowest attenuation has to be found before the beams are realigned. This will have to happen within less than 100 ms. Finally, the return to the initial transmission path has to be ensured once the person has left the path, which can be done via an occasional scanning. Such complex functionalities dictated by the specialties of THz wave propagation reach far beyond the concepts of existing systems, which again stresses the necessity to understand a newly opened radio channel in all facets. Hence, the channel will finally have to be modeled as a whole.

17.4.9 Modeling of THz Indoor Radio Channels

In the final step, all individual propagation phenomena have to be brought together to model the overall THz indoor radio channel.

This has been done implicitly in several parts of the section already whenever a scenario has been simulated on the basis of ray tracing. Due to the quasi-optical wave propagation, ray tracing, that is, the geometrical ray search according to optical principles combined with electromagnetic models, becomes the method of choice to simulate the THz indoor channel with high accuracy and an acceptable runtime. Figure 17.13 demonstrates the results of geometric ray-tracing simulations conducted in a small-office scenario. Moreover, the corresponding electromagnetic computations have been validated against indoor radio channel measurements around 300 GHz in Ref. [48]. With further calibration and extension of ray tracing to channel simulation in the frequency domain, very accurate channel predictions can be achieved [47].

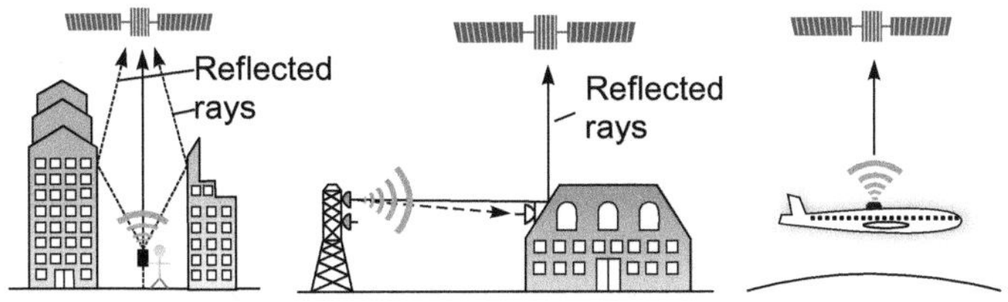

Figure 17.13 Exemplary possible interference scenarios of THz communications with Earth exploration satellite services.

In summary, the comprehensive conveyed basic knowledge on propagation mechanisms in combination with an appropriate ray-tracing simulator at hand enables the inclined reader to accurately predict the THz wave propagation behavior for further studies.

17.5 Regulation and Standardization Activities

The commercial introduction and success of a communication system requires the system to be standardized and a dedicated spectrum to be available. The corresponding activities are standardization and regulation. In both areas, first activities regarding THz communications have been started in parallel to the research described in the previous sections. Since the processes for standard development and for spectrum management can be expected to be running over a couple of years, an early beginning of these activities is essential.

Standardization of THz communications has been initiated within IEEE 802, where the Interest Group THz (802.15 IG THz) has been established within its working group 802.15 in 2008 [49]. The goal of the IEEE 802.15 IG THz is to explore the potential of the THz spectrum as a frequency band for future wireless communication systems, both through studies and through own activities. In its first years of existence, the focus of this group has been threefold. The first aspect is to follow the technological progress of semiconductor research in this frequency range to assess the maturity of this technology with regard to the development of a concrete standard. Second, the work on channel characterization is in the focal point to have appropriate channel models available. The third aspect is the identification of possible applications and usage scenarios for ultrafast wireless THz data transmission [50].

The first and second aspects have been addressed comprehensively in other parts of this section. The latter includes categorization in terms of specific propagation conditions and the requirements with respect to the antenna alignment accuracy, which distinctly determine the necessary complexity of the system concept to be developed. In each of these categories, a dedicated standard addressing the specifics of the requirements for the physical layer becomes appropriate so that it is likely that a family of standards based on the THz communications idea will emerge in the future. As soon as technology matures sufficiently to master the requirements in one of these categories, it is intended that a corresponding study group will be initiated, which will potentially be followed by a task group developing a concrete standard.

On the active side, the 802.15 IG THz has influenced the spectrum discussions with respect to the World Radiocommunication Conference (WRC) 2012 and has linked the standardization to the spectrum regulation activities. Spectrum allocations of frequencies to specific applications are defined by the International Telecommunications Union in the Radio Regulations [51], which are subject to revision at the WRC, which takes place ever three years. Depending on whether active transmission is involved or not, the services are subdivided into active and passive services. Apart from active communications potentially transmitting at THz frequencies, radio astronomy (RAS) as well as Earth exploration satellite services (EESS) are passive services already under operation in the THz range [38, 52]. Although the spectrum beyond 275 GHz has not been assigned to any specific service yet, Footnote 5.565 of Radio Regulations states that passive services like the RAS and the EESS have to be protected from harmful interference, should active services be operated in the same band.

This leaves THz communications with two options: Either only the remaining fragmented parts of the spectrum are used, or THz radio systems are operated coexistently with passive services throughout the entire THz range. In fact, the first option cannot be considered as a real choice because the nonprotected bands provide continuous bandwidths of no more than about 10 GHz, which completely counteract the idea of THz communications with bandwidths of multiple 10 GHz up to 100 GHz and beyond. For coexistent spectrum usage, Footnote 5.565 effectively puts the burden on active THz systems, demanding that the passive services not be interfered with. Hence, interference studies between active and passive THz services have shifted into the spotlight of THz communications.

Regarding the two passive applications named above, entirely different operational conditions necessitate independent investigations of both. Radio astronomy telescopes are mainly located on several 1000 m high mountains, looking skyward with extremely focusing dish antennas. Such locations in combination with the pointing direction of the antennas make any interference from ground-based transceivers extremely unlikely. Even air- or spaceborne transmitters passing over the dishes sufficiently fast can be tolerated due to long integrational times of the telescopes. A corresponding report has been published by the ITU-R, reasoning the negligible interference probabilities [38].

Second, the EESS and space research services operate passive THz spaceborne sensors, whereas the latter does not become relevant because of the deep-space antenna orientation. On the other hand, the highly directive EESS antennas may be oriented groundward, picking up interference power from any devices transmitting in the immediate direction of the satellite. Thinkable scenarios for this can be manifold and comprise accidentally skyward-pointed nomadic devices, fixed links with reflecting or scattering objects close to the propagation path, and airborne THz transceivers. These and more setups have been subject to thorough investigations in Ref. [6]. There, it has been found that interference may occur under worst-case conditions but can be avoided with the help of transmit power limitations. Despite such transmit power constraints, THz WLAN cells, fixed links, and airborne systems remain operable with radii of the order of a few tens of meters, a few kilometers, and 10–20 m within the maximum-allowed transmit power ranges, respectively.

In the end, it could be shown that interference is extremely improbable at any rate and that interference freedom can be ensured with appropriate countermeasures under avoidance of a transmit power reduction so that the coexistent operation of active THz communications and passive THz services can be guaranteed.

17.6 Conclusion and Outlook

Clearly, chances of the THz frequency range for communication purposes are as numerous as the challenges. Not only can well-established use cases for wireless technology like backhaul links, WLANs, or WPANs be enhanced with superior performance of several tens up to several hundreds of gigabits per second, but also new applications like intradata center links or chip-to-chip connections can be envisioned on the basis of THz radio.

The most recent feasibility studies have clearly illustrated the potential of THz communications. Now that output powers of components are climbing up to a few megawatts, the first commercial systems move within reach in the next years. Still, several crucial open points remain to be addressed in the future, even under the assumption of readily available RF hardware. These can be divided into four main topics:

1. *Antenna architecture*: Steerable antennas will form one central component of THz radio systems. Their layout and performance will become crucial for the system performance, depending on the respective radio channel. Whereas fixed links can employ highly focusing antennas with 50 dBi and more, WLANs will build upon dynamically redirectable antennas with around 20–30 dBi. Discrete phase shifters are one option to be considered, being difficult to realize though. Alternatively, switching networks for smart antennas with a limited number of preselectable beam-pointing directions are likely to be limited in precision so that the optical steering

method bases on, for example, liquid lenses or metamaterials will also have to be considered as an option.

2. *Baseband processing*: Coding schemes, modulation, and the sampling of data streams with extremely high rates will demand dedicated baseband processing units designed in hardware. Software-based processing is extremely unlikely to fulfill the throughput demands. Digital-to-analog (D/A) as well as analog-to-digital (A/D) converters will have to be developed, and hardware baseband components, including appropriate interfaces to existing systems, will have to be designed.
3. *Procedure and protocol definition*: Data transmission format protocols of current systems have been designed for use with omnidirectional antenna concepts, and these do not comply with connection establishment procedures of radio systems based on smart directive antennas. The first corresponding steps toward the integration of a beam alignment process in the medium access control (MAC) layer have been undertaken in the framework of the IEEE standards 802.15.3c and 802.11ad for 60 GHz WPANs and WLANs, respectively. However, those are based on in-band signaling at 60 GHz with a beam refinement procedure, which cannot work in the THz range because of the need for high gains, straight from the antenna beam alignment phase onward. Out-of-band signaling based on conventional WLANs or ultrawide band (UWB) systems becomes a viable option to realize the beam alignment and access procedure. Appropriate MAC protocols will have to be designed in the future.
4. *Standardization and spectrum allocations*: Along the lines of propagation studies for different use cases, appropriate application-specific physical layer as well as protocol definitions will have to be specified, for example, within the IEEE standardization forum. Moreover, the ITU spectrum allocation process in the framework of the WRC will have to be followed. In-depth interference studies between active and passive services complementing those introduced in this chapter will have to be conducted, ensuring interference-free simultaneous operations of active and passive devices

in the THz range. Ultimately, such coexistent spectrum usage without any fixed allocations appears far more attractive than dedicated assignments as it leaves potentially the entire THz range for wireless data transmission.

Beyond these rather technological aspects, economic feasibility, that is, the production and overall system costs, will come into the immediate focus of consideration soon. Whereas high-performance components can already be manufactured in compound semiconductor technology, such devices are rather costly. Once CMOS technology matures and reaches a performance comparable with InP or GaAs, this will give way for cheap, broad implementation and distribution of the end-user THz radio. Until then, THz waves will most likely be employed for less cost-sensitive professional use cases only.

Given a future progress comparable to the past and taking into account cycle times in standardization, the wide availability of THz communication systems can be expected roughly until 2020. Then, they will help satisfy the tremendous bandwidth hunger of today's information society.

References

1. S. Cherry, Edholm's law of bandwidth, *IEEE Spectrum* **41**, 58 (2004).
2. H.-J. Song and T. Nagatsuma, Present and future of terahertz communications, *IEEE Trans. THz Sci. Technol.* **1**, 256 (2011).
3. G. J. Foschini and M. J. Gans, On limits of wireless communications in a fading environ-ment when using multiple antennas, *Wireless Personal Commun.* **6**, 311 (1998).
4. A. J. Paulraj, D. A. Gore, R. U. Nabar, and H. Bolcskei, An overview of MIMO communications: a key to gigabit wireless, *Proc. IEEE* **92**, 198 (2004).
5. ITU-R Recommendation P.676-8 (2009). *Attenuation by Atmospheric Gases.*
6. R. Piesiewicz, T. Kleine-Ostmann, N. Krumbholz, D. M. Mittleman, M. Koch, J. Schoebel, and T. Kürner, Short-range ultra-broadband terahertz communications: concepts and perspectives, *IEEE Antennas Propagat. Mag.* **49**(6), 24–39 (2007).
7. A. Myles and R. Vegt, *Wi-Fi (WFA) VHT Study Group Usage Models*, IEEE 802.11 document, 802.11-07/2988r4.

8. D. Britz, *Evolution of Extreme Bandwidth Personal and Local Area Terahertz Wireless Networks*, IEEE 802.15 document IEEE 802.15-10/162r0.
9. M. Koch, Terahertz communications: A 2020 vision, in *Terahertz Frequency Detection and Identification of Materials and Objects* (Springer, the Netherlands, 2007), 325–338.
10. J. Federici and L. Moeller, Review of terahertz and subterahertz wireless communications, *J. Appl. Phys.* **107**, 111101 (2011).
11. R. Lai, X. B. Mei, W. R. Deal, W. Yoshida, Y. M. Kim, P. H. Liu, and A. Fung, Sub 50 nm InP HEMT device with fmax greater than 1 THz, *IEEE Int. Electron. Devices Meet. 2007,* 609–611 (2007).
12. D. H. Kim, J. A. del Alamo, P. Chen, W. Ha, M. Urteaga, and B. Brar, 50-nm E-mode In 0.7 Ga 0.3 As PHEMTs on 100-mm InP substrate with f max> 1 THz, *IEEE Int. Electron. Devices Meet. 2010,* 30.6.1–30.6.4 (2010).
13. V. Radisic, D. Sawdai, D. Scott, W. R. Deal, L. Dang, D. Li, and R. Lai, Demonstration of a 311-GHz fundamental oscillator using InP HBT technology, *IEEE Trans. Microwave Theory Tech.* **55**, 2329–2335 (2007).
14. I. Kallfass, H. Massler, A. Leuther, A. Tessmann, and M. Schlechtweg, A 210 GHz dual-gate FET mixer MMIC with >2 dB conversion gain, high LO-to-RF isolation, and low LO-drive requirements, *IEEE Microwave Wireless Comp. Lett.* **18**, 557–559 (2008).
15. ITU-R Report RS.2189 (2010). *Sharing Between the Radio Astronomy Service and Active Services in the Frequency Range 275 to 3000 GHz. ITU-R Report RS.2189 (2010).*
16. H. Takahashi, T. Kosugi, A. Hirata, K. Murata, and N. Kukutsu, 10-Gbit/s quadrature phase-shift-keying modulator and demodulator for 120-GHz-band wireless links, *IEEE Trans. Microwave Theory Tech.* **58**, 4072–4078 (2010).
17. H.-J. Song, J.-Y. Kim, K. Ajito, M. Yaita, and N. Kukutsu, Fully integrated ASK receiver MMIC for terahertz communications at 300 GHz, *IEEE Trans. THz Sci. Technol.* **3**, 445–452 (2013).
18. V. Radisic, K. M. Leong, X. Mei, S. Sarkozy, W. Yoshida, P. Liu, J. Uyeda, R. Lai, and W. R. Deal, A 50 mW 220 GHz power amplifier module, *Int. Microwave Symp. Dig. (IMS)* **TU1C-1** (2010).
19. V. Radisic, W. R. Deal, K. M. Leong, X. Mei, W. Yoshida, P. H. Liu, J. Uyeda, A. Fung, L. Samoska, and T. Gaier, A 10-mW submillimeter-wave solid-state power-amplifier module, *IEEE Trans. Microwave Theory Tech.* **58**, 1903–1909 (2010).

20. S. Sankaran, M. Chuying, S. Eunyoung, S. Dongha, C. Changhua, H. Ruonan, D. J. Arenas, D. B. Tanner, S. Hill, H. Chih-Ming, and K. K. O, Towards terahertz operation of CMOS, *Proc. Int. Solid-State Circuits Conf.*, 202–203 (2009).

21. H. Daquan, T. R. LaRocca, M. C. F. Chang, L. Samoska, A. Fung, R. L. Campbell, and M. Andrews, Terahertz CMOS frequency generator using linear superposition technique, *IEEE J. Solid-State Circuits* **43**, 2730–2738 (2008).

22. E. Seok, C. Cao, D. Shim, D. Arenas, D. Tanner, C. Hung, and K. K. O, A 410 GHz CMOS push-push oscillator with an on-chip patch antenna, *Proc. Int. Solid-State Circuits Conf.* 472–629 (2008).

23. U. R. Pfeiffer, E. Ojefors, and Z. Yan, A SiGe quadrature transmitter and receiver chipset for emerging high-frequency applications at 160GHz, *Proc. Int. Solid-State Circuits Conf.* 416–417 (2010).

24. S. Suzuki, M. Asada, A. Teranishi, H. Sugiyama, and H. Yokoyama, Fundamental oscillation of resonant tunneling diodes above 1 THz at room temperature, *Appl. Phys. Lett.* **97**, 242102 (2010).

25. H.-J. Song, K. Ajito, Y. Muramoto, A. Wakazuki, T. Nagatsuma, and N. Kukutsu, Uni-travelling-carrier photodiode module generating 300 GHz power greater than 1 mW 1mW UTC-PD at 300 GHz, *IEEE Trans. THz Sci. Technol.* **3**, 158–164 (2013).

26. A. Hirata, T. Kosugi, N. Meisl, T. Shibata, and T. Nagatsuma, High-directivity photonic emitter using 6 module integrated with HEMT amplifier for 10-Gbit/s wireless link, *IEEE Trans. Microwave Theory* **52**, 1843–1850 (2004).

27. A. Hirata, T. Kosugi, H. Takahashi, R. Yamaguchi, F. Nakajima, T. Furuta, H. Ito, H. Sugahara, Y. Sato, and T. Nagatsuma, 120-GHz-band millimeter-wave photonic wireless link for 10-Gb/s data transmission, *IEEE Trans. Microwave Theory Tech.* **54**, 1937–1944 (2006).

28. A. Hirata, R. Yamaguchi, T. Kosugi, H. Takahashi, K. Murata, T. Nagatsuma, N. Kukutsu, Y. Kado, N. Iai, S. Okabe, S. Kimura, H. Ikegawa, H. Nishikawa, T. Nakayama, and T. Inada, 10-Gbit/s wireless link using InP HEMT MMICs for generating 120-GHz-band millimeter-wave signal, *IEEE Trans. Microwave Theory Tech.* **57**, 1102–1109 (2009).

29. J. Takeuchi, A. Hirata, H. Takahashi, and N. Kukutsu, 10-Gbit/s bi-directional and 20-Gbit/s uni-directional data transmission over a 120-GHz-Band wireless link using a finline ortho-mode transducer, *Proc. Asia-Pacific Microwave Conf.* (2010).

30. H. J. Song, K. Ajito, A. Hirata, A. Wakatsuki, Y. Muramoto, T. Furuta, N. Kukutsu, T. Nagatsuma, and Y. Kado, 8 Gbit/s wireless data transmission at 250 GHz, *Electron. Lett.* **45**, 1121-U32 (Oct. 22 2009).

31. H.-J. Song, K. Ajito, A. Wakatsuki, Y. Muramoto, N. Kukutsu, Y. Kado, and T. Nagatsuma, Terahertz wireless communication link at 300 GHz, *Int. Topical Meet. Microwave Photon.*, 42–45 (2010).

32. H.-J. Song, K. Ajito, Y. Muramoto, A. Wakatsuki, T. Nagatsuma, and N. Kukutsu, 24 Gbit/s data transmission in 300 GHz band for future terahertz communications, *Electron. Lett.* **48**, 953–954 (2012).

33. I. Kallfass, J. Antes, T. Schneider, F. Kurz, D. Lopez-Diaz, S. Diebold, H. Massler, A. Leuther, and A. Tessmann, All active MMIC-based wireless communication at 220 GHz, *IEEE Trans. THz Sci. Technol.* **1**, 477–487 (2011).

34. C. Jastrow, K. Munter, R. Piesiewicz, T. Kurner, M. Koch, and T. Kleine-Ostmann, 300 GHz transmission system, *Electron. Lett.* **44**, 213–215 (2008).

35. C. Jastrow, S. Priebe, B. Spitschan, J. Hartmann, M. Jacob, T. Kurner, T. Schrader, and T. Kleine-Ostmann, Wireless digital data transmission at 300 GHz, *Electron. Lett.* **46**, 661–662 (2010).

36. H.T. Friis, A note on a simple transmission formula, *Proc. IRE* **34**(5), 254–256 (1946).

37. T. H. Lee, H. Samavati, and H. R. Rategh, 5-GHz CMOS wireless LANs, *IEEE Trans. Microwave Theory Tech.* **50**(1), 268–280 (2002).

38. A. Gasiewski, *Applications and Requirements for Environmental Remote Sensing*, IEEE 802.15 document 15-11-0765-00-0thz (2011).

39. T. Schneider, A. Wiatrek, S. Preussler, M. Grigat, and R.-P. Braun, Link budget analysis for terahertz fixed wireless links, *IEEE Trans. THz Sci. Technol.* **2**(2), 250–256 (2012).

40. R. Piesiewicz, C. Jansen, S. Wietzke, D. M. Mittleman, M. Koch, and T. Kürner, Properties of building and plastic materials in the THz range, *Int. J. IR mm Waves* **28**(5), 363–371 (2007).

41. R. Piesiewicz, C. Jansen, D. M. Mittleman, T. Kleine-Ostmann, M. Koch, and T. Kürner, Scattering analysis for the modeling of THz communication systems, *IEEE Trans. Antennas Propagat.* **55**(11), 3002–3009 (2007).

42. C. Jansen, R. Piesiewicz, D. M. Mittleman, T. Kürner, and M. Koch, The impact of reflections from stratified building materials on the wave propagation in future indoor terahertz communication systems, *IEEE Trans. Antennas Propagat.* **56**(5), 1413–1419 (2008).

43. C. Jansen, S. Priebe, C. Möller, M. Jacob, H. Dierke, M. Koch, and T. Kürner, Diffuse scattering from rough surfaces in THz communication channels, *IEEE Trans. THz Sci. Technol.* **2**(1), 462–472 (2011).

44. S. Priebe, M. Jacob, C. Jansen, and T. Kürner, Non-specular scattering modeling for THz propagation simulations, *Proc. 5th Eur. Conf. Antennas Propagat. (EuCAP)*, 1–5 (2011).

45. M. Jacob, S. Priebe, R. Dickhoff, T. Kleine-Ostmann, T. Schrader, and T. Kürner, Diffraction in mm and sub-mm wave indoor propagation channels, *IEEE Trans. Microwave Theory Tech.* **60**(3), 833–844 (2012).

46. W. Q. Malik, D. J. Edwards, and C. J. Stevens, Frequency-dependent pathloss in the ultrawideband indoor Channel, *Proc. Intl. Conf. Commun. ACC* **12**, 5546–5551 (2006).

47. S. Priebez, M. Jacob, and T. Kürner, Calibrated broadband ray tracing for the simulation of wave propagation in mm and sub-mm wave indoor propagation channels, *Proc. 18th Eur. Wireless Conf., EW*, 1–10 (2012).

48. S. Priebez, M. Jacob, C. Jastrow, T. Kleine-Ostmann, T. Schrader, and T. Kürner, A comparison of indoor channel measurements and ray tracing simulations at 300 GHz, *Proc. 35th Int. Conf. IR mm THz Waves, IRMMW-THz*, 1–2 (2010).

49. T. Kürner, I. Kallfass, K. Ajito, A. Kasamatsu, M. Britz, and S. Priebez, *What's Next? Wireless Communication beyond 60 GHz (Tutorial IG THz)*, IEEE 802.15 document 15-12-0320-01-0thz (2012).

50. T. Kürner, Towards future THz communications, *THz Sci. Technol.* **5**(1), 11–17 (2012).

51. ITU (2012). *Radio Regulations.*

52. A. Clegg, *Sharing Between Radio Astronomy and Active Services at THz Frequencies*, IEEE 802.15 document 15-10-0829-00-0thz (2010).

53. ITU-R Recommendation P.838-3 (2005). *Specific Attenuation Model for Rain for Use in Prediction Methods.*

Chapter 18

Terahertz Remote Gas Sensing System

Naofumi Shimizu,[a] Atsushi Wakatsuki,[b] Ken'ichi Kikuchi,[c] Satoshi Kohjiro,[c] Tomofumi Ikari,[d] Ryoichi Fukasawa,[d] and Ken Matsuyama[e]
[a]*NTT Device Technology Laboratories, NTT Corporation, 3-1, Morinosato-Wakamiya, Atsugi, Kanagawa 243-0198, Japan.*
[b]*NTT Device Innovation Center, NTT Corporation, 3-1, Morinosato-Wakamiya, Atsugi, Kanagawa 243-0198, Japan.*
[c]*National Institute of Advanced Industrial Science and Technology, Tsukuba Central 2, 1-1-1 Umezono, Tsukuba-shi, Ibaraki 305-8568, Japan*
[d]*Spectra Design Corporation, 285-1 Yuzukami, Ohtawara-shi, Tochigi 324-0403, Japan*
[e]*Center for Fire Science and Technology, Tokyo University of Science, 2641 Yamazaki, Noda-shi, Chiba 278-8510, Japan*
shimizu.naofumi@lab.ntt.co.jp

This chapter presents a remote gas-sensing system with an operation band between 200 GHz and 500 GHz, developed for detecting toxic gases at the scene of a fire. The system employs a photonic terahertz wave generator and a superconductor-insulator-superconductor tunneling mixer, which provides both fast frequency sweeping and high receiver sensitivity at the same time. In a performance test of the system with a full-scale simulated fire, we succeeded in detecting hydrogen cyanide (HCN), which is produced by incomplete

Handbook of Terahertz Technologies: Devices and Applications
Edited by Ho-Jin Song and Tadao Nagatsuma

ISBN 978-981-4613-08-8 (Hardcover), 978-981-4613-09-5 (eBook)
www.panstanford.com

combustion of urethane. The coincidence of the HCN concentration obtained from the developed system with the result of chemical analysis of a sampled gas demonstrates the potential of the system for use as a remote toxic gas sensor at disaster sites.

18.1 Introduction

Gas molecules have a lot of absorption lines over a wide frequency range between microwaves and light. These lines come from energy transition in a molecule, and the peak frequencies, spacing, shape, and other characteristics of absorption spectra depend on the composition and structure of a molecule itself. It is not necessary to investigate all lines spreading in the wide frequency range to identify what gaseous substances are presented there. Observing just a few absorption lines among them is sufficient. For gas sensing and analysis, absorption lines in the light region are commonly used because of the strong absorption intensity of gas molecules and the wide variety of mature measurement technologies in this region.

In contrast, the absorption intensity of gas molecules is generally weaker for the terahertz (THz) and millimeter-wave regions than for the light region. Moreover, technologies available for the THz region are limited. However, there is an advantage in sensing gaseous substances with THz and millimeter waves. The size of particles suspended in a dusty space or in smoke is typically a few microns or less. The wavelength of light is comparable to them and is easily scattered by dust, soot, and smoke. Therefore, using light for sensing gaseous substances isn't suitable in such dusty space; on the other hand, these tiny floating particles hardly affect the transmission characteristic of THz and millimeter waves at all. This is because their wavelengths, ranging from 100 microns to a few millimeters, are much larger than the size of the floating particles. Consequently, using THz and millimeter waves for sensing gaseous substances is suitable in a terrestrial area [1].

Gopalsami et al. reported environmental monitoring in the field with millimeter waves with frequencies between 225 GHz and 315 GHz. They developed an active sensing system and succeeded in detecting heavy water vapor, methyl chloride, and butanol from a distance [2, 3]. They also demonstrated remote detection of nitric oxide with a passive system with the 146–154 GHz band [4]. Brown et al. carried out a feasibility study of remote detection of bioparticles suspended in air with an active technique in the THz region [5, 6]. Concerning the analysis of chemical compounds in flame and smoke with THz waves, several groups have proved that THz waves have good transmission characteristics for this purpose, and they reported the observation of absorption lines of various gaseous substances in smoke generated by combustion [7–11].

These results suggest that it should be possible to detect typical hazardous gases formed at fire sites, such as carbon monoxide (CO) and hydrogen cyanide (HCN), from a distance using electromagnetic waves in the THz and millimeter-wave bands and to warn of possible secondary disasters in advance at disaster sites such as at the scene of a fire. On the basis of the concept [12], we have developed a remote gas-sensing system with an operation band between 200 GHz and 500 GHz and have succeeded in detecting HCN molecules produced by the combustion of urethane in a simulated full-scale fire [13].

The next section discusses the considerations for a remote gas-sensing system. Then, the hardware of the system—THz wave generator, photomixer, heterodyne receiver, and THz optics—is presented. Finally, we describe the experimental results of gas sensing with the developed system in a full-scale simulated fire.

18.2 System Considerations

Selecting a suitable band is very important for realizing a remote sensing system for use in a terrestrial area because attenuation of the THz and millimeter waves due to water molecules in the atmosphere is very strong at some frequencies. Figure 18.1 shows the transmission loss of THz and millimeter waves calculated from the HITRAN database [14] under the condition that they propagate 50 m in air at 23°C with a relative humidity of 50%. The transmission characteristics are not very good above 1 THz because the absorption of water molecules appears as a continuous level rather than a discrete one at around this frequency and the absorption spectra of other substances will be mostly masked. This region would, therefore, be very difficult to use. The loss is comparatively low between 200 GHz and 1 THz, except for the frequencies of water absorption lines of around 380 GHz, 448 GHz, 557 GHz, and 753 GHz. In particular, there is no frequency mask below 500 GHz. We can, therefore, measure the spectrum effectively with a simple frequency sweep in this frequency band.

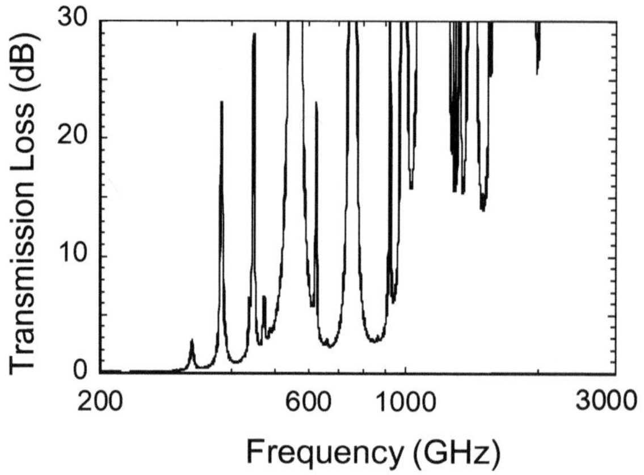

Figure 18.1 Transmission loss of THz waves in air at 23°C with a relative humidity of 50%.

Next, the fingerprint spectra of gas molecules are considered. Figure 18.2 shows the absorption spectra of CO and HCN between 30 GHz and 1 THz, calculated from the HITRAN database. Here, we assume that the THz waves travel 10 m in the space filled with each gas in uniform concentration and that the concentrations are lethal levels for CO and one-tenth of the lethal levels for HCN. These two toxic gases have unique absorption spectra in the frequency region below 1 THz, but the absorption intensity becomes weak in lower frequencies. Considering the loss of the THz waves and fingerprint spectra of gas molecules, we set the operation band of our sensing system between 200 GHz and 500 GHz.

There are two ways to measure an absorption spectrum of a gaseous substance. One is to use blackbody radiation of the ambience as a probing signal. The temperature difference between the target material and the background induces a rise or drop in the radiation intensity at certain frequencies in the THz or millimeter-wave region. From the frequency of the peak, we can identify existing materials. This method, called passive detection, needs no active source for the THz and millimeter waves and is widely used in radio astronomy, atmospheric sensing, and earth resource mapping. The other is to use the THz or millimeter waves generated in the system as probing signals. The transmitter of the system emits THz or millimeter waves to a target. The waves are scattered there, and the receiver detects part of the returning waves. This is called active detection, and a radar is a typical example of an active system. The drawback of the passive method for remote gas sensing at the scene of a fire is that the temperature difference between the target material and the background is very difficult to predict.

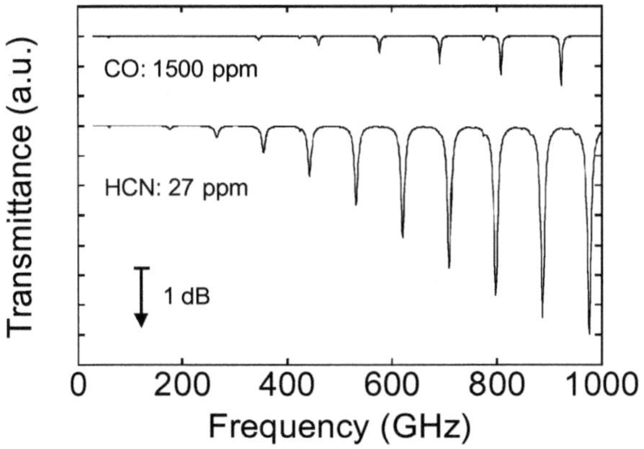

Figure 18.2 Absorption spectra calculated from the HITRAN database.

In the active system, there should be something that works as a reflector of the THz waves around a target. Then, we investigated the possible use of the walls of buildings as reflectors in disaster situations. The reflectance of 30 building materials was measured at THz frequencies. Figure 18.3a shows the reflectance of several wood samples. The reflectance level was measured as 1%–10%, depending on the species. Figure 18.3b shows results for metallic and ceramic sidings. The high reflectance of metallic sidings indicates these might function well as a reflector. The reflection level of different glasses was measured, as shown in Fig. 18.3c, and no frequency dependence was apparent. In our measurements, tatami and carpet had very low reflection levels of waves, as shown in Fig. 18.3d. One reason for low reflection is the scattering at the surface. The scattering loss depends on the surface roughness. When the surface roughness meets the following requirement, the surface is generally regarded as a smooth surface for incident waves:

$$\Delta h < \frac{\lambda}{8\cos\theta}$$

where Δh is the standard deviation of the surface height, λ is the wavelength of the incident waves, and θ is the angle of incident and reflection. The above condition is known as the Rayleigh criterion. The maximum surface height at which a surface is regarded as smooth for 300 GHz, λ = 1 mm, was calculated, respectively, as 0.12 mm and 0.25 mm for the incident angles for 0° and 60°. We were unable to measure the precise surface roughness of materials. However, the tatami and carpet surface height can be estimated as much larger than 0.12 mm from microscopic measurements. For wallpaper and plaster, detailed scattering analysis results were reported by Piesiewicz et al. [15], who found that the standard deviation of the surface height is 0.15 mm for wallpaper and 0.05 mm for plaster, indicating that these surfaces can be regarded as smooth for frequencies less than 300 GHz.

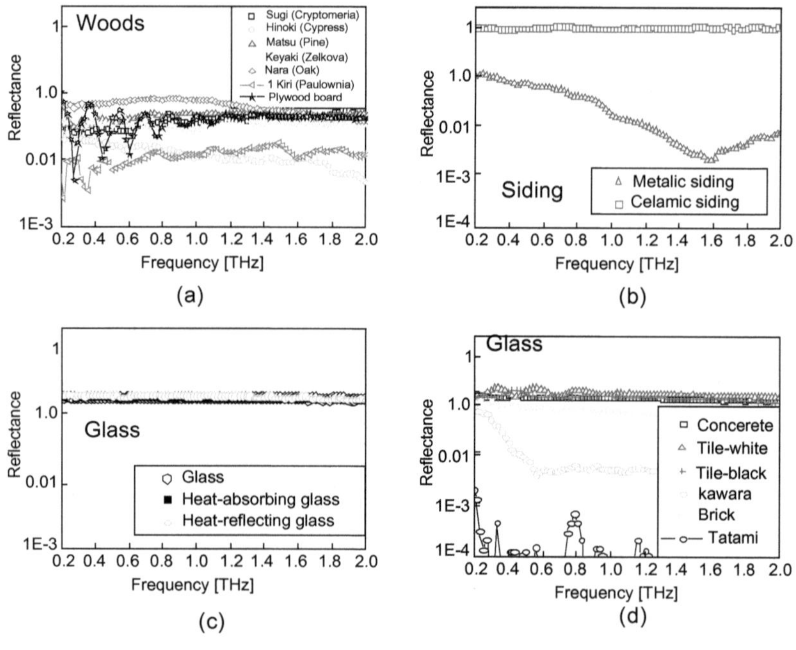

Figure 18.3 Reflection spectra of building material samples: (a) wood, (b) siding, (c) glass, and (d) concrete, tile, and tatami.

Figure 18.4 summarizes the reflectance levels around 300 GHz. The results indicate that the materials commonly used in the outer walls of buildings, such as metallic siding, glass, tile, concrete, and mortar, have a reflectance of more than 10% and they can be expected to work as effective reflectors of the THz waves. Moreover, their reflectance spectra are very flat. In view of these considerations, the active method is employed in our system.

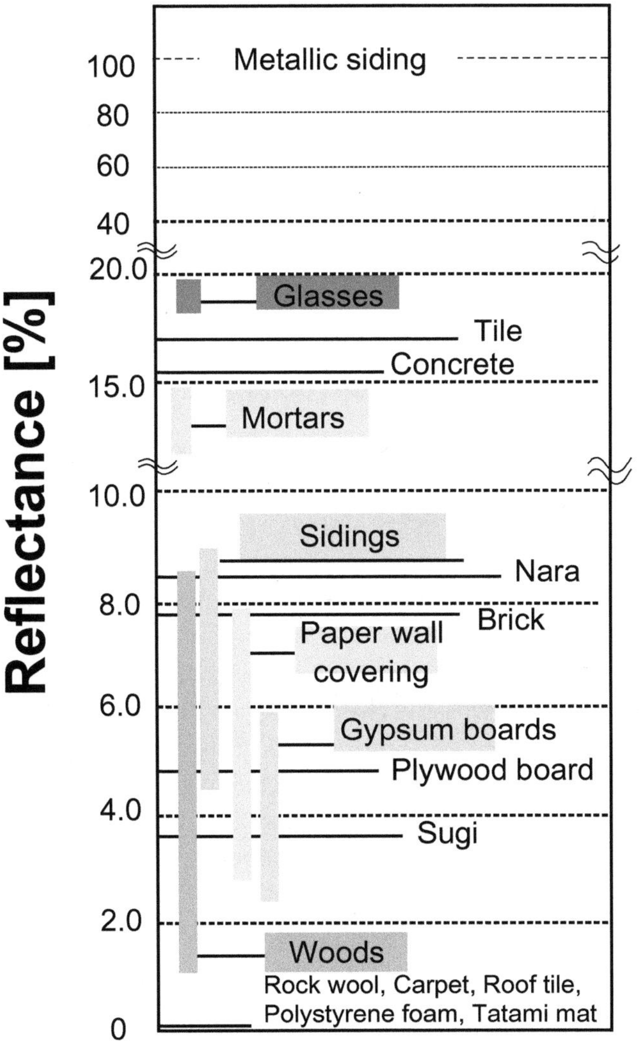

Figure 18.4 Summary of reflectance levels of building materials at 0.2–1 THz.

In an active system, there are also two ways to arrange the transmitter and receiver: One is a bistatic arrangement, in which the transmitter and receiver use their own antennas and detach them from each other in space; the other is a monostatic arrangement in which, the transmitter and receiver share an antenna and are integrated at one place. One disadvantage of the bistatic arrangement is that it requires fine alignment of the position and direction of each antenna before measurement. In contrast, the relative position and direction of the transmitter and receiver are fixed in the monostatic configuration, which means we don't have to adjust them according to time and circumstances. The ability to instantly start up the system is indispensable for remote toxic gas detection at the scene of a fire. Therefore, we adopted the monostatic arrangement for the transmitter and receiver.

18.3 Hardware

18.3.1 Photonic THz Wave Generator

Figure 18.5 shows the configuration of the photonic THz wave generator used in this research [16]. It consists of two wavelength-fixed laser sources (FLS 1 and 2), a wavelength-tunable laser source (TLS), two optical couplers, two erbium-doped optical fiber amplifiers (EDFAs), and two photomixers. With this configuration, the generator outputs two THz waves at once. The operation principle is as follows: the laser light from the TLS is divided into two, one is coupled with the laser light from FLS 1 and the other with that from FLS 2. Then, each mixed optical signal is amplified by each EDFA and converted into electrical signal by a photomixer. Here we denote THz waves generated at photomixers 1 and 2 as THz waves 1 and 2, respectively. Small parts of the laser lights from two FLSs are picked up and coupled, and their beat signal is obtained. This beat signal is mixed with the reference signal, and its frequency error is extracted. The FM–AM converter then changes this frequency error into an error signal with its amplitude proportional to the frequency error. Finally, the driving current of FLS 2 is adjusted with the error signal, and the frequency difference of the FLSs is locked.

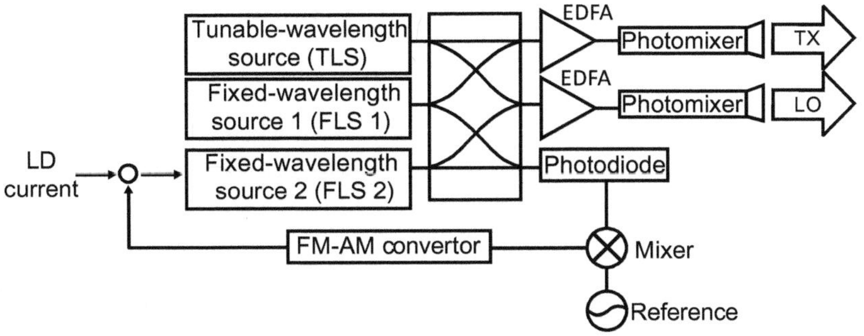

Figure 18.5 Photonic THz wave generator.

This photonic THz wave generator has a clear advantage in building a broadband continuous-wave (CW) THz spectrometer with a heterodyne receiver. This is because the frequency difference between THz wave 1 and THz wave 2 is only determined by the wavelengths of FLS 1 and 2; it is independent of wavelengths of the TLS or the frequency of the THz waves themselves. When the frequency difference between the received signal and the local oscillator (LO) signal of the heterodyne receiver is unstable, the operation band of the intermediate-frequency (IF) circuit of the heterodyne receiver has to cover all the possible minimum and maximum frequencies of the IF signal: This results in an increase of the thermal noise of the receiver, which degrades the receiver sensitivity. When we use THz waves 1 and 2 for the transmitter signal and photonic LO (Ph-LO) signal of the heterodyne receiver, we don't have to widen the bandwidth of the IF circuit needlessly, because their frequency difference is fixed. Therefore, we can exploit the low-noise performance of superconductor-insulator-superconductor (SIS) mixers, as shown in Section 18.3.3, to build a broadband CW THz spectrometer.

18.3.2 Photomixer

Photomixer modules operating in the W-, F-, and D-bands were previously developed [17, 18]. On the basis of these technologies, we have successfully developed a high-output-power photomixer module operating at 200–500 GHz [19].

Figure 18.6 shows the band diagram of the photodiode (PD) used in the photomixer module. It is a variation of a uni-traveling-carrier photodiode (UTC-PD), which has a p-type photoabsorption layer and a wide-gap electron collection layer. Because of the high electron velocity in the collection layer, this device can operate at high current densities, while keeping its high-frequency response. We designed the structure so that the UTC-PD has higher responsivity than that of one with the same neutral absorption layer thickness. Due to the thin undoped absorption layer between the p-doped absorption layer and the wide-gap electron collection layer, THz waves with high output power are obtained with relatively lower optical input power. To increase the output power at around the designed center frequency of 350 GHz, we integrated a short stub with the UTC-PD chip. The fabricated PD was mounded in a metal package that has a PD–waveguide mode conversion coupler and an output rectangular waveguide (WR-3). The coupler, made on a quartz substrate, acts as a circuit to match the impedance between the PD chip and the waveguide. For operation at frequencies beyond the fundamental mode frequency range of WR-3, the coupler is designed to suppress the propagation of higher-order modes below –30 dB up to 530 GHz. Figure 18.7 shows a photograph of the fabricated photomixer module.

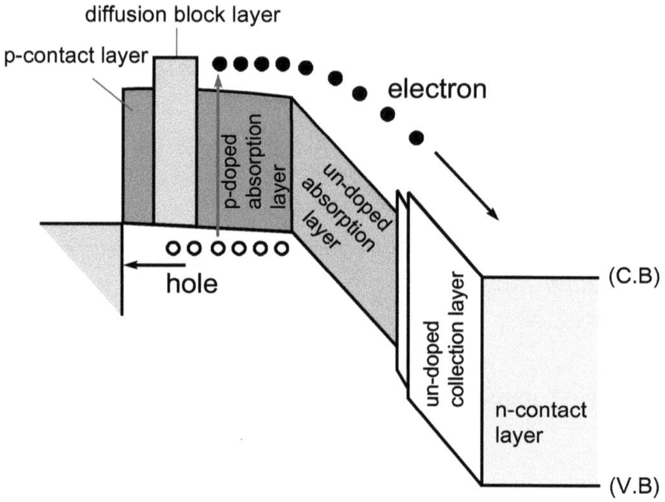

Figure 18.6 Band diagram of a photomixer.

The responsivity of the fabricated module is about 0.22 A/W, and the dark current is 20 nA at a DC bias of –1 V. At 350 GHz, the output power starts to saturate above the photocurrent of 20 mA at a DC bias of −2.5 V. The maximum output power is −2.7 dBm at around 350 GHz for a photocurrent of 20 mA, as shown in Fig. 18.8. The output power of more than −20 dBm is available from the transmitter between 225 GHz and 600 GHz. This demonstrates that the rectangular-waveguide output scheme is applicable for broadband THz wave generation at frequencies beyond the fundamental mode frequency range.

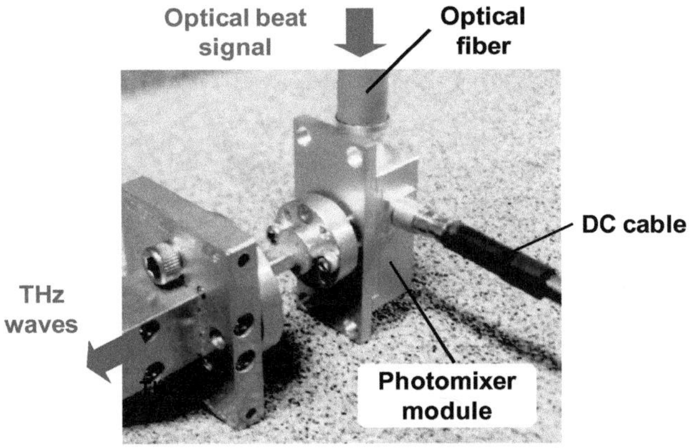

Figure 18.7 Photograph of a fabricated photomixer module.

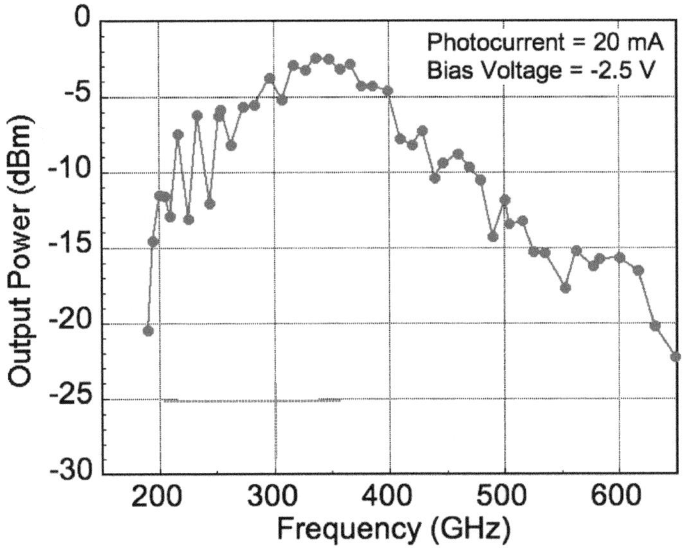

Figure 18.8 Output characteristics of a fabricated photomixer module.

18.3.3 Heterodyne Receiver

18.3.3.1 Advantage of SIS mixers

In contrast to millimeter-wave heterodyne receivers, a mixer determines the sensitivity of THz receivers because of the lack of low-noise preamplifiers operating above 200 GHz. The LO power requirement of <10 μW is also a key issue for mixer choice since the reported output powers of a Ph-LO are 400 μW at 350 GHz and 50 μW at 500 GHz [19]. SIS mixers exhibit quantum-limited sensitivity and satisfy this LO power requirement [20]. To operate SIS mixers, however, a relatively large system is necessary because SIS operation requires cryogenic equipment, such as a cryostat and a cryocooler, which consists of cold stages and compressors. Although there have been some precedents for a compact receiver with SIS mixers (SIS receivers), their applications have been limited to astronomical or upper-atmospheric observations (e.g., Ref. [21]). One of the issues related to a compact SIS receiver for ground-level observation is how to achieve a wide bandwidth because the absorption line is broadened to several gigahertz or more under atmospheric pressure. Besides, wide-frequency monitoring is necessary for identifying dangerous gases from the complex absorption spectra caused by many different gas molecules.

For this, we have developed a very wide-band SIS mixer that shows excellent performance in the 200–500 GHz range [22, 23] and succeeded in laboratory demonstration of absorption spectrum measurement [11]. This wide-band mixer can decrease the volume and heat load of the cryogenic stage, making it possible to use the smallest commercially available 4 K cryocooler. Since the Atacama large millimeter/submillimeter array (ALMA) [24] project will cover the 0.2–0.5 THz range with three separate receiver bands, it could be said that our receiver has a very good "size–bandwidth" performance [25].

18.3.3.2 Design and fabrication of SIS mixers

A quasi-optical mixer we developed for the fractional bandwidth (FBW) of FBW > 60% consists of a distributed junction array (DJA) of eight SIS junctions, a twin-slot antenna (TSA), and a three-stage transformer, as shown in the inset of Fig. 18.9. An approach for wide-RF-band SIS mixers is to adopt SIS junctions with a smaller time constant $C_J R_N$ originating from a thinner tunnel barrier for higher frequency, where C_J is the junction capacitance and R_N is the normal tunneling resistance of a junction. Above 400 GHz, only a few institutes with excellent junction fabrication technology have succeeded in this approach by changing the barrier material from mature AlO_X to AlN [26]. Another approach is the DJA [27], where N junctions with a relatively thick tunnel barrier are connected in parallel through a superconducting microstrip line (MSL). The DJA mixer shown in Fig. 18.10a incorporates N-1 unit cells, which provide N-1 resonant frequencies at which the junction capacitance C_J is tuned out by an MSL inductance L_{MS}. If these frequencies are positioned closely, C_J is almost tuned out even in their separation, which ensures efficient and wide-band coupling between the DJA and an antenna. We found numerically this frequency separation is approximately given by $f_S = 1/[2N(C_J L_{MS})^{1/2}]$. In contrast, the DJA becomes too capacitive to couple to the antenna for $f > f_C = 1/[\pi(C_J L_{MS})^{1/2}]$. Optimum N, C_J, and L_{MS} should be determined for satisfying $f_{op} > f_S$, $f_{op} < f_C$, $R_{ANT} \approx \eta R_N/N$, and $N(hf_{op}/e)^2/(4R_N) < P_{LOav}$, where f_{op} is the operation frequency, R_{ANT} the radiation resistance of the antenna, η the impedance conversion ratio of the transformer, P_{LOav} the available LO power, h Planck's constant, and e the unit charge. On the basis of these conditions and our fabrication technology of Nb-based devices [28], we determined the parameters as $N = 8$, junction area $A_J = 2.0\ \mu m^2$, $C_J = 0.12$ pF, and $R_N \approx 32\ \Omega$. From these, related characteristic frequencies are $f_S = 120$ GHz and $f_C = 610$ GHz when the length of L_{MS} is 14 μm.

Self-complementary antennas, such as spiral and log-periodic ones, are known for their wide-band operation: the radiation resistance is independent of the frequency and given by $R_{ANT} = 60\pi/[(\varepsilon_{rS}+1)/2]^{1/2}$, where ε_{rS} is the dielectric constant of the substrate. However, $R_{ANT} = 75\ \Omega$ for a Si substrate ($\varepsilon_{rS} = 11.7$) is too large for wide-band coupling to a DJA with $R_N/N \approx 4\ \Omega$. Moreover, their polarization is frequency dependent, which prevents wide-band coupling to an LO. In contrast, a TSA is a linearly polarized and its main beam efficiency can be almost independent of frequency from its radiation patterns of 246 GHz and 500 GHz [29]. Though its impedance depends on frequency, our electromagnetic simulation [30] indicates that its imaginary part is smaller than R_{ANT} in the frequency range wider than 80% of its center. In addition, R_{ANT} decreases with the slot width W_S and $9 < R_{ANT} < 25\ \Omega$ for $1.5 < W_S < 15$ μm at the full-wave resonance around 350 GHz. For these reasons, we adopted a TSA followed by a three-stage MSL transformer designed for 250–450 GHz.

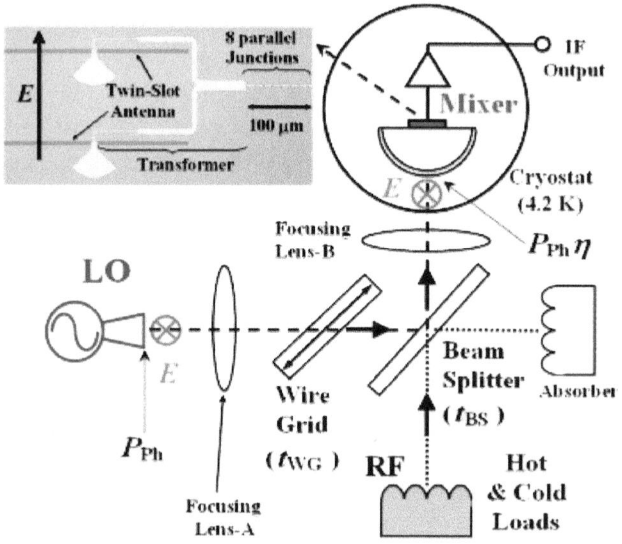

Figure 18.9 Setup for characterization of an SIS mixer. Inset shows the central part of the mixer chip. E denotes the electric field of an LO signal. P_{Ph}, t_{WG}, t_{BS}, and h are defined in the text.

The device consists of six thin-film layers on a 3-inch-diameter Si wafer: Nb/AlO_X/Nb junction sandwiches attached to an Nb ground plane, a SiO_2 insulation layer with thickness t_I, and a dielectric constant of ε_{rI}, Nb wirings, and Au contact pads. The fabrication process established for Nb-based rapid-single-flux-quantum (RSFQ) circuits [27] were adopted, where the Al oxidation time and pressure are modified to realize $C_J R_N \approx 3.8$ ps.

The chip was cooled at 4.2 K in a liquid helium cryostat after it had been glued on a 5 mm radius silicon elliptical lens coated with an epoxy antireflection film whose designed center frequency was 350 GHz. To evaluate radio frequency (RF) transmission from the TSA to the DJA, the current responsivity of a voltage-biased DJA was measured by Fourier transform spectroscopy (FTS) prior to heterodyne measurement.

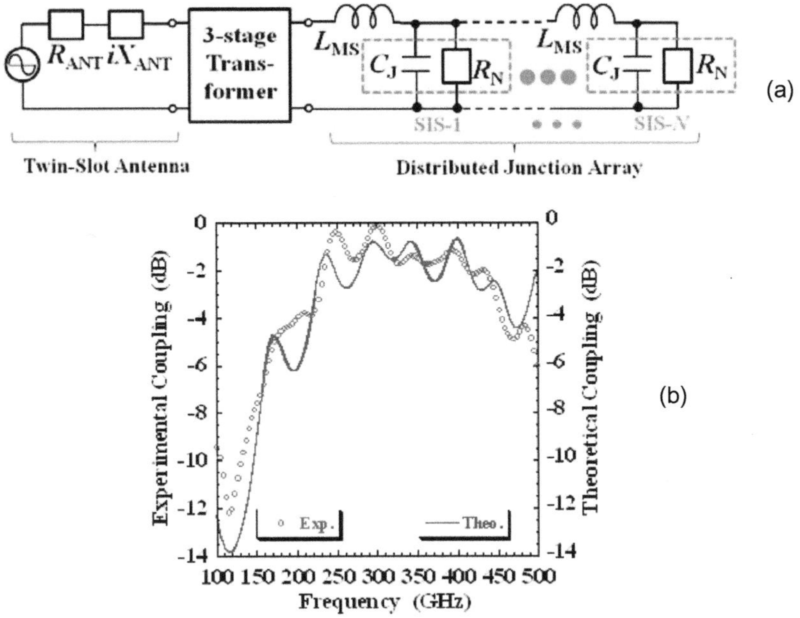

Figure 18.10 (a) Equivalent circuit of a mixer chip. (b) Calculated (solid line) and experimental (circle) power coupling between an antenna and a DJA.

Figure 18.10b shows experimental and theoretical power transmissions from the antenna to the DJA. An experimental transmission C_{EX} is calculated as $C_{EX} > -2.5$ dB in 231–444 GHz from the FTS response as $C_{EX} = 10\log_{10}$(FTS response) + C_O, where C_O is adjusted so that the maximum C_{EX} in the 100–500 GHz range is 0 dB. From a systematic comparison between experimental and theoretical transmissions for various mixers, we found the difference in their frequencies that gives local maxima and minima on the transmission is less than 6% in the band of 250–450 GHz, confirming the validity of our design. Their discrepancy out of the band may be due to approximations that were not valid at a frequency far from its center, that is, zero impedance of radial stubs positioned at the TSA and infinite RF impedance of the IF port of the mixer.

18.3.3.3 Characterization of the SIS receiver

The uncorrected double-sideband (DSB) receiver noise temperature T_{RX} was measured by the conventional Y-factor method, where T_{RX}

is derived from the ratio of the IF output power for the hot antenna load to the cold one. For the single-ended mixer adopted in this experiment, the sideband noise of the LO is also down-converted to the IF output of the mixer and degrades T_{RX}. Thus, comparison of T_{RX} of an identical mixer pumped by two kinds of LOs makes clear whether these two LOs produce different amounts of noise. Figure 18.11 shows T_{RX}-f characteristics with two kinds of LOs. One is a Ph-LO with a 16 μm or 25 μm thick beam splitter (BS). The efficiency of LO–mixer coupling of the latter is about two times as large as that of the former, with the sacrifice of the RF coupling of the mixer. The other LO device is composed of commercial multiplier-based CW oscillators driven by a microwave synthesizer or Gunn diodes. As shown in Fig. 18.11, T_{RX}-f characteristics of both LOs agree quantitatively with the inverse of the FTS responsivity with a fitting coefficient of $K = 0.07$. This indicates that T_{RX} of the Ph-LO is almost equal to that of conventional LOs, demonstrating that the Ph-LO does not produce remarkable additional noise in comparison with the conventional multiplier-based ones. The origin of the slightly higher T_{RX} for the Ph-LO at particular frequency points, such as 312 GHz, 339 GHz, and 460 GHz, is now under investigation. The origin may be related to the dynamics of the source lasers, since the T_{RX} of Ph-LOs is found to be sensitive to the choice and operating conditions of the lasers. In this case, the optimization of source lasers will contribute to the reduction of T_{RX} at these frequency points. Figure 18.11 indicates the clear difference of bandwidth between the Ph-LO and conventional LOs. A Ph-LO covers 200–500 GHz completely, while three conventional LOs with different frequency bands cover a part of this range. $T_{RX} \leq 20hf/k\text{B}$ is achieved in FBW = 74%. This bandwidth is dominated by that of the present SIS mixer. The pumping power of this wide-band and low-noise SIS mixer was roughly estimated from the frequency-dependent P_{Ph} [19]. The coupling coefficient between the LO and the mixer, η, is given by $\eta = \zeta \times (1 - t_{BS}) \times t_{WG}$, where t_{BS} is the BS transmission, t_{WG} the WG transmission, and ζ the beam-coupling efficiency in terms of other factors. The product of $P_{Ph}\eta$ is the pumping power at the mixer input. From the frequency-dependent P_{Ph}, taking theoretical values of t_{BS} and t_{WG} and assuming the ideal coupling of $\zeta = 1$, $0.7 \leq P_{Ph}\eta \leq 5$ μW is calculated [31] (Fig. 18.12). The resulting $P_{Ph}\eta$ is 2 to 3 orders of magnitude lower than the typical LO power requirement P_{LO} for

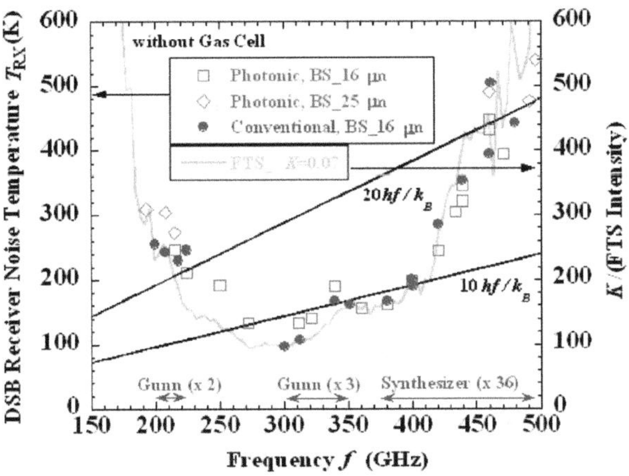

Figure 18.11 Uncorrected DSB receiver noise temperature vs. frequency. The SIS mixer is pumped either by a Ph-LO or three multiplier-based LOs (filled circles) driven by a microwave synthesizer or Gunn diodes with different bands (arrows; ×2, ×3, and ×36 denote each harmonic number). For the Ph-LO, two kinds of beam splitters (BSs) with different thicknesses are used. The solid curve denotes the inverse of FTS responsivity with K = 0.07 (right axis).

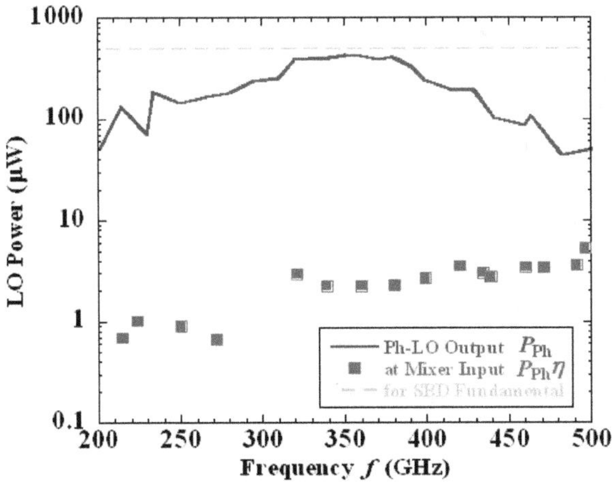

Figure 18.12 Frequency dependence of the output power of a photonic LO P_{Ph} (solid line) and the maximum [31] pumping power at the mixer input $P_{Ph}h$ (filled squares) calculated from t_{WG} and t_{BS}. The dashed line denotes the reported LO power requirement for an SBD fundamental mixer at room temperature.

an SBD operating at room temperature, that is, $P_{LO} \geq 0.5$ mW [32, 33] for the fundamental mixing and $P_{LO} \geq 2$ mW for the second-harmonic mixing [32, 34]. The combination of the SIS mixer with such a small P_{LO} and the Ph-LO with 50–400 μW output achieves wide-band and low-T_{RX} heterodyne detection.

18.3.3.4 Compact mechanical cryocooler and system stability

Figure 18.13 shows a cross-sectional view of a cryostat.

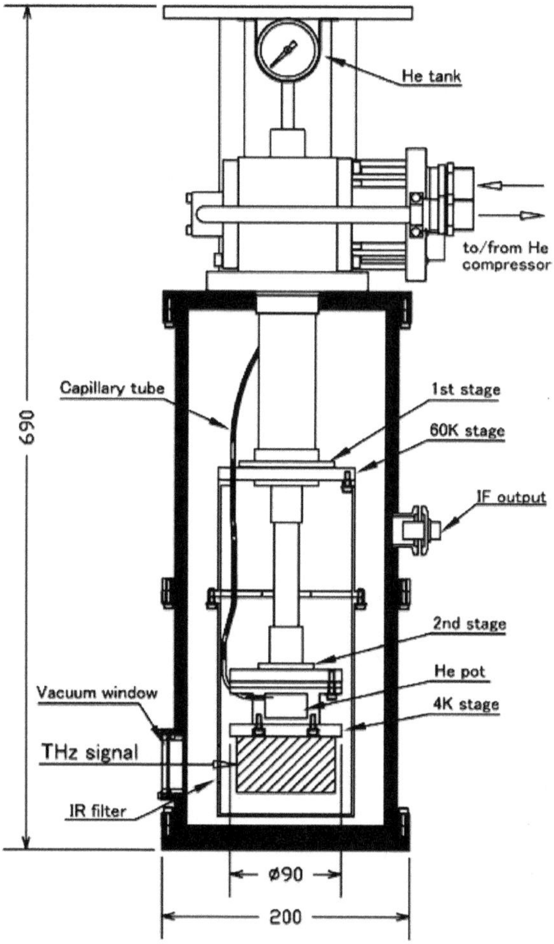

Figure 18.13 Cross-sectional drawing of a cryostat. The SIS mixer and the HEMT amplifier are in the hatching region attached below the 4-K stage.

A cryostat is a self-standing, compact commercially available Gifford–McMahon cryocooler (model SRDK-101D, Sumitomo Heavy Industries, Ltd.). The cryocooler is operated by an air-cooled helium (He) compressor (model CNA-11B), which circulates refrigerant He gas through a pair of flexible tubes. Table 18.1 summarizes the receiver specifications. The whole receiver system could be installed on a small vehicle to transport it somewhere for measurement. The cooling capacities of the first and second stages are 3 W at 60 K and 0.1 W at 4.2 K, respectively. The SIS mixer and a high-electron-mobility transistor (HEMT) amplifier are located at the 4 K stage. To cope with large thermal radiation through the THz window, which is estimated to be about a 165 mW heat load at the 4 K stage, we attached two-ply-type G-108 Zitex [35], a porous polytetrafluoroethylene (PTFE) film, to the hole of the 60 K shield as an infrared (IR) filter. It reduces the radiation heat load by more than 80% and makes the temperature at the 4 K stage less than 4 K without degrading the receiver's noise performance. To stabilize the 4 K stage temperature, a He pot [36, 37], which reduces temperature fluctuations owing to the large heat capacity of He, is mounted between the second stage and the 4 K stage and connected to the external He tank by a capillary tube. As a result, a large temperature fluctuation of ~200 $mK_{p\text{-}p}$ at the second stage caused by periodic motion of a displacer in the cryocooler does not directly affect the 4 K stage, the fluctuation of which is kept as small as 5 $mK_{p\text{-}p}$.

Table 18.1 Summary of receiver specifications

Parameter	
Frequency:	
Input	0.2–0.5 THz
Output	0.2–1.0 GHz
Weight:	
Compressor	42 kg
Cryostat	27 kg
Dimensions:	
Compressor	$400 \times 385 \times 529$ mm^3
Compressor	$200 \times 270 \times 690$ mm^3
Power Consumption:	
Compressor	<1.5 kW (100 V AC)
Electronics	About 5 W

For heterodyne measurement, an external LO is necessary. Both LO and THz signals reach the SIS mixer through a 125 μm thick vacuum window made of a polyimide film and IR filter. The SIS mixer down-converts the incoming THz signal into a 0.2–1.0 GHz IF signal, the following HEMT amplifies the IF signal, and finally the IF signal is extracted from the cryostat. To evaluate the stability, we measured the Allan variance [38, 39] of the IF output (Fig. 18.14). The fact that the Allan minimum time is ~70 s indicates measured data accumulated over a period of 1 minute will not be degraded due to the receiver system instability such as gain drift.

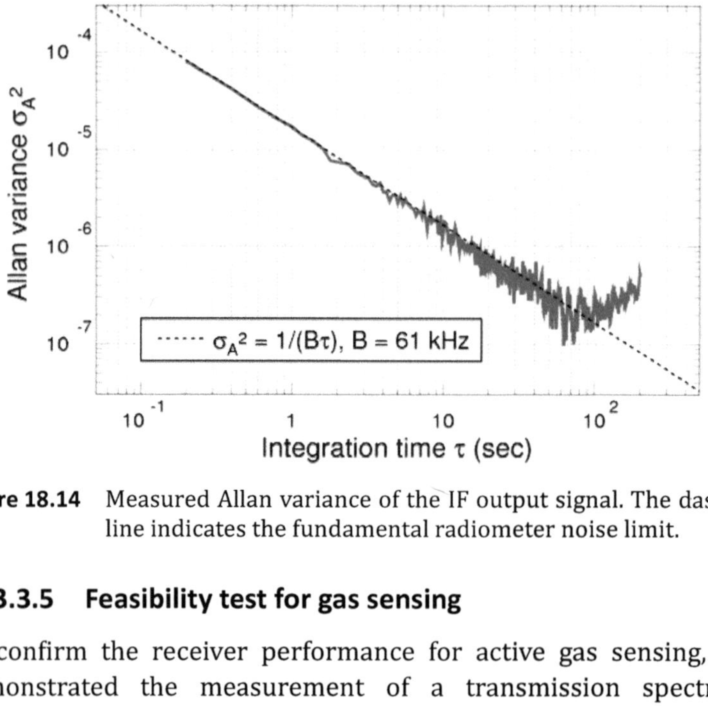

Figure 18.14 Measured Allan variance of the IF output signal. The dashed line indicates the fundamental radiometer noise limit.

18.3.3.5 Feasibility test for gas sensing

To confirm the receiver performance for active gas sensing, we demonstrated the measurement of a transmission spectrum through a gas cell by passive method (Fig. 18.15). A blackbody radiator behind the gas cell emitted a broadband THz spectrum. Since the molecular gas inside the gas cell absorbed THz energy at characteristic frequencies, the receiver observed absorption lines in the spectrum. A photomixer using a J-band uni-traveling-carrier photodiode (J-UTC-PD) [19] was used as an LO, and the total IF power was measured with a fast square-law detector. Synchronizing between the LO frequency sweep and power detection [40], we could obtain a full 0.2–0.5 THz spectrum. Figure 18.16 shows the measured transmission spectra of acetonitrile (CH_3CN) diluted 0.25%, 0.5%, and 1.0% with nitrogen (N_2) at 1 atm. Each spectrum is the average of sixteen 0.2–0.5 THz sweeps with a frequency step of 0.5 GHz. Since a single frequency sweep takes less than 1 s, the total measurement time was much shorter than the Allan minimum time. The frequency resolution is dominated by LO stability. The LO has a line width of ~10 MHz [25] and a frequency accuracy of ~100 MHz [40], which is small enough to identify the gas contents whose lines are broadened to a few gigahertz at atmospheric pressure. The measured absorption peaks agree well with predicted model calculations.

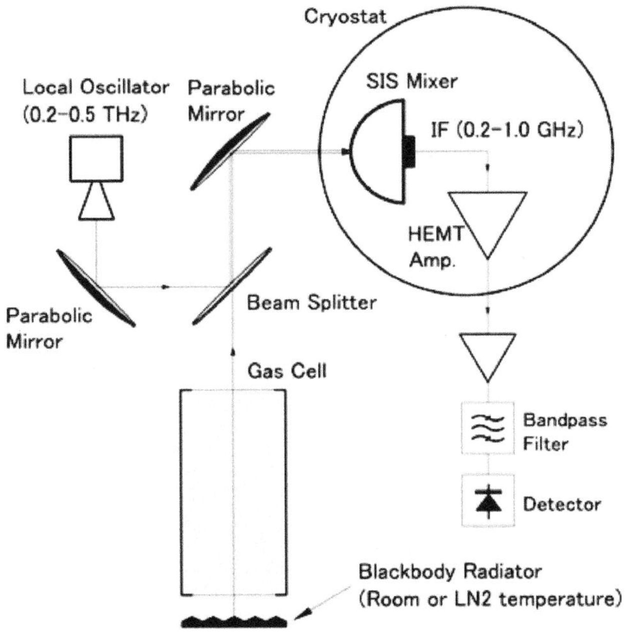

Figure 18.15 Schematic view of the gas cell measurement setup.

On the basis of the Dicke radiometer equation, the root-mean square uncertainty of brightness temperature δT_{RMS}, and the corresponding minimum detectable power of the receiver δP_{RMS} are defined as

$$\delta T_{RMS} = \frac{T_{RX} + T_{ANT}}{\kappa\sqrt{(B\tau)}} \tag{18.1}$$

and

$$\delta P_{\mathrm{RMS}} = k_{\mathrm{B}} B \delta T_{\mathrm{RMS}} \tag{18.2}$$

respectively, where T_{ANT} is the radiation temperature of the antenna load, κ is the coupling efficiency between the receiver and the object being measured, B is the final detection bandwidth, and τ is the total signal integration time. For $T_{\mathrm{RX}} = 150$ K, $T_{\mathrm{ANT}} = 77$ K, $\kappa = 1$, $B = 0.5$ GHz, $\tau = 0.17 \times 16 = 2.7$ ms, one finds the theoretical power resolution of $\delta P_{\mathrm{RMS}} = 1.4$ fW. The experimental δP_{RMS} is about 5–10 fW from the passive gas spectroscopy we carried out [40], which is of nearly the same order as the theoretical prediction. In a rough estimation for active gas spectroscopy, the receiver detects the reflected THz waves with power $P_{\mathrm{k}} = -55$ dBm = 3 nW: the minimum output power of the photomixer is −15 dBm (Fig. 18.8); the round-trip transmission loss of optics and free space are both 10 dB; and the reflection loss of building material is 20 dB (Figs. 18.3 and 18.4). Thus, the ratio of detected THz wave power to noise is estimated as $P_{\mathrm{k}}/\delta P_{\mathrm{RMS}} = 3[\mathrm{nW}]/10[\mathrm{fW}] = 3 \times 10^5 = 55$ dB.

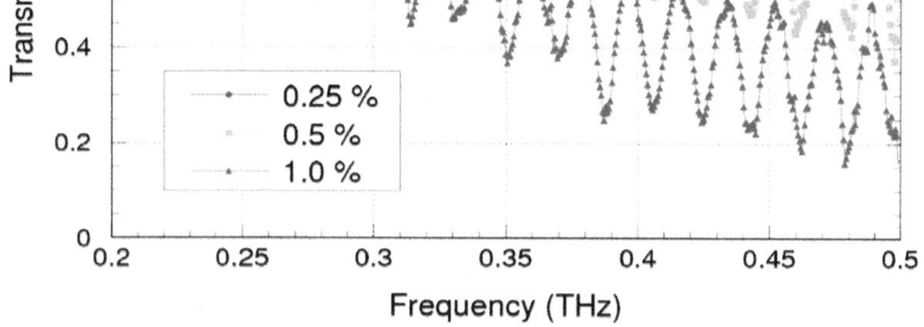

Figure 18.16 Measured transmission spectra of CH_3CN diluted 0.25%, 0.5%, and 1.0% with N_2.

Closed circles in Fig. 18.17 show the measured absorbance of the smoke emitted from heated nylon fabric by active method. In this experiment, we placed a 1 m long gas cell filled with smoke between the THz transmitter and receiver and measured absorbance. As a THz transmitter and LO signal of the receiver, each output of the photonic THz wave generator was used. To clarify the origin of these peaks, we carried out a spectral analysis, in which we took into account HCN, CH_3CN, and H_2O. We assumed that the temperature and total pressure in the gas cell were 25°C and 1013 mbar, respectively. Besides distinct absorption peaks due to these three gas molecules, broad absorption was observed in measured absorption spectra when the gas cell was filled with smoke. This might be attributable to the attenuation of soot or gaseous organic compounds generated due to the combustion of nylon fabric. In our spectral analysis, we treated such broad absorption as a frequency-independent loss. We determined the contributions of these four components by the least-squares method. The best-fit result for the spectrum is shown as a solid line in Fig. 18.17. The contributions of HCN, CH_3CN, and H_2O and broad absorption in the fitted absorbance are shown as broken lines. The frequencies of the measured and fitted absorption peaks agree quite well, indicating that the experimentally observed distinct peaks are mainly due to these three gas molecules.

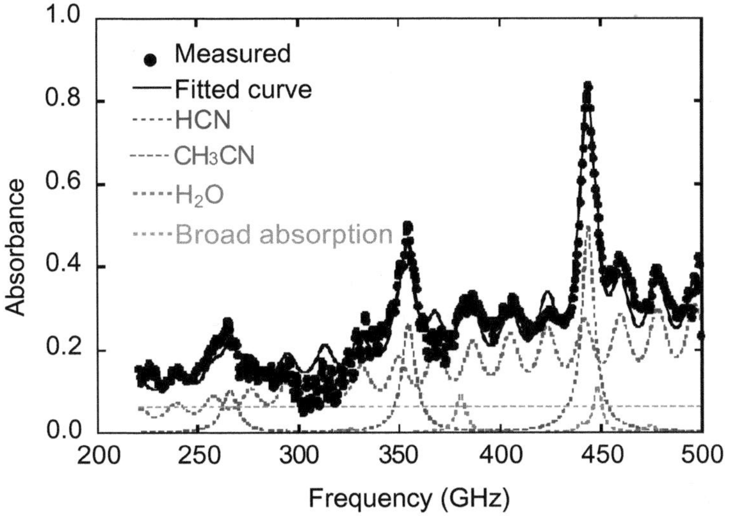

Figure 18.17 Measured absorbance of smoke emitted from heated nylon.

These results indicate the functions and sensitivity of the developed receiver are appropriate for an active spectrometer.

18.3.4 Optics

Figure 18.18 shows a monostatic arrangement of the components used to transmit and collect THz waves [13]. The output from the J-UTC-PD with a feed horn antenna becomes a Gaussian beam of THz waves, whose electric field is horizontally polarized. A flat mirror redirects this beam to an ellipsoidal mirror, which makes the beam into a quasi-parallel one. Then, the beam passes between two wire grid polarizers (WG1 and WG2). The polarization directions of WG1 and WG2 are horizontal and tilted 45°, respectively. The transmission loss of the beam at the first grid is 0, whereas it is 3 dB when the beam passes through the second grid. The polarization direction of the beam is tilted to 45° after the second grid. Next, the beam is reflected by a subreflector to a 40 cm square hyperbolic main reflector. Here, the beam expands in size. The main reflector forms the beam into a quasi-parallel one and projects it to the target. The full width at 1/e power of the projected beam was measured to be 19 cm at 375 GHz at a distance of 7.5 m from the reflector, as shown in Fig. 18.19.

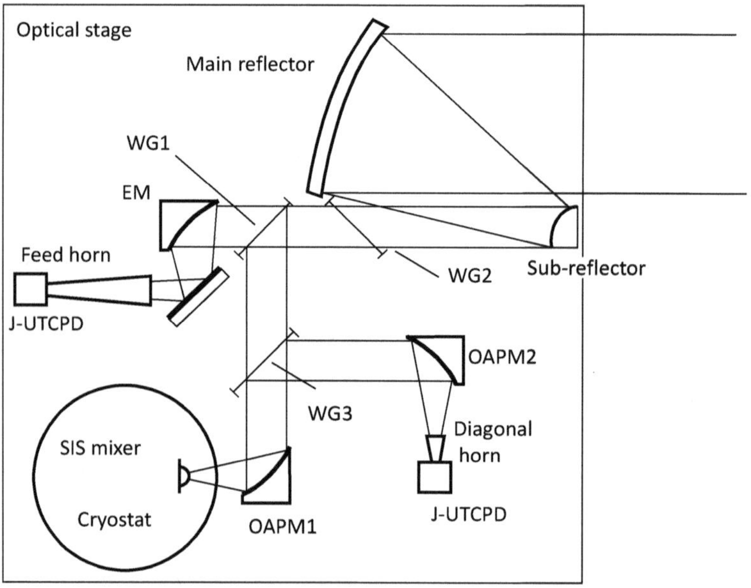

Figure 18.18 Optics of the gas-sensing system. FM, EM, WG, and OAPM stand for flat mirror, ellipsoidal mirror, wire grid filter, and off-axis parabolic mirror, respectively.

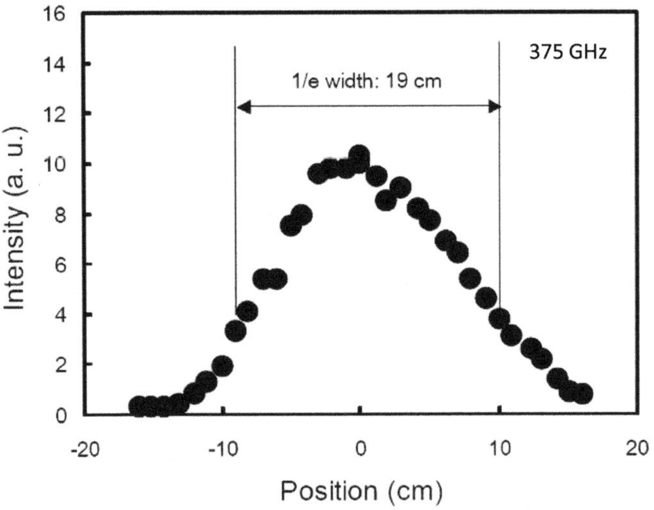

Figure 18.19 Profile of a THz beam measured at 7.5 m from the reflector.

Some of the THz waves that reach the target return to the main mirror. This is because an uneven part of an exterior wall formed by a window, entrance, eave, or craftwork or furnishings in the room becomes an effective scatterer and reflects the beam of THz waves. The returned beam is refocused by the main reflector and the subreflector into a quasi-parallel beam. It passes through WG2 and is divided into two beams of THz waves, whose electric fields are polarized in the vertical and horizontal directions, respectively. The vertically polarized beam is directed to the wire grid polarizer WG3, while the horizontally polarized one passes through WG1. The transmission loss of the beam at WG3 is as small as 3% because the polarization direction of WG3 is tilted only 10° from the vertical axis. The beam is then redirected by off-axis parabolic mirror 1 (OAPM1) into the SIS mixer in a mechanical cooler.

OAPM2 focuses the output from the other J-UTC-PD with the diagonal horn antenna into a quasi-parallel beam of THz waves. Its electric field is horizontally polarized. This beam is used as the LO signal of the SIS mixer. As the electric field direction of the LO beam and the polarization direction of WG3 form an angle of 80°, 97% of the LO beam is redirected to the receiver at WG3. Then, the LO beam is also reflected by OAPM1 and fed into the SIS mixer. The SIS mixer is sensitive only to the vertical electric field of incident THz waves because of the direction of the antenna integrated with the SIS mixer. Therefore, 97% of the beam from WG1 and 3% of the LO beam are received by the SIS mixer. All of these THz optics and the mechanical cooler are arranged on a mobile cart, as shown in Fig. 18.20.

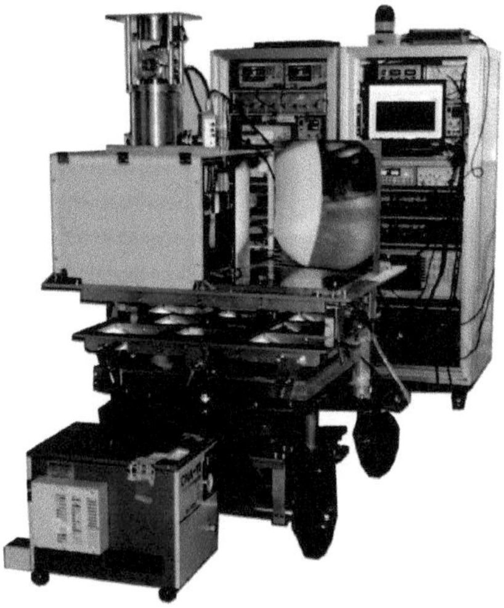

Figure 18.20 Developed active remote gas-sensing system.

18.4 Experiment

18.4.1 Measurement Setup

To confirm the performance of the developed system, we carried out an experiment in the full-scale fire test hall at the Center for Fire Science and Technology, Tokyo University of Science [13, 41]. Figure 18.21 shows the arrangement. The test room in the hall has heat-resistant walls, and its internal dimensions are 3.6 m in depth, 2.4 m in width, and 2.4 m in height, respectively. In it, we put 12 foam urethane cubes, 25 cm long on a side, in 2 lines in the longitudinal direction of the room. Then, we burned the cubes, which filled with the room with fire and smoke, as shown in Fig. 18.22. The temperature at a height between 20 cm and 220 cm from the floor was monitored by thermocouples placed in a corner of the testing room. The area of the opening is 0.06 m^2 or 0.12 m^2. As a scatterer of the THz waves, we used a metal corner reflector, which was set at the opposite end of the testing room facing the sensing system. The THz waves transmitted from the sensing system propagated horizontally at a height of 95 cm in the test room to the metal corner reflector. Then, the metal corner reflector reflected the THz waves back to the system. The photocurrent of the J-UTC-PD was kept at 10 mA during the measurement. The corresponding output power of the J-UTC-PD was around –12 dBm at 350 GHz.

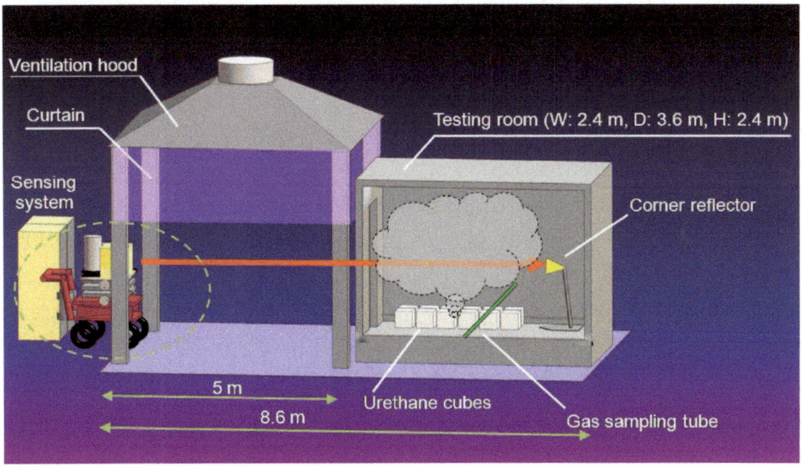

Figure 18.21 Arrangement of the test room and sensing system in a full-scale fire test hall.

Figure 18.22 Opening of the testing room as seen from the sensing system.

We sampled gases just beside the beam path of the THz waves simultaneously through a thin glass tube inserted through the wall of the testing room. In the hall, a ventilation hood is installed at the upper part of the area between the sensing system and the testing room. The smoke emitted from the opening of the testing room is collected by this hood and burned completely by a following secondary combustion furnace before exhaust. The area under the ventilation hood is electromagnetically shielded by aluminum sheets.

18.4.2 Result

We begin by presenting the results of the experiment when the testing room had a 0.06 m^2 opening. The urethane cubes were set on fire at $t = 0$ (s), and we started recording the power of the received THz waves at each frequency. The path of the THz beam was completely covered by fire and smoke in the first 30 s. The ambient temperature near the THz beam path increased rapidly from 40 s to 100 s, as shown in Fig. 18.23. Figure 18.24 shows the transmission spectra of the THz waves that propagated to and from the metal corner cube in the testing room. Here, the spectra were normalized by dividing the measured frequency-dependent received power by that before the fire had been started. Each spectrum is the average of seven sweeps.

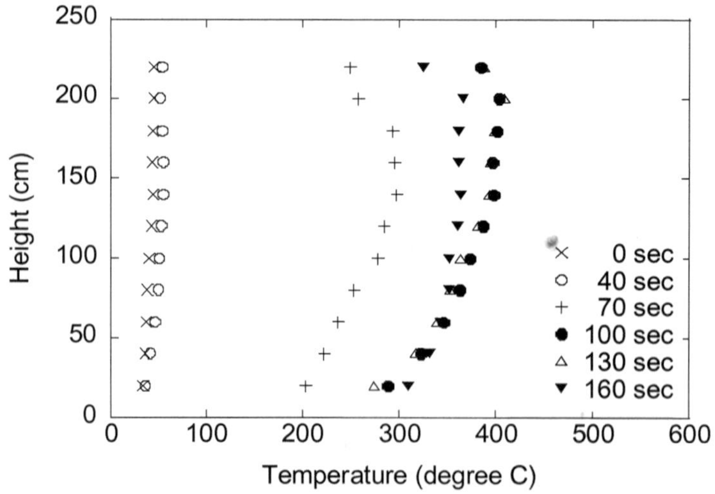

Figure 18.23 Temperature distribution in the testing room.

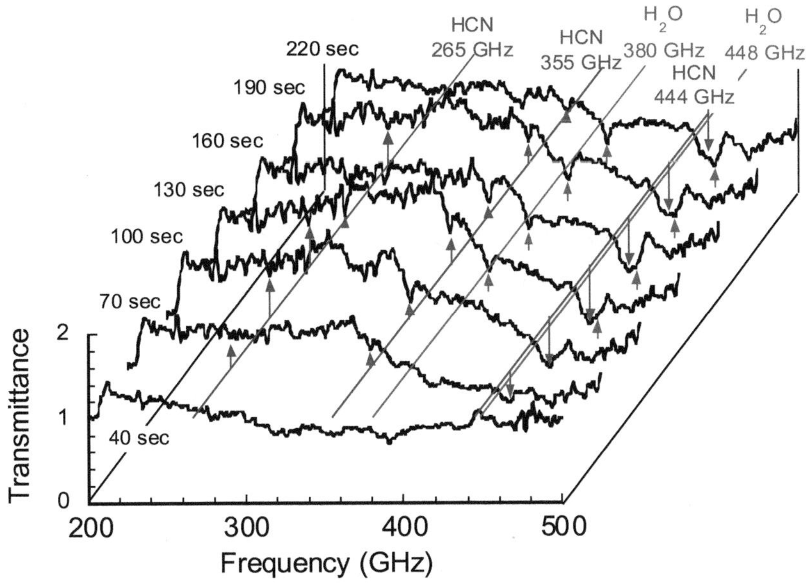

Figure 18.24 Transmission spectra of the THz waves that propagated to and from the metal corner cube in the testing room. The testing room has a 0.06-m^2 opening.

Absorption peaks were observed at 265 GHz, 355 GHz, and 444 GHz in the measured spectra at 70 s, 100 s, 130 s, and 160 s. The peaks are more pronounced at higher frequencies. These characteristics agree with those for the absorption spectrum of HCN. Therefore, these peaks are assigned to HCN in smoke. Peaks at 380 GHz and 448 GHz are identified as absorption of water molecules. These molecules were generated due to the combustion of foam urethane cubes.

Noteworthy is that the baseline of the measured spectrum becomes larger than 1. This doesn't indicate that THz waves radiated from high-temperature substances in the testing room are detected in addition to those transmitted from the sensing system. The three metal plates that form the metal corner reflector warp when the ambient temperature in the testing room increases. The warpage changes the shape of the metal corner reflector and can result in an increase in its back reflection efficiency.

We did the same experiment for the testing room with a 0.12 m^2 opening and measured the transmission spectra. The result is shown in Fig. 18.25. The absorption intensity of HCN for this condition is weaker than that for the 0.06 m^2 opening.

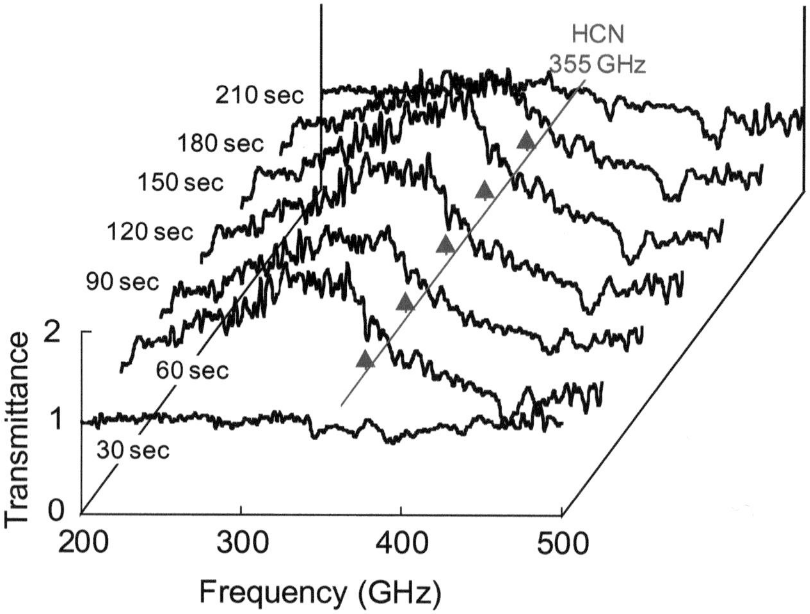

Figure 18.25 Transmission spectra of THz waves that propagated to and from the metal corner cube in the testing room. The testing room has a 0.12 m^2 opening.

The HCN concentration of smoke in the testing room was estimated from the measured absorption intensity. We used the HITRAN database [10] and prepared a calibration curve that relates the absorption intensity at the frequency of 355 GHz to the HCN content. Here, we assumed an interaction length and temperature of HCN of 6.6 m and 350°C, respectively. To compensate for the effect of the baseline shift of the measured spectra due to the deformation of the metal corner reflector, we used the absorption intensity corrected for offsets. The result shown in Fig. 18.26 indicates that the HCN concentration reached a maximum around 100 s and then gradually decreased after that. The maximum HCN concentrations are 1300 ppm and 950 ppm for the room openings of 0.06 m^2 and 0.12 m^2, respectively.

To chemically analyze the HCN concentration in the testing room, a sufficient amount of gas should be collected from it. So, it is impossible to sample the gas in a moment with a thin tube. In this experiment, we used gases collected from the testing room between t = 90 s and 300 s. The HCN concentration in the collected gas was determined by using photometric detection with 4-pyridinecarboxylic acid-pyrazolone. The chemical analysis shows that the average HCN concentrations during this period were 640 ppm and 470 ppm for the room having the openings of 0.06 m^2 and 0.12 m^2, respectively.

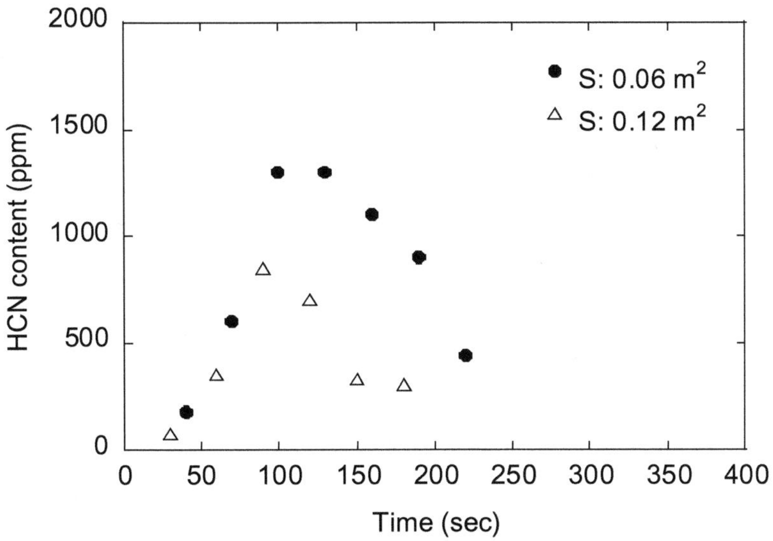

Figure 18.26 HCN concentrations estimated from the measured absorption intensity. Closed circles and open triangles represent the area of the opening of the testing room of 0.06 m^2 and 0.12 m^2, respectively.

18.5 Discussion

We examined the validity of our sensing system by comparing the HCN concentration obtained from it with that obtained from chemical analysis. The result is shown in Fig. 18.27. Here, the x axis and y axis of the graph indicate values of the chemical analysis and those obtained from our sensing system, respectively. For the opening of 0.06 m^2, the average HCN concentration between 100 s and 250 s was used for the value obtained from our sensing system. The average between 95 s and 220 s was used for that of 0.12 m^2.

The agreement for both measurements reveals the accuracy of HCN concentration estimation from our sensing system. The production of HCN strongly depends on the amount of oxygen available. A larger opening improves ventilation and ensures a plentiful supply of oxygen. As a result, less HCN is produced. Our sensing system picked up this difference.

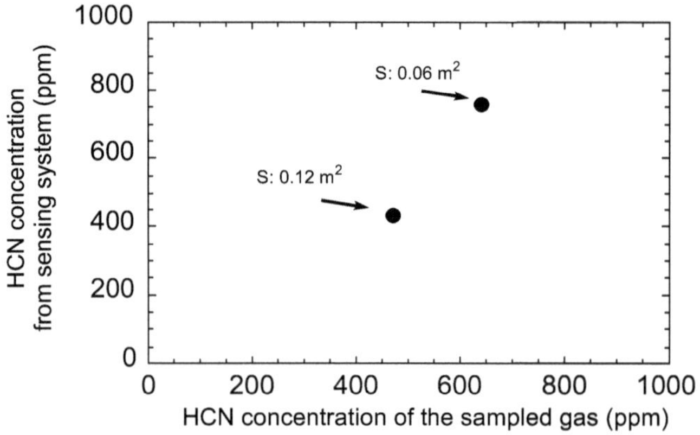

Figure 18.27 Averaged HCN concentration obtained from the THz spectrum and chemical analysis. *S*: the area of the opening of the testing unit.

We used a metal corner reflector to efficiently receive THz waves from the fire site in this study. However, the power of THz waves required for sensing was 9.3 dB lower than the maximum of the photomixer output. Therefore, our system operated with a sufficient power margin in this experiment. This suggests that instead of an artificial metal corner reflector, various things in a room could be used as a reflector. These results indicate the potential of our system as a remote toxic gas sensor available for use at disaster sites.

In this research, we couldn't detect CO both in the gas and the full-scale fire experiments. This is mainly due to small absorbance intensity at frequencies between 200 GHz and 500 GHz. Recently, we measured absorption spectra of smoke produced by incomplete combustion of molded charcoal using a 1 m long gas cell and succeeded in observing absorption peaks of CO at 692 GHz and 921 GHz [42]. Therefore, extending the measurement frequency range up to 1000 GHz will provide us simultaneous detection and quantification of CO and HCN at the scene of a fire.

18.6 Conclusion

We developed a remote gas sensor with an operation bandwidth between 200 GHz and 500 GHz. The system employs a photonic THz wave generator and an SIS tunneling mixer, which provide both fast frequency sweeping and high receiver sensitivity at the same time. The performance of the developed system was investigated in a full-scale simulated fire. The measured transmission spectra of smoke showed the existence of HCN molecules, which were generated due to incomplete combustion of foam urethane blocks. The HCN concentration estimated from the absorption intensity agreed quite well with that obtained from the chemical analysis of a sampled gas. These results demonstrate the potential of the developed sensing system as a remote toxic gas sensor at disaster sites.

Acknowledgments

This work was supported in part by the National Institute of Information and Communications Technology, Japan. We carried out our experiments with a full-scale simulated fire with the support of the Center of Fire Science of Technology, Tokyo University of Science. Satoshi Kohjiro and Ken'ichi Kikuchi acknowledge M. Maezawa for fabrication of SIS mixers and T. Yamada for characterization of the cryocooler.

References

1. H. J. Hansen, Standoff detection using millimter and submillimeter wave spectroscopy, *Proc. IEEE* **95**, 1691–1703 (2007).
2. N. Gopalsami, S. Bakhtiari, A. C. Raptis, S.L. Dieckman, and F. C. DeLucia, Millimeter-wave measurements of molecular spectra with application to environmental monitoring, *IEEE Trans., Instrum. Meas.* **45**, 225–230 (1996).
3. N. Gopalsami and A. C. Raptis, Millimeter-wave radar sensing of airborne chemicals, *IEEE MTT* **49**, 646–653 (2001).
4. N. Gopalsami and S. Bakhtiari, T. W. Elmer II, and A. C. Raptis, Application of millimeter-wave radiometry for remote chemical detection, *IEEE MTT* **56**, 700–709 (2008).
5. E. R. Brown, D. L, Woolard, A. C. Samuels, T. Globus, and B. Gelmont, Remote detection of bioparticles in the THz region, *2002 IEEE Int. Microwave Symp. Digest* **3**, 1591–1594 (2002).

6. E. R. Brown, J. E. Bjarnason, T. L. J. Chan, A. M. M. Lee, and M. A. Celis, Optical attenuation signatures of Bacillus subtillis in the THz region, *Appl. Phys. Lett.* **84**, 3438–3440 (2004).
7. R. A. Cheville and D. Grischkowsky, Far-infrared terahertz time-domain spectroscopy of flames, *Opt. Lett.* **20**, 1646–1648 (1995).
8. D. Bigourd A. Cuisset, F. Hindle, S. Maton, E. Fertein, R. Bocquet, and G. Mouret, Detection and quantification of multiple molecular species in mainstream cigarette smoke by continuous-wave terahertz spectroscopy, *Opt. Lett.* **31**, 2356–2358 (2006).
9. T. Uno and H. Tabata, *In situ* measurement of combustion gas using terahertz time domain spectroscopy setup for gas phase spectroscopy and measurement of solid sample, *Jpn. J. Appl. Phys.* **49**, 04DL17 1–7 (2010).
10. M. R. Stringer, J. Bassi, R. E. Miles, Y. Zhang, and K. Ozanyan, Terahertz spectroscopy through a high-pressure combustion system, *33rd Int. Conf. IR mm THz Waves*, W4B2 1702 (2008).
11. N. Shimizu, K. Kikuchi, T. Ikari, K. Mtsuyama, A. Wakatsuki, S. Kohjiro, and R. Fukasawa, Absorption spectra of smoke emitter from heated nylon fabric measured with a continuous-wave sub-terahertz spectrometer, *Appl. Phys. Express* **4**, 032401-1–032401-3 (2011).
12. N. Shimizu, H. J. Song, Y. Kado, T. Furuta, A. Wakatsuki, and Y. Muramoto, Gas detection using terahertz waves, *NTT Tech. Rev.* **7**, 37–40 (2009).
13. N Shimizu, T Ikari, K Kikuchi, K Matsuyama, A Wakatsuki, S Kohjiro, and R Fukasawa, *2011 IEEE MTT-S International Microwave Symposium Digest*, Baltimore, WE2G-5 (2011).
14. HITRAN, http://cfa-www.harvard.edu/hitran/welcometop.html.
15. R. Piesiewicz, C. Jansen, D. Mittleman, T. Kleine-Ostmann, M. Koch, and T. Kurner, Scattering analysis for the modeling of THz communication systems, *IEEE Trans. Antennas Propagat.* **55**(11), 3002–3009 (2007).
16. K.-H. Oh, N. Shimizu, S. Kohjiro, K. Kikuch, A. Wakatsuki, N. Kukutsu, and Y. Kado, High-sweeping-speed optically synchronized dual-channel terahertz-signal generator for driving a superconducting tunneling mixer and its application to active gas sensing, *Opt. Express* **17**, 18455–18461 (2009).
17. H. Ito, T. Ito, Y. Muramoto, T. Furuta, and T. Ishibashi, Rectangular waveguide output unitraveling-carrier photodiode module for high-power photonic millimeter-wave generation in the F-band, *J. Lightwave Technol.* **21**, 3456–3461 (2003).
18. H. Ito, T. Furuta, Y. Muramoto, T. Ito, and T. Ishibashi, Photonic millimeter- and sub-millimetre-wave generation using J-band

rectangular-waveguide-output uni-traveling-carrier photodiode module, *Electron. Lett.* **42**, 1424–1425 (2006).

19. A. Wakatsuki, T. Furuta, Y. Muramoto, T. Yoshimatsu, and H. Ito, High-power and broadband sub-terahertz wave generation using a J-band photomixer module with rectangular-waveguide output port, *Proc. 33rd Int. Conf. IR mm THz Waves*, M4K2 1199 (2008).
20. R. Tucker and M. J. Feldman, Quantum detection at millimeter wavelength, *Rev. Mod. Phys.* **57**, 1055–1113 (1985).
21. T. Oka, K. Kamegai, M. Hayashida, M. Nagai, M. Ikeda, N. Kuboi, K. Tanaka, N. Bronfman, and S. Yamamoto, Atomic carbon in the southern Milky Way, *Astrophys. J.* **623**, 889–896 (2005).
22. S. Kohjiro, S.-C. Shi, J. Inatani, M. Maezawa, Y. Uzawa, Z. Wang, and A. Shoji, An octave bandwidth SIS mixer for accurate and compact terahertz spectrometers, *IEEE Trans. Appl. Supcond.* **17**, 355–358 (2007).
23. S. Kohjiro, K. Kikuchi, M. Maezawa, T. Furuta, A. Wakatsuki, H. Ito, N. Shimizu, T. Nagatsuma, and Y. Kado, A 0.2-0.5 THz single-band heterodyne receiver based on a photonic local oscillator and a superconductor-insulator-superconductor mixer, *Appl. Phys. Lett.* **93**, 093508 (2008).
24. A. Wootten and A. R. Thompson, The Atacama large millimeter/submillimeter array, *Proc. IEEE* **97**, 1463–1471 (2009).
25. K. Kikuchi, S. Kohjiro, T. Yamada, N. Shimizu, and A. Wakatsuki, Compact terahertz passive spectrometer with wideband superconductor-insulator- superconductor mixer, *Rev. Sci. Instrum.* **83**, 023110 (2012).
26. J. Kawamura, D. Miller, J. Chen, J. Zmuidzinas, B. Bumble, H. G. LeDuc, and J. A. Stern, Very high-current-density Nb/AlN/Nb tunnel junctions for low-noise submillimeter mixers, *Appl. Phys. Lett.* **76**, 2119–2121 (2000).
27. S. C. Shi, T. Noguchi, J. Inatani, Y. Irimajiri, and T. Saito, Characterization of the bandwidth performance of distributed junction array, *IEEE Trans. Appl. Supercond.* **9**, 3777–3779 (1999).
28. M. Maezawa, F. Hirayama, and M. Suzuki, Design and fabrication of RSFQ cell library for middle-scale applications, *Physica C*, **412–414**, 1591–1596 (2004).
29. D. F. Filipovic, S. S. Gearhart, and G. M. Rebeiz, Double-slot antennas on extended hemispherical and elliptical silicon dielectric lenses, *IEEE Trans. Microwave Theory Tech.* **41**, 1738–1749 (1993).

30. http://www.mwoffice.com. The antenna impedance agrees with that in Kominami et al., Dipole and slot elements and arrays on semi-infinite substrates, *IEEE Trans. Ant. Propag.* **AP-33**, 600–607 (1985).
31. Because of the assumption of $\zeta = 1$, the calculated $P_{\mathrm{Ph}}\eta$ gives the maximum value of the real pumping power.
32. http://www.virginiadiodes.com/.
33. J. L. Hesler, W. R. Hall, T. W. Crowe, R. M. Weikle, II, B. S. Deaver, Jr., R. F. Bradley, and S. K. Pan, Fixed-tuned submillimeter wavelength waveguide mixers using planar Schottky-barrier diodes, *IEEE Trans. Microwave Theory Tech.* **45**, 653–658 (1997).
34. N. R. Erikson, Low-noise submillimeter receivers using single-diode harmonic mixers, *Proc. IEEE* **80**, 1721–1728 (1992).
35. Saint-Gobain Performance Plastics; see http://www.norton-films.com/norton-zitex-films.aspx.
36. R. Li Onishi, A. T. Satoh, and Y. Kanazawa, A simple method of temperature stabilization for 4 K GM cryocooler, in *Proceedings of the 16th International Cryogenic Engineering Conference/International Cryogenic Materials Conference*, eds. T. Haruyama, T. Mitsui, and K. Yamafuji (Elsevier, Kitakyusyu, Japan, 1997), 355–358.
37. K. Kikuchi, Y. Yamada, and S. Kohjiro, Development of superconductor-insulator-superconductor (SIS) terahertz receiver with mechanical and thermal vibration-reduced cryocooler, *IEEE Trans. Appl. Supcond.* **21**, 649–653 (2011).
38. D. W. Allan, Statistics of atomic frequency standards, *Proc. IEEE* **54**, 221–230 (1969).
39. J. W. Kooi, G. Chattopadhyay, M. Thielman, and T. G. Phillips, Noise stability of SIS receivers, *Int. J. IR mm Waves* **21**, 689–716 (2000).
40. K.-H. Oh, N. Shimizu, N. Kukutsu, Y. Kado, S. Kohjiro, K. Kikuchi, T. Yamada, and A. Wakatsuki, Heterodyne THz-wave receiver with a superconducting mixer driven by a high sweeping-speed photonic-based THz-wave local oscillator, *IEICE Electron. Exp.* **6**, 601–606 (2009).
41. K. Matsuyama, N. Shimizu, K. Kikuchi, S. Kohjiro, T. Ikari, R. Fukasawa, and A. Wakatsuki, Advanced detection system for toxic gases in fire by using terahertz electromagnetic waves, *10th Int. Symp. Fire Safety Sci., 2011*, Poster Session II-25 (2011).
42. N. Shimizu, I. Hosako, and K. Matsuyama, Absorption spectra of hydrogen chloride and carbon monoxide in smoke, *Proc. 37rd Int. Conf. IR mm THz Waves*, Fri-A-2-2 (2012).

Chapter 19

Terahertz 3D Tomography

Motoki Imamura and Shigeki Nishina
Advantest Corporation, 48-2 Matsubara, Kamiayashi, Aoba-ku, Sendai, Miyagi 989-3124, Japan
motoki.imamura@jp.advantest.com, shigeki.nishina@jp.advantest.com

19.1 Introduction

Terahertz (THz) waves, which are located in the band between infrared and millimeter waves, offer advantages such as the ability to obtain fingerprint absorption spectra and the ability to pass through materials that are opaque to visible light. These features of THz waves are particularly useful for imaging techniques for nondestructive analysis. Kawase et al. performed demonstration experiments on using 2D spectral imaging to distinguish and identify the fingerprint spectra of (a) methamphetamine and 3,4-methylenedioxy-*N*-methylamphetamine (MDMA), two of the most widely consumed illicit drugs, which are often hidden in mail envelopes, and (b) aspirin as a reference. In this way, Kawase et al. demonstrated the possibility of new applications of THz waves [1].

Handbook of Terahertz Technologies: Devices and Applications
Edited by Ho-Jin Song and Tadao Nagatsuma

ISBN 978-981-4613-08-8 (Hardcover), 978-981-4613-09-5 (eBook)
www.panstanford.com

This chapter describes the THz wave imaging techniques that are used in nondestructive analysis technology. From among these, the THz computed tomography (CT) method, which excels as a 3D imaging technique, is described in detail, and examples of its practical application are presented.

The latter part of this chapter describes the technology used in the TAS7000 3D imaging analysis system produced by Advantest Corporation; the TAS7000 is the world's first commercial device to make practical use of this new nondestructive analysis technology. During the process of creating a THz CT device suitable for practical applications, increasing the measurement speed was particularly challenging. This chapter describes in detail the high-speed measurement technique and laser control technology for addressing this problem.

19.2 Imaging Techniques Using THz Waves

Table 19.1 shows a summary of the THz imaging techniques used in nondestructive analysis technology. Although microscopy techniques such as scanning near-field optical microscopy [2] and laser THz emission microscopy [3], which measures THz waves that have been excited optically, are also used for THz imaging, they are omitted from this chapter, which focuses on nondestructive analysis techniques.

THz imaging techniques can be broadly divided into active and passive. Active techniques use a THz emitter, such as a quantum cascade laser (QCL) or photoconductive antenna (PCA), to illuminate the object being measured with THz waves and measure these THz waves using a detector. Passive techniques have found practical application in security applications such as body-scanning systems, which utilize the differences in the energies of the THz waves emitted from an object to determine its shape [4].

The first 2D imaging in the THz band was performed in 1976 by Hartwick et al., who performed external shape imaging of a key hidden inside a box by ray mapping using an infrared laser and infrared detector [5]. Since the term "terahertz" was not in general use at the time, the technique was introduced as an infrared imaging technique.

Table 19.1 Various THz imaging methods

Mode	Imaging parameter	2D/3D	Main methods
Passive	Power	2D	THz camera
Active transmission	Power	2D	QCL with THz camera Mapping by QCL with detector
		3D	CT by QCL with detector
	Spectroscopy	2D	Mapping by TDS Mapping by FTIR Mapping by tunable terahertz lightsource
		3D	CT by TDS
	Phase	2D	Mapping TDS
		3D	CT by TDS
Active reflection	Power	2D	QCL with THz camera Mapping by QCL with detector
	Spectroscopy	2D	Mapping by TDS Mapping by FTIR Mapping by tunable terahertz lightsource
	Differential time	1D (depth)	TOF
		2D/3D	Mapping TOF

QCL: quantum cascade laser; TDS: time domain spectroscopy; FTIR: Fourier transform infrared spectroscopy; TOF: time of flight

2D THz spectral imaging can be performed through a combination of ray mapping and THz spectral techniques. Although THz spectroscopic techniques that use Fourier transform infrared spectroscopy (FTIR) and a parametric tunable THz light source [6] also exist, the current mainstream method is THz time domain spectroscopy (THz-TDS) [7, 8]. The first 2D imaging by THz-TDS was transmission imaging of the lead frame of an integrated circuit in a plastic package and of tree leaves by Hu and Nuss in 1995 [9]. Figure 19.1 shows a block diagram of mapping measurement by THz-TDS. This method has become popular with advances in femtosecond lasers and offers advantages such as the ease of deducing the complex optical constants, since phase information can be obtained in addition to amplitude information, the ability to handle data in the time domain, and the ability to measure signals with high signal-to-noise (S/N) ratios over short periods of time.

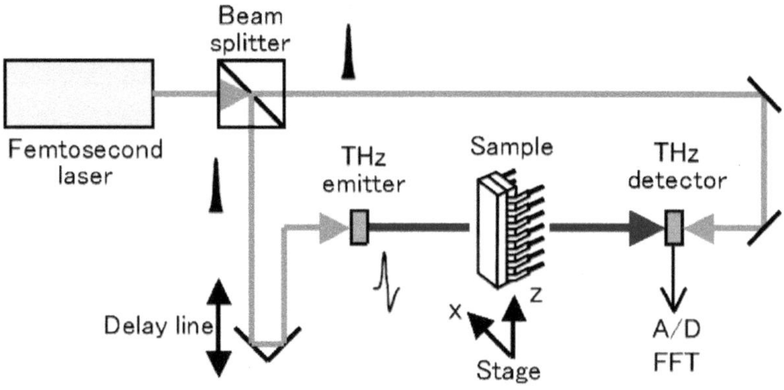

Figure 19.1 Schematic diagram of 2D mapping imaging by THz-TDS.

though mapping measurement in which rays are scanned spatially in two dimensions over the object being measured is a spectral imaging technique that can be used both with transmission and reflection, it cannot be used as is in the 3D imaging techniques focused on in this chapter. To obtain 3D information using a THz wave beam, structural information about the object being measured in the propagation direction of the THz waves is needed. The methods for obtaining this information are the CT transmission technique and the time-of-flight (TOF) reflection technique [10].

As described in detail in the next section, the CT method provides cross-sectional information about an object by measuring the transmission properties from various directions with respect to the object. From the temporal waveform of THz reflection pulses in a TDS measurement system, the TOF method provides information such as structure in the depth direction and differences in density. The following sections describe in detail the THz CT 3D imaging technique from among these THz imaging techniques.

19.3 Overview of the THz CT Method

A leading example of a current 3D imaging technique is X-ray CT, which uses electromagnetic waves (like THz waves but at a shorter wavelength). X-ray CT was developed by Hounsfield in 1972 as a nondestructive cross-sectional scanning technique for medical diagnosis. CT analysis provides cross-sectional images that are reconstructed by performing the inverse Radon transform on acquired projection data [11, 12]. Details on the formative period of X-ray CT development can be found in an article by Beckmann [13].

Although THz CT is the same as X-ray CT in that each is a 3D imaging technique that provides cross-sectional images through inverse projection calculations, a new feature of THz CT is that it provides THz temporal waveforms and spectral information, including the phase as the source data for transformation into an image; furthermore, THz CT offers the ability to perform analysis that not only determines the structure of the object being measured but also offers new features based on exploiting the spectral information. The following sections describe the principles of reconstructing images using the CT method and the features of CT using the TDS method.

19.3.1 Principles of CT

THz CT is an excellent imaging technique for determining the internal characteristics of an object. An object with an unknown internal structure is illuminated with THz waves from different directions, and the transmitted THz waves are measured as projection data that forms a 1D mapping image as shown in Fig. 19.2. The cross-sectional images of the object are found by performing inverse projection calculations on this data, as shown below.

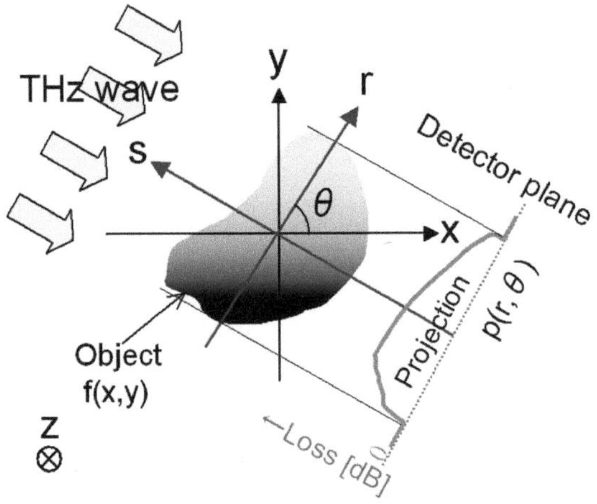

Figure 19.2 Projection data $p(r,\theta)$.

During the propagation of THz waves, attenuation occurs due to absorption by the medium. These electromagnetic waves pass through an object that has a spatial distribution $f(x,y)$ of attenuation per unit length of propagation (dB). The projection data for the amount of attenuation of the electromagnetic waves can be represented as follows:

$$\begin{aligned} p(r,\theta) &= \int_{-\infty}^{+\infty} f(x,y)ds \\ &= \int_{-\infty}^{+\infty} f(r\cos\theta - s\sin\theta, r\sin\theta + s\cos\theta)ds \end{aligned} \quad (19.1)$$

This is called the Radon transform.

The 2D Fourier transform $F(X,Y)$ of $f(x,y)$ can be expressed as

$$F(X,Y) = \int_{-\infty}^{+\infty}\int_{-\infty}^{+\infty} f(x,y)e^{-i(xX+yY)}dxdy \quad (19.2)$$

Expressing this in polar coordinates $X = R\cos\theta$, $Y = R\sin\theta$, we can write the following because $r = x\cos\theta + y\sin\theta$ and $dxdy = drds$:

$$\begin{aligned} F(R\cos\theta, R\sin\theta) &= \int_{-\infty}^{+\infty}\int_{-\infty}^{+\infty} f(x,y)e^{-iRr}drds \\ &= \int_{-\infty}^{+\infty}\left[\int_{-\infty}^{+\infty} f(x,y)ds\right]e^{-iRr}dr = \int_{-\infty}^{+\infty} p(r,\theta)e^{-iRr}dr \end{aligned} \quad (19.3)$$

In other words, if we can obtain $p(r,\theta)$, then the Fourier transform $F(X,Y)$ of $f(x,y)$ is completely determined and the distribution $f(x,y)$ of the attenuation coefficient can be found from the following equation, which is the 2D Fourier transform of $F(X,Y)$:

$$f(x,y) = \frac{1}{(2\pi)^2}\int_{-\infty}^{+\infty}\int_{-\infty}^{+\infty} F(X,Y)e^{-2\pi i(xX+yY)}dXdY \quad (19.4)$$

The projection data is the values of the Radon transform of $f(x,y)$ for all (r,θ) and is called a sinogram. The cross-sectional images, which are the original distribution $f(x,y)$ of attenuation coefficients, can then be reconstructed by performing an inverse projection calculation on the sinogram.

During the acquisition of actual measurement data, a single $f(x,y)$ slice is measured by spatially scanning the object being measured over the r and θ directions at some z position, where z is the axis perpendicular to the x- and y-axes. Spatial scans are then also performed at other z positions, and a 3D cross-sectional image is obtained by stacking these slices.

19.3.2 THz CT Using TDS

The use of QCLs for CT measurement by using THz waves has been reported [15], and the first report on using the TDS method was by Ferguson et al. [16]. CT measurement by THz-TDS acquires the projection data $p(r,\theta)$ by adding the rotation angle θ to the XY stage in the experimental system shown in Fig. 19.1 and finds the reconstructed image $f(x,y)$ by performing the inverse projection calculation shown in Section 19.3.1 above. Reconstruction using the CT method of Ferguson et al. demonstrated not only that the shape of the sample can be obtained from the loss characteristics when the THz waves passed through the sample but also that the distribution of the refractive indices of polystyrene resin can be found from the delay in the THz wave temporal waveform [17].

Since spectral information that includes temporal information and phase information can be found by the THz-TDS method, the 3D distribution of the complex optical constants of the object being measured can be analyzed by CT measurement. Since this means that more information can be obtained about the object being measured than by X-ray CT using transmission power information, THz CT makes it possible to perform analysis with a wider range of parameters and has attracted attention as a new nondestructive analysis technique for a wide range of applications.

19.4 Examples of THz CT Measurement

This section gives several examples of THz CT measurement using the TDS method and discusses the usefulness and applicability of these measurements.

The instrument used in the analysis is the TAS7000 3D imaging analysis system produced by Advantest Corporation [18]. This instrument nondestructively visualizes and analyzes the internal shape and characteristics of the object being measured as 2D and 3D images and displays the results in three dimensions. The TAS7000 also has a track record of being used by some automobile manufacturers for evaluating ceramic automobile components. The technology used in this device is described in detail in Section 19.5.

19.4.1 Plastic Samples

Although plastics are difficult to analyze with X-rays due to their extremely low absorption, they can be measured with good sensitivity by THz waves and are expected to be one of the targets of nondestructive analysis using THz CT. Figures 19.3 and 19.4 show images of plastic industrial components measured on the TAS7000. The 3D images were found by obtaining the 3D distribution data of the spatial THz wave absorption coefficient from the transmittance of THz waves through the samples and by further applying a volume shape-rendering algorithm to the distribution of absorption coefficients.

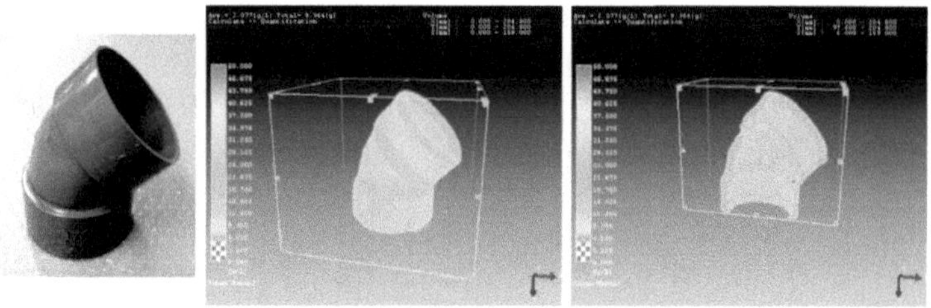

(a) (b) (c)

Figure 19.3 THz CT images of a ø75 mm PVC pipe fitting. (a) Photograph of the pipe fitting, (b) 3D image by THz CT, and (c) cross-sectional 3D image of (b).

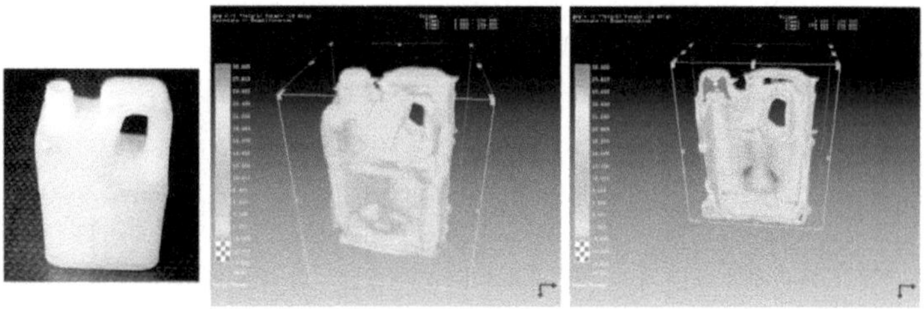

(a) (b) (c)

Figure 19.4 THz CT image of an empty polyethylene container (140 mm [W] × 215 mm [H]). (a) Photograph of the polyethylene container, (b) 3D image by THz CT, and (c) cross-sectional 3D image of (b).

In the figures, (a) shows a photograph of the object being measured, (b) shows the 3D image found by THz CT in which areas of higher THz wave transmission losses are shaded darker, and (c) shows a 3D cross-sectional image in which part of the structure from (b) has been cut away so that the internal structure is visible. This shows that not only the external shape but also the internal shape is reproduced.

19.4.2 Dummy Reagent Samples

To demonstrate the usefulness of spectral CT, the results of performing THz CT analysis using samples of substances that have a fingerprint spectrum are shown. The samples that were measured were sugar lactose and the amino acid tyrosine, which were diluted to 5 wt.% with polyethylene powder. The lactose and tyrosine powders were, respectively, packed into 8 mm and 10 mm diameter holes made in a 32 mm diameter polyethylene cylinder (Fig. 19.5a).

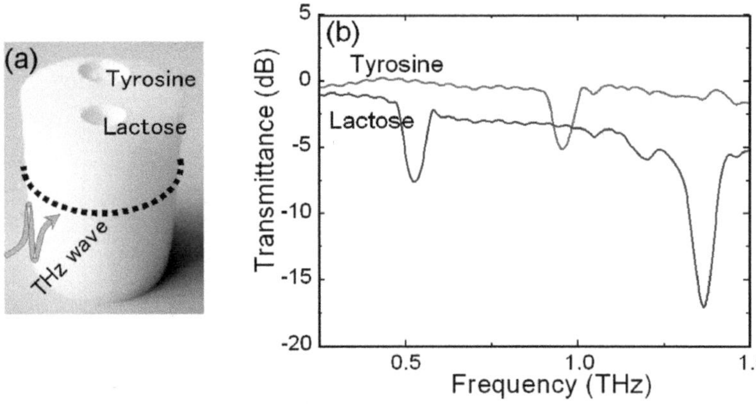

Figure 19.5 Analysis of powders. (a) Photograph of a test sample and (b) transmission spectra of tyrosine and lactose.

The transmission spectra of the unmixed samples of the lactose and tyrosine reagents are shown in Fig. 19.5b. It is clear that lactose and tyrosine exhibit sharp absorption peaks at 0.53 THz and 0.96 THz, respectively. These can be treated as material-specific spectra called fingerprint spectra [1] and can be used to identify the constituents.

Since THz CT gives information originating from the spectra for each pixel in the cross-sectional images by changing the frequency band for which the sinogram is found, this technique allows analysis of the separate constituents of the object being measured, as shown below [19].

Figure 19.6a,b shows the sinogram $p(r,\theta)$ acquired at the absorption frequency of tyrosine (0.96 THz) and the corresponding re-created cross-sectional image $f(x,y)$. Figure 19.6c,d shows the sinogram acquired at the absorption frequency of lactose (0.53 THz) THz and the corresponding re-created cross-sectional image. The lactose and tyrosine were distinguished and detected.

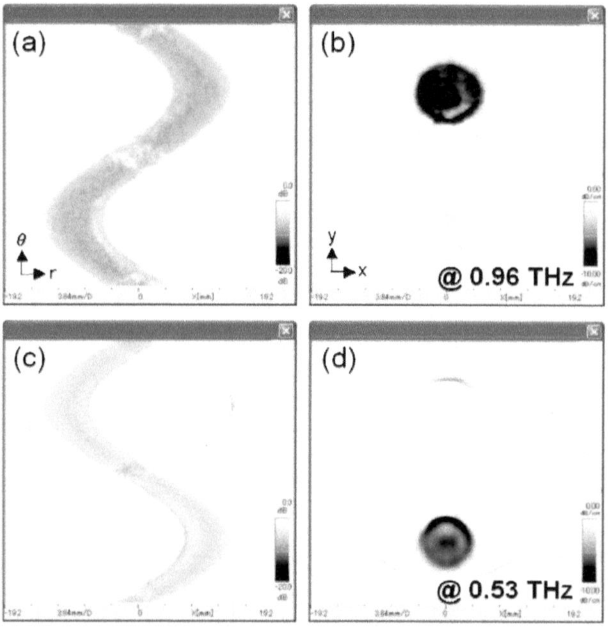

Figure 19.6 Sinogram and reconstructed cross sections of a test sample. (a) Sinogram for 0.96 THz (tyrosine), (b) reconstructed cross section calculated from (a), (c) sinogram for 0.53 THz (lactose), and (d) reconstructed cross section calculated from (c).

Figure 19.7 shows 3D images constructed by stacking each of the 2D slices. Figure 19.7a shows an image obtained by integrating across the entire bandwidth without using any particular absorption frequency and can be used to observe the 3D structure of the samples, and Fig. 19.7b shows an image of the superposition of the images for each of the absorption frequencies (Fig. 19.6b,d), showing each of the components separately.

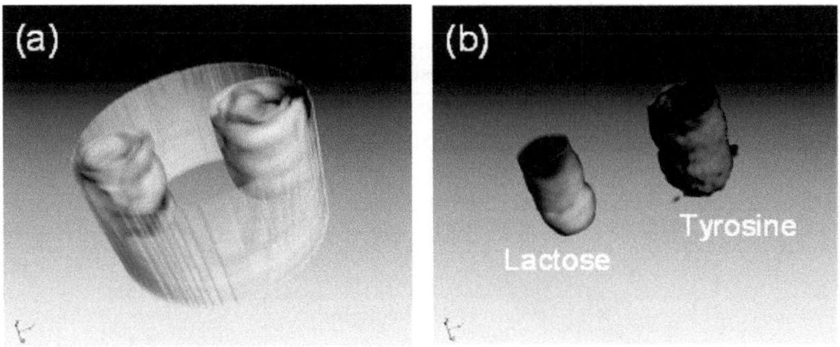

Figure 19.7 Reconstructed 3D spectroscopic images of a test sample. (a) 3D image for integrated transmission data over the whole spectral range and (b) superimposed 3D images of each powder constituent.

19.4.3 Quantitative Analysis of the Distribution of PM Accumulated in a DPF

We next give an introduction to nondestructive imaging analysis of the ceramic filter in a diesel engine as an example of a practical application of THz CT analysis [20].

The exhaust gases of diesel engines contain particulate matter (PM), of which the main constituent is soot. The main component that is used for cleaning exhaust gases that contain PM is called the diesel particulate filter (DPF), and this has a porous honeycomb ceramic structure in which both ends are blocked alternatingly (Fig. 19.8). When exhaust gas passes through the DPF, the side walls act as a filter and clean the exhaust gas. However, since PM accumulates inside the DPF the longer the engine is run, it is necessary to intermittently clear away the PM by forced regeneration, which burns (oxidizes) the PM.

To improve the reliability of forced regeneration systems, it is necessary to gain an understanding of the distribution of PM accumulated in the DPF, and the main analysis method in the past was to cut the DPF into blocks and measure the mass of each block. Since THz CT analysis allows the 3D density distribution of PM accumulated inside the DPF to be visualized nondestructively and the soot density to be analyzed quantitatively, the causes of failures can be identified more efficiently.

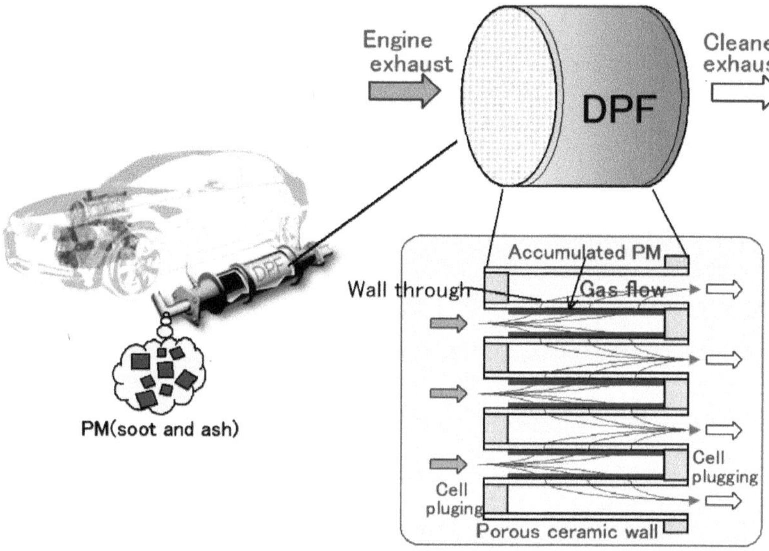

Figure 19.8 DPF structure.

The following paragraph explains how quantitative analysis is performed by THz CT [20]. As THz waves propagate inside the DPF, they are affected by interactions with the accumulated PM, which changes the transmission spectrum. At this time, the density of the accumulated matter can be found from the difference in the total mass of the DPF before and after accumulation. A calibration curve can be plotted, as shown in Fig. 19.9, by taking the change in the measured THz wave transmission parameter as the horizontal axis and the density of PM as the vertical axis. The 3D density distribution of accumulated matter can then be found in units of grams per liter by applying the calibration curve to all of the data from the THz CT.

To demonstrate the capabilities of quantitative analysis by THz CT, PM was deposited in only the left half of a DPF (diameter: 144 mm) by using a diesel engine and attaching a semicircular mask to the DPF input side, and analysis was performed on the DPF sample. The main constituent of the accumulated PM was carbon in the form of soot, and the expected value of the amount of accumulated soot was found to be 8.0 g/L from the difference in weight of the

DPF before and after accumulation. Furthermore, the 3D density distribution of soot accumulated inside the DPF was obtained by constructing a calibration curve from the CT measurement results from the TAS7000 before and after accumulation and applying it to the CT measurement results (Fig. 19.10). The results match the sample fabrication specifications and confirm that this method can be used for quantitative analysis.

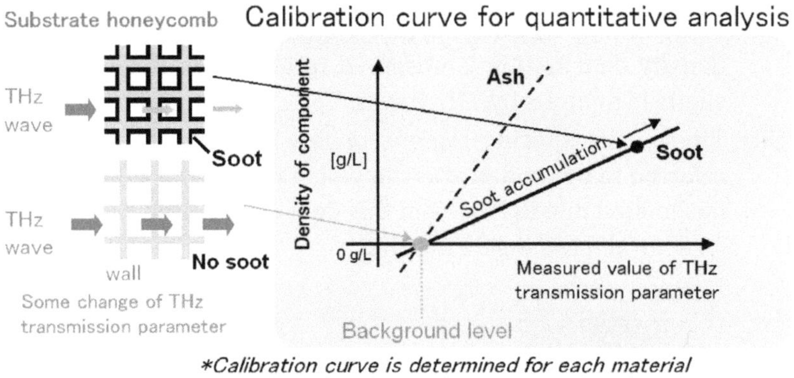

Figure 19.9 Quantitative analysis using THz CT and the calibration curve.

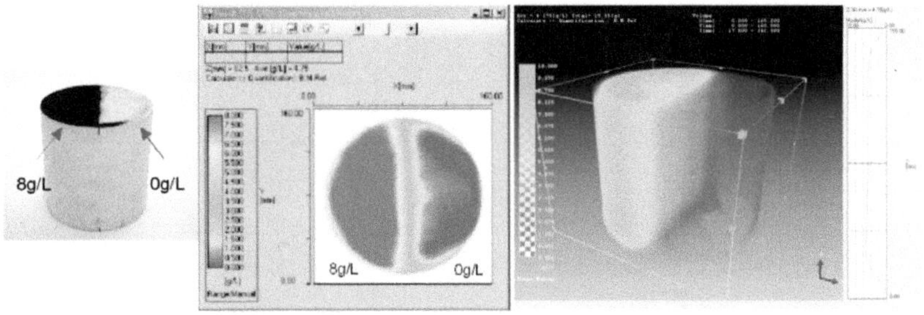

Figure 19.10 3D image of a masked DPF.

As an example of actual analysis, Fig. 19.11 shows the analysis results for a DPF that was fitted in a truck that was driven approximately 70,000 km. The DPF is assumed to have undergone several repetitions of force regeneration. This visualizes the distribution of the PM components that were trapped inside the DPF as exhaust gas flowed from the top to the bottom of the DPF. From these analysis results, several pieces of characteristic information can be obtained:

(i) There is a layer of ash that is not incinerated by forced regeneration and that remains at the bottom of the DPF. Since the ash builds up with regular driving distance and leads to drops in performance, it is necessary to know the amount.

(ii) PM density variation attributed to thermal variability in the DPF extends in the vertical direction. In contrast virtually no contrast is obtained by nondestructive scanning by X-rays for PM of which the main constituent is carbon, because it is a light element that exhibits extremely low absorption [21], and the density distribution is observed to have a clear distribution shape in analysis by THz waves.

(iii) There exist specific regions of low PM density. These are inferred to be eroded areas in which part of the DPF material has melted due to heat, and this data can be used to check for structural flaws inside the DPF.

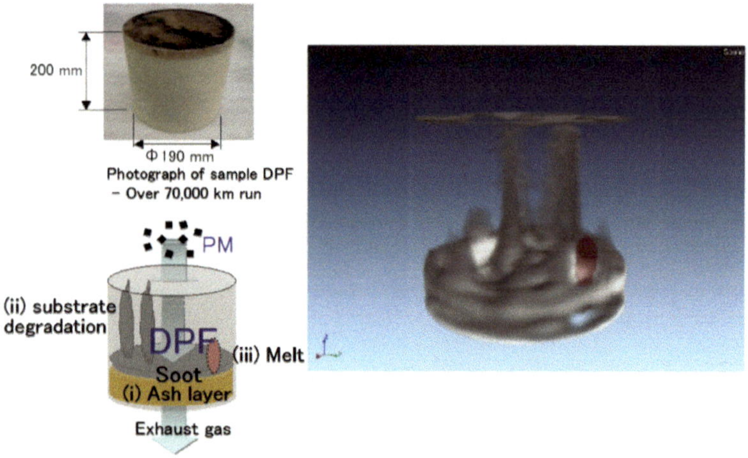

Figure 19.11 3D image of a DPF recovered from the field.

19.5 Technology Needed for Practical Application of THz CT

The biggest challenge in terms of putting CT measurement into practical use is increasing the measurement speed [17]. An extremely large amount of measurement data is needed in order to perform 3D imaging. For example, the measurement shown in Fig. 19.11 in the previous section was based on dividing the X, θ, and Z axes into 64, 60, and 64 divisions, respectively, giving a total of 246,000 pieces of projection data. THz pulse temporal waveforms were obtained by the conventional TDS method by changing the amount of optical delay mechanically using a translation stage. However, since this takes several minutes to measure a single temporal waveform [17], calculation shows that several days of measurement time would be needed to obtain the 3D internal image of an object if applied to CT measurement.

The TAS7000 3D imaging analysis system used in the example analysis in the previous section is the first commercial THz CT system in the world and delivers the throughput performance and functionality required for making practical use of CT measurement. Since there are limitations on increasing the speed when using the conventional TDS method, the TAS7000 implements increased speed through a unique optical sampling method that provides a CT analysis system with throughput sufficient for practical applications. This section describes this technology of the TAS7000.

19.5.1 TAS7000 3D Imaging Analysis System

The main specifications of the TAS7000 (Fig. 19.12) are shown below.

Specifications [18]:

- Analytical functions
  2D/3D CT density distribution analysis and THz spectroscopic transmission analysis
- Analytical frequency range
  High dynamic range module: 0.02–0.60 THz
  Broadband module: 0.05–3.00 THz
- Spatial resolution of CT
  ≤3 mm (using wire phantom; modulation transfer function, 10%)
- Throughput
  Maximum waveform acquisition speed:
  250 Hz (spectroscopic measurement mode)

Acquisition time for CT cross section: $\leq$15minutes
(1 CT cross section divided into 64 × 64)

- Maximum sample size and weight
  $\leq\Phi$310 mm (12.2 in.) × H310 mm (12.2 in.), $\leq$20 kg

Figure 19.12 Photograph of the TAS7000 3D imaging analysis system, Advantest, Japan, © 2013.

Figure 19.13 shows a block diagram of the components of the TAS7000. The TDS method is used as the basic spectroscopic principle. The system is equipped with two ultrashort pulse fiber lasers for generating THz waves and detecting them and employs an optical sampling method that obtains the THz temporal waveform by electrically controlling the timing between the two light sources. The ultrashort pulses are irradiated from a PCA, and the temporal waveform in the THz band is acquired by sampling. The generated pulses contain a broad spectrum of components up to the THz band, and the transmission characteristics (spectra) of the object being measured are obtained by fast Fourier transform (FFT) of the THz waveform transmitted through the object being measured. Because this data simultaneously contains not only amplitude information but also phase information, properties such as the complex refraction index of the material can be found.

Furthermore, the object being measured can be spatially scanned, and cross-sectional images can be obtained by reconstruction using the CT method. The ultrashort pulse fiber laser and optical sampling technology used in this system are described in detail in the following sections.

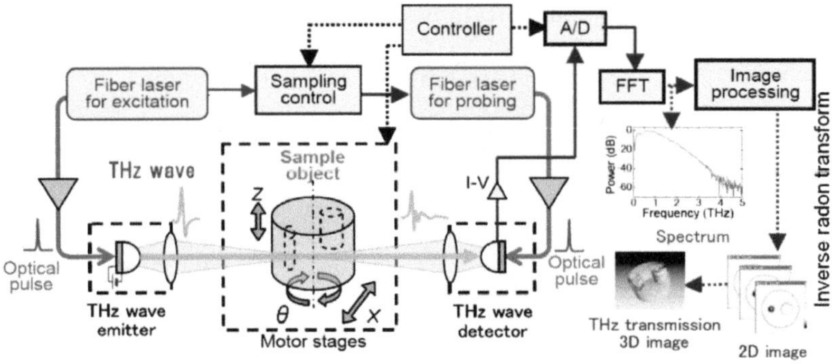

Figure 19.13 Schematic diagram of the TAS7000 3D imaging analysis system.

19.5.2 High-Speed Spectroscopic Measurement Using the THz Optical Sampling Method

The TAS7000 uses the principles of an optical sampling oscilloscope that performs electrostatic control for high-throughput waveform measurement without relying on the mechanical optical delay method. This THz optical sampling method [24] creates a slight difference Δf in the repetition frequency of two femtosecond pulse lasers and is able to acquire the waveform at time intervals of the inverse of this $1/\Delta f$ [22, 23]. Although measurement jitter occurs in the conventional THz optical sampling method due to the difference in timing between the two pulsed lasers, the TAS7000 has achieved high throughput and wideband measurement by newly adding a jitter reduction circuit to eliminate this source of jitter. Figure 19.14 shows the sampling principles [25], and Table 19.2 lists the features of the device.

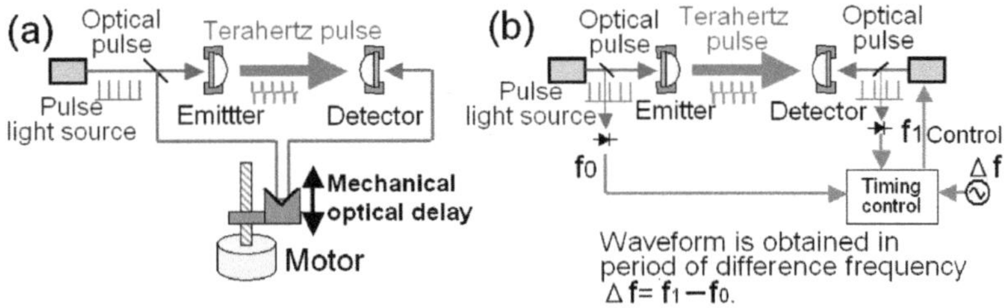

Figure 19.14 Methods of THz spectroscopy. (a) THz-TDS and (b) THz optical sampling.

Table 19.2 Features of the THz optical sampling method

	Conventional technology	TAS7000
Method	THz-TDS (Time Domain Spectroscopy)	THz optical sampling
Time sweep	Mechanical optical delay	Electrical control sweep
Light source	Branches of one light source	Two light sources
Throughput	1 min/sweep 68 h/image	Min. 4 ms/sweep 15 min/image
Affected by light source jitter	No	Yes → overcame using low-jitter light source

By this method, a maximum of 250 THz wave pulse waveforms can be captured per second, the time to capture a cross-sectional image by CT can be reduced to under 15 minutes, and the time to capture a 3D image of an entire sample can be reduced to under 5 hours, making the THz CT system suitable for practical applications.

19.5.3 High-Power Ultrashort Pulse Fiber Laser Technology

To measure the temporal waveform of THz waves, the pulse width of the THz waves that are measured is extremely short at approximately 300 fs and the measurement is found with the uncertainty that the time is less than the pulse width. In the optical sampling method, laser control is required that generates pulses with low phase noise so as to limit considerably the variation in phase difference (jitter) between the two fiber lasers, which is the source of uncertainty. An effective means of accomplishing this is to control the resonator length to synchronize the oscillation frequency with an external signal [26].

We, therefore, next explain the ultrashort pulse fiber laser technology with low phase noise that is used in this system (Fig. 19.15).

The light source is a fiber laser that has a ring-shaped resonator structure [28] and that performs passive mode synchronization by using a semiconductor saturable absorber mirror; the light source obtains self-starting oscillations with a pulse width of less than 80 fs, an average power of 200 mW, and a repetition frequency of 50 MHz at 1.55 μm. Part of the output from the fiber laser is diverted out by a branch coupler and used to stabilize the repetition frequency through a controller that varies the resonator length.

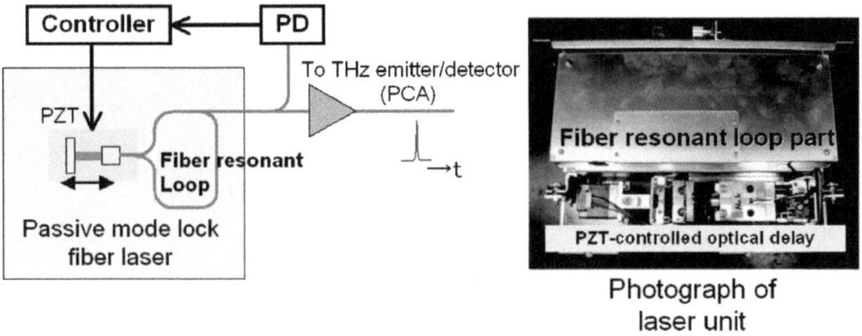

Figure 19.15 Passive mode-lock fiber laser (PD is the photodetector, and PZT is lead zirconate titanate).

Figure 19.16 shows the phase noise spectrum of the fiber laser. It is clear that the low-end phase noise is reduced by resonator length control. The root-mean-square jitter calculated by integrating this phase noise spectrum was (a) 471 fs for the free-running laser versus (b) 55 fs when controlled, which shows that the phase noise (jitter) was significantly reduced. In THz sampling, this method is used to control the system such that the light source–phase relationship at the detector has a constant frequency difference with respect to the light source oscillator phase at the THz pulse generator, removing the effect of jitter.

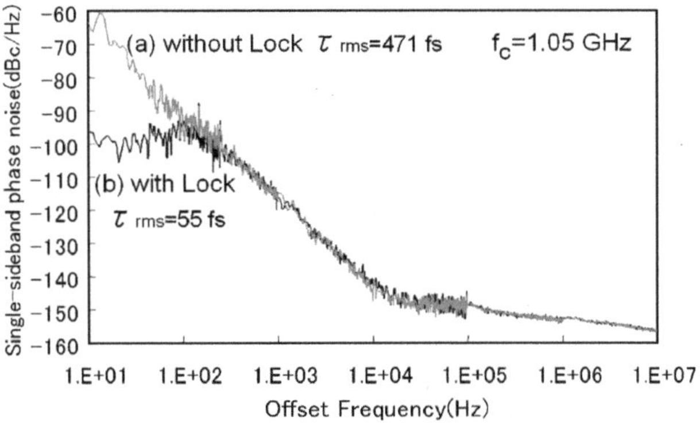

Figure 19.16 Phase noise spectrum.

Next, Fig. 19.17 shows an example of a temporal waveform measurement of a THz pulse measured using the THz optical sampling method and the power spectrum of the waveform obtained by FFT. From these results, it is clear that a broadband THz wave spectrum spanning across 4 THz can be observed.

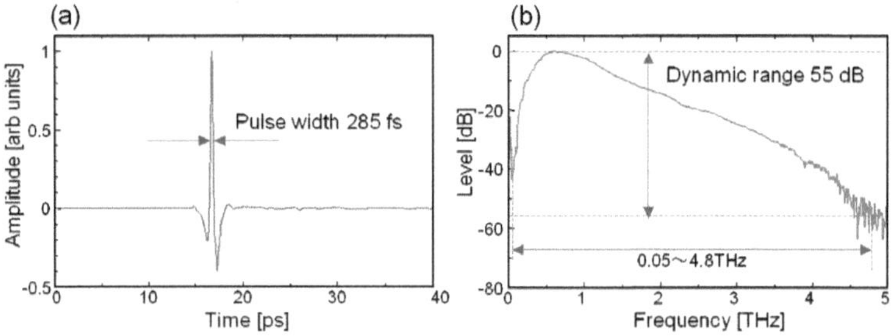

Figure 19.17 (a) THz waveform of a sampled signal and (b) FFT power spectrum (bandwidth at S/N = 1: 0.05–4.8 THz).

19.6 Summary

This chapter has described THz CT, which is a THz imaging technique that is highly anticipated as a 3D imaging technology, and has focused, in particular, on nondestructive analysis applications, THz CT high-speed measurement technology, and laser control technology.

THz CT using TDS is a new technique for 3D visualization based on spectral information that is not available in conventional X-ray CT systems. We have demonstrated some of the usefulness of spectral CT through the example of taking CT measurement of the absorption spectrum information of a sugar and an amino acid and showed that they could be identified and their distributions measured. Quantitative analysis of the distribution of PM inside the ceramic filter for a diesel engine was also shown as an example of a practical application of THz CT analysis that is already being used.

THz imaging using the TDS method is expected to find applications in fields such as analysis of crystal structures and other aspects of pharmaceutical products [29] and biorelated applications [30]. The THz time-of-flight (TOF) imaging technique, which is a reflection method, is also expected to be effective in these applications [10]. The high-speed measurement technique using the

THz optical sampling method introduced in this chapter offers great benefits not only for CT measurement but also for THz TOF and THz spectral measurement. Because of this, the THz optical sampling method has become an extremely useful technique for making nondestructive analysis applications feasible. Another equipment, Advantest's TAS7500 THz spectroscopic imaging system, has already been created for these applications and uses the THz optical sampling method [18]. In the future, we hope that with advances in THz technology, nondestructive analysis applications using THz waves will find use in a wide variety of fields.

Acknowledgments

The authors thank Professor Kodo Kawase of Nagoya University for guidance regarding the CT technology in the TAS7000.

References

1. K. Kawase, Y. Ogawa, Y. Watanabe, and H. Inoue, Non-destructive terahertz imaging of illicit drugs using spectral fingerprints, *Opt. Express* **11**, 2549–2554 (2003).
2. B. Gompf and M. Dressel, THz-micro-spectroscopy, *IEEE J. Sel. Top. Quantum Electron.* **14**, 470–475 (2008).
3. H. Murakami, N. Uchida, R. Inoue, S. Kim, T. Kiwa, and M. Tonouchi, Laser terahertz emission microscope, *Proc. IEEE* **95**, 1646–1657 (2007).
4. http://www.sds.l-3com.com/advancedimaging/provision.htm.
5. T. S. Hartwic, D. T. Hodgrs, D. H. Barker, and F. B. Foote, Far infrared imagery, *Appl. Opt.* **15**, 1919–1922 (1976).
6. A. Dobroiu, C. Otani, and K. Kawase, Terahertz-wave sources and imaging, *Meas. Sci. Technol.* **17**, 161–174 (2006).
7. D. H. Auston, K. P. Cheung, J. A. Valdmanis, and D. A. Kleinman, Cherenkov radiation from femtosecond optical pulses in electro-optic media, *Phys. Rev. Lett.* **53**, 1555–1558 (1984).
8. Ch. Fattinger and D. Grischkowsky, Point source terahertz optics, *Appl. Phys. Lett.* **53**, 1480–1482 (1988).
9. B. B. Hu and M. C. Nuss, Imaging with terahertz waves, *Opt. Lett.* **20**, 1716–1718 (1995).

10. J. L. Johnson, T. D. Dorney, and D. M. Mittleman, Enhanced depth resolution in terahertz imaging using phase-shift interferometry, *Appl. Phys. Lett.* **78**, 835–837 (2001).

11. J. Radon, Mathematical and physical section, Berichte Saechsiche Akademie Wissenschaften **69**, 262–277 (1917).

12. A. C. Kak and M. Slaney, *Principles of Computerized Tomographic Imaging* (IEEE Press, New York, 1988).

13. E. C. Beckmann, CT scanning the early days, *Br. J. Radiol.* **79**, 5–8 (2006).

14. T. Hashimoto and H. Shinohara, *Basis for Image Reconstruction in C Language* (Iryo Kagaku-sya Press, Japan, 2006) (in Japanese).

15. K. L. Nguyen, M. L. Johns, and L. F. Gladden, Three-dimensional imaging with a terahertz quantum cascade laser, *Opt. Express* **14**, 2123–2129 (2006).

16. B. S. Ferguson, S. Wang, D. Gray, D. Abbot, and X.-C. Zhang, BT-ray computed tomography, *Opt. Lett.* **27,** 1312–1314 (2002).

17. X.-C. Zhang, Three-dimensional terahertz wave imaging, *Phil. Trans. R. Soc. Lond. A* **362**, 283–299 (2004).

18. http://www.advantest.co.jp/products/3DIAS/en-index.shtml.

19. M. Imamura, S. Nishina, A. Irisawa, T. Yamashita, and E. Kato, 3D spectroscopic computed tomography imaging using terahertz waves, infrared millimeter and terahertz waves (IRMMW-THz), 2010, *35th Int. Conf.* 1–3 (2010).

20. S. Nishina, K. Takeuchi, M. Shinohara, M. Imamura, M. Shibata, Y. Hashimoto, and F. Watanabe, Novel nondestructive imaging analysis for catalyst washcoat loading and DPF soot distribution using terahertz wave computed tomography, SAE Paper 2011-01-2064, 2011 PF&L Conference, Kyoto, 2011.

21. P. A. Elzinga, R. J. Kneisler, F. E. Lytle, Y. Jiang, G. B. King, and N. M. Laurendeau, A pump/probe method for fast analysis of visible spectral signatures utilizing asynchronous optical sampling, *Appl. Opt.* **26,** 4303 (1987).

22. H. Takara, S. Kawanishi, A. Yokoo, S. Tomaru, T. Kitoh, and M. Saruwatari, 100 Gbit/s optical signal eye diagram measurement with optical sampling using organic nonlinear optical crystal, *Electron. Lett.* **32**, 2256–2258 (1996).

23. C. Janke, M. Forst, M. Nagel, H. Kurz, and A. Bartels, Asynchronous optical sampling for high-speed characterization of integrated resonant terahertz sensors, *Opt. Lett.* **30**, 1405–1407 (2005).

24. T. Yamashita and A. Irisawa, Japanese Patent No. 4782889 (2011).

25. J. B. Schlager, R. P. Mirin, and N. A. Sanford, Passively mode-locked waveguide laser with low residual jitter, *IEEE Photon. Technol. Lett.* **14**, 1351–1353 (2002).

26. E. A. De Souza, C. E. Soccolich, W. Pleibel, R. H. Stolen, J. R. Simpson, and D. J. DiGiovanni, Saturable absorber mode-locked polarization maintaining erbium-doped fiber laser, *Electron. Lett.* **29**, 447–449 (1993).

27. S. Masuda, S. Niki, and M. Nakazawa, Environmentally stable, simple passively mode-locked fiber ring laser using a four-port circulator, *Opt. Express* **17**, 6613–6622 (2009).

28. A. Strzelec, H. Bilheux, C. Finney, C. Daw, D. Foster, C. Rutland, B. Schillinger, and M. Schulz, Neutron imaging of diesel particulate filters, SAE Paper 2009-01-2735, 2009.

29. S. Byrn, R. Pfeiffer, M. Ganey, C. Hoiberg, and G. Poochikian, Pharmaceutical solids: a strategic approach to regulatory considerations, *Pharm. Res.* **12**, 945–954 (1995).

30. S. Mickan, A. Menikh, H. Liu, C. Mannella, R. MacColl, D. Abbott, J. Munch, and Xi Zhang, Label-free bioaffinity detection using terahertz technology, *Phys. Med. Biol.* **47**, 3789 (2002).

Index